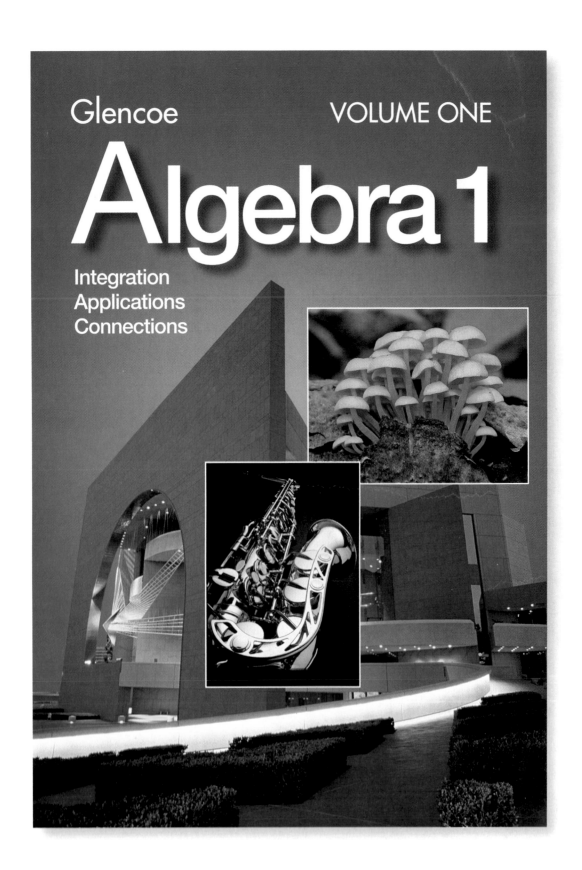

Glencoe

VOLUME ONE

Algebra 1

Integration
Applications
Connections

GLENCOE
McGraw-Hill

New York, New York   Columbus, Ohio   Woodland Hills, California   Peoria, Illinois

Glencoe/McGraw-Hill

A Division of The **McGraw·Hill** *Companies*

Copyright © 1998 by The McGraw-Hill Companies, Inc. All rights reserved.
Printed in the United States of America. Except as permitted under the United
States Copyright Act, no part of this publication may be reproduced or distributed
in any form or by any means, or stored in a database or retrieval system, without
prior written permission from the publisher.

Send all inquiries to:
Glencoe/McGraw-Hill
936 Eastwind Drive
Westerville, OH 43081-3329

ISBN: 0-02-825333-7

6 7 8 9 10 071/043 05 04 03 02 01 00 99

WHY IS ALGEBRA IMPORTANT?

Why do I need to study algebra? When am I ever going to have to use algebra in the real world?

Many people, not just algebra students, wonder why mathematics is important. *Algebra 1* is designed to answer those questions through **integration, applications,** and **connections.**

INTEGRATION
Geometry

Did you know that algebra and geometry are closely related? Topics from all branches of mathematics, like geometry and statistics, are integrated throughout the text.

You'll learn how to find the complement and supplement of an angle and how to find the measure of the third angle of a triangle. (Lesson 3–4*, pages 162–164)

APPLICATION
Nutrition

I can't believe that a double cheeseburger has that much fat! Real-world uses of mathematics are presented.

The number of grams of fat in a double cheeseburger is determined by solving an open sentence. (Lesson 1–5*, page 32)

Seymour, Wisconsin
"The Home of the Hamburger"

CONNECTION
Biology

What does biology have to do with mathematics? Mathematical topics are connected to other subjects that you study.

Punnett squares, which are models that show the possible ways that genes combine, are connected to squaring a binomial. (Lesson 9–8*, page 542)

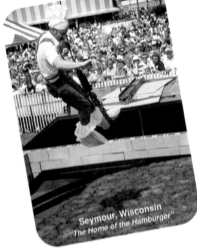

*Chapters 1–7 are contained in Volume One. Chapters 7–13 are contained in Volume Two.

iii

Authors

WILLIAM COLLINS teaches mathematics at James Lick High School in San Jose, California. He has served as the mathematics department chairperson at James Lick and Andrew Hill High Schools. Mr. Collins received his B.A. in mathematics and philosophy from Herbert H. Lehman College in Bronx, New York, and his M.S. in mathematics education from California State University, Hayward. Mr. Collins is a member of the Association of Supervision and Curriculum Development and the National Council of Teachers of Mathematics, and is active in several professional mathematics organizations at the state level. He is also currently serving on the Teacher Advisory Panel of the *Mathematics Teacher.*

"In this era of educational reform and change, it is good to be part of a program that will set the pace for others to follow. This program integrates the ideas of the NCTM Standards with real tools for the classroom, so that algebra teachers and students can expect success every day."

GILBERT CUEVAS is a professor of mathematics education at the University of Miami in Miami, Florida. Dr. Cuevas received his B.A. in mathematics and M.Ed. and Ph.D., both in educational research, from the University of Miami. He also holds a M.A.T. in mathematics from Tulane University. Dr. Cuevas is a member of many mathematics, science, and research associations on the local, state, and national levels and has been an author and editor of several National Council of Teachers of Mathematics (NCTM) publications. He is also a frequent speaker at NCTM conferences, particularly on the topics of equity and mathematics for all students.

ALAN G. FOSTER is a former mathematics teacher and department chairperson at Addison Trail High School in Addison, Illinois. He obtained his B.S. from Illinois State University and his M.A. in mathematics from the University of Illinois. Mr. Foster is a past president of the Illinois Council of Teachers of Mathematics (ICTM) and was a recipient of the ICTM's T.E. Rine Award for Excellence in the Teaching of Mathematics. He also was a recipient of the 1987 Presidential Award for Excellence in the Teaching of Mathematics for Illinois. Mr. Foster was the chairperson of the MATHCOUNTS question writing committee in 1990 and 1991. He frequently speaks and conducts workshops on the topic of cooperative learning.

BERCHIE GORDON is the mathematics/science coordinator for the Northwest Local School District in Cincinnati, Ohio. Dr. Gordon has taught mathematics at every level from junior high school to college. She received her B.S. in mathematics from Emory University in Atlanta, Georgia, her M.A.T. in education from Northwestern University in Evanston, Illinois, and her Ph.D. in curriculum and instruction at the University of Cincinnati. Dr. Gordon has developed and conducted numerous inservice workshops in mathematics and computer applications. She has also served as a consultant for IBM, and has traveled throughout the country making presentations on graphing calculators to teacher groups.

"Using this textbook, you will learn to think mathematically for the 21st century, solve a variety of problems based on real-world applications, and learn the appropriate use of technological devices so you can use them as tools for problem solving."

BEATRICE MOORE-HARRIS is an educational specialist at the Region IV Education Service Center in Houston, Texas. She is also the Southwest Regional Director of the Benjamin Banneker Association. Ms. Moore-Harris received her B.A. from Prairie View A&M University in Prairie View, Texas. She has also done graduate work there, at Texas Southern University in Houston, Texas, and at Tarleton State University in Stephenville, Texas. Ms. Moore-Harris is a consultant for the National Council of Teachers of Mathematics (NCTM) and serves on the Editorial Board of the NCTM's *Mathematics Teaching in the Middle School.*

"This program will bring algebra to life by engaging you in motivating, challenging, and worthwhile mathematical tasks that mirror real-life situations. Opportunities to use technology, manipulatives, language, and a variety of other tools are an integral part of this program, which allows all students full access to the algebra curriculum."

JAMES RATH has 30 years of classroom experience in teaching mathematics at every level of the high school curriculum. He is a former mathematics teacher and department chairperson at Darien High School in Darien, Connecticut. Mr. Rath earned his B.A. in philosophy from The Catholic University of America and his M.Ed. and M.A. in mathematics from Boston College. He has also been a Visiting Fellow in the mathematics department at Yale University in New Haven, Connecticut.

DORA SWART is a mathematics teacher and department chairperson at W.F. West High School in Chehalis, Washington. She received her B.A. in mathematics education at Eastern Washington University in Cheney, Washington, and has done graduate work at Central Washington University in Ellensburg, Washington, and Seattle Pacific University in Seattle, Washington. Ms. Swart is a member of the National Council of Teachers of Mathematics, the Western Washington Mathematics Curriculum Leaders, and the Association of Supervision and Curriculum Development. She has developed and conducted numerous inservices and presentations to teachers in the Pacific Northwest.

"Glencoe's algebra series provides the best opportunity for you to learn algebra well. It explores mathematics through hands-on learning, technology, applications, and connections to the world around us. Mathematics can unlock the door to your success— this series is the key."

LESLIE J. WINTERS is the former secondary mathematics specialist for the Los Angeles Unified School District and is currently supervising student teachers at California State University, Northridge. Mr. Winters received bachelor's degrees in mathematics and secondary education from Pepperdine University and the University of Dayton, and master's degrees from the University of Southern California and Boston College. He is a past president of the California Mathematics Council-Southern Section, and received the 1983 Presidential Award for Excellence in the Teaching of Mathematics and the 1988 George Polya Award for being the Outstanding Mathematics Teacher in the state of California.

Consultants, Writers, and Reviewers

Consultants

Robbie Bonneville
Consultant, 2-Volume Edition
Mathematics Coordinator
La Joya ISD
Alamo, Texas

David Foster
Glencoe Author and Mathematics
    Consultant
Morgan Hill, California

Bonnie Coulter Leech
Consultant, 2-Volume Edition
Mathematics Teacher
Marcus High School
Flower Mound, Texas

Melissa McClure
Consultant, Tech Prep and 2-Volume
    Edition
Mathematics Consultant
Teaching for Tomorrow
Fort Worth, Texas

Cindy J. Boyd
Mathematics Teacher
Abilene High School
Abilene, Texas

Eva Gates
Independent Mathematics
    Consultant
Pearland, Texas

Carol Malloy
Consultant, 2-Volume Edition
Assistant Professor of Mathematics
    Education
University of North Carolina at
    Chapel Hill
Chapel Hill, North Carolina

Dr. Luis Ortiz-Franco
Consultant, Diversity
Associate Professor of Mathematics
Chapman University
Orange, California

Gail Burrill
National Center for Research/
    Mathematical & Science Education
University of Wisconsin
Madison, Wisconsin

Joan Gell
Mathematics Department Chairman
Palos Verdes High School
Palos Verdes Estates, California

Daniel Marks
Consultant, Real-World Applications
Associate Professor of Mathematics
Auburn University at Montgomery
Montgomery, Alabama

Pamela Summers
Consultant, 2-Volume Edition
Secondary Mathematics and Science
    Coordinator
Lubbock ISD
Lubbock, Texas

Writers

David Foster
Writer, Investigations
Glencoe Author and Mathematics
    Consultant
Morgan Hill, California

Jeri Nichols-Riffle
Writer, Graphing Technology
Assistant Professor of Teacher Education/Mathematics and Statistics
Wright State University
Dayton, Ohio

Reviewers

Susan J. Barr
Mathematics Department Chairperson
Dublin Coffman High School
Dublin, Ohio

Wayne Boggs
Mathematics Supervisor
Ephrata High School
Ephrata, Pennsylvania

William A. Brinkman
Mathematics Department Chairperson
Paynesville High School
Paynesville, Minnesota

Kenneth Burd, Jr.
Mathematics Teacher
Hershey Senior High School
Hershey, Pennsylvania

Kimberly C. Cox
Mathematics Teacher
Stonewall Jackson High School
Manassas, Virginia

Sabine Goetz
Mathematics Teacher
Hewitt-Trussville Junior High
Trussville, Alabama

William Biernbaum
Mathematics Teacher
Platteview High School
Springfield, Nebraska

Donald L. Boyd
Mathematics Teacher
South Charlotte Middle School
Charlotte, North Carolina

Louis A. Bruno
Mathematics Department Chairman
Somerset Area Senior High School
Somerset, Pennsylvania

Todd W. Busse
Mathematics/Science Teacher
Wenatchee High School
Wenatchee, Washington

Janis Frantzen
Mathematics Department Chairman
McCullough High School
The Woodlands, Texas

Dee Dee Hays
Mathematics Teacher
Henry Clay High School
Lexington, Kentucky

John R. Blickenstaff
Mathematics Teacher
Martinsville High School
Martinsville, Indiana

Judith B. Brigman
Mathematics Teacher
South Florence High School
Florence, South Carolina

Luajean Nipper Bryan
Mathematics Teacher
McMinn County High School
Athens, Tennessee

Esther Corn
Mathematics Department Chairperson
Renton High School
Renton, Washington

Vicki Fugleberg
Mathematics Teacher
May-Port CG School
Mayville, North Dakota

Ralph Jacques
Mathematics Department Chairperson
Biddeford High School
Biddeford, Maine

Dixie T. Johnson
Mathematics Department Chairperson
Paul Blazer High School
Ashland, Kentucky

Jane Housman Jones
Mathematics/Technology Teacher
du Pont Manual High School
Louisville, Kentucky

Rebecca A. Luna
Mathematics Teacher
Austin Junior High School
San Juan, Texas

Sandra A. Nagy
Mathematics Resource Teacher
Mesa Schools
Mesa, Arizona

Linda Palmer
Mathematics Teacher
Clark High School
Plano, Texas

Cynthia H. Ray
Mathematics Teacher
Ravenswood High School
Ravenswood, West Virginia

Hazel Russell
Mathematics Teacher
Riverside Middle School
Fort Worth, Texas

Mabel S. Sechrist
Mathematics Teacher
Mineral Springs Middle School
Winston-Salem, North Carolina

Diane Baish Stamy
Mathematics Chairperson
Big Spring School District
Newville, Pennsylvania

Robert L. Thompson
Mathematics Teacher
Raines High School
Jacksonville, Florida

John J. Turechek
Mathematics Department Chairperson
Bunnell High School
Stratford, Connecticut

Jan S. Wessell
Mathematics Supervisor
New Hanover County Schools
Wilmington, North Carolina

Kathy Zwanzig
Middle School Algebra Resource
    Teacher
Jefferson County Public Schools
Louisville, Kentucky

Bernita Jones
Mathematics Department Chairperson
Turner High School
Kansas City, Kansas

Nancy W. Kinard
Mathematics Teacher
Palm Beach Gardens High School
Palm Beach Gardens, Florida

Frances J. MacDonald
Area Mathematics Coordinator, K–12
Pompton Lakes School District
Pompton Lakes, New Jersey

Roger O'Brien
Mathematics Supervisor
Polk County Schools
Bartow, Florida

Deborah Patonai Phillips
Mathematics Department Head
St. Vincent-St. Mary High School
Akron, Ohio

Carrol W. Rich
Mathematics Teacher
High Point Central
High Point, North Carolina

Judith Ryder
Mathematics Teacher
South Charlotte Middle School
Charlotte, North Carolina

Richard K. Sherry
Mathematics Department Chairman
Roosevelt Junior High School
Altoona, Pennsylvania

Carolyn Stanley
Mathematics Department Chairperson
Tidewater Academy
Wakefield, Virginia

Linda K. Tinga
Mathematics Teacher
Laney High School
Wilmington, North Carolina

Tonya Dessert Urbatsch
Mathematics Curriculum Coordinator
Davenport Community Schools
Davenport, Iowa

Ted Wiecek
Mathematics Department Chairperson
Wells Community Academy
Chicago, Illinois

Cheryl M. Jones
Mathematics Department Coordinator
Winnetonka High School
Kansas City, Missouri

Margaret M. Kirk
Mathematics Teacher
Welborn Middle School
High Point, North Carolina

George R. Murphy
Mathematics Department Chairperson
Pine-Richland High School
Gibsonia, Pennsylvania

Barbara S. Owens
Mathematics Teacher
Davidson Middle School
Davidson, North Carolina

Kristine Powell
Mathematics Teacher
Lehi Junior High School
Lehi, Utah

Mary Richards
Mathematics Teacher
Clyde A. Erwin High School
Asheville, North Carolina

Donna Schmitt
Reviewer, Technology
Mathematics and Computer Science
    Teacher
Dubuque Senior High School
Dubuque, Iowa

Catherine C. Sperando
Mathematics Teacher
Shades Valley High School
Birmingham, Alabama

Richard P. Strausz
Math/Computer Coordinator
Farmington Schools
Farmington, Michigan

Kimberly Troise
Mathematics Teacher
A. C. Reynolds High School
Asheville, North Carolina

Anthony J. Weisse
Mathematics Teacher
Central High School
LaCrosse, Wisconsin

Rosalyn Zeid
Mathematics Director
Union High School
Union, New Jersey

Table of Contents

Getting Acquainted with the Graphing Calculator 2

CHAPTER **1** **Exploring Expressions, Equations, and Functions**   4

1–1	Variables and Expressions	6
1–2	Patterns and Sequences	12
1–3	Order of Operations	19
1–4	**Integration:** Statistics Stem-and-Leaf Plots	25
1–5	Open Sentences	32
	Self Test	36
1–6	Identity and Equality Properties	37
1–7A	**Modeling Mathematics:** The Distributive Property	44
1–7	The Distributive Property	45
1–8	Commutative and Associative Properties	51
1–9	A Preview of Graphs and Functions	56
	Chapter 1 Highlights	63
	Chapter 1 Study Guide and Assessment	64

Technology
    Explorations **8, 21**
Modeling Mathematics
    Activities **15**
Mathematics and Society
    International Mathematical
        Olympiad **11**

CHAPTER 2 Exploring Rational Numbers 70

2–1   Integers and the Number Line 72
2–2   Integration: Statistics
       Line Plots 78
2–3A   Modeling Mathematics: Adding and Subtracting Integers 84
2–3   Adding and Subtracting Integers 85
2–4   Rational Numbers 93
       Self Test 99
2–5   Adding and Subtracting Rational Numbers 100
2–6A   Modeling Mathematics: Multiplying Integers 105
2–6   Multiplying Rational Numbers 106
2–7   Dividing Rational Numbers 112
2–8A   Modeling Mathematics: Estimating Square Roots 118
2–8   Square Roots and Real Numbers 119
2–9   Problem Solving: Write Equations and Formulas 126
       Chapter 2 Highlights 133
       Chapter 2 Study Guide and Assessment 134
       Cumulative Review, Chapters 1–2 138

Long-Term Investigation
     The Greenhouse Effect   68–69
     • Working on the Investigation   83, 111
Modeling Mathematics
     Activities   74, 79, 128
Mathematics and Society
     Mayan Mathematics   77

CHAPTER 3 Solving Linear Equations 140

3–1A   Modeling Mathematics: Solving One-Step Equations 142
3–1   Solving Equations by Using Addition and Subtraction 144
3–2   Solving Equations by Using Multiplication and Division 150
3–3A   Modeling Mathematics: Solving Multi-Step Equations 155
3–3   Solving Multi-Step Equations 156
       Self Test 161
3–4   Integration: Geometry
       Angles and Triangles 162
3–5   Solving Equations with the Variable on Both Sides 168
3–6   Solving Equations and Formulas 173
3–7   Integration: Statistics
       Measures of Central Tendency 178
       Chapter 3 Highlights 185
       Chapter 3 Study Guide and Assessment 186

Long-Term Investigation
     The Greenhouse Effect
     • Working on the Investigation   154, 177
     • Closing the Investigation   184
Technology
     Explorations   147, 157, 180
Modeling Mathematics
     Activities   164
Mathematics and Society
     Teen Talk Barbies   183

INTEGRATION

Content Integration

What does geometry have to do with algebra? Believe it or not, you can study most math topics from more than one point of view. Here are some examples.

Music Sales by Categories

Sales (millions of dollars)

= Rock
= Rap
= Country

1985   '87   '89   '91
Year

Source: The Recording Industry Association of America

Probability You'll model the roll of two dice with relations and line plots. (Lesson 5–2*, Exercise 45)

▲ **Problem Solving** You'll interpret graphs and learn how to sketch graphs of real-world situations. (Lesson 1–9*, page 62)

LOOK BACK

You can refer to Lesson 1-7 for information on the distributive property.

Look Back features refer you to skills and concepts that have been taught earlier in the book.

Source: Lesson 9–6*, page 529

Discrete Mathematics ▶ You'll investigate patterns and sequences and use them to solve a variety of problems. (Lesson 1–2*, pages 12–18)

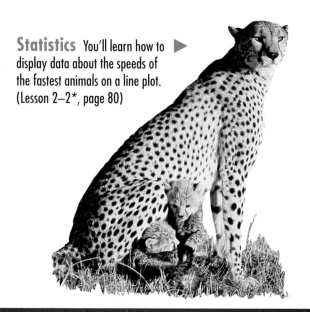

Statistics You'll learn how to ▶ display data about the speeds of the fastest animals on a line plot. (Lesson 2–2*, page 80)

Geometry You'll use your algebra skills to find the missing measures of similar triangles. (Lesson 4–2*, pages 201–205)

Chapters 1–7 are contained in Volume One.
Chapters 7–13 are contained in Volume Two.

x

CHAPTER **4** Using Proportional Reasoning 192

4–1A	**Modeling Mathematics:** Ratios	194
4–1	Ratios and Proportions	195
4–2	**Integration: Geometry** Similar Triangles	201
4–3	**Integration: Trigonometry** Trigonometric Ratios	206
4–4	Percents	215
	Self Test	221
4–5	Percent of Change	222
4–6	**Integration: Probability** Probability and Odds	228
4–7	Weighted Averages	233
4–8	Direct and Inverse Variation	239
	Chapter 4 Highlights	245
	Chapter 4 Study Guide and Assessment	246
	Cumulative Review, Chapters 1–4	250

Long-Term Investigation
    Go Fish! **190–191**
       • Working on the Investigation **200, 227**
Technology
    Exploration **224**
Modeling Mathematics
    Activities **209, 216, 229**
Mathematics and Society
    Morgan's Conjecture **214**

CHAPTER **5** Graphing Relations and Functions 252

5–1	The Coordinate Plane	254
5–2A	**Graphing Technology:** Relations	260
5–2	Relations	262
5–3A	**Graphing Technology:** Equations	270
5–3	Equations as Relations	271
5–4A	**Graphing Technology:** Linear Relations and Functions	278
5–4	Graphing Linear Equations	280
	Self Test	286
5–5	Functions	287
5–6	Writing Equations from Patterns	295
5–7A	**Graphing Technology:** Measures of Variation	303
5–7	**Integration: Statistics** Measures of Variation	305
	Chapter 5 Highlights	315
	Chapter 5 Study Guide and Assessment	316

Long-Term Investigation
    Go Fish!
       • Working on the Investigation **269, 302**
       • Closing the Investigation **314**
Technology
    Explorations **273, 297**
Modeling Mathematics
    Activities **265**
Mathematics and Society
    Record Cold Temperatures **313**

CHAPTER **6** Analyzing Linear Equations 322

6-1A	**Modeling Mathematics:** Slope	324
6-1	Slope	325
6-2	Writing Linear Equations in Point-Slope and Standard Forms	332
6-3	Integration: **Statistics**	
	Scatter Plots and Best-Fit Lines	339
6-4	Writing Linear Equations in Slope-Intercept Form	346
	Self Test	353
6-5A	Graphing Technology: Parent and Family Graphs	354
6-5	Graphing Linear Equations	356
6-6	Integration: **Geometry**	
	Parallel and Perpendicular Lines	362
6-7	Integration: **Geometry**	
	Midpoint of a Line Segment	369
	Chapter 6 Highlights	375
	Chapter 6 Study Guide and Assessment	376
	Cumulative Review, Chapters 1–6	380

Long-Term Investigation
  Smoke Gets In Your Eyes   **320–321**
  • Working on the Investigation   **338, 361**
Technology
  Explorations   **327, 342**
Modeling Mathematics
  Activities   **365, 370**
Mathematics and Society
  People Around the World   **374**

CHAPTER **7** Solving Linear Inequalities 382

7-1	Solving Inequalities by Using Addition and Subtraction	384
7-2A	**Modeling Mathematics:** Solving Inequalities	391
7-2	Solving Inequalities by Using Multiplication and Division	392
7-3	Solving Multi-Step Inequalities	399
7-4	Solving Compound Inequalities	405
	Self Test	412
7-5	Integration: **Probability**	
	Compound Events	413
7-6	Solving Open Sentences Involving Absolute Value	420
7-7	Integration: **Statistics**	
	Box-and-Whisker Plots	427
7-7B	Graphing Technology: Box-and-Whisker Plots	433
7-8A	Graphing Technology: Graphing Inequalities	435
7-8	Graphing Inequalities in Two Variables	436
	Chapter 7 Highlights	443
	Chapter 7 Study Guide and Assessment	444

Long-Term Investigation
  Smoke Gets In Your Eyes
  • Working on the Investigation   **398, 426**
  • Closing the Investigation   **442**
Technology
  Explorations   **401, 437**
Mathematics and Society
  Computers and Chess   **419**

Chapter B
To the Student **B1**
Pretest **B2**
Chapter 1 Review
    and Test **B4**
Chapter 2 Review
    and Test **B14**
Chapter 3 Review
    and Test **B24**
Chapter 4 Review
    and Test **B32**
Chapter 5 Review
    and Test **B41**
Chapter 6 Review
    and Test **B49**
Posttest **B57**

APPLICATIONS

Real-Life Applications

Have you ever wondered if you'll ever actually use math? Every lesson in this book is designed to help show you where and when math is used in the real world. Since you'll explore many interesting topics, we believe you'll discover that math is relevant and exciting. Here are some examples.

Top Five List, FYI, and **Fabulous Firsts** contain interesting facts that enhance the applications.

Selling Prices of Paintings

1. *Portrait du Dr. Gachet* by van Gogh, $75,000,000
2. *Au Moulin de la Galette* by Renoir, $71,000,000
3. *Les Noces de Pierrette* by Picasso, $51,700,000
4. *Irises* by van Gogh, $49,000,000
5. *Yo Picasso* by Picasso, $43,500,000

Source: Lesson 9–4*, page 514

fabulous FIRSTS

Elmer Simms Campbell

(1906–1971)

Elmer Simms Campbell was the first African-American cartoonist to work for national publications. He contributed cartoons and other art work to *Esquire, Cosmopolitan,* and *Redbook,* as well as syndicated features in 145 newspapers.

Source: Lesson 8–3*, page 469

Football You'll see how punting a football is related to angle measure. (Lesson 3–4*, page 162)

Carpentry You'll study an industrial technology application that involves computing with rational numbers. (Lesson 2–7*, Exercise 47)

◀ **Recreation** Games from several world cultures that are similar to hopscotch will illustrate multiplying a polynomial by a monomial. (Lesson 9–6*, page 529)

F Y I

In September 1995, blue M&M's® completely replaced the tan ones. The ratios of colors for plain M&M's are as follows.

brown	30%
red	20%
yellow	20%
orange	10%
green	10%
blue	10%

Source: Lesson 2–2*, page 79

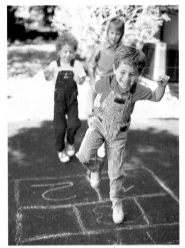

▲ **World Cultures** The Great Pyramid of Khufu is the setting for an application involving systems of linear equations. (Lesson 8–3*, Exercise 42)

Space Science You'll discuss the extreme temperatures encountered during a walk in space as you learn about absolute value. (Lesson 2–3*, page 85)

Mathematics and SOCIETY

Teen Talk Barbies

What do Barbie dolls have to do with mathematics? Actual reprinted articles illustrate how mathematics is a part of our society. (Lesson 3–7*, page 183)

*Chapters 1–7 are contained in Volume One.
Chapters 7–13 are contained in Volume Two.*

CHAPTER **8** **Solving Systems of Linear Equations and Inequalities** **450**

8-1A **Graphing Technology:** Systems of Equations **452**
8-1 Graphing Systems of Equations **454**
8-2 Substitution **462**
8-3 Elimination Using Addition and Subtraction **469**
Self Test **474**
8-4 Elimination Using Multiplication **475**
8-5 Graphing Systems of Inequalities **482**
Chapter 8 Highlights **487**
Chapter 8 Study Guide and Assessment **488**
Cumulative Review, Chapters 1–8 .... **492**

Long-Term Investigation
Ready, Set, Drop! **448–449**
• Working on the Investigation **461**
Technology
Explorations **471, 484**
Modeling Mathematics
Activities **464**
Mathematics and Society
High-Tech Checkout Lanes **481**

CHAPTER **9** **Exploring Polynomials** **494**

9-1 Multiplying Monomials **496**
9-2 Dividing by Monomials **501**
9-3 Scientific Notation **506**
9-4A **Modeling Mathematics:** Polynomials **513**
9-4 Polynomials **514**
9-5A **Modeling Mathematics:** Adding and Subtracting Polynomials **520**
9-5 Adding and Subtracting Polynomials **522**
Self Test **527**
9-6A **Modeling Mathematics:** Multiplying a Polynomial by a Monomial **528**
9-6 Multiplying a Polynomial by a Monomial **529**
9-7A **Modeling Mathematics:** Multiplying Polynomials **534**
9-7 Multiplying Polynomials **536**
9-8 Special Products **542**
Chapter 9 Highlights **549**
Chapter 9 Study Guide and Assessment **550**

Long-Term Investigation
Ready, Set, Drop!
• Working on the Investigation **519, 541**
• Closing the Investigation **548**
Technology
Explorations **502, 507**
Modeling Mathematics
Activities **545**
Mathematics and Society
Measurements Great and Small **512**

CHAPTER **10** Using Factoring   556

10–1   Factors and Greatest Common Factor.. 558
10–2A   **Modeling Mathematics:** Factoring Using the Distributive Property..... 564
10–2   Factoring Using the Distributive Property... 565
10–3A   **Modeling Mathematics:** Factoring Trinomials 572
10–3   Factoring Trinomials... 574
        Self Test 580
10–4   Factoring Differences of Squares... 581
10–5   Perfect Squares and Factoring.. 587
10–6   Solving Equations by Factoring 594
        Chapter 10 Highlights.. 601
        Chapter 10 Study Guide and Assessment 602
        Cumulative Review, Chapters 1–10... 606

Long-Term Investigation
    The Brickyard   **554–555**
    • Working on the Investigation   **586, 600**
Technology
    Exploration   **578**
Modeling Mathematics
    Activity   **581**
Mathematics and Society
    Number Sieve   **563**

CHAPTER **11** Exploring Quadratic and
             Exploring Exponential Functions   608

11–1A   Graphing Technology: Quadratic Functions 610
11–1   Graphing Quadratic Functions 611
11–1B   Graphing Technology: Parent and Family Graphs............................. 618
11–2   Solving Quadratic Equations by Graphing... 620
11–3   Solving Quadratic Equations by Using the Quadratic Formula.......... 628
        Self Test 633
11–4A   Graphing Technology: Exponential Functions 634
11–4   Exponential Functions 635
11–5   Growth and Decay.. 643
        Chapter 11 Highlights.. 651
        Chapter 11 Study Guide and Assessment 652

Long-Term Investigation
    The Brickyard
    • Working on the Investigation   **617, 627**
    • Closing the Investigation   **650**
Technology
    Explorations   **622, 646**
Modeling Mathematics
    Activities   **612, 636**
Mathematics and Society
    Minimizing Computers   **649**

CONNECTIONS

Interdisciplinary Connections

Did you realize that mathematics is used in biology? in history? in geography? Yes, it may be hard to believe, but mathematics is frequently connected to other subjects that you are studying.

GL⊕BAL CONNECTIONS

About 15,000 years ago, an important application of the lever appeared as hunters used an *atlatl,* the Aztec word for spear-thrower. This simple device was a handle that provided extra mechanical advantage by adding length to the hunter's arm. This enabled the hunter to use less motion to get greater results when hunting.

Source: Lesson 7–2*, page 392

◀ **Global Connections** features introduce you to a variety of world cultures.

Health You'll write inequalities to model target heart rates and exercise. (Lesson 7–8*, Exercise 47)

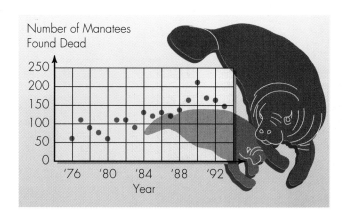

Number of Manatees Found Dead

▲ **Biology** You'll study the mortality rates of Florida's manatees, an endangered species. (Lesson 5–2*, page 262)

CAREER CHOICES

A **zoologist** studies animals, their origin, life process, behavior, and diseases. They are usually identified by the animal group in which they specialize, for example herpetology (reptiles), ornithology (birds), and ichthyology (fish).

Zoology requires a minimum of a bachelor's degree. Many college positions in this field require a Ph.D. degree.

For more information, see:

Occupational Outlook Handbook, 1994–95, U.S. Department of Labor.

Source: Lesson 5–2*, page 264

◀ **Career Choices** features include information on interesting careers.

Geography In 1994, the population of New York was surpassed by Texas. This situation is modeled as a system of linear equations. (Lesson 8–2*, page 462)

▲ **Art** A Mondrian painting entitled *Composition with Red, Yellow, and Blue* is used to model polynomials. (Lesson 9–3*, page 514)

Math Journal exercises give you the opportunity to assess yourself and write about your understanding of key math concepts. (Lesson 7–1*, Exercise 6)

MATH JOURNAL

6. Sometimes statements we make can be translated into inequalities. For example, *In some states, you have to be at least 16 years old to have a driver's license* can be expressed as $a \geq 16$, and *Tomás cannot lift more than 72 pounds* can be translated into $w \leq 72$. Following these examples, write three statements that deal with your everyday life. Then translate each into a corresponding inequality.

*Chapters 1–7 are contained in Volume One.
Chapters 7–13 are contained in Volume Two.*

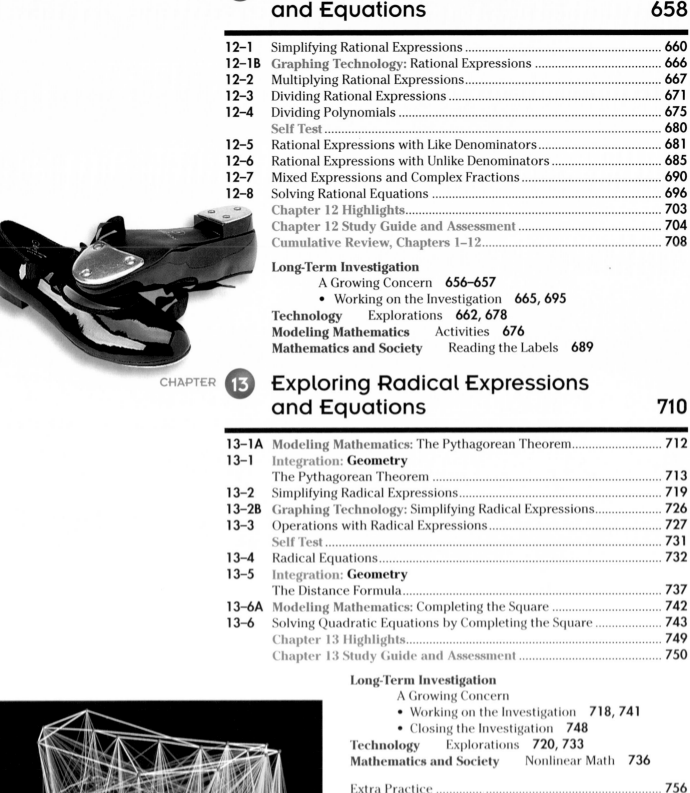

CHAPTER 12 **Exploring Rational Expressions and Equations** 658

12–1	Simplifying Rational Expressions	660
12–1B	Graphing Technology: Rational Expressions	666
12–2	Multiplying Rational Expressions	667
12–3	Dividing Rational Expressions	671
12–4	Dividing Polynomials	675
	Self Test	680
12–5	Rational Expressions with Like Denominators	681
12–6	Rational Expressions with Unlike Denominators	685
12–7	Mixed Expressions and Complex Fractions	690
12–8	Solving Rational Equations	696
	Chapter 12 Highlights	703
	Chapter 12 Study Guide and Assessment	704
	Cumulative Review, Chapters 1–12	708

Long-Term Investigation
   A Growing Concern **656–657**
   • Working on the Investigation **665, 695**
Technology   Explorations **662, 678**
Modeling Mathematics   Activities **676**
Mathematics and Society   Reading the Labels **689**

CHAPTER 13 **Exploring Radical Expressions and Equations** 710

13–1A	Modeling Mathematics: The Pythagorean Theorem	712
13–1	Integration: **Geometry**	
	The Pythagorean Theorem	713
13–2	Simplifying Radical Expressions	719
13–2B	Graphing Technology: Simplifying Radical Expressions	726
13–3	Operations with Radical Expressions	727
	Self Test	731
13–4	Radical Equations	732
13–5	Integration: **Geometry**	
	The Distance Formula	737
13–6A	Modeling Mathematics: Completing the Square	742
13–6	Solving Quadratic Equations by Completing the Square	743
	Chapter 13 Highlights	749
	Chapter 13 Study Guide and Assessment	750

Long-Term Investigation
   A Growing Concern
   • Working on the Investigation **718, 741**
   • Closing the Investigation **748**
Technology   Explorations **720, 733**
Mathematics and Society   Nonlinear Math **736**

Extra Practice	756
Chapter Tests	787
Glossary	794
Spanish Glossary	803
Selected Answers	816
Photo Credits	849
Applications & Connections Index	851
Index	854

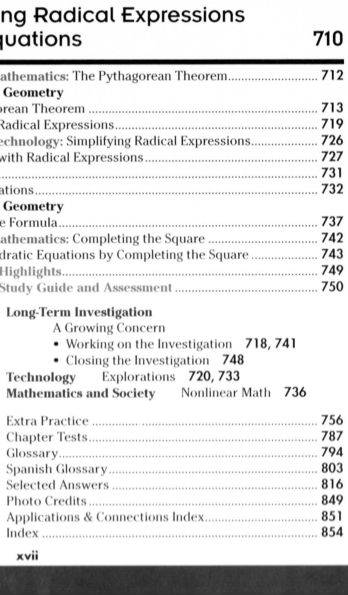

TECHNOLOGY

D o you know how to use computers and graphing calculators? If you do, you'll have a much better chance of being successful in today's high-tech society and workplace.

GRAPHING CALCULATORS

There are several ways in which graphing calculators are integrated.

- **Getting Acquainted with the Graphing Calculator**   On pages 2–3, you'll get acquainted with the basic features and functions of a graphing calculator.

- **Graphing Technology Lessons**   In Lesson 5–2A*, you'll learn how to plot points using a graphing calculator.

- **Graphing Calculator Explorations**   You'll learn how to use a graphing calculator to find the mean and median in Lesson 3–7*.

- **Graphing Calculator Programs**   In Lesson 3–5*, Exercise 39 includes a graphing calculator program that can be used to solve equations of the form $ax + b = cx + d$.

- **Graphing Calculator Exercises**   Many exercises are designed to be solved using a graphing calculator. For example, see Exercises 40–42 in Lesson 7–8*.

COMPUTER SOFTWARE

- **Spreadsheets**   On page 297 of Lesson 5–6*, a spreadsheet is used to help write an equation that models a relation.

- **BASIC Programs**   The program on page 8 of Lesson 1–1* is designed to evaluate expressions.

- **Graphing Software**   The graphing software Exploration on page 469 in Lesson 8–3* involves graphing and solving systems of linear equations.

Technology Tips, such as this one on page 216 of Lesson 4–4*, are designed to help you make more efficient use of technology through practical hints and suggestions.

TECHNOLOGY
Tip

Most calculators have a key labeled %. To find 28% of 58.4 on a TI-34, enter:

28 [2nd] [%] [×]

58.4 [=]   16.352.

*Chapters 1–7 are contained in Volume One.
Chapters 7–13 are contained in Volume Two.

xviii

To The Student

Chapter A contains three sections: pretest, concept review, and posttest. The pretest is a mixture of twelve concepts that are generally acknowledged as necessary for success in Algebra 1. You should take the pretest to determine which concepts you need to work on. The concept review section allows you to develop, and eventually master, the individual skills the pretest identified as needing to be reinforced. You should take the posttest to make sure you understand all of the concepts and to measure your progress.

The chart on the next page contains commonly used formulas and measurement conversions. You may use it on the pretest and posttest.

FORMULA CHART AND MEASUREMENT CONVERSIONS

Formula Chart		
Perimeter	square rectangle	$P = 4s$ $P = 2(\ell + w)$
Circumference	circle	$C = 2\pi r$
Area	square rectangle triangle trapezoid circle	$A = s^2$ $A = \ell w$ or $A = bh$ $A = \frac{bh}{2}$ or $\frac{1}{2}bh$ $A = \frac{1}{2}h(b_1 + b_2)$ $A = \pi r^2$
Surface Area	cube cylinder (lateral)	$S = 6s^2$ $S = 2\pi rh$
Volume	rectangular prism cylinder cube	$V = \ell wh$ $V = \pi r^2 h$ $V = s^3$
Pythagorean Theorem	right triangle	$a^2 + b^2 = c^2$
Slope of a Line	$m = \dfrac{y_2 - y_1}{x_2 - x_1}$	
Quadratic Formula	$x = \dfrac{-b \pm \sqrt{b^2 - 4ac}}{2a}$	
Slope-Intercept Form of an Equation	$y = mx + b$	
Point-Slope Form of an Equation	$y - y_1 = m(x - x_1)$	
Standard Form of an Equation	$Ax + By = C$	

Measurement Conversions		
Measure	**Metric**	**Customary**
Length	1 kilometer = 1000 meters 1 meter = 100 centimeters 1 centimeter = 10 millimeters	1 mile = 1760 yards 1 mile = 5280 feet 1 yard = 3 feet 1 foot = 12 inches
Volume and Capacity	1 liter = 1000 milliliters	1 gallon = 4 quarts 1 gallon = 128 ounces 1 quart = 2 pints 1 pint = 2 cups 1 cup = 8 ounces
Weight and Mass	1 kilogram = 1000 grams 1 gram = 1000 milligrams	1 ton = 2000 pounds 1 pound = 16 ounces
Time		1 year = 12 months 1 year = 52 weeks 1 year = 365 days 1 week = 7 days 1 day = 24 hours 1 hour = 60 minutes 1 minute = 60 seconds

1. The graph below shows the percent of responses in a survey of families in the United States that say they depend on teens to earn money. If 750 people were surveyed, how many of the responses were "Not at all"?

Dependent On Teen's Income?

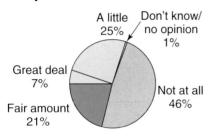

- A little 25%
- Don't know/no opinion 1%
- Great deal 7%
- Not at all 46%
- Fair amount 21%

A 5

B 75

C 157.5

D 345

E Not Here

2. The graph below shows a store's monthly sales of CDs.

CD Sales

The average price of a CD was $12. What would be a reasonable sales total, with tax, for the CDs sold during the 4-month period?

F More than $12,000

G Between $9600 and $12,000

H Between $7200 and $9600

J Between $4800 and $7200

K Less than $4800

3. The graph below shows the percent of men and women who held two jobs from 1970 to 1995.

Working Twice a Day

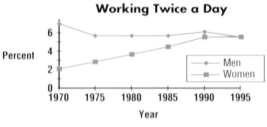

(Continued at top of next column.)

Which is a reasonable conclusion from the information on the graph?

A The number of women working two jobs increases as the number of men working two jobs decreases.

B The number of women working two jobs will continue to increase due to inflation.

C Men are retiring.

D Women are more ambitious.

E The number of men and women working two jobs was the same in 1995.

4. Akikta is making banana bread. The recipe for one loaf calls for $1\frac{3}{4}$ cups of flour. How many cups of flour does he need to make 3 loaves?

F $4\frac{1}{2}$ c

G $4\frac{3}{4}$ c

H $4\frac{9}{16}$ c

J $5\frac{1}{4}$ c

K Not Here

5. Jorge laid 18 bricks end to end to make a border for a flower bed. Each brick is $8\frac{1}{2}$ inches long. How long is the row of bricks?

A $26\frac{1}{2}$ in.

B 153 in.

C $144\frac{1}{2}$ in.

D $9\frac{1}{2}$ in.

E Not Here

6. If a bookshelf is 28 inches wide, how many books each $1\frac{3}{4}$ inches wide can be used to fill the shelf?

F 49

G 36

H 21

J 16

K Not Here

7. A developer has $43\frac{1}{2}$ acres of land that she plans to sell in $\frac{3}{8}$-acre lots. How many lots can she get from this land?

A 16.3125

B 87

C 116

D 174

E Not Here

8. A recipe calls for $\frac{3}{8}$ cup of milk and $\frac{3}{8}$ cup of syrup. What is the total amount of milk and syrup in the recipe?

F $\frac{3}{8}$ c

G $\frac{3}{4}$ c

H $1\frac{1}{8}$ c

J $\frac{9}{64}$ c

K Not Here

9. The width of a column of a magazine is $5\frac{7}{8}$ inches. The left margin is $\frac{7}{8}$ inches, and the right margin is $1\frac{1}{2}$ inches. What is the total width of the page?

A $6\frac{3}{4}$ in.

B $7\frac{3}{4}$ in.

C $8\frac{1}{4}$ in.

D $8\frac{1}{8}$ in.

E Not Here

10. Mr. Baker had $5\frac{1}{2}$ yards of curtain material. He used $2\frac{7}{8}$ yards. How much material did he have left?

F $8\frac{3}{8}$ yd

G $7\frac{3}{8}$ yd

H $3\frac{5}{8}$ yd

J $2\frac{5}{8}$ yd

K Not Here

11. Jane and Julie are identical twins. At birth, Jane weighed $5\frac{5}{8}$ pounds, and Julie weighed 7 pounds. What was the difference in their weight?

A $\frac{3}{4}$ lb

B $1\frac{1}{4}$ lb

C $2\frac{1}{8}$ lb

D $2\frac{3}{4}$ lb

E Not Here

12. What is the approximate width of a door?

F 1.2 meters

G 12 centimeters

H 12 millimeters

J 1.2 millimeters

13. Ken's recipe for chili calls for $3\frac{1}{2}$ cups of tomato sauce. He pours two 14-ounce cans of tomato sauce into a pan. How many more *ounces* does Ken need to add?

A 0 oz.

B 4 oz.

C 6 oz.

D 8 oz.

E Not Here

14. Light travels approximately 1.86 million miles in 10 seconds. The distance from the sun to Earth is about 93 million miles. Which proportion could be used to find x, the time it takes for light to travel from the sun to Earth?

F $\frac{1.86}{x} = \frac{93}{10}$

G $\frac{1.86}{10} = \frac{x}{93}$

H $\frac{1.0}{x} = \frac{93}{1.86}$

J $\frac{1.0}{93} = \frac{x}{1.86}$

K $\frac{1.86}{10} = \frac{93}{x}$

15. The ratio of your weight on the moon to your weight on Earth is about 1:6. If you weigh 72 kilograms on Earth, about how much would you weigh on the moon?

   A 36 kg

   B 28 kg

   C 14 kg

   D 12 kg

   E Not Here

16. In the figure below, what is the total number of rays in which the endpoint is *P*?

   F 1

   G 2

   H 3

   J 4

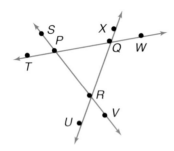

17. The sides of the triangle are best described as:

   A rays

   B line segments

   C angles

   D lines

18. Tashema wants to frame a rectangular picture that is 12 inches by 18 inches. What is the perimeter in *feet*?

   F 60 ft   H 5 ft

   G 24 ft   J $2\frac{1}{2}$ ft

19. The area of a basketball court is 4032 square feet. If the length of the court is 84 feet, what is the *approximate* width of the court?

   A 40 ft   C 80 ft

   B 50 ft   D 90 ft

20. What is the surface area of the prism?

   E 31.5 sq cm

   F 297 sq cm

   G 316 sq cm

   H 346.5 sq cm

21. What is the volume of the cylinder? $\left(\text{Use } \pi \approx \frac{22}{7}.\right)$

   A 88 in^3

   B 176 in^3

   C $276\frac{4}{7}$ in^3

   D 352 in^3

   E Not Here

22 in.

2 in.

22. Jamonte hikes 6 kilometers north and then 8 kilometers west. How far is he from his starting point?

   F 10 km

   G 12 km

   H 14 km

   J 100 km

23. *ABCD* is a rectangle. *AD* = 7 centimeters and *AC* = 25 centimeters. What is the length of $\overline{CD}$?

   A 576 cm

   B 175 cm

   C 64 cm

   D 32 cm

   E 24 cm

24. The school store has sweatshirts in 4 sizes and in 3 different colors. If the store has one of every possible size and color sweatshirt, how many different sweatshirts do they have?

   F 7

   G 8

   H 12

   J 14

25. If a fair coin is tossed 4 times, the probability that 4 heads will occur is $\frac{1}{16}$. What is the probability that 4 heads will *not* occur?

   A $\frac{7}{8}$

   B $\frac{1}{16}$

   C $\frac{15}{16}$

   D $\frac{3}{4}$

Statistics: Graphs

What YOU'LL LEARN

- To interpret graphs in real-world settings.

Why IT'S IMPORTANT

Graphs are a useful way to display data so it can be understood quickly and easily.

When data is collected and displayed in a graph, you can look for trends and often make decisions or predictions based on the data.

The most common statistical graphs include *line graphs*, *circle graphs*, and *bar graphs*. A **line graph** is the usual choice to show how values change over a period of time.

Example **1**

The double-line graph at the right shows the percent of young people who go out on dates and who play pickup sports games.

a. At what age does dating become as popular as sports?

b. What is the percent of young people who date and play pickup sports when both are equally popular?

Ready, Set, Shift

Source: Youth Sports Institute, Michigan

a. By looking at the graph, you can see that the lines appear to intersect at or shortly after 12 years of age. This is the age where dating becomes as popular as sports.

b. about 35%

Circle graphs may be the most popular type of visual display of data because they allow for more creativity than the other types. Circle graphs are used to show how parts are related to the whole.

Example **2**

The graph below shows how people describe their health. If 1500 people were surveyed to determine the graph, how many people responded good or very good?

Look at the section representing responses of good and very good.

A total of 29.9% + 36.1% or 66% responded good or very good. Find 66% of 1500 to find the number of people surveyed who answered good or very good.

1500 ☒ .66 ☐ *990*

990 people responded good or very good.

How's Your Health?

Don't Know
0.5%

Fair/Poor
13.7%

Very Good
36.1%

Good
29.9%

Excellent
19.8%

Source: NFO Research, Inc., HealthMed Division

A **bar graph** is used to compare quantities.

Example **3** The double-bar graph at the right shows how long men and women think it will be before a woman is elected U.S. president.

  a. During which time interval is the difference between men and women the greatest?

  b. What is the size of this gap?

  a. By looking at the graph, the greatest gap between men and women is in the interval labeled 11–20 years.

  b. The gap is 35% − 28% or 7%.

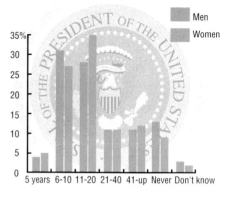

When Will She Win?

Men
Women

Source: Yankelovich for Gazelle Productions

EXERCISES

Practice **For each set of data, determine whether a bar graph, circle graph, or line graph is the best way to display the data.**

  1. number of viewers of *Monday Night Football* each year from 1970 to 1996

  2. how much different age groups in the U.S. spend on sporting apparel

  3. how U.S. consumers pay for goods and services

  4. the percent of men and women who held two jobs from 1977 to 1996

  5. the publishing budget for your school yearbook

  6. number of U.S. shipments of CDs and cassette tapes from 1985 to 1994

Solve. Use the line graph.

  7. About how many more women ran for congress in 1992 than in 1988?

  8. In what two-year span did the largest increase in the number of female candidates running for congress occur?

Female Candidates for U.S. Congress

Applications **Draw a graph that you might use to accurately represent the data. You may want to use computer software.**

  9. Olympics The table at the right shows the number of gold medals earned by athletes from the United States in the six summer Olympic Games from 1976 to 1996.

Year	Gold Medals
1976	34
1980	0
1984	83
1988	36
1992	37
1996	44

  10. Advertising The table at the right shows what TV viewers did when a commercial came on in 1985 and in 1991.

Activity	1985	1991
Change channel/mute	13%	22%
Stop paying attention	25%	27%
Watch the commercial	33%	33%
Leave the room or do something else	29%	18%

Multiplying Fractions and Mixed Numbers

What YOU'LL LEARN

- To multiply fractions and mixed numbers.

Why IT'S IMPORTANT

You can multiply fractions to help you solve problems involving statistics and economics.

You can draw an **area model** on grid paper to show multiplication of fractions. The diagram at the right models the product $\frac{2}{3} \cdot \frac{3}{4}$. The rectangle is three units wide. Two columns are colored yellow to represent $\frac{2}{3}$. The rectangle is four units long. Three rows are colored blue to represent $\frac{3}{4}$. Notice that twelve 1-by-1 squares are inside the rectangle and six of them overlap the blue and yellow, which results in green. The product of $\frac{2}{3}$ and $\frac{3}{4}$ is $\frac{6}{12}$, or $\frac{1}{2}$.

An area model shows what happens when you multiply fractions. You will get the same results if you multiply fractions as shown below.

$\frac{2}{3} \cdot \frac{3}{4} = \frac{2 \cdot 3}{3 \cdot 4}$ *Multiply the numerators and multiply the denominators.*

$\phantom{\frac{2}{3} \cdot \frac{3}{4}} = \frac{6}{12}$ or $\frac{1}{2}$ *Simplify by dividing the numerator and denominator by 6.*

Multiplying Fractions and Mixed Numbers	**To multiply fractions, multiply the numerators and then multiply the denominators. Simplify if necessary.** **To multiply mixed numbers, rename each mixed number as an improper fraction. Multiply the fractions.**

You can use the following shortcut to simplify a product.

$\frac{3}{4} \cdot \frac{8}{9} = \frac{3 \cdot 8}{4 \cdot 9}$ *Divide both the numerator and the denominator by 3 and 4.*

$\phantom{\frac{3}{4} \cdot \frac{8}{9}} = \frac{\overset{1}{3} \cdot \overset{2}{8}}{\underset{1}{4} \cdot \underset{3}{9}}$ *3 is the greatest common factor (GCF) of 3 and 9.*
   4 is the GCF of 4 and 8.

$\phantom{\frac{3}{4} \cdot \frac{8}{9}} = \frac{2}{3}$ ← *simplest form*

Example ❶ Solve $x = \frac{1}{3} \cdot 5\frac{1}{4}$.

$x = \frac{1}{3} \cdot 5\frac{1}{4}$

$x = \frac{1}{3} \cdot \frac{21}{4}$ *Rename $5\frac{1}{4}$ as $\frac{21}{4}$.*

$x = \frac{1 \cdot \overset{7}{21}}{3 \cdot 4}$ *The GCF of 21 and 3 is 3.*
$ \underset{1}{}$ *Divide 21 and 3 by 3.*

$x = \frac{7}{4}$ or $1\frac{3}{4}$

Practice

1. **Draw** an area model that demonstrates that the product of $\frac{3}{4}$ and $\frac{1}{2}$ is $\frac{3}{8}$.

Solve each equation. Write the solution in simplest form.

2. $\frac{1}{7}\left(\frac{1}{3}\right) = x$

3. $v = \frac{2}{3}\left(\frac{1}{5}\right)$

4. $\frac{5}{6}\left(\frac{3}{10}\right) = y$

5. $w = \frac{7}{12}\left(\frac{4}{5}\right)$

6. $\frac{7}{10}\left(\frac{5}{28}\right) = z$

7. $s = \frac{9}{15}\left(\frac{5}{9}\right)$

8. What is the product of 9 and $\frac{2}{3}$?

9. What is $\frac{4}{5}$ of 30?

10. What is $\frac{2}{5}$ of $10\frac{1}{2}$?

Solve each equation. Write the solution in simplest form.

11. $k = 2\left(1\frac{5}{18}\right)$

12. $3\frac{2}{5}(10) = m$

13. $n = 2\left(3\frac{5}{8}\right)$

14. $2\frac{2}{3}\left(\frac{4}{5}\right) = n$

15. $t = \frac{7}{8}\left(4\frac{1}{4}\right)$

16. $d = 2\frac{3}{10}\left(\frac{5}{12}\right)$

17. $9\frac{1}{3}\left(3\frac{3}{4}\right) = f$

18. $3\frac{1}{3}\left(4\frac{1}{2}\right) = w$

19. $v = 6\frac{1}{4}\left(1\frac{7}{16}\right)$

20. $\left(1\frac{7}{8}\right)4\frac{2}{7}\left(3\frac{4}{8}\right) = c$

21. $3\frac{1}{8}\left(2\frac{4}{5}\right)\left(\frac{5}{7}\right) = p$

22. $3\frac{2}{3}\left(\frac{1}{8}\right)\left(1\frac{1}{11}\right) = h$

Applications

23. **Statistics** The graph at the right shows which fast-food restaurants kids prefer according to a 1995 poll. Suppose 600 kids participated in the survey.
    a. How many kids preferred Taco Bell?
    b. How many kids preferred Pizza Hut?
    c. How many more kids preferred McDonald's over Burger King?

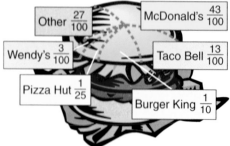

Kid's Fast-Food Favorites

Other $\frac{27}{100}$ McDonald's $\frac{43}{100}$

Wendy's $\frac{3}{100}$ Taco Bell $\frac{13}{100}$

Pizza Hut $\frac{1}{25}$ Burger King $\frac{1}{10}$

Source: *Sports Illustrated for Kids*

24. **Economics** The sales tax in Franklin County is $7\frac{3}{4}\%$. This means for every dollar you spend, you pay $7\frac{3}{4}$¢ in tax. If you purchase an item for $24, how much will you owe in tax?

25. **Economics** A personal AM/FM radio and cassette player that usually sells for $69 is on sale at $\frac{1}{3}$ off. This means that it is on sale for $\frac{2}{3}$ of the regular price. What is the sale price?

Dividing Fractions and Mixed Numbers

Dana is planning a pizza party. She estimates that each person will eat an average of $\frac{2}{3}$ of a pizza. If she buys four pizzas, will there be enough for five people?

To solve this problem, you need to find how many $\frac{2}{3}$ pizzas are in 4 pizzas.

Method 1

You can model the problem above using an area model. Divide 4 by $\frac{2}{3}$.

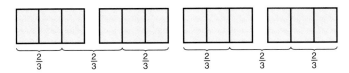

So, $4 \div \frac{2}{3} = 6$.

Method 2

You can also divide by a fraction or mixed number. To do this, multiply by its multiplicative inverse.

You can also use the GCF of 2 and 4, which is 2.

$$4 \div \frac{2}{3} = \frac{\overset{2}{\cancel{4}}}{1} \cdot \frac{3}{\underset{1}{\cancel{2}}}$$
$$= \frac{6}{1} \text{ or } 6$$

$$4 \div \frac{2}{3} = \frac{4}{1} \div \frac{2}{3} \quad \textit{Rename 4 as } \frac{4}{1}.$$
$$= \frac{4}{1} \cdot \frac{3}{2} \quad \textit{Dividing by } \frac{2}{3} \textit{ is the same as multiplying by } \frac{3}{2}.$$
$$= \frac{12}{2} \text{ or } 6$$

Four pizzas will be enough to feed 6 people.

Dividing Fractions and Mixed Numbers	**To divide by a fraction, multiply by its multiplicative inverse. Simplify if necessary.** **To divide by a mixed number, rename the mixed number as an improper fraction. Multiply by its multiplicative inverse.**

Example **1** Solve $x = \frac{2}{3} \div 2\frac{6}{7}$.

$$x = \frac{2}{3} \div 2\frac{6}{7}$$

$$x = \frac{2}{3} \div \frac{20}{7} \quad \textit{Rename } 2\frac{6}{7} \textit{ as } \frac{20}{7}.$$

$$x = \frac{2}{3} \cdot \frac{7}{20} \quad \textit{Dividing by } \frac{20}{7} \textit{ is the same as multiplying by } \frac{7}{20}.$$

$$x = \frac{\overset{1}{\cancel{2}} \cdot 7}{3 \cdot \underset{10}{\cancel{20}}} \quad \textit{The GCF of 2 and 20 is 2. Divide 2 and 20 by 2.}$$

$$x = \frac{7}{30}$$

Practice

1. **Draw** an area model that demonstrates $\frac{1}{2} \div 6 = \frac{1}{12}$.

Solve each equation. Write the solution in simplest form.

2. $2 \div \frac{1}{3} = j$

3. $g = \frac{1}{2} \div 8$

4. $3 \div \frac{1}{2} = y$

5. $q = 7 \div 4$

6. $n = 5 \div 11$

7. $\frac{5}{6} \div \frac{1}{6} = f$

8. What is the quotient of $\frac{2}{3}$ and $\frac{4}{9}$?

9. Solve the equation $c = \left(\frac{1}{4} \cdot \frac{5}{8} \right) \div 3\frac{1}{5}$.

Solve each equation. Write the solution in simplest form.

10. $r = \frac{7}{9} \div \frac{1}{9}$

11. $\frac{7}{16} \div \frac{7}{11} = s$

12. $\frac{6}{13} \div \frac{5}{7} = t$

13. $e = 1\frac{3}{4} \div \frac{7}{12}$

14. $f = \frac{3}{5} \div 11\frac{2}{5}$

15. $4\frac{2}{5} \div \frac{11}{15} = g$

16. $d = 1 \div 2\frac{3}{5}$

17. $2\frac{5}{8} \div 7\frac{1}{2} = x$

18. $5\frac{3}{5} \div 4\frac{1}{5} = y$

Applications

19. **Carpentry** Builders Square has two stacks of $\frac{5}{8}$-in. plywood sheets. One stack is 40 inches high. The other is 25 inches high. How many sheets of plywood are there in both stacks?

20. **Home Economics** The graph below shows the cost per year for different uses of energy in the average household.

Energy Expenses

Other $\frac{11}{100}$

Home heating $\frac{13}{25}$

Air conditioning $\frac{2}{25}$

Cooking/refrigeration $\frac{11}{100}$

Water heating $\frac{9}{50}$

Source: *The Family Handyman, Reader's Digest Special Interest Magazine Group*

a. Suppose a family's only use of natural gas is for water heating. If their cost for gas in a year is $172.80, what was their total household energy expense for the year?

b. How many times as much does it cost for water heating than it does for air conditioning?

c. How many times as much does it cost for heating the home than it does for air conditioning?

Adding Fractions and Mixed Numbers

WhaT YOU'LL LEARN

- To add fractions and mixed numbers.

Why IT'S IMPORTANT

You can divide fractions to help you solve problems involving cooking, carpentry, and consumer spending.

Since a fraction can be represented using an area model, you can model the addition of fractions by using different colors of shading. The model at the right represents $\frac{3}{5} + \frac{1}{5}$. Since 4 of the 5 sections are shaded, the sum is $\frac{4}{5}$.

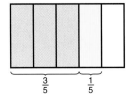

| **Adding Like Fractions** | **To add fractions with like denominators, add the numerators. Then simplify.** |

Example Solve each equation. Write the solution in simplest form.

a. $\frac{2}{3} + \frac{2}{3} = x$

$\frac{4}{3} = x$ *Since the denominators are the same, add the numerators.*

$1\frac{1}{3} = x$ *Rename $\frac{4}{3}$ as a mixed number.*

b. $t = 2\frac{3}{8} + 3\frac{5}{8}$

$t = (2 + 3) + \left(\frac{3}{8} + \frac{5}{8}\right)$ *Use the associative and commutative properties to add the whole numbers and fractions separately.*

$t = 5 + \frac{8}{8}$ *Add the numerators of the fractions.*

$t = 5 + 1 \text{ or } 6$ *Simplify.*

The model at the left represents the sum of $\frac{1}{3}$ and $\frac{1}{6}$. This leads us to the following rule for adding unlike fractions.

| **Adding Unlike Fractions** | **To find the sum of two fractions with unlike denominators, rename the fractions with a common denominator. Then add and simplify.** |

Example Solve each equation. Write the solution in simplest form.

a. $y = \frac{1}{8} + \frac{3}{4}$ *The least common multiple (LCM) of 8 and 4 is 8.*

$y = \frac{1}{8} + \frac{6}{8}$ *Rename $\frac{3}{4}$ as $\frac{6}{8}$.*

$y = \frac{7}{8}$

b. $9\frac{1}{6} + 7\frac{7}{9} = d$ *Use the LCM of 6 and 9 to rename $\frac{1}{6}$ as $\frac{3}{18}$ and $\frac{7}{9}$ as $\frac{14}{18}$.*

$9\frac{3}{18} + 7\frac{14}{18} = d$

$16\frac{17}{18} = d$ *Add the whole numbers. Then add the fractions.*

EXERCISES

Practice   **Write the addition sentence shown by each model.**

1. ⊕ + ⊕ → ○ ◐   **2.** ⬡ + ⬡ → ⬡

Draw an area model to show each sum.

3. $\frac{5}{8} + \frac{5}{8} = 1\frac{1}{4}$

4. $\frac{1}{2} + \frac{1}{4} = \frac{3}{4}$

Solve each equation. Write each solution in simplest form.

5. $x = \frac{1}{3} + \frac{1}{3}$

6. $\frac{4}{7} + \frac{5}{7} = a$

7. $\frac{2}{9} + \frac{4}{9} = t$

8. $y = \frac{1}{8} + \frac{3}{8}$

9. $\frac{1}{6} + \frac{5}{6} = f$

10. $\frac{7}{12} + \frac{11}{12} = g$

11. $\frac{1}{3} + \frac{1}{2} = b$

12. $r = \frac{1}{2} + \frac{1}{8}$

13. $h = \frac{2}{3} + \frac{3}{4}$

14. $\frac{1}{2} + \frac{1}{8} + \frac{1}{4} = z$

15. $c = \frac{7}{8} + \frac{3}{4} + \frac{1}{2}$

16. $m = \frac{1}{4} + \frac{1}{3} + \frac{1}{2}$

17. $4\frac{1}{2} + \frac{5}{8} = k$

18. $s = 73\frac{1}{2} + \frac{2}{3}$

19. $3\frac{1}{4} + 5\frac{1}{3} = d$

20. $j = 9\frac{1}{3} + 5\frac{5}{6}$

21. $v = 4\frac{1}{2} + 7\frac{5}{8}$

22. $1\frac{2}{3} + 5\frac{7}{8} = n$

Applications   **23. Home Economics**   Mr. Sopher used $\frac{1}{3}$ cup of sour cream and $\frac{1}{4}$ cup of mayonnaise as the base for a dip. How much dip does he have so far?

24. Carpentry   The students in Advanced Cabinetry made a desktop by glueing a sheet of $\frac{1}{16}$-inch walnut veneer to both sides of a sheet of $\frac{5}{8}$-inch plywood. What was the total thickness of the desktop?

25. Consumer Spending
The graph at the right shows the fraction of each dollar a family spends on different expenses.
  a. What fraction of each dollar is spent on shelter and transportation?
  b. What fraction of each dollar is spent on apparel and medical/personal care?
  c. What fraction of each dollar is spent on things not mentioned in parts a and b?

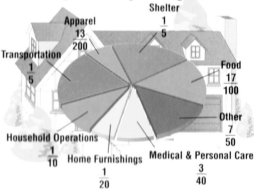

Average Family Budget, 1990

Shelter $\frac{1}{5}$
Apparel $\frac{13}{200}$
Transportation $\frac{1}{5}$
Food $\frac{17}{100}$
Other $\frac{7}{50}$
Household Operations $\frac{1}{10}$
Home Furnishings $\frac{1}{20}$
Medical & Personal Care $\frac{3}{40}$

Source: The Conference Board

Subtracting Fractions and Mixed Numbers

- To subtract fractions and mixed numbers.

Why IT'S IMPORTANT

You can subtract fractions to help you solve problems involving sewing and business.

As you can see in the model below, subtracting fractions with like denominators is similar to adding fractions with like denominators. The model represents $\frac{3}{4} - \frac{1}{4}$. Since 2 of the 4 sections are shaded, the difference is $\frac{2}{4}$ or $\frac{1}{2}$.

| **Subtracting Like Fractions** | **To subtract fractions with like denominators, subtract the numerators. Then simplify.** |

Example ① **Solve each equation. Write the solution in simplest form.**

a. $\frac{3}{5} - \frac{1}{5} = m$

$\frac{2}{5} = m$ *Since the denominators are the same, subtract the numerators.*

b. $n = 4\frac{1}{8} - 2\frac{5}{8}$

$n = 3\frac{9}{8} - 2\frac{5}{8}$ *Replace $4\frac{1}{8}$ with $3\frac{9}{8}$.*

$n = (3 - 2) + \left(\frac{9}{8} - \frac{5}{8}\right)$ *Use the associative and commutative properties to subtract the whole numbers and fractions separately.*

$n = 1 + \frac{4}{8}$ *Subtract the numerators of the fractions.*

$n = 1 + \frac{1}{2}$ or $1\frac{1}{2}$ *Simplify.*

The model at the left represents the difference of $\frac{1}{3}$ and $\frac{1}{6}$. This leads us to the following rule for subtracting unlike fractions.

| **Subtracting Unlike Fractions** | **To find the difference of two fractions with unlike denominators, rename the fractions with a common denominator. Then subtract and simplify.** |

Example ② **Solve each equation. Write the solution in simplest form.**

a. $p = \frac{6}{7} - \frac{2}{3}$

$p = \frac{18}{21} - \frac{14}{21}$ *The LCM of 7 and 3 is 21. Rename $\frac{6}{7}$ as $\frac{18}{21}$ and $\frac{2}{3}$ as $\frac{14}{21}$.*

$p = \frac{4}{21}$

b. $9\frac{1}{6} - 7\frac{7}{9} = q$

$9\frac{3}{18} - 7\frac{14}{18} = q$ *Use the LCM of 6 and 9 to rename $\frac{1}{6}$ as $\frac{3}{18}$ and $\frac{7}{9}$ as $\frac{14}{18}$.*

$8\frac{21}{18} - 7\frac{14}{18} = q$ *Rename $9\frac{3}{18}$ as $8\frac{21}{18}$.*

$1\frac{7}{18} = q$ *Subtract the whole numbers. Then subtract the fractions.*

EXERCISES

Practice   **Write the subtraction sentence shown by each model.**

1.   **2.**

Draw an area model to show each difference.

3. $\frac{3}{5} - \frac{2}{5} = \frac{1}{5}$ **4.** $\frac{1}{2} - \frac{1}{4} = \frac{1}{4}$

Solve each equation. Write the solution in simplest form.

5. $\frac{5}{8} - \frac{1}{8} = w$ **6.** $\frac{5}{6} - \frac{1}{6} = p$ **7.** $\frac{9}{10} - \frac{7}{10} = n$

8. $a = \frac{7}{11} - \frac{3}{11}$ **9.** $1\frac{1}{4} - \frac{3}{4} = g$ **10.** $x = \frac{7}{12} - \frac{1}{12}$

11. $\frac{1}{2} - \frac{1}{8} = j$ **12.** $\frac{2}{3} - \frac{1}{6} = m$ **13.** $k = \frac{5}{6} - \frac{5}{12}$

14. $\frac{1}{2} - \frac{1}{3} = v$ **15.** $\frac{5}{6} - \frac{3}{4} = d$ **16.** $b = \frac{8}{9} - \frac{1}{6}$

17. $z = 7\frac{5}{8} - 4\frac{5}{8}$ **18.** $c = 15 - 2\frac{1}{3}$ **19.** $f = 4\frac{5}{8} - 3\frac{1}{2}$

20. $8\frac{7}{8} - 4\frac{2}{3} = r$ **21.** $4\frac{1}{2} - 2\frac{3}{4} = q$ **22.** $h = 5\frac{5}{16} - 2\frac{3}{8}$

23. Subtract $\frac{11}{16}$ from 20.

Applications   **24. Sewing**   Becky is making a suit. She needs $2\frac{5}{8}$ yards of material for the jacket and $\frac{7}{8}$ yard of material for the skirt. How much more material does she need for the jacket than the skirt?

25. Business   The chart at the right shows the closing price of Nike stock for each day of the week before the 1996 New York City Marathon.

   a. How much did the price of the stock drop from Tuesday to Wednesday?

   b. Between which two consecutive days did the largest drop occur?

Day	Closing Price
Monday	$62\frac{1}{4}$
Tuesday	$60\frac{7}{8}$
Wednesday	$59\frac{3}{4}$
Thursday	$59\frac{1}{8}$
Friday	$54\frac{3}{4}$

Measurement

The **meter (m)** is the standard unit of length in the metric system. All units of length in the metric system are defined in terms of the meter. In the customary system, the basic unit of length is the **inch (in.)**.

Metric Length	Customary Length
1000 millimeters (mm) = 1 meter	12 inches = 1 foot (ft)
100 centimeters (cm) = 1 meter	3 feet = 1 yard (yd)
1000 meters = 1 kilometer (km)	1760 yards = 1 mile (mi)

The **liter (L)** is the standard unit of capacity in the metric system. Liters, milliliters, kiloliters, and so on, are related in a manner similar to the units of length. In the customary system, the basic unit of capacity for liquids is the **fluid ounce (fl oz)**.

Metric Capacity	Customary Capacity
1000 milliliters (mL) = 1 liter	8 fluid ounces = 1 cup (c)
1000 liters = 1 kiloliter (kL)	2 cups = 1 pint (pt)
	2 pints = 1 quart (qt)
	4 quarts = 1 gallon (gal)

The *mass* of an object is the amount of matter that it contains. The **gram (g)** is the standard unit of mass in the metric system. Units of mass are also related in a manner similar to the units of length. In the customary system, the basic unit of *weight* is the **ounce (oz)**.

Metric Mass	Customary Weight
1000 milligrams (mg) = 1 gram	16 ounces = 1 pound (lb)
1000 grams = 1 kilogram (kg)	2000 pounds = 1 ton (T)

To convert units within the metric system, you follow the same procedure as when you multiply or divide by powers of ten.

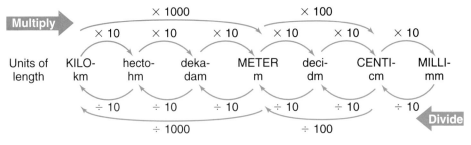

The prefixes in capital letters are more commonly used.

Example **Complete each sentence.**

a. 200 cm = _?_ m

Centimeters are smaller than meters. Converting to larger units means there will be fewer units. Divide by 100.

$200 \div 100 = 2$

$200 \text{ cm} = 2 \text{ m}$

b. 0.354 L = _?_ mL

Liters are larger than milliliters. Converting to smaller units means there will be more units. Multiply by 1000.

$0.354 \times 1000 = 354$

$0.354 \text{ L} = 354 \text{ mL}$

Converting from one unit to another in the customary system may involve more than one step. The rules about when to multiply or divide are the same as those used with the metric system.

Example **Complete each sentence.**

a. 1035 ft = _?_ yd

Feet are smaller than yards. Smaller to larger means fewer units. Divide by 3 since there are 3 feet in a yard.

1035 ÷ 3 = 345
1035 ft = 345 yd

b. 8 qt = _?_ pt

Quarts are larger than pints. Larger to smaller means more units. Multiply by 2.

8 × 2 = 16
8 qt = 16 pt

EXERCISES

Practice

1. Suppose your best friend misses school the day measurement was covered. Write a few sentences telling her how to determine when to multiply or divide when converting measures.

Complete each sentence.

2. 6.9 mm = _?_ cm

3. 2.5 km = _?_ m

4. 9.1 kg = _?_ mg

5. 34 cm = _?_ m

6. 642 mL = _?_ L

7. 7 g = _?_ mg

8. 12 ft = _?_ in.

9. 4 T = _?_ lb

10. 9 pt = _?_ c

11. 31,680 ft = _?_ mi

12. 3 lb = _?_ oz

13. 15 qt = _?_ gal

Write which metric unit you would probably use to measure each item. Then write which customary unit you would probably use.

14. water in a swimming pool

15. width of a door

16. weight of a 15-year old male

17. soup spoon

Applications

18. **Archaeology**   The dimensions of an arrowhead are shown in the illustration at the right. Will this arrowhead fit in a box that is 0.04 m long and 2.5 cm wide? Explain your reasoning.

34 mm   18 mm   14 mm

19. **Biology**   About 200 million years ago, the largest living animal was a dinosaur called a brachiosaurus. It weighed about 170,000 pounds. Today the largest living animal is the blue whale. It weighs about 220 tons. How many more tons does today's blue whale weigh than the brachiosaurus?

20. **Ecology**   Every ton of recycled office paper saves about 17 trees. Glencoe/McGraw-Hill's office in Westerville, Ohio, recycled 32,634 pounds of paper in 1995. How many trees did this save?

Using Proportions

***What* YOU'LL LEARN**

- To use proportions to solve problems.

***Why* IT'S IMPORTANT**

You can use proportions to solve many real-life problems involving cooking, sports, and photography.

If you are planning a trip to a town you have never visited, you probably would look at a road map to see what roads to take. On the map, you might find a scale that says one inch equals approximately 18.2 miles or 29.2 kilometers.

The scale on a map is the ratio of the distance on the map to the actual distance. A **ratio** is a comparison of two numbers by division. The ratio that compares 5 to 146 can be written as follows.

$$5 \text{ to } 146 \qquad 5{:}146 \qquad 5 \text{ out of } 146 \qquad \frac{5}{146}$$

When you know the scale of a map, you can find the actual distances by writing and solving proportions. A **proportion** is an equation that shows that two ratios are equivalent. The **cross products** in the proportion below are $9 \cdot 24$ and $12 \cdot 18$. In a proportion, the two cross products are equal.

$$\frac{9}{12} = \frac{18}{24} \qquad \begin{array}{l} 12 \cdot 18 \\ 9 \cdot 24 \end{array} \qquad \begin{array}{l} 12 \cdot 18 = 216 \\ 9 \cdot 24 = 216 \end{array}$$

You can use the fact that cross products are equal to solve proportions.

Example **Solve each proportion.**

a. $\dfrac{c}{36} = \dfrac{15}{24}$

$c \cdot 24 = 36 \cdot 15$ *Cross products are equal.*

$24c = 540$ *Multiply.*

$\dfrac{24c}{24} = \dfrac{540}{24}$ *Divide each side by 24.*

$c = 22.5$

The solution is 22.5.

b. $\dfrac{7}{12} = \dfrac{1.68}{d}$

$7 \cdot d = 12 \cdot 1.68$ *Cross products are equal.*

$d = \dfrac{12 \cdot 1.68}{7}$

12 ☒ 1.68 ÷ 7 = *2.88*

Thus, $d = 2.88$.

The solution is 2.88.

Proportions can be used to solve many real-life problems.

Example **The distance between Houston and San Antonio on a map is 6.2 inches. The scale on the map is 1 inch:30 miles. Find the distance between Houston and San Antonio.**

$$\begin{array}{l} \textit{map distance} \rightarrow \\ \textit{actual distance} \rightarrow \end{array} \frac{1 \text{ inch}}{30 \text{ miles}} = \frac{6.2 \text{ inches}}{n} \begin{array}{l} \leftarrow \textit{map distance} \\ \leftarrow \textit{actual distance} \end{array}$$

$$1 \cdot n = 6.2 \cdot 30$$

$$n = 186$$

The actual distance between Houston and San Antonio is about 186 miles.

Example ③ A tree casts a shadow 19.5 meters long at the same time a 2-meter post casts a shadow 3 meters long. How tall is the tree?

$$\begin{array}{cc} \textit{tree height} & \to \\ \textit{tree shadow} & \to \end{array} \quad \frac{x \text{ meters}}{19.5 \text{ meters}} = \frac{2 \text{ meters}}{3 \text{ meters}} \quad \begin{array}{c} \leftarrow \\ \leftarrow \end{array} \begin{array}{c} \textit{post height} \\ \textit{post shadow} \end{array}$$

$$\frac{x}{19.5} = \frac{2}{3}$$

$3x = 39$ *Find the cross products.*

$x = 13$ *Divide each side by 3.*

The tree is 13 meters tall.

EXERCISES

Practice

1. **Write** the cross products for $\frac{1}{2} = \frac{x}{8}$.

2. **Write** two ratios that do not form a proportion. Explain why they do not.

Solve each proportion.

3. $\frac{m}{4} = \frac{7}{14}$ 4. $\frac{a}{3} = \frac{10}{15}$ 5. $\frac{28}{35} = \frac{8}{r}$ 6. $\frac{14}{t} = \frac{4}{6}$

7. $\frac{8}{14} = \frac{12}{y}$ 8. $\frac{10}{6} = \frac{h}{26}$ 9. $\frac{c}{2} = \frac{7}{4.2}$ 10. $\frac{0.7}{n} = \frac{2.1}{18}$

Write a proportion that could be used to solve for each variable. Then solve the proportion.

11. 5 quarts fill c cups   12. 625 bushels for 5 acres

    2 quarts fill 8 cups b bushels for 2 acres

13. n boxes in 5 crates   14. 60 sliced mushrooms on 4 pizzas

    8 boxes in 2 crates m sliced mushrooms on 15 pizzas

15. 2 liters at \$0.98   16. 18 plums weigh 3.375 pounds

    3 liters at \$$d$ 27 plums weigh z pounds

Applications

17. **Geometry** In geometry, figures that have the same shape but differ in size are called **similar figures**. Also, if two triangles are similar, then their corresponding sides are proportional. In the figure at the right, $\triangle ABC$ is similar to $\triangle ADE$. Find the value of x.

18. **Cooking** Miguel is baking cookies for the class bake sale. The recipe calls for 3 cups of flour for 48 cookies. How many cups of flour are needed for 72 cookies?

19. **Sports** Ken Griffey, Jr. got 3 hits in his first 8 at-bats. How many hits must he get in his next 200 at-bats to maintain this ratio?

20. **Photography** In simple cameras, like the one shown at the right, light from a subject passes through a lens and makes an image on film. The image and subject are always in proportion.

$$\frac{\textit{image size}}{\textit{subject size}} \quad \frac{\textit{image distance from lens}}{\textit{subject distance from lens}}$$

Find the distance represented by d in the diagram.

Geometry

A **point** is the simplest figure studied in geometry. A point is an exact location in space with no size or shape. A point can be represented by a dot and named with a capital letter.

Two points and all points on a straight path between them is called a **line segment**. A line segment is named by its endpoints. When two line segments have the same length, we say they are **congruent**.

If a segment connecting two points is extended indefinitely in both directions, it is called a **line**. Arrowheads are used to show that a line has no endpoints. A line can be named by two points on the line or with a single lowercase letter.

In geometry, a **ray** is defined as a segment that is extended indefinitely in one direction. A ray is named using the endpoint first and then any other point on the ray.

A•

point B •

C • ———————— • D
segment *CD* or $\overline{CD}$ or
segment *DC* or $\overline{DC}$

E   F
ℓ ←—•————•—→
line *EF* or $\overleftrightarrow{EF}$ or
line *FE* or $\overleftrightarrow{FE}$ or
line ℓ

G   H
•————•—→
ray *GH* or $\overrightarrow{GH}$

Example ❶ **Name two points, a line, two rays, and a line segment in the figure at the right.**

Two of the points are *A* and *B*.

You can use a single lowercase letter or any two points labeled on the line. So you can name it *n*, $\overleftrightarrow{AB}$, $\overleftrightarrow{AC}$, or $\overleftrightarrow{BC}$.

Two of the rays are $\overrightarrow{BC}$ and $\overrightarrow{BA}$.

One of the line segments shown is $\overline{AC}$.

When two rays have a common endpoint, they form an **angle**. The common endpoint is called the **vertex**, and the rays are called the **sides**. An angle is named using three points with the vertex as the middle letter. If no other angle shares the same vertex, the angle can be named with just the vertex letter or with a numeral. Two angles that have the same measure are also said to be congruent.

To say the measure of angle DEF is 120 degrees, we write m∠DEF = 120°.

Another basic geometric figure is a **plane**. A plane is a flat surface with no edges, or boundaries. It can be named by a single uppercase script letter or by using any three points on the plane.

Two lines that lie in the same plane and never intersect are called **parallel lines**.

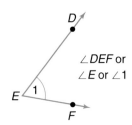

∠*DEF* or
∠*E* or ∠1

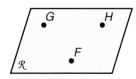

Plane *FGH* or plane ℛ

ℓ ∥ m

A line that intersects two other lines is called a **transversal**.

Congruent Angles with Parallel Lines	**If a pair of parallel lines is intersected by a transversal, these pairs of angles are congruent.** **alternate interior angles:** $\angle 3 \cong \angle 5, \angle 4 \cong \angle 6$ **alternate exterior angles:** $\angle 1 \cong \angle 7, \angle 2 \cong \angle 8$ **corresponding angles:** $\angle 1 \cong \angle 5, \angle 2 \cong \angle 6, \angle 3 \cong \angle 7, \angle 4 \cong \angle 8$

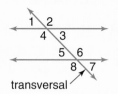

transversal

Example ② In the figure, $m \parallel n$ and $m\angle 1 = 110°$.

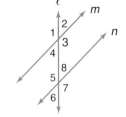

    a. Find $m\angle 7$.

        $\angle 1$ and $\angle 7$ are alternate exterior angles.

        They are congruent so their measures are the same. $m\angle 7 = 110°$

    b. Find $m\angle 3$.

        $\angle 3$ and $\angle 7$ are corresponding angles.

        Their measures are congruent, so $m\angle 3 = 110°$.

EXERCISES

Practice **Draw and label a diagram to represent each of the following.**

1. point A
2. line b
3. plane XYZ
4. $\overline{EF}$
5. $\overrightarrow{GH}$
6. $\angle D$
7. $\overline{AB} \parallel \overline{CD}$
8. $\angle TUV$
9. $\overrightarrow{RS}$

Draw $\overleftrightarrow{AB} \parallel \overleftrightarrow{CD}$ and transversal $\overleftrightarrow{EF}$.

10. Name the interior angles.
11. Name the exterior angles.

In the figure below, $\ell \parallel m$ and $m\angle 1 + m\angle 2 = 180°$. If $m\angle 1 = 49°$, find the measure of each angle.

12. $\angle 2$
13. $\angle 7$
14. $\angle 8$
15. $\angle 5$
16. $\angle 3$
17. $\angle 6$

Applications 18. **Carpentry** Mrs. Kwan wants to build some stools that won't wobble. Explain why she should make the stools with three legs.

19. **Traffic** The shape of a stop sign is a regular octagon.
    a. What geometric figure does each edge of a stop sign represent?
    b. How many pairs of edges in a stop sign are parallel?

20. **Carpentry** A carpenter uses a protractor and a plumb line (a string with a weight attached) to measure the angle between a cathedral ceiling and the supporting wall. If $m\angle CXA = 115°$, find $m\angle CWB$.

Perimeter and Area

The distance around a geometric figure is called its **perimeter**. Below are all the different rectangles that have a perimeter of 16 and sides with measures that are whole numbers. You can find the perimeter of each by counting.

Perimeter can also be found by doubling the sum of the length and the width.

$P = 2(1 + 7)$	$P = 2(2 + 6)$	$P = 2(3 + 5)$	$P = 2(4 + 4)$
$= 16$	$= 16$	$= 16$	$= 16$

The measure of the surface enclosed by a geometric figure is called its **area**. You can find the area of a rectangle by counting the number of 1-by-1 unit squares or by multiplying the measures of the length and the width.

$$A = 1 \cdot 7 \text{ or } 7 \qquad A = 2 \cdot 6 \text{ or } 12 \qquad A = 3 \cdot 5 \text{ or } 15 \qquad A = 4 \cdot 4 \text{ or } 16$$

Perimeter and Area of a Rectangle

> The perimeter of a rectangle is twice the sum of its length ℓ and width w.
>
> $$P = 2(\ell + w)$$
>
> The area of a rectangle is the product of its length ℓ and width w.
>
> $$A = \ell w$$

A square is a special rectangle in which the lengths of all the sides are equal. For this reason, the formula for the perimeter of square is $P = 4s$ and the formula for the area of a square is $A = s^2$.

Example **1** **Find the perimeter and area of rectangle *ABCD*.**

$P = 2(\ell + w)$
$P = 2(6 + 4)$
$P = 2(10) \text{ or } 20$ The perimeter is 20 inches.

$A = \ell w$
$A = 6 \cdot 4 \text{ or } 24$ The area is 24 square inches.

Use the formulas below to find the areas of other plane figures.

	Parallelograms	Triangles	Trapezoids	Circles
Area Formulas for Parallelograms, Triangles, Trapezoids, and Circles	The area of a parallelogram is the base times the height. $A = bh$	The area of a triangle is one-half the base times the height. $A = \frac{1}{2}bh$	The area of a trapezoid is one-half the height times the sum of the bases. $A = \frac{1}{2}h(b_1 + b_2)$	The area of a circle is π times the square of the radius. $A = \pi r^2$

Example **2** Find the area of each figure.

a.

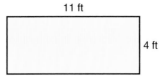
11 ft

4 ft

$A = bh$

$A = 11 \cdot 4$ or 44

The area is 44 ft².

b.

5 in.

9 in.

$A = \frac{1}{2}bh$

$A = \frac{1}{2}(5)(9)$ or $22\frac{1}{2}$

The area is $22\frac{1}{2}$ in².

c.

8 in.

6 in.

19 in.

$A = \frac{1}{2}h(b_1 + b_2)$

$A = \frac{1}{2} \cdot 6(8 + 19)$

$A = \frac{1}{2} \cdot \overset{3}{\underset{1}{\cancel{6}}} \cdot 27$ or 81

The area is 81 in².

d.

4 cm

$A = \pi r^2$

$A \approx 3.14(4^2)$

$A \approx 3.14(16)$

$A \approx 50.24$

The area is about 50 cm².

EXERCISES

Practice

1. **Draw and label** a rectangle that has a perimeter of 12 inches.

2. **Draw and label** a triangle that has an area of 30 square centimeters.

Find the perimeter and area of each figure.

3.

13 in.

21 in.

4.

4 cm

5.

4 ft   5 ft

3 ft

Find the area of each figure.

6.

3 m

5 m

7.

5 mi

6 mi

12 mi

8.

10 m

Applications

9. **Home Maintenance** How much will it cost to tile the floor of a rectangular room if the tiles cost $1.29/ft² and the room is 12 ft by 18 ft?

10. **Gardening** Each participant in a community vegetable garden is allotted a rectangular-shaped plot that covers 1620 ft². Each plot is 18 ft wide.
    a. What is the length of each plot?
    b. What is the perimeter of each plot?

Surface Area and Volume

In geometry, we study three-dimensional figures. A three-dimensional figure that has two parallel congruent sides is called a prism. A **prism** is named by the shape of its parallel bases. The figure at the right is an example of a **rectangular prism**.

The amount of material that it would take to cover a three-dimensional figure is called the **surface area** of the figure. To find the surface area of a three-dimensional figure, find the area of each side of the figure. One way to find the surface area of a three-dimensional figure is to "unfold" it. The result is called a **net**. The surface area of a three-dimensional figure is the area of the net.

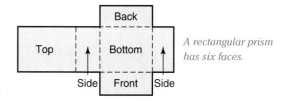

A rectangular prism has six faces.

Example ❶ **Find the surface area of a CD case that is 14.3 cm long, 12.4 cm wide, and 1.1 cm thick.**

Draw a net for the CD case and label the dimensions of each face. Use the formula $A = \ell w$ to find the areas of the faces.

Top and Bottom: 14.3 ⊠ 12.4 ⊠ 2 ⊟ *354.64*

Two Sides: 12.4 ⊠ 1.1 ⊠ 2 ⊟ *27.28*

Front and Back: 14.3 ⊠ 1.1 ⊠ 2 ⊟ *31.46*

Add to find the total surface area.

Total: 354.64 ⊞ 27.28 ⊞ 31.46 ⊟ *413.38*

The surface are of the CD case is 413.38 cm².

A can of soup is an example of a **cylinder**. The bases of a circular cylinder are two parallel, congruent circular regions. As with prisms, you find the surface area of a cylinder by finding the area of the two bases and adding the area of the curved side.

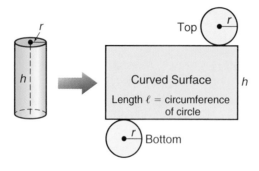

A net of a cylinder is shown at the left. Notice that the curved side is a rectangle that has the same height as the cylinder and the length is the same as the circumference of the base. Thus, the formula for the surface area of a circular prism is $2\pi r^2 + 2\pi rh$.

The amount a container will hold is called its capacity, or **volume**. Volume is usually measured in cubic units. To find the volume of a prism, multiply the area of the base times the height. You can use the formulas below to find the volumes of prisms and cylinders.

	Prisms	Circular Cylinders
Volume Formulas for Prisms and Circular Cylinders	**The volume of a rectangular prism is the area of the base times the height.** $V = Bh$	**The volume of a circular cylinder is the product of π, the square of the radius, and the height.** $V = \pi r^2 h$

Example **Find the volume of the cylinder. Use 3.14 for π.**

$V = \pi r^2 h$ *Formula for volume*

$V = 3.14 \cdot 7^2 \cdot 20$ *The diameter of the cylinder is 14 cm so*

$V \approx 3077$ *the radius is 7 cm.*

The volume is about 3077 cm^3.

EXERCISES

Practice

1. **Draw and label** a net for a circular cylinder that has a diameter of 3 inches and a height of 8.5 inches.

2. **Draw and label** a net for a cube with sides that have a length of 6 inches.

Find the surface area and volume of each three-dimensional figure. Use 3.14 for π. Round to the nearest tenth.

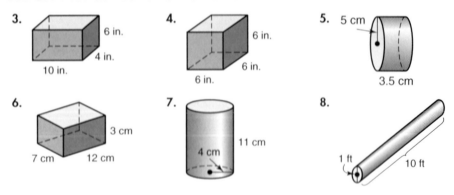

3.
6 in.
4 in.
10 in.

4.
6 in.
6 in.
6 in.

5. 5 cm
3.5 cm

6.
3 cm
7 cm   12 cm

7.
11 cm
4 cm

8.
1 ft
10 ft

Applications

9. **Manufacturing** A cylindrical tank is being built for a tank truck. The tank is supposed to be 25 feet long and have a radius of 34 inches.
   a. How much material is needed to build the tank?
   b. What will be the volume of the tank when it is completed?

10. **Pet Care** Jeremy has an end table that doubles as a fish tank. The base of the table is a rectangle that is 10 inches wide and 26 inches long. The table is 25 inches tall.
    a. What is the surface area of the table?
    b. How many cubic feet of water does it hold?

The Pythagorean Theorem

What YOU'LL LEARN

• To use the Pythagorean theorem.

Why IT'S IMPORTANT

You can use the Pythagorean theorem to solve problems involving sports, safety, and surveying.

On grid paper, the length of each square is defined as one unit. Therefore, a square with sides 1 unit long has an area of 1 square unit.

The model at the right shows what the Greek mathematician Pythagoras learned about right triangles. A **right triangle** has one angle that measures 90°. This angle is called a **right angle**. The side opposite the right angle is called the **hypotenuse**. The sides that form the right angle are called the **legs**. Note that the sum of the squares of the lengths of the legs is equal to the square of the length of the hypotenuse.

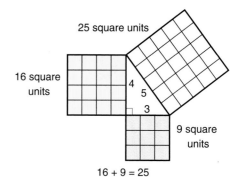

25 square units

16 square units

9 square units

$16 + 9 = 25$

Today we call this relationship the **Pythagorean theorem**. It is true for *any* right triangle.

Pythagorean Theorem	If *a* and *b* are the measures of the legs of a right triangle and *c* is the measure of the hypotenuse, then $c^2 = a^2 + b^2$.

Example **1** *ABCD* is a rectangle. What is the length of $\overline{AC}$?

Segments *AC*, *AB*, and *BC* form a right triangle. The legs of the right triangle are $\overline{AB}$ and $\overline{BC}$, and the hypotenuse of the right triangle is $\overline{AC}$.

$a^2 + b^2 = c^2$ *Pythagorean theorem*

$9^2 + 12^2 = c^2$ *Replace a with 9 and b with 12.*

$81 + 144 = c^2$

$225 = c^2$

$\sqrt{225} = c$ *Take the square root of each side.*

$15 = c$

The length of $\overline{AC}$ is 15 inches.

You can use the Pythagorean theorem to find the length of a leg of a right triangle if you know the length of the other leg and the hypotenuse.

Example **2** **Find the value of _b_ in the triangle below.**

$$a^2 + b^2 = c^2$$
$$15^2 + b^2 = 39^2$$
$$225 + b^2 = 1521$$
$$225 - 225 + b^2 = 1521 - 225$$
$$b^2 = 1296$$
$$b = \sqrt{1296}$$
$$b = 36$$

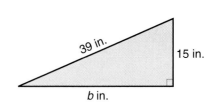

39 in.
15 in.
_b_ in.

The value of _b_ is 36.

EXERCISES

Practice

1. **Write** an equation that describes the relationship among the three large squares in the figure at the right.

2. **Draw and label** a right triangle with a hypotenuse of 20 units and legs of 12 units and 16 units.

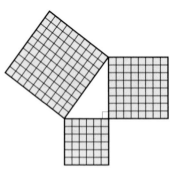

Find the missing lengths. Round decimal answers to the nearest tenth.

3.
9 yd 15 yd
_b_

4.
27 cm
_b_ cm 13 cm

5.
13 in. _a_ in.
18 in.

In a right triangle, if _a_ and _b_ are the measures of the legs and _c_ is the measure of the hypotenuse, find each missing measure. Round decimal answers to the nearest tenth.

6. $a = 6, b = 5$

7. $a = 12, b = 12$

8. $a = 8, b = 16$

9. $a = 20, c = 25$

10. $a = 9, c = 14$

11. $b = 15, c = 20$

12. $a = 5, b = 50$

13. $b = 54, c = 68$

14. $a = 4.5, c = 8.5$

Applications

15. **Sports**   In football, receivers often run a pass route called the "down and out." The route runs down the field, parallel to the sideline, then turns 90° toward the sideline. How far from his original position is a receiver who runs downfield 12 yards, turns and runs toward the sideline for another 5 yards?

16. **Safety**   The National Safety Council recommends placing the base of a ladder one foot from the wall for every three feet of the ladder's length.
    a. How far away from a wall should you place a 30-foot ladder?
    b. How high can the ladder reach safely?

17. **Surveying**   To determine the distance across a pond, surveyors made the diagram at the right. Write an equation they can use to determine how far it is across the lake. Then solve.

200 ft
450 ft

Counting Outcomes and Probability

What YOU'LL LEARN

- To count outcomes using a tree diagram or the fundamental counting principle, and
- to find the probability of a simple event.

Why IT'S IMPORTANT

You can use probability to calculate the chance that something will happen. This is especially helpful in games.

Suppose you decide to guess on the answers to a quiz with four true-false questions on it. How many different sets of answers, or **outcomes**, are possible? You can draw a diagram to find all the possible outcomes.

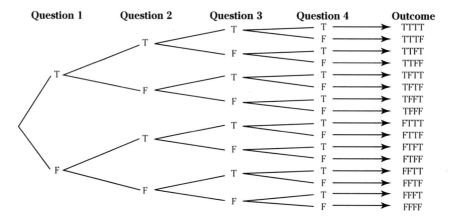

There are 16 different sets of guesses, or outcomes.

The diagram above is called a **tree diagram**. You can also find the total number of outcomes by multiplying. This principle is known as the **fundamental counting principle**.

Fundamental Counting Principle	**If event *M* can occur in *m* ways and is followed by event *N* that can occur in *n* ways, then the event *M* followed by the event *N* can occur in *m · n* ways.**

The number of possible outcomes for the quiz can be determined by using this principle. Each question has two outcomes and there are four questions. So, the number of outcomes is $2 \cdot 2 \cdot 2 \cdot 2$ or 16.

Example **A number cube is rolled once, and a coin is tossed once. How many outcomes are possible?**

number of outcomes for number cube	×	*number of outcomes for coin*	=	*number of possible outcomes*
6	×	2	=	12

There are 12 possible outcomes.

When all outcomes have an *equally likely* chance of happening, we say that the outcomes happen at **random**.

Probability is the chance that an event will happen. The probability of an outcome is written as a number from 0 to 1.

Definition of Probability	$\text{Probability} = \dfrac{\text{number of ways an event can occur}}{\text{number of possible outcomes}}$

Example 2 A spinner like the one at the right is used in a game. What is the probability that the spinner will stop on blue when you spin it?

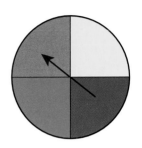

There is one way to stop on blue.

There are 4 possible outcomes.

$$P(\text{blue}) = \frac{\text{number of ways to stop on blue}}{\text{number of possible outcomes}} = \frac{1}{4}$$

EXERCISES

Practice

1. **Draw** a tree diagram to list all the outcomes for Example 2.

2. **Write** a problem that corresponds to the tree diagram at the right.

3. **Explain** why the probability of stopping on white on the spinner in Example 2 is 0.

4. **Draw** a spinner where the probability of an outcome is $\frac{1}{8}$.

Find the number of possible outcomes for each event.

5. two number cubes are rolled   6. a coin is tossed three times

7. a spinner with 5 equal sections is spun twice

A drawer contains one red sock, three blue socks, two black socks, and two green socks. One sock is removed. Find each probability.

8. $P(\text{red})$ 9. $P(\text{blue})$ 10. $P(\text{not black})$

11. $P(\text{green})$ 12. $P(\text{orange})$ 13. $P(\text{red or blue})$

The spinner at the right is spun once. Find each probability.

14. $P(\text{even})$ 15. $P(\text{odd})$

16. $P(\text{less than one})$ 17. $P(\text{prime})$

Applications

18. **Fashion** Paul has 7 shirts and 3 pairs of slacks. How many different outfits can he make using 1 shirt and 1 pair of slacks?

19. **Weather** The TV weather reporter said that the probability of rain on Friday is 40%. What is the probability that it will not rain?

20. **Retail Sales** Mercedes surveyed 75 students to determine how many of each size of school jackets to order. The results of the survey are shown in the table at the right.

    a. If one of the students is chosen at random, what is the probability that the student wears a medium?

    b. If one of the students is chosen at random, what is the probability that the student wears a large or extra large?

Size	Tally
Small	ⅢⅢ ⅢⅢ ⅢⅢ ⅢⅢ
Medium	ⅢⅢ ⅢⅢ ⅢⅢ
Large	ⅢⅢ ⅢⅢ ⅢⅢ ⅢⅢ III
Extra Large	ⅢⅢ ⅢⅢ ⅢⅢ II

1. In a survey, parents were asked where their children prefer to do homework. The graph shows the results of the survey. If 1600 parents were surveyed, how many of them said their children preferred the kitchen?

Where Do Your Kids Do Their Homework?

- **A** 25
- **B** 40
- **C** 400
- **D** 40,000
- **E** Not Here

2. The graph below shows a store's monthly sales of video games.

The average price of a video game was $40. What would be a reasonable sales total, without tax, for the video games sold during the 3-month period?

- **F** Less than $28,000
- **G** Between $28,000 and $32,000
- **H** Between $32,000 and $36,000
- **J** Between $36,000 and $40,000
- **K** More than $40,000

3. The graph below shows school enrollment in the U.S. from 1966 to 1996.

(Continued at top of next column)

Which is a reasonable conclusion from the information on the graph?

- **A** A record high will be reached in 2001.
- **B** Enrollment was at a record low in 1981.
- **C** Enrollment reached its peak in 1971.
- **D** Enrollment decreased between 1966 and 1976.
- **E** Enrollment increased between 1986 and 1996.

4. A recipe that makes 12 servings calls for $4\frac{2}{3}$ cups of flour. How much flour is needed to make 24 servings?

- **F** $2\frac{1}{3}$ c
- **G** $8\frac{2}{3}$ c
- **H** $9\frac{1}{3}$ c
- **J** $16\frac{2}{3}$ c
- **K** Not Here

5. In the United States flag, the length is $1\frac{9}{10}$ times the width. What is the length of a flag if its width is $4\frac{1}{2}$ feet?

- **A** $2\frac{3}{5}$ ft
- **B** $4\frac{9}{20}$ ft
- **C** $6\frac{2}{5}$ ft
- **D** $8\frac{11}{20}$ ft
- **E** Not Here

6. Brian and three friends bought $\frac{3}{4}$ pound of trail mix. If they spilt the mix equally, how much will each person get?

- **F** $\frac{1}{16}$ lb
- **G** $\frac{3}{16}$ lb
- **H** $\frac{3}{8}$ lb
- **J** $\frac{1}{4}$ lb
- **K** Not Here

7. How many slices of pepperoni, each $\frac{1}{16}$-inch thick, can be cut from a stick of pepperoni $4\frac{1}{2}$ inches long?

   A 72

   B 80

   C 128

   D 144

   E Not Here

8. After galloping $1\frac{7}{8}$ miles, a racehorse trotted an additional $\frac{7}{8}$ miles to cool down. How far did the horse travel altogether?

   F $1\frac{3}{4}$ mi

   G $2\frac{1}{8}$ mi

   H $2\frac{1}{4}$ mi

   J $2\frac{3}{4}$ mi

   K Not Here

9. A recipe calls for $2\frac{1}{3}$ cups of biscuit mix, $\frac{1}{2}$ cup of milk, $\frac{1}{4}$ cup of sugar, and $\frac{1}{4}$ cup of margarine. How many cups of ingredients are in the recipe?

   A $3\frac{1}{4}$ c

   B $3\frac{1}{3}$ c

   C $3\frac{5}{12}$ c

   D $3\frac{1}{2}$ c

   E Not Here

10. The tank of Zina's car holds 12 gallons of gas. It took $6\frac{1}{2}$ gallons to fill the tank. How much gasoline was already in the tank?

    F $4\frac{1}{2}$ gal

    G $5\frac{1}{2}$ gal

    H $6\frac{1}{2}$ gal

    J $18\frac{1}{2}$ gal

    K Not Here

11. Margurite weighed $7\frac{1}{2}$ pounds at birth. Her brother weighed $7\frac{3}{4}$ pounds when he was born. What was the difference in their birth weights?

    A $\frac{1}{8}$ lb

    B $\frac{1}{4}$ lb

    C $\frac{3}{8}$ lb

    D $\frac{1}{2}$ lb

    E Not Here

12. What is the approximate length of a large paper clip?

    F 4.8 millimeters

    G 48 millimeters

    H 48 centimeters

    J 4.8 meters

    K 0.048 kilometers

13. Roberta is driving a truck that weighs 55,600 pounds when fully loaded. A highway sign states that the maximum weight allowed is 28 tons. If Roberta's truck is fully loaded, how far is it away from the maximum?

    A 400 pounds

    B 6000 pounds

    C 3 tons

    D 400 tons

    E Not Here

14. According to the scale on a map, 1 inch represents 19.9 kilometers. If two cities are 180 kilometers apart, which proportion could be used to find x, the distance between the two cities on the map?

    F $\frac{1}{19.9} = \frac{x}{180}$

    G $\frac{180}{1} = \frac{x}{19.9}$

    H $\frac{19.9}{1} = \frac{x}{180}$

    J $\frac{x}{19.9} = \frac{1}{180}$

    K $\frac{1}{x} = \frac{180}{19.9}$

15. The ratio of your weight on the moon to your weight on Earth is 1:6. If your weight on Earth is 144 pounds, how much would you weigh on the moon?

  A 864 lb

  B 72 lb

  C 24 lb

  D 12 lb

  E Not Here

16. Which figure best represents ray *XY*?

  F

  G

  H

  J

17. The sides of the trapezoid are best described as:

  A rays

  B points

  C angles

  D lines

  E Not Here

18. The length of 1 side of an equilateral triangle is 2.2 meters. What is its perimeter in *centimeters*?

  F 6.6 cm

  G 66 cm

  H 660 cm

  J 6600 cm

19. What is the area of the trapezoid?

  A 10 sq ft

  B 30 sq ft

  C 40 sq ft

  D 95 sq ft

20. What is the surface area of the cylinder? (Use $\pi \approx 3.14$.)

  F 401.92 sq cm

  G 427.04 sq cm

  H 452.16 sq cm

  J 854.08 sq cm

21. What is the volume of the cube?

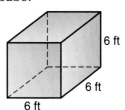

  A 200 ft^3

  B 175 ft^3

  C 36 ft^3

  D 216 ft^3

  E Not Here

22. An airplane flies 3 miles due south, then 4 miles due west. How far is the airplane from its starting point?

  F 5 mi   **J** 25 mi

  G 6 mi   **K** Not Here

  H 7 mi

23. A 15-foot ladder is leaning against a house. The base of the ladder is 5 feet from the house. How high does the ladder reach?

  A 40 ft   **D** 10 ft

  B 14.1 ft   **E** Not Here

  C 15.8 ft

24. A number cube is tossed three times. How many outcomes are possible?

  F 18   **J** 216

  G 36   **K** Not Here

  H 108

25. A bag contains 5 blue, 10 black, 5 red, and 5 yellow marbles. If you select one marble at random, what is the probability of selecting a black marble?

  A $\frac{2}{5}$

  B $\frac{1}{5}$

  C $\frac{2}{9}$

  D $\frac{1}{6}$

  E Not Here

SYMBOLS AND MEASURES

Symbols

$=$	is equal to		π	pi
$\neq$	is not equal to		$\%$	percent
$>$	is greater than		$0.1\overline{2}$	decimal 0.12222...
$<$	is less than		$\circ$	degree
$\geq$	is greater than or equal to		$f(x)$	f of x, the value of f at x
$\leq$	is less than or equal to		(a, b)	ordered pair a, b
$\approx$	is approximately equal to		$\overline{AB}$	line segment $\overline{AB}$
$\sim$	is similar to		$\overset{\frown}{AB}$	arc AB
$\times$ or $\cdot$	times		$\overrightarrow{AB}$	ray AB
$\div$	divided by		$\overleftrightarrow{AB}$	line AB
$-$	negative or minus		AB	measure of $\overline{AB}$
$+$	positive or plus		$\angle$	angle
$\pm$	positive or negative		Δ	triangle
$-a$	opposite or additive inverse of a		$\cos A$	cosine of A
$|a|$	absolute value of a		$\sin A$	sine of A
$a \overset{?}{=} b$	Does a equal b?		$\tan A$	tangent of A
$a : b$	ratio of a to b		$(\ )$	parentheses; *also* ordered pairs
$\sqrt{a}$	square root of a		$[\ ]$	brackets; *also* matrices
$P(A)$	probability of A		$\{\ \}$	braces; *also* sets
O	origin		$\varnothing$	empty set

Measures

mm	millimeter		in.	inch
cm	centimeter		ft	foot
m	meter		yd	yard
km	kilometer		mi	mile
g	gram		in^2 or sq in.	square inch
kg	kilogram		s	second
mL	milliliter		min	minute
L	liter		h	hour

GETTING ACQUAINTED WITH THE GRAPHING CALCULATOR

What is it?

What does it do?

How is it going to help me learn math?

These are just a few of the questions many students ask themselves when they first see a graphing calculator. Some students may think, "Oh, no! Do we *have* to use one?", while others may think, "All right! We get to use these neat calculators!" There are as many thoughts and feelings about graphing calculators as there are students, but one thing is for sure: a graphing calculator *can* help you learn mathematics.

So what is a graphing calculator? Very simply, it is a calculator that draws graphs. This means that it will do all of the things that a "regular" calculator will do, *plus* it will draw graphs of simple or very complex equations. In algebra, this capability is nice to have because the graphs of some complex equations take a lot of time to sketch by hand. Some are even considered impossible to draw by hand. This is where a graphing calculator can be very useful.

But a graphing calculator can do more than just calculate and draw graphs. You can program it, work with matrices, and make statistical graphs and computations, just to name a few things. If you need to generate random numbers, you can do that on the graphing calculator. If you need to find the absolute value of numbers, you can do that, too. It's really a very powerful tool—so powerful that it is often called a pocket computer. But don't let that intimidate you. A graphing calculator can save you time and make doing mathematics easier.

As you may have noticed, graphing calculators have some keys that other calculators do not. The Texas Instruments TI-83 will be used throughout this text. The keys located on the bottom half of the calculator are probably familiar to you as they are the keys found on basic scientific calculators. The keys located just below the screen are the graphing keys. You will also notice the up, down, left, and right arrow keys. These allow you to move the cursor around on the screen and to "trace" graphs that have been plotted. The other keys located on the top half of the calculator access the special features such as statistical and matrix computations.

There are some keystrokes that can save you time when using the graphing calculator. A few of them are listed below.

- Any yellow commands written above the calculator keys are accessed with the 2nd key, which is also yellow. Similarly, any green characters above the keys are accessed with the ALPHA key, which is also green.

- 2nd ENTRY copies the previous calculation so you can edit and use it again.

- Pressing ON while the calculator is graphing stops the calculator from completing the graph.

- 2nd QUIT will return you to the home (or text) screen.

- 2nd A-LOCK locks the ALPHA key, which is like pressing "shift lock" or "caps locks" on a typewriter or computer. The result is that all caps will be typed and you do not have to hold the shift key down. (This is handy for programming.)

- 2nd OFF turns the calculator off.

Some commonly used mathematical functions are shown in the table below. As with any scientific calculator, the graphing calculator observes the order of operations.

Mathematical Operation	Examples	Keys	Display
evaluate expressions	Find 2 + 5.	2 + 5 ENTER	2+5 7
exponents	Find 3^5.	3 ∧ 5 ENTER	3^5 243
multiplication	Evaluate 3(9.1 + 0.8).	3 × (9.1 + .8) ENTER	3(9.1+.8) 29.7
roots	Find $\sqrt{14}$.	2nd √ 14 ENTER	√14 3.741657387
opposites	Enter −3.	(−) 3	−3

Graphing on the TI–83

Before graphing, we must instruct the calculator how to set up the axes in the coordinate plane. To do this, we define a **viewing window.** The viewing window for a graph is the portion of the coordinate grid that is displayed on the **graphics screen** of the calculator. The viewing window is written as [left, right] by [bottom, top] or [Xmin, Xmax] by [Ymin, Ymax]. A viewing window of [−10, 10] by [−10, 10] is called the **standard viewing window** and is a good viewing window to start with to graph an equation. The standard viewing window can be easily obtained by pressing ZOOM 6. Try this. Move the arrow keys around and observe what happens. You are seeing a portion of the coordinate plane that includes the region from -10 to 10 on the x-axis and from -10 to 10 on the y-axis. Move the cursor, and you can see the coordinates of the points for the current position of the cursor.

Any viewing window can be set manually by pressing the WINDOW key. The window screen will appear and display the current settings for your viewing window. First press ENTER . Then, using the arrow and ENTER keys, move the cursor to edit the window settings. Xscl and Yscl refer to the x-scale and y-scale. This is the number of tick marks placed on the x- and y-axes. Xscl=1 means that there will be a tick mark for every unit of one along the x-axis. The standard viewing window would appear as follows.

$$\text{Xmin} = -10$$
$$\text{Xmax} = 10$$
$$\text{Xscl} = 1$$
$$\text{Ymin} = -10$$
$$\text{Ymax} = 10$$
$$\text{Yscl} = 1$$

Graphing equations is as simple as defining a viewing window, entering the equations in the Y= list, and pressing GRAPH . It is often important to view enough of a graph so you can see all of the important characteristics of the graph and understand its behavior. The term **complete graph** refers to a graph that shows all of the important characteristics such as intercepts or maximum and minimum values.

Example: Graph $y = x - 14$ in the standard viewing window.

Enter: Y= X,T,θ,n — 14 GRAPH

We see only a small portion of the graph. Why?

The graph of $y = x - 14$ is a line that crosses the x-axis at 14 and the y-axis at -14. The important features of the graph are plotted off the screen and so the graph is not complete. A better viewing window for this graph would be $[-20, 20]$ by $[-20, 20]$, which includes both intercepts. This is considered a complete graph.

Programming on the TI–83

The TI–83 has programming features that allow us to write and execute a series of commands to perform tasks that may be too complex or cumbersome to perform otherwise. Each program is given a name. Commands begin with a colon (:), followed by an expression or an instruction. Most of the features of the calculator are accessible from program mode.

When you press PRGM , you see three menus: EXEC, EDIT, and NEW. EXEC allows you to execute a stored program by selecting the name of the program from the menu. EDIT allows you to edit or change an existing program and NEW allows you to create a new program. To break during program execution, press ON . The following example illustrates how to create and execute a new program that stores an expression as Y and evaluates the expression for a designated value of X.

1. Enter PRGM ▶ ▶ ENTER to create a new program.

2. Type EVAL ENTER to name the program. (Make sure that the caps lock is on.) You are now in the program editor, which allows you to enter commands. The colon (:) in the first column of the line indicates that this is the beginning of the command line.

3. The first command lines will ask the user to designate a value for x. Enter PRGM ▶ 3 2nd A-LOCK " ENTER THE VALUE FOR X " ALPHA ENTER PRGM ▶ 1 X,T,θ,n ENTER .

4. The expression to be evaluated for the value of x is $x - 7$. To store the expression as Y, enter X,T,θ,n — 7 STO▶ ALPHA Y ENTER .

5. Finally, we want to display the value for the expression. Enter PRGM ▶ 3 ALPHA Y ENTER .

6. Now press 2nd QUIT to return to the home screen.

7. To execute the program, press PRGM . Then press the down arrow to locate the program name and press ENTER , or press the number or letter next to the progam name. The program asks for a value for x. You will input any value for which the expression is defined and press ENTER . To immediately re-execute the program, simply press ENTER when Done appears on the screen.

While a graphing calculator cannot do everything, it can make some things easier. To prepare for whatever lies ahead, you should try to learn as much as you can. The future will definitely involve technology, and using a graphing calculator is a good start toward becoming familiar with technology. Who knows? Maybe one day you will be designing the next satellite, building the next skyscraper, or helping students learn mathematics with the aid of a graphing calculator!

Exploring Expressions, Equations, and Functions

Objectives

In this chapter, you will:

- translate verbal expressions into mathematical expressions,
- solve problems by looking for a pattern,
- use mathematical properties to evaluate expressions,
- solve open sentences, and
- use and interpret stem-and-leaf plots, tables, graphs, and functions.

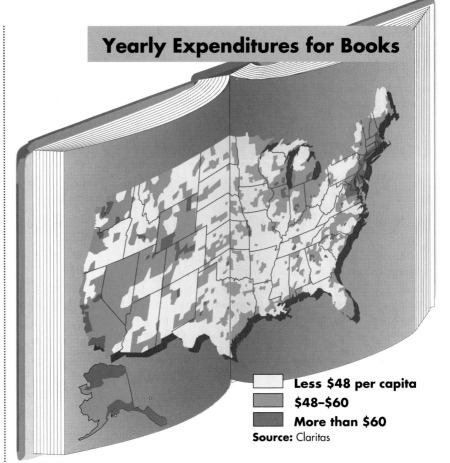

Yearly Expenditures for Books

Less $48 per capita

$48–$60

More than $60

Source: Claritas

The writing and publishing of books is an important part of our economy. Do you like to express yourself? Do you share your thoughts by writing poems or short stories? Would you like to eventually make a living putting words on paper?

TIME *Line*

1650 B.C. The Rhind Papyrus shows solutions of simple equations.

A.D. 786 The Grand Mosque of Cordoba, Spain, is completed.

1900 B.C. | 1800 | 1700 | 1600 | 500 | 400 | 300 | 200 | 100 | A.D. 500 | 600 | 700 | 800

400 B.C. Theodorus of Cyrene demonstrates that the square roots of some numbers are irrational.

Chapter Project

Choose a poet and read 10 of his or her poems.

- Describe some patterns that you see in these poems.

- Count the number of words in each poem. Make a stem-and-leaf plot of the number of words in the poems. Does the poet seem to write poems with about the same number of words? Are one or more of the poems exceptionally long or short for this poet?

- Make a graph that models one of the poems. Write a paragraph that describes the relationship between the graph and the poem.

When **Wendy Isdell** was in her eighth grade Algebra 1 class, she got an idea for a short story. Later, she wrote an outline and completed a first draft. The following year, she entered the finished story in the Virginia Young Authors' Contest and won first place in the state competition. Wendy sent her story to Free Spirit Publishing in Minneapolis, and her book *A Gebra Named Al* was published during her senior year in high school.

Wendy's Algebra 1 and 2, geometry, trigonometry, chemistry, Earth science, and physical science classes have provided background for the information in her book. She hopes to make a living as a professional writer.

1455 The Gutenberg Bible is printed in Mainz, Germany.

1881 French impressionist painter Auguste Renoir completes the *Boating Party Lunch*.

1663 The first known work on probability is published.

1993 Toni Morrison becomes the first African American to receive the Nobel Prize for Literature.

900 | 1000 | 1100 | 1200 | 1300 | 1400 | 1500 | 1600 | 1700 | 1800 | 1900 | 2000

Variables and Expressions

What YOU'LL LEARN

• To translate verbal expressions into mathematical expressions and vice versa.

Why IT'S IMPORTANT

You can use expressions to solve problems involving health, recycling, and geometry.

10 × 5 and 10 × 15 are numerical expressions.

APPLICATION
Health

Many people enjoy going to the beach in the summer. Unfortunately, the ultraviolet radiation in direct sunlight causes skin cancer. In recent years, the number of skin cancer cases has increased. Many sunbathers use sunscreen lotions to protect themselves in the sun.

The Sun Protection Factor (SPF) scale gives numbers that represent the length of time you can stay in the sun without burning if you have lotion on. Let's say you can stay in the sun with no sunscreen for 10 minutes without burning. If you put on SPF 5 lotion, you can stay in the sun for 10 minutes × 5 or 50 minutes, or a little less than one hour. Use the table below to see the pattern.

No Sunscreen	With Sunscreen	
Minutes in Sun Without Burning	SPF Number	Minutes in Sun Without Burning
10	5	10 × 5
10	15	10 × 15
5	4	5 × 4
15	8	15 × 8
m	s	$m × s$

The letters m and s are called **variables,** and $m × s$ is an **algebraic expression.** In algebra, variables are symbols that are used to represent unspecified numbers. Any letter may be used as a variable. We selected m because it is the first letter of the word "minutes" and s because it is the first letter of "SPF".

An algebraic expression consists of one or more numbers and variables along with one or more arithmetic operations. Here are some other examples of algebraic expressions.

$$x - 2 \qquad \frac{a}{b} + 3 \qquad t × 2s \qquad 7mn ÷ 3k$$

The symbol "×" as shown in x × y is often avoided because it may be confused with the variable "x."

In algebraic expressions, a raised dot or parentheses are often used to indicate multiplication. Here are ways to represent the product of x and y.

$$xy \qquad x \cdot y \qquad x(y) \qquad (x)(y) \qquad (x)y \qquad x × y$$

In each of the multiplication expressions, the quantities being multiplied are called **factors,** and the result is called the **product.**

It is often necessary to translate verbal expressions into algebraic expressions.

Verbal Expression	Algebraic Expression
7 less than the product of 3 and a number x	$3x - 7$
the product of 7 and s divided by the product of 8 and y	$7s \div 8y$
four years younger than Sarah (S = Sarah's age)	$S - 4$
half as big as last night's crowd (c = size of last night's crowd)	$\frac{c}{2}$

Example **1** Write an algebraic expression for each verbal expression.

  a. three times a number x subtracted from 24

  $24 - 3x$

  b. 5 greater than half of a number t

  $\frac{t}{2} + 5$ or $\frac{1}{2}t + 5$

Another important skill is translating algebraic expressions into verbal expressions.

Example **2** Write a verbal expression for each algebraic expression.

  a. $(3 + b) \div y$

  the sum of 3 and b divided by y

  b. $5y + 10x$

  the product of 5 and y plus the product of 10 and x

An expression like x^n is called a **power.** The variable x is called the **base,** and n is called the **exponent.** The exponent indicates the number of times the base is used as a factor.

$$x^2 \text{ means } x \cdot x. \qquad 12^4 \text{ means } 12 \cdot 12 \cdot 12 \cdot 12.$$

Symbols	Words	Meaning
5^1	5 to the first power	5
5^2	5 to the second power or 5 squared	$5 \cdot 5$
5^3	5 to the third power or 5 cubed	$5 \cdot 5 \cdot 5$
5^4	5 to the fourth power	$5 \cdot 5 \cdot 5 \cdot 5$
$3a^5$	three times a to the fifth power	$3 \cdot a \cdot a \cdot a \cdot a \cdot a$
x^n	x to the nth power	$\underbrace{x \cdot x \cdot x \cdot \ldots \cdot x}_{n \text{ factors}}$
	An expression of the form x^n is read as "x to the nth power."	

Example **3**

INTEGRATION

Geometry

Write a power that represents the number of smallest squares in the large square.

There are 8 squares on each side.

The total number of squares is $8^2 = 8 \cdot 8$ or 64.

You can use the x^2 key on a calculator to square a number.

Enter: 8 $\boxed{x^2}$ *64*

To **evaluate** an expression means to find its value.

Example **4** **Evaluate 3^4.**

$$3^4 = 3 \cdot 3 \cdot 3 \cdot 3$$

$$= 81$$

You can use the $\boxed{y^x}$ key on a calculator to raise a number to a power.

Enter: 3 $\boxed{y^x}$ 4 $\boxed{=}$ *81*

EXPLORATION

PROGRAMMING

BASIC is a computer language. The symbols used in BASIC are similar to those used in algebra.

+ means add.   / means divide.

− means subtract.   ↑ means exponent.

* means multiply.

Numerical variables in BASIC are represented by capital letters. The program at the right can be used to add, subtract, multiply, divide, and find powers.

```
10 INPUT A,B
20 PRINT "A + B ="; A + B
30 PRINT "A − B ="; A − B
40 PRINT "A * B ="; A * B
50 PRINT "A / B ="; A / B
60 PRINT "A ↑ B ="; A ↑ B
```

Your Turn

a. Let $a = 10$ and $b = 3$. Use the program to find $10 + 3$, $10 - 3$, 10×3, $10 \div 3$, and 10^3.

b. Let $a = 9$ and $b = 3$. Use the program to find $9 + 3$, $9 - 3$, 9×3, $9 \div 3$, and 9^3.

c. Let $a = 5.2$ and $b = 2$. Use the program to find $5.2 + 2$, $5.2 - 2$, 5.2×2, $5.2 \div 2$, and 5.2^2.

Communicating Mathematics

Study the lesson. Then complete the following.

1. **Describe** how you would compute the length of time you could sunbathe using an SPF 20 sunscreen lotion if you can stay in the sun for 15 minutes without burning.

2. Can you find the volume of a cube if you only know the length of a side, s? How?

3. **Describe** what x^7 means.

4. **Explain** the difference between numerical expressions and algebraic expressions.

5. **You Decide** Darcy and Tonya are studying for an exam. Darcy says, "The square of a number is always greater than the number." Tonya disagrees. Who is correct? Explain your answer.

Guided Practice

Write an algebraic expression for each verbal expression.

6. the product of the fourth power of a and the second power of b

7. six less than three times the square of y

Write a verbal expression for each algebraic expression.

8. 7^5

9. $3x^2 + 4$

Write each expression as an expression with exponents.

10. $4 \cdot 4 \cdot 4$

11. $a \cdot a \cdot a \cdot a \cdot a \cdot a \cdot a$

Evaluate each expression.

12. 6^2

13. 2^5

14. **Geometry** The area of a circle can be found by multiplying the number π times the square of the radius. If the radius of the circle is r, write an expression that represents the area of the circle.

Practice

Write an algebraic expression for each verbal expression.

15. the sum of k and 20

16. the product of 16 and p

17. a to the seventh power

18. 49 increased by twice a number

19. two thirds of the square of a number

20. the product of 5 and m plus half of n

21. 8 pounds heavier than his brother (b = brother's weight)

22. 3 times as many wins as last season (w = last season's wins)

Write a verbal expression for each algebraic expression.

23. $4m^5$

24. $\dfrac{x^2}{2}$

25. $c^2 + 23$

26. $\dfrac{1}{2}n^3$

27. $2(4)(5)^2$

28. $3x^2 - 2x$

Write each expression as an expression with exponents.

29. $8 \cdot 8$

30. $10 \cdot 10 \cdot 10 \cdot 10 \cdot 10$

31. $4 \cdot 4 \cdot 4 \cdot 4 \cdot 4 \cdot 4 \cdot 4$

32. $t \cdot t \cdot t$

33. $z \cdot z \cdot z \cdot z \cdot z$

34. $d \cdot d \cdot d \cdot d \cdot d \cdot d \cdot d \cdot d \cdot d$

Evaluate each expression.

35. 7^2

36. 9^2

37. 4^3

38. 5^3

39. 2^4

40. 3^5

Write an algebraic expression for each verbal expression.

41. triple the difference of 55 and the cube of w

42. four times the sum of r and s, increased by twice the difference of r and s

43. the sum of a and b, increased by the quotient of a and b

44. the perimeter of a square if the length of a side is s

45. the amount of money in Karen's savings account if she has y dollars and adds x dollars per week for 10 weeks

46. Geometry Write an expression that represents the total number of small cubes in the large cube shown at the right. Then evaluate the expression.

Critical Thinking

47. Evaluate 4^2 and 2^4.

   a. What do you notice? From this example, can you say that the same relationship exists between 2^5 and 5^2? What about between 3^4 and 4^3?

   b. In general, do you think that $a^b = b^a$ for positive whole numbers a and b? Support your answer.

Applications and Problem Solving

48. Babysitting According to *Smart Money* magazine, in New York City, the average teenage babysitter makes $7 per hour for a Saturday night. In Atlanta, the average teenage babysitter makes only $3 per hour.

   a. Write an expression representing the amount of money that the average teenage babysitter in New York City makes for working x hours on a Saturday night.

   b. Write an expression representing the amount of money that the average teenage babysitter in Atlanta makes for working x hours on a Saturday night.

   c. Write an expression representing the difference between the amount that the New York babysitter makes and the Atlanta babysitter makes for working x hours on a Saturday night.

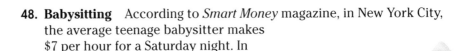

49. Recycling The alarming amount of garbage produced by our society requires that everyone work to recycle. According to *Vitality* magazine, each person in the United States produces an average of 3.5 pounds of trash a day.

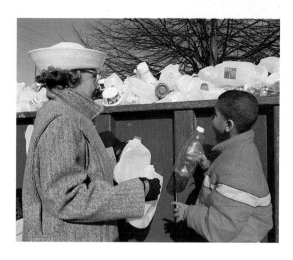

    a. Write an expression representing the pounds of trash produced in a day by a family that has x members.

    b. Write an expression representing the pounds of trash produced in a day by a family that has y members.

    c. Write an expression representing the pounds of trash produced in one day by both of the families in parts a and b.

50. Geometry The surface area of a rectangular prism is the sum of the product of twice the length ℓ and the width w and the product of twice the length and the height h and the product of twice the width and the height. Write an expression that represents the surface area of this type of prism.

Mathematics and SOCIETY

International Mathematical Olympiad

The excerpt below appeared in an article in *TIME* magazine on August 1, 1994.

IF THE UNITED STATES SENDS SIX KIDS TO THE International Mathematical Olympiad in Hong Kong, where a perfect individual score is 42, and together they score 252, does the country have reason to cheer? If you can't answer that one, then you desperately need a remedial course in arithmetic (or perhaps just new batteries in your calculator). The U.S. team members, all public high-school students, started out by competing against 350,000 of their peers on the American High School Mathematics Examination, aced two tougher exams, and prepped for a month at the U.S. Naval Academy. Only then did they board a plane and became the first squad in the Math Olympiad's 35-year history to get perfect scores across the board, outstripping 68 other nations to win the competition. . . . "They showed the world!" suggests the justifiably proud U.S. coach, Walter Mientka, a math professor at the University of Nebraska at Lincoln. ■

1. Do you think this type of international contest is a good idea? Why or why not?

2. Would you like to represent your country in this competition? Why or why not?

3. How do you feel about the math courses you have had in school? What factors have influenced the way you feel?

Patterns and Sequences

APPLICATION

Handicrafts

The photograph at the right shows a Hopi woman completing a basket. Baskets are important in the Hopi culture since they can serve either a functional or a spiritual purpose. Looking at the inside of the basket, we can see a pattern. Following this pattern in a clockwise direction, what color would you expect to find on the first diamond shape that cannot be seen in the photo?

Example ① **Study the pattern below.**

    1   2   3   4   5   6

a. Draw the next three figures in the pattern.
b. Draw the 35th square in the pattern.

a. The pattern consists of squares with one corner shaded. The corner that is shaded is rotated in a clockwise direction. The next three figures are drawn below.

    7   8   9

You may find a different pattern. For example, the seventh square may be the same as the fifth square, the eighth square may be the same as the fourth square, and the ninth square may be the same as the third square.

b. The pattern repeats after every 4 designs. Since 32 is the greatest number less than 35 that is divisible by 4, the 33rd square will be the same as the first square.

    33   34   35

The 35th square will have its upper right corner shaded.

The numbers 2, 4, 6, 8, 10, and 12 form a pattern called a **sequence.** A sequence is a set of numbers in a specific order. The numbers in the sequence are called **terms.**

Example **Find the next three terms in each sequence.**

a. 7, 13, 19, 25, . . .

Study the pattern in the sequence.

7,　13,　19,　25,...

$+6$　$+6$　$+6$

Each term is 6 more than the term before it.

$25 + 6 = 31$

$31 + 6 = 37$

$37 + 6 = 43$

The next three terms are 31, 37, and 43.

b. 243, 81, 27, 9, . . .

Study the pattern in the sequence.

243,　81,　27,　9,...

$\times \frac{1}{3}$　$\times \frac{1}{3}$　$\times \frac{1}{3}$

Each term is $\frac{1}{3}$ of the term before it.

$9 \times \frac{1}{3} = 3$

$3 \times \frac{1}{3} = 1$

$1 \times \frac{1}{3} = \frac{1}{3}$

The next three terms are 3, 1, and $\frac{1}{3}$.

One of the most-used strategies in problem solving is **look for a pattern.** When using this strategy, you will often need to make a table to organize the information.

Example **What is the number of diagonals in a 10-sided polygon?**

INTEGRATION

Geometry

Drawing a 10-sided polygon and all its diagonals would be difficult. Another way to solve the problem is to study the number of diagonals for polygons with fewer sides and then look for a pattern.

triangle	quadrilateral	pentagon	hexagon
3 sides	4 sides	5 sides	6 sides
0 diagonals	2 diagonals	5 diagonals	9 diagonals

(continued on the next page)

Use a chart to see the pattern.

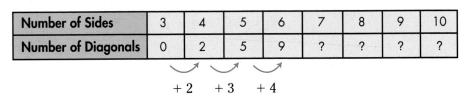

Number of Sides	3	4	5	6	7	8	9	10
Number of Diagonals	0	2	5	9	?	?	?	?

+ 2   + 3   + 4

Use the pattern to complete the chart.

Number of Sides	3	4	5	6	7	8	9	10
Number of Diagonals	0	2	5	9	14	20	27	35

+ 2   + 3   + 4   + 5   + 6   + 7   + 8

A 10-sided polygon has 35 diagonals.

Recognizing patterns is important, but it does not always guarantee finding the correct answer. You may wish to check this answer by drawing a 10-sided polygon and its diagonals.

Sometimes a pattern can lead to a general rule that can be written as an algebraic expression.

Example **Study the following pattern. Write an algebraic expression for the perimeter of the pattern consisting of *n* trapezoids.**

1 trapezoid   2 trapezoids   3 trapezoids   4 trapezoids
$P = 5$ units $P = 8$ units $P = 11$ units $P = 14$ units

Use a chart to look for a pattern.

Number of Trapezoids	1	2	3	4	5	6	n
Perimeter	5	8	11	14	?	?	?

+ 3   + 3   + 3

Notice that each trapezoid adds 3 units to the perimeter.
   For 1 trapezoid, the perimeter is $3 \cdot 1 + 2$, or 5 units.
   For 2 trapezoids, the perimeter is $3 \cdot 2 + 2$, or 8 units.
   For 3 trapezoids, the perimeter is $3 \cdot 3 + 2$, or 11 units.
   For 4 trapezoids, the perimeter is $3 \cdot 4 + 2$, or 14 units.

Extend this pattern.
   For 5 trapezoids, the perimeter should be $3 \cdot 5 + 2$, or 17 units.
   For 6 trapezoids, the perimeter should be $3 \cdot 6 + 2$, or 20 units.
   For *n* trapezoids, the perimeter should be $3 \cdot n + 2$, or $3n + 2$ units.

The algebraic expression $3n + 2$ represents the perimeter of this pattern with *n* trapezoids.

MODELING MATHEMATICS

Looking for Patterns

Materials: string ✂ scissors

If you use a pair of scissors to cut a piece of string in the normal way, you will have 2 pieces of string. What happens if you loop the string around one of the cutting edges of the scissors and cut?

0 loop   1 loop

Your Turn

a. Make a piece of string loop once around your scissors as shown above. Cut the string. How many pieces do you have?

b. Make 2 loops and cut. How many pieces do you have?

c. Continue making loops and cutting until you see a pattern. Describe the pattern and write the sequence.

d. How many pieces would you have if you made 20 loops?

e. Now tie the ends of the string together before you loop the string around the scissors. Investigate to determine how many pieces you would have if you made 10 loops with this string.

0 loop   1 loop

CHECK FOR UNDERSTANDING

Communicating Mathematics

Study the lesson. Then complete the following.

1. **Explain** how looking for a pattern can help to solve some problems.

2. **Write** a sequence that has 22 as its fourth term.

3. **You Decide**   Chi-Yo studies the sequence 1, 2, and 4. She notices that the second number is 1 more than the first, and the third number is 2 more than the second. She concludes that the next number in the sequence is $4 + 3$ or 7. Alonso disagrees. He notices that each number is twice the number before it. He says the next number in the sequence is 8. Who is correct? Explain.

MODELING MATHEMATICS

4. Refer to the Modeling Mathematics activity above. Suppose the ends of a string are *not* tied together.

   a. The string is looped around the scissors 50 times, and the string is cut. How many pieces will you have?

   b. The string is looped around the scissors y times, and the string is cut. How many pieces will you have?

Guided Practice

Give the next two items for each pattern.

5.

6. 85, 76, 67, 58,...

7. $1x + 1, 2x + 1, 3x + 1, 4x + 1,...$

Lesson 1–2 Patterns and Sequences **15**

8. Consider the following pattern.

    a. Suppose the length of each side of the triangles is 1 unit. What is the perimeter of each figure in the pattern?

    b. Draw the next figure in the pattern. What is the perimeter of this figure?

    c. What is the perimeter of the tenth figure in this pattern?

    d. What is the perimeter of the *n*th figure in this pattern?

9. a. Copy and complete the following table.

4^1	4^2	4^3	4^4	4^5
4	16			

    b. If you found the value of 4^6, what number do you think will be in the ones place? Check your conjecture.

    c. Find the number in the ones place for the value of 4^{225}. Explain your reasoning.

EXERCISES

Practice

Give the next two items for each pattern.

10.

11.

12.

13. 3, 6, 12, 24, ...

14. 4, 5.5, 7, 8.5, ...

15. 1, 4, 9, 16, ...

16. 9, 7, 10, 8, 11, 9, 12, ...

17. $a + 1, a + 3, a + 5, ...$

18. $x - 1y, x - 2y, x - 3y, ...$

19. a. Draw the next three figures in the following pattern.

    b. What is the color of the 38th figure? Explain your reasoning.

    c. How many sides will the 19th figure have? Explain your reasoning.

20. a. Copy and complete the following table.

3^1	3^2	3^3	3^4	3^5	3^6
3	9				

b. Write the sequence of numbers representing the numbers in the ones place.

c. What are the next six numbers in this sequence?

d. Find the number in the ones place for the value of 3^{100}. Explain your reasoning.

21. a. Copy and find each sum.

$1 = 1$

$1 + 3 = ?$

$1 + 3 + 5 = ?$

$1 + 3 + 5 + 7 = ?$

$1 + 3 + 5 + 7 + 9 = ?$

b. Pythagoras is credited with a discovery about the sum of consecutive odd numbers such as those shown in part a. What did he discover?

c. What is the sum of the first 100 odd numbers?

d. What is the sum of the first x odd numbers?

22. If y represents the counting numbers 1, 2, 3, 4, ..., then the algebraic expression $2y$ represents the terms of a sequence. To find the first term of the sequence, replace y with 1 and find $2 \cdot 1$.

a. How could you find the second term of the sequence?

b. Write the first ten terms of the sequence.

c. Describe the terms of this sequence.

23. a. Use a calculator to find each product.

i. $999,999 \times 2$

ii. $999,999 \times 3$

iii. $999,999 \times 4$

iv. $999,999 \times 5$

b. Without using a calculator, find the product of 999,999 and 9.

24. a. Copy and complete the following table.

10^1	10^2	10^3	10^4	10^5
10	100			

b. Use the pattern to find the value of 10^0.

25. If n represents the counting numbers 1, 2, 3, 4, ..., then the algebraic expression $3n + 1$ represents the terms of a sequence. Write the first five terms of the sequence.

Critical Thinking

26. A special sequence is called the *Fibonacci sequence*. It is named after Leonardo Fibonacci from Italy who presented it in 1201. This sequence is very intriguing because the numbers in the sequence are often found in nature. The first six terms of the sequence are listed below. Give the next six numbers in the sequence. 1, 1, 2, 3, 5, 8, ...

Applications and Problem Solving

27. Transportation Olga has part of a bus schedule. She wishes to take the bus to go to the mall, but she cannot leave until after 1:00 P.M. What is the earliest time Olga can catch the bus?

Bus Schedule
Departures
8:25 A.M.
9:13 A.M.
10:01 A.M.
10:49 A.M.

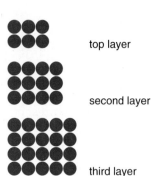

top layer

second layer

third layer

28. Sales Paula needs to make a tower of soup cans as a display in a supermarket. Each layer of the tower will be in the shape of a rectangle as shown at the left. The length and width of each layer will be one less than the layer below it.

a. How many cans will be needed for the fourth layer?

b. What is the total number of cans needed for an 8-layer tower?

29. Recreation The house of cards at the right used 26 cards to build four stories.

a. How many cards are needed to build a similar house that is eight stories high?

b. Plan another way to construct a house of cards. Make a drawing showing four stories of your house.

c. Write a sequence that represents the number of cards needed to build 1, 2, 3, and 4 stories of your house.

d. How many cards will you need to build eight stories of your house?

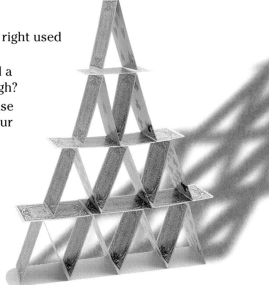

Mixed Review

30. Write an algebraic expression for *eight less than the square of q*. (Lesson 1–1)

31. Write a verbal expression for $\frac{x^3}{9}$. (Lesson 1–1)

32. Write $m \cdot m \cdot m \cdot m \cdot m \cdot m \cdot m \cdot m$ as an algebraic expression with exponents. (Lesson 1–1)

33. Geometry Write an expression that represents the total number of small cubes in the large cube at the right. Then evaluate the expression. (Lesson 1–1)

34. Write an algebraic expression for the amount of mileage on Seth's car if the car initially has 20,000 miles and Seth adds an average of x miles per month for 2 years. (Lesson 1–1)

35. Science When water freezes, its volume is increased by one eleventh. In other words, the volume of the ice would equal the sum of the volume of the water and the product of $\frac{1}{11}$ and the volume of the water. Suppose x cubic inches of water is placed in a freezer. Write an expression for the volume of the ice that is formed. (Lesson 1–1)

Order of Operations

APPLICATION

Investments

Bobbie Jackson is investing money in stocks to help pay for her child's college education. She buys one share of Nike stock at $16. She also purchases five shares of Disney stock at $35 each. The expression below represents the amount of money Ms. Jackson spends for the stock purchase.

cost of 1 share of Nike stock ————┐   ┌———— *cost of 1 share of Disney stock*

$$16 + 5 \cdot 35$$

————— *number of Disney stocks*

Numerical and algebraic expressions often contain more than one operation. A rule is needed to let you know which operation to perform first. This rule is called the **order of operations.**

To find the total amount of money Ms. Jackson invests, evaluate the expression $16 + 5 \cdot 35$. Which of the following methods is correct?

Method 1		**Method 2**	
$16 + 5 \cdot 35 = 16 + 175$	*Multiply first.*	$16 + 5 \cdot 35 = 21 \cdot 35$	*Add first.*
$= 191$	*Then add.*	$= 735$	*Then multiply.*

The answers are not the same because a different order of operations was used in each method. Since numerical expressions must have only one value, the following order of operations has been established.

Order of Operations	1. **Simplify the expressions inside grouping symbols, such as parentheses, brackets, and braces, and as indicated by fraction bars.** 2. **Evaluate all powers.** 3. **Do all multiplications and divisions from left to right.** 4. **Do all additions and subtractions from left to right.**

Based on the context of the problem and the order of operations, Method 1 is correct. Therefore, Ms. Jackson invested $16 + 5 \cdot 35$ or $191 in the stock market.

Other expressions can be evaluated by using the order of operations.

Example **Evaluate $5 \times 7 - 6 \div 2 + 3^2$.**

Evaluate $5 \times 7 - 6 \div 2 + 3^2 = 5 \times 7 - 6 \div 2 + 9$ *Evaluate 3^2.*

$= 35 - 6 \div 2 + 9$ *Multiply 5 by 7.*

$= 35 - 3 + 9$ *Divide 6 by 2.*

$= 32 + 9$ *Subtract 3 from 35.*

$= 41$ *Add 32 and 9.*

In mathematics, grouping symbols such as parentheses ( ), brackets [ ], and braces { } are used to clarify or change the order of operations. They indicate that the expression within the grouping symbol is to be evaluated first. When more than one grouping symbol is used, start evaluating within the innermost grouping symbols.

Example **Evaluate $8[6^2 - 3(2 + 5)] \div 8 + 3$.**

$8[6^2 - 3(2 + 5)] \div 8 + 3 = 8[6^2 - 3(7)] \div 8 + 3$ *Add $2 + 5$, the innermost group.*

$= 8[36 - 3(7)] \div 8 + 3$ *Evaluate 6^2.*

$= 8[36 - 21] \div 8 + 3$ *Multiply 3 by 7.*

$= 8[15] \div 8 + 3$ *Subtract 21 from 36.*

$= 120 \div 8 + 3$ *Multiply 8 by 15.*

$= 15 + 3$ *Divide 120 by 8.*

$= 18$ *Add 15 and 3.*

Algebraic expressions can be evaluated when the values of the variables are known. First, replace the variables by their values. Then, calculate the value of the numerical expression.

Example **The figure at the right is a rectangle.**

$s + 3$ mm

s
mm

a. Find the perimeter of the rectangle when $s = 5$.
b. Find the area of the rectangle.

INTEGRATION

Geometry

a. The perimeter of the rectangle is the sum of 2 times the measure of the width (s) and 2 times the measure of the length ($s + 3$).

$P = 2s + 2(s + 3)$

$= 2(5) + 2(5 + 3)$ *Replace the variable s with 5.*

$= 2(5) + 2(8)$ *Add 5 and 3.*

$= 10 + 2(8)$ *Multiply 2 and 5.*

$= 10 + 16$ *Multiply 2 and 8.*

$= 26$ *Add 10 and 16.*

The perimeter is 26 mm.

b. The area is the product of the width (s) and the length ($s + 3$).

$A = s(s + 3)$

$\quad = 5(5 + 3)$ *Replace the variable s with 5.*

$\quad = 5(8)$ *Add 5 and 3.*

$\quad = 40$ *Multiply 5 and 8.*

The area is 40 mm².   *Area is expressed in units squared.*

EXPLORATION

GRAPHING CALCULATORS

You can use a graphing calculator to evaluate algebraic expressions. Use a calculator to evaluate $\frac{0.25x^2}{7x^3}$ when $x = 0.75$.

Enter: .75 [STO▸] [X,T,θ,n] [2nd] [:] [(] .25 [X,T,θ,n] [x^2] [)]

[÷] [(] 7 [X,T,θ,n] [∧] 3 [)] [ENTER] *0.0476190476*

Your Turn

a. Evaluate the expression when $x = 24.076$.

b. Evaluate $\frac{2x^2}{(x^2 - x)}$ when $x = 27.89$.

c. Work through some of the examples in this lesson using a graphing calculator.

The fraction bar is another grouping symbol. It indicates that the numerator and denominator should each be treated as a single value.

$$\frac{2 \times 3}{1 + 2} \text{ means } (2 \times 3) \div (1 + 2) \text{ or } 2.$$

Example **Evaluate** $\frac{x^3 + y^3}{x^2 - y^2}$ **when** $x = 4.2$ **and** $y = 1.8$.

$$\frac{x^3 + y^3}{x^2 - y^2} = \frac{(4.2)^3 + (1.8)^3}{(4.2)^2 - (1.8)^2}$$

Estimate: $\dfrac{4^3 + 2^3}{4^2 - 2^2} = \dfrac{64 + 8}{16 - 4} = \dfrac{72}{12}$ *or 6*

Use a scientific calculator.

Enter: [(] 4.2 [y^x] 3 [+] 1.8 [y^x] 3 [)] [÷]

[(] 4.2 [x^2] [−] 1.8 [x^2] [)] [=] *5.55*

Is the answer reasonable?

Communicating Mathematics

Study the lesson. Then complete the following.

1. When evaluating the expression $4 + 7 \cdot 2$, what would you do first?

2. **Name** two types of grouping symbols and explain how you may use them.

3. **Explain** how you would evaluate the expression $2[5 + (30 \div 6)]^2$.

Guided Practice

Evaluate each expression.

4. $15 + 3 \cdot 2$

5. $5^3 + 3(4^2)$

6. $\dfrac{38 - 12}{2 \cdot 13}$

7. $12 \div 3 \cdot 5 - 4^2$

Evaluate each expression when $w = 12$, $x = 5$, $y = 6$, and $z = 4$.

8. $wx - yz$

9. $2w + x^2 - yz$

10. Use two or more of the variables in Exercises 8–9 and their assigned values to write an expression whose value is 18.

11. If $a = 1.5$, which is less, $a^2 - a$ or 1?

12. **Geometry**   Find the area of the rectangle when $t = 6$ inches.

Rectangle with top side labeled $2t + 1$ and left side labeled t.

Practice

Evaluate each expression.

13. $3 + 2 \cdot 3 + 5$

14. $(4 + 5)7$

15. $29 - 3(9 - 4)$

16. $50 - (15 + 9)$

17. $15 \div 3 \cdot 5 - 4^2$

18. $4(11 + 7) - 9 \cdot 8$

19. $\dfrac{(4 \cdot 3)^2 \cdot 5}{(9 + 3)}$

20. $[7(2) - 4] + [9 + 8(4)]$

21. $\dfrac{6 + 4^2}{3^2(4)}$

22. $(5 - 1)^3 + (11 - 2)^2 + (7 - 4)^3$

23. $\dfrac{2 \cdot 8^2 - 2^2 \cdot 8}{2 \cdot 8}$

24. $7(0.2 + 0.5) - 0.6$

Evaluate each expression when $v = 5$, $x = 3$, $a = 7$, and $b = 5$.

25. $(v + 1)(v^2 - v + 1)$

26. $\dfrac{x^2 - x + 6}{x + 3}$

27. $[a + 7(b - 3)]^2 \div 3$

28. $v^2 - (x^3 - 4b)$

29. $(2v)^2 + ab - 3x$

30. $\dfrac{a^2 - b^2}{v^3}$

Write an algebraic expression for each verbal expression. Then, evaluate the expression when $r = 2$, $s = 5$, and $t = \frac{1}{2}$.

31. the square of r increased by $3s$

32. t times the sum of four times s and r

33. the sum of r and s times the square of t

34. r to the fifth power decreased by t

Geometry

A formula for the perimeter of each figure is given. Find the perimeter when $a = 5$, $b = 6$, and $c = 8.5$.

35. triangle

$P = a + b + c$

36. square

$P = 4c$

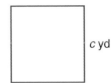

37. parallelogram

$P = 2(a + b)$

Graphing Calculator

Use a graphing calculator to evaluate each expression to the nearest hundredth.

38. $5(2)^4 + 3$

39. $\dfrac{(5 \cdot 7)^2 + 5}{(9 \cdot 3^2) - 7}$

40. $4.79 \, (0.05)^2 + 0.375 \, (6.34)^3$

41. $1 - 2u + 3u^2$ when $u = 1.35$

Critical Thinking

42. Patterns Consider the value of the expression $\dfrac{2x - 1}{2x}$ for various values of x.

   a. Copy and complete the chart below.

x	$\frac{1}{2}$	1	10	50	100
$\frac{2x-1}{2x}$					

   b. Describe the pattern that is emerging.

   c. What would the approximate value of the expression be if x were a very large number?

43. Consider the following sequence of numbers.

       4 2 5 3 2

   a. Insert operation symbols and parentheses in the sequence so that its value is 2.

   b. Insert operation symbols and parentheses in the sequence so that the answer is the largest possible number.

Applications and Problem Solving

44. Accounting Alicia and Travis are selling tickets for a school talent show. Bleacher seats cost $3.00, and floor seats cost $4.00. Alicia sells 30 bleacher seat tickets and 25 floor seat tickets. Travis sells 65 bleacher seat tickets.

   a. Write an expression to show how much money Alicia and Travis have collected for the tickets.

   b. How much money have they collected?

45. World Landmarks   The Great Pyramid of Cheops in Egypt is considered to be one of the Seven Wonders of the World. It is also the largest pyramid in the world. The area of its base is 4050 square meters. The volume of any pyramid is one-third the product of the area of the base B and its height h.

Pyramid	Height (meters)
Great Pyramid of Cheops in Egypt	147
Bent Pyramid in Egypt	101
Inca Pyramid in Peru	75
Pyramid of the Sun in Mexico	60
Step Pyramid of Djoser in Egypt	60
Pyramid of the Sun in Peru	50

a. Write an expression that represents the volume of a pyramid.

b. Find the volume of the Great Pyramid of Cheops.

Mixed Review

46. Find the next two terms in the sequence 2, 4, 8, 16, ... (Lesson 1–2)

47. Find the next two terms in the sequence 2, 5.5, 9, 12.5, ... (Lesson 1–2)

48. Find the next two items in the pattern a, a^2b, a^3b^2c, $a^4b^3c^2d$, ... (Lesson 1–2)

49. Ishi wants to save her money to buy a drum set from her uncle. Her uncle is willing to sell her the drums for $525. She has already saved $357, and she plans to add $1 on June 1, $2 on June 2, $3 on June 3, and so on. On what day will Ishi be able to buy the drums? (Lesson 1–2)

50. Write an algebraic expression for *h to the fifth power*. (Lesson 1–1)

51. The price of tickets for the spring concert is $3 more than last year. If last year's tickets cost t dollars, write an expression for the cost of the tickets this year. (Lesson 1–1)

52. Evaluate 11^2. (Lesson 1–1)

53. Write a verbal expression for $9 + 2y$. (Lesson 1–1)

Integration: Statistics
Stem-and-Leaf Plots

1-4

What YOU'LL LEARN

- To display and interpret data on a stem-and-leaf plot.

Why IT'S IMPORTANT

Stem-and-leaf plots are a useful way to display data.

APPLICATION
Consumerism

Carmela Perez sells artificial joints to hospitals. She does a lot of highway driving and frequently works with the doctors who perform operations to replace joints. Her present car is old, and she needs to buy a new one. Before deciding which car to buy, she wants to find out the miles-per-gallon (MPG) ratios of the cars she likes. This information will help her decide which cars to investigate in detail before she chooses one to buy. The MPG ratios for 25 cars are listed below.

31	30	28	26	22	31	26	34	47
32	18	33	26	23	18	29	13	40
31	42	17	22	50	12	41

Which MPG ratio occurs most frequently? Which is the highest MPG ratio? Which is the lowest?

Each day when you read newspapers or magazines, watch television, or listen to the radio, you are bombarded with numerical information about food, sports, the economy, politics, and so on. Interpreting this numerical information, or **data,** is important to your understanding of the world around you. A branch of mathematics called **statistics** helps provide you with methods of collecting, organizing, and interpreting data.

Graphs are often used to display data. The misuse of graphs can lead to false assumptions. One way that graphs are often used to mislead the reader is by labeling the vertical and horizontal scales inconsistently. All of the interval marks should represent the same units. If either of the scales does not begin at zero, this should be indicated by a broken or jagged line. For example, suppose Ana wants to convince her parents that her math grades are improving. Which graph below seems to show the greatest improvement in Ana's grades?

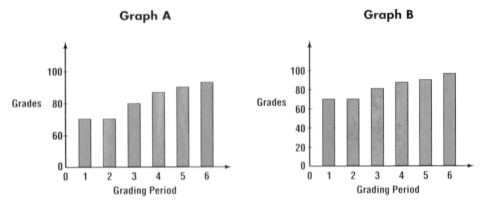

Graph A seems to show the greatest improvement. However, it is misleading. Notice the vertical axis. The distance from 0 to 60 is the same as the distance from 80 to 100. This is an incorrect representation of the data and can lead to the wrong conclusions.

A stem may have one or more digits. A leaf usually has just one digit.

Another way to organize and display data is by using a **stem-and-leaf plot.** In a stem-and-leaf plot, the greatest common place value of the data is used to form the *stems.* The numbers in the next greatest place-value position are then used to form the *leaves.* In the list on the previous page, the greatest place value is tens. Thus, 31 miles per gallon would have stem 3 and leaf 1.

To make the stem-and-leaf plot, first make a vertical list of the stems. Since the mileage data range from 12 to 50, the stems range from 1 to 5. Then, plot each number by placing the units digit (leaf) to the right of its correct stem. Thus, the mileage 31 is plotted by placing leaf 1 to the right of stem 3. Include a key with the plot. The complete stem-and-leaf plot is shown below.

Stem	Leaf
1	8 8 3 7 2
2	8 6 2 6 6 3 9 2
3	1 0 1 4 2 3 1
4	7 0 2 1
5	0 *3	1 = 31* ← *Key*

A second stem-and-leaf plot can be made to arrange the leaves in numerical order from least to greatest as shown below. This will make it easier for Ms. Perez to analyze the data.

Stem	Leaf
1	2 3 7 8 8
2	2 2 3 6 6 6 8 9
3	0 1 1 1 2 3 4
4	0 1 2 7
5	0 *3	1 = 31*

Example ➊

APPLICATION
Consumerism

Use the information in the stem-and-leaf plot shown above to answer each question.

a. **Which MPG ratio did Ms. Perez plot most frequently?**
26 and 31 (each three times)

b. **What are the highest and lowest MPG ratios?**
50 and 12

c. **Each line of the stem-and-leaf plot represents an interval of the data. In which mileage interval did Ms. Perez find the most cars?**
20–29 miles per gallon (8 cars)

d. **How many cars have a ratio of between 20 and 39 miles per gallon?**
15 cars

e. **If you were Ms. Perez, which cars might you investigate more closely? Why?**
Ms. Perez might want to investigate the cars represented in the last two lines of the stem-and-leaf plot, since she should get more miles per gallon with these cars than the other cars.

Sometimes the data for a stem-and-leaf plot are numbers that mostly begin with the same digit. In this case, use the digits in the first two places to form the stems.

Example 2

APPLICATION

Health

The weights (in pounds) of the students in a health and nutrition class are listed below. Make a stem-and-leaf plot of the student weights and answer the questions.

102	117	119	147	135	148	122	137	103
116	147	152	117	149	108	123	130	123
147	112	133	99	101	135	138	155	118
142	103	159	131	137	156	149	120	98

Since the data range from 159 to 98 pounds, the stems range from 15 to 9.

Notice that the numbers in the stem can be placed in ascending or descending order.

Stem	Leaf
15	9 6 5 2
14	9 9 8 7 7 7 2
13	8 7 7 5 5 3 1 0
12	3 3 2 0
11	9 8 7 7 6 2
10	8 3 3 2 1
9	9 8 *12|3 = 123*

Dr. Antonia C. Novello (1944–)

In 1989, Dr. Antonia C. Novello became the first woman and the first Hispanic American to become U.S. Surgeon General. She was appointed by President George Bush.

a. **What does 14 | 8 represent on the plot?**
   It represents 148 pounds.

b. **Which interval has the most students in it?**
   130–139 pounds (8 students)

c. **What is the difference between the lowest and highest weight?**
   159 – 98 or 61 pounds

d. **Which weight occurred most frequently?**
   147 pounds (3 students)

A *back-to-back stem-and-leaf plot* can also be used to compare two related sets of data.

Example 3

Kyle and Mikito wanted to compare boys' and girls' heights. They measured the height (in inches) of every student in their class. The data they collected and stem-and-leaf plot they made are shown below.

Boys' Heights (in.)		Girls' Heights (in.)	
65	60	72	64
63	70	57	60
69	72	61	63
71	66	65	62
73	71	59	61
59	58	61	71

(continued on the next page)

In this case, the heights of the boys and the heights of the girls are to be compared. To compare these numbers most effectively, use a back-to-back stem-and-leaf plot.

Boys	Stem	Girls
3 2 1 1 0	7	1 2
9 6 5 3 0	6	0 1 1 1 2 3 4 5
9 8	5	7 9 *5│7 = 57*

a. What is the height of the shortest boy? the shortest girl?
The shortest boy is 58 inches tall. The shortest girl is 57 inches tall.

b. What is the difference in height between the shortest boy and the tallest girl?
$72 - 58$ or 14 inches

c. What does 6│3 represent in each plot?
63 inches

d. What is the greatest number of boys who are the same height? What is the greatest number of girls who are the same height?
There are 2 boys who are 71 inches tall and 3 girls who are 61 inches tall.

e. What patterns, if any, do you see in the data?
The boys appear to be slightly taller than the girls. Seven boys are 65 inches or taller, while only 2 of the girls are that tall.

CHECK FOR UNDERSTANDING

Communicating Mathematics

Study the lesson. Then complete the following.

1. **Describe** the information you can determine by looking at a stem-and-leaf plot.

2. **Name** two or three ways in which a graph can be misleading.

3. **List** the steps used to make a stem-and-leaf plot.

4. **Assess Yourself** Write a paragraph about your use of stem-and-leaf plots. Begin the paragraph by completing this sentence: "I can use stem-and-leaf plots to _____ ." Think of the ways in which you deal with numbers: earnings from a job, number of assignments turned in, test scores, attendance at club meetings, and so on.

Guided Practice

Suppose the number 25,678 is rounded to 25,700 and plotted using stem 25 and leaf 7. Write the stem and leaf for each number below if the numbers are part of the same set of data.

5. 12,221   6. 6323   7. 126,896

8. Write the stems that would be used for a plot of the following set of data.

57, 43, 34, 12, 29, 8

9. Use the stem-and-leaf plot below to answer each question.

Number of Prom Tickets Sold During a 20-Day Period

Stem	Leaf
3	1 3 5
4	5 5 6 7
5	0 0 1 1 2 3 3
6	3 4 4 4
7	1 5 5|1 = 51

   a. What does the entry 3│5 represent?

   b. What was the least number of tickets sold?

   c. What was the greatest number of tickets sold during the 20-day period?

   d. What was the total number of tickets sold?

10. Architecture The *World Almanac* lists 23 tall buildings in Denver, Colorado. Each number below represents the number of floors for one of these buildings.

54	52	43	41	40	36	35	31
32	34	42	32	29	26	28	33
31	30	30	29	27	26	56	

   a. Make a stem-and-leaf plot of these data.

   b. What was the least number of floors?

   c. What was the greatest number of floors?

   d. How many buildings had 35 floors or more?

   e. How many buildings were in the 20–29 floor range?

EXERCISES

Practice

Suppose the number 178,651 is rounded to 179,000 and plotted using stem 17 and leaf 9. Write the stem and leaf for each number below if the numbers are part of the same set of data.

11. 133,271   **12.** 44,589   **13.** 442,672

14. 99,278   **15.** 1,112,750   **16.** 8443

Suppose the number 0.0478 is rounded to 0.048 and plotted using stem 4 and leaf 8. Write the stem and leaf for each number below if the numbers are part of the same set of data.

17. 0.14278   **18.** 0.00997   **19.** 1.114

Write the stems that would be used for a plot of each set of data.

20. 123, 436, 507, 449, 278, 489, 134, 770, 98, 110, 398

21. 12,367; 24,003; 27,422; 9447; 39,550; 38,045; 40,196

22. 37.2, 8.9, 12.74, 33.5, 27, 17.001, 13.5, 29.6

23. Manufacturing A video game company tests its products before selling them in stores. Its latest product is an inexpensive space detective adventure. The company would like to know whether to market the game to teens or young adults. A group of 25 teens and 25 young adults rated the game after playing it. The players rated the game for features like graphics, level of difficulty, and player interaction. The ratings consisted of a score between 1 (Trash it, man!) and 60 (Excellent, dude!). The players' ratings are listed below.

Teens (13–19 years old)	19 45 22 44 30 35 41 43 18 35 21 43 30 57 17 27 20 35 15 41 57 22 33 41 55
Young Adults (20–25 years old)	51 43 36 51 42 33 27 48 42 31 48 26 48 31 38 26 45 13 37 25 52 12 44 12 37

a. Should this game be sold to teens or young adults (or to both groups)?

b. Write a memo to the company with a recommendation. Be sure to give reasons for your decision.

24. Geology According to the *Universal Almanac,* there were 40 major earthquakes from June, 1990, to June, 1994. The stem-and-leaf plot below shows the magnitude of each earthquake in Richter scale units.

Stem	Leaf
5	1 4 4 5 6 9 9
6	0 0 1 2 2 2 2 4 4 5 8 8 8 8 8 8
7	0 0 1 2 2 2 2 3 4 5 5 5 6 7 7
8	0 7|3 = 7.3

a. What was the most frequent magnitude of the earthquakes?

b. What was the highest Richter scale value recorded for an earthquake?

c. How many more earthquakes had magnitudes in the 6.0–6.9 range than in the 5.0–5.9 range?

d. What was the magnitude of the weakest earthquake?

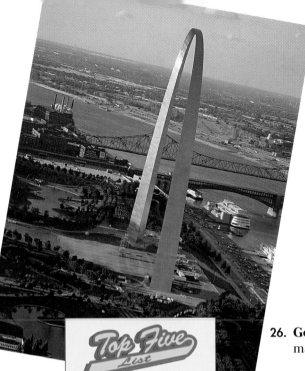

25. Economics The table at the right shows the average acreage per farm for six western states in 1980 and 1993 according to the *Universal Almanac*.

State	1980	1993
Arizona	5080	4557
Colorado	1358	1286
Montana	2601	2445
Nevada	3100	3708
New Mexico	3467	3274
Wyoming	3846	3742

a. Round each number to the nearest hundred. Then make a back-to-back stem-and-leaf plot of the average farm sizes for 1980 and 1993 for these six states.

b. Which interval shows the most common farm size found in 1980 for these six states? in 1993?

c. Write a statement about the information the stem-and-leaf plot presents.

26. Geography The numbers below show the lengths (in miles) of 30 major rivers in North America as listed in the *World Almanac*.

424	313	444	301	659	652	314	538
377	800	883	525	360	512	500	722
865	360	390	425	309	336	430	692
540	610	800	350	300	420		

a. Round each number to the nearest ten. Then make a stem-and-leaf plot of the data.

b. What is the difference in length between the shortest and longest rivers?

c. How many rivers are less than 400 miles long?

Top Five List

Longest rivers in the world
1. Nile, 4145 miles
2. Amazon, 4007 miles
3. Yangtze-Kiang, 3915 miles
4. Mississippi-Missouri-Red Rock, 3710 miles
5. Yenisey-Angara-Selenga, 3442 miles

Mixed Review

27. Evaluate $3 \cdot 6 - \dfrac{12}{4}$. (Lesson 1–3)

28. Evaluate $9a - 4^2 + b^2 \div 2$ when $a = 3$ and $b = 6$. (Lesson 1–3)

29. **Weather** The time between seeing lightning flash and hearing the bang of thunder it produces can be used to approximate the distance from the lightning. The distance in miles from lightning can be approximated by dividing the number of seconds between seeing the lightning and hearing the thunder by 5. Lightning within three miles can be dangerous and is a warning to take shelter.

a. Suppose there are s seconds between seeing the lightning and hearing the thunder. Write an algebraic expression for the distance from the lightning. (Lesson 1–1)

b. Fernando counted 10 seconds between seeing lightning and hearing thunder. How far away was the lightning? Is Fernando in danger? (Lesson 1–3)

30. Find the next two terms in the sequence $\dfrac{1}{2}, \dfrac{3}{4}, \dfrac{5}{8}, \dfrac{7}{16}, \dfrac{9}{32}, \ldots$ (Lesson 1–2)

31. Give the next two items for the following pattern. (Lesson 1–2)

Open Sentences

APPLICATION
Nutrition

The grams of fat found in cheeseburgers sold by various fast food chains are listed in the stem-and-leaf plot at the right.

Stem	Leaf
1	3 3 3 6 9
2	0 5 8 9
3	6 9
4	6
6	3

$4 \mid 6 = 46$

Obviously, some of the fast food chains are selling burgers with more than the usual amount of fat. Some students at Middletown High School have formed a consumer awareness club and are concerned about the high content of fat in this particular cheeseburger. They decide to write a letter to the company recommending a way to reduce the fat of their cheeseburger to 37 grams.

The students know that a one-ounce slice of American cheese has about 7 grams of fat and that ground beef has about 6 grams of fat per ounce. The students need to determine the number of ounces of ground beef that can be used to make a cheeseburger with a fat content of 37 grams.

Let *b* represent the number of ounces of beef that will go into the cheeseburger. This problem can be represented by the equation below.

$$6b + 7 = 37$$

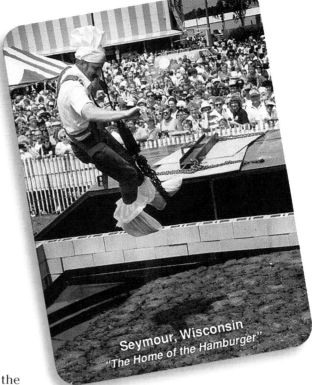

Seymour, Wisconsin
"The Home of the Hamburger"

The number 6 represents the amount of fat in one ounce of ground beef. The 7 represents the number of grams of fat in one slice of cheese. The 37 represents the total number of grams of fat for a cheeseburger.

Mathematical statements with one or more variables, or unknown numbers, are called **open sentences.** An open sentence is neither true nor false until the variable has been replaced by a value. Finding a replacement for the variable that results in a true sentence is called **solving the open sentence.** This replacement is called a **solution** of the open sentence.

Replace b in $6b + 7 = 37$ with the values 3, 4, 5, and 6. Then see whether each replacement results in a true or false sentence.

Replace b with:	$6b + 7 = 37$	True or False?
3	$6(3) + 7 \overset{?}{=} 37 \rightarrow 25 \neq 37$	false
4	$6(4) + 7 \overset{?}{=} 37 \rightarrow 31 \neq 37$	false
5	$6(5) + 7 \overset{?}{=} 37 \rightarrow 37 = 37$	true
6	$6(6) + 7 \overset{?}{=} 37 \rightarrow 43 \neq 37$	false

Since $b = 5$ makes the sentence $6b + 7 = 37$ true, the solution for $6b + 7 = 37$ is 5. The students can write a letter suggesting that the chain reduce the amount of fat to 37 grams by using 5 ounces of ground beef in their cheeseburger.

A set of numbers from which replacements for a variable may be chosen is called a **replacement set.** A **set** is a collection of objects or numbers. Sets are often shown by using braces { }. Each object or number in a set is called an **element,** or member. Sets are usually named by capital letters. Set A has three elements; they are 1, 3, and 5.

$$A = \{1, 3, 5\} \qquad B = \{2, 4, 5\} \qquad C = \{1, 2, 3, 4, 5\}$$

The **solution set** of an open sentence is the set of all replacements for the variable that make the sentence true.

Example **Find the solution set for $y + 5 \leq 7$ if the replacement set is {0, 1, 2, 3, 4}.**

Replace y with:	$y + 5 \leq 7$	True or False?
0	$0 + 5 \overset{?}{\leq} 7 \rightarrow 5 \leq 7$	true
1	$1 + 5 \overset{?}{\leq} 7 \rightarrow 6 \leq 7$	true
2	$2 + 5 \overset{?}{\leq} 7 \rightarrow 7 \leq 7$	true
3	$3 + 5 \overset{?}{\leq} 7 \rightarrow 8 \not\leq 7$	false
4	$4 + 5 \overset{?}{\leq} 7 \rightarrow 9 \not\leq 7$	false

The symbol $\leq$ means "less than or equal to." The symbol $\geq$ means "greater than or equal to."

Therefore, the solution set for $y + 5 \leq 7$ is {0, 1, 2}.

A sentence that contains an equals sign, $=$, is called an **equation.** A sentence having the symbols $<$, $\leq$, $>$, or $\geq$ is called an **inequality.** Which of the open sentences below are equations? Which ones are inequalities?

Mathematical Sentence	Equation or Inequality?
$2x + 10 = 50$	equation
$3a \leq 43$	inequality
$y - 6 > 12$	inequality

Sometimes you can solve an equation by simply applying the order of operations.

Example **Solve** $\dfrac{5(3+5)}{3 \cdot 2 + 2} = d.$

$$\frac{5(3+5)}{3 \cdot 2 + 2} = d$$

$$\frac{5(8)}{6+2} = d$$

$$\frac{40}{8} = d \quad \textit{Evaluate the numerator and denominator.}$$

$$5 = d \quad \textit{Divide.}$$

The solution is 5.

Example ③ Refer to the application at the beginning of the lesson. Solve for the number of grams of fat *f* in a McDonald's Quarter Pounder® with Cheese. This cheeseburger has one slice of cheese and four ounces of ground beef. Assume that the bun and toppings contain no fat.

APPLICATION
Nutrition

$f = 6b + 7 \qquad$ *f = grams of fat and b = ounces of beef*

$f = 6(4) + 7 \quad$ *Replace b with 4.*

$f = 24 + 7 \qquad$ *Evaluate using the order of operations.*

$f = 31$

A Quarter Pounder® with Cheese has about 31 grams of fat.

CHECK FOR UNDERSTANDING

Communicating Mathematics

Study the lesson. Then complete the following.

1. **Explain** why an open sentence always has at least one variable.

2. **Explain** the difference between an expression and an open sentence.

3. **Define** in your own words the phrase *solution set of an open mathematical sentence.*

4. **Explain** how to find the solution set for $7 + 2n > 31$ if the replacement set for *n* is {10, 11, 12, 13}.

5. Make up an inequality with a replacement set. Find the solution set and explain how you obtained it.

Guided Practice

State whether each equation is *true* or *false* for the value of the variable given.

6. $7(x^2) - 15 \div 5 = 25, x = 2$

7. $\dfrac{7a + a}{(9 \cdot 3) - 7} = 4, a = 5$

Find the solution set for each inequality, given the replacement set.

8. $3x + 2 > 2; \{0, 1, 2\}$

9. $2y^2 - 1 > 0; \{1, 3, 5\}$

Solve each equation for *y* if *x* is replaced with 6.

10. $x - 2 = y$

11. $2x^2 + 3 = y$

12. **Diet** During a lifetime, the average American drinks 15,579 glasses of milk, 6220 glasses of fruit juice, and 18,995 glasses of soft drinks.
    a. Write an equation for the total number of glasses of milk, juice, and soft drinks that the average American drinks in a lifetime.
    b. What is the total number of glasses of milk, juice, and soft drinks consumed by the average American in a lifetime?

EXERCISES

Practice

State whether each equation is *true* or *false* for the value of the variable given.

13. $a + \frac{3}{4} = \frac{3}{2} + \frac{1}{4}, a = \frac{1}{2}$

14. $\frac{3 + 15}{x} = \frac{1}{2}(x), x = 6$

15. $y^6 = 4^3, y = 2$

16. $3x^2 - 4(5) = 6, x = 3$

17. $\frac{5^2 - 2y}{5^2 - 6} \leq 1, y = 3$

18. $a^5 \div 8 \div a^2 \div a < \frac{1}{2}, a = 2$

Find the solution set for each inequality if the replacement sets are $x = \left\{\frac{1}{2}, \frac{3}{4}, 1, \frac{5}{4}\right\}$ and $y = \{5, 10, 15, 20\}$.

19. $y - 2 < 6$

20. $y + 2 > 7$

21. $8x + 1 < 8$

22. $2x > 1$

23. $\frac{y}{5} \geq 2$

24. $3x \leq 4$

Solve each equation.

25. $y = \frac{14 - 8}{2}$

26. $4(6) + 3 = a$

27. $\frac{21 - 3}{12 - 3} = x$

28. $d = 3\frac{1}{2} \div 2$

29. $s = 4\frac{1}{2} + \frac{1}{3}$

30. $x = 5^2 - 2^3$

Critical Thinking

31. Find five pairs of values for p and q such that the open sentence $q + 2 > 3p$ is true.

Applications and Problem Solving

32. **Weather Forecasting** The graph at the right shows the states with the most tornadoes.

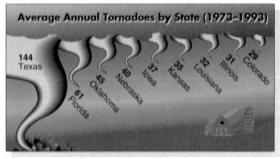

Average Annual Tornadoes by State (1973–1993)

144 Texas · 61 Florida · 45 Oklahoma · 40 Nebraska · 37 Iowa · 35 Kansas · 32 Louisiana · 31 Illinois · 29 Colorado

Source: National Severe Storm Forecast Center

    a. Write an equation that estimates the number of tornadoes Texas will have in the next three years. Justify why your equation will provide a good estimate.
    b. At this rate, how many tornadoes will Texas have in the next three years?
    c. Make up another problem using the data in the graph.

33. **Nutrition** A person must burn 3500 calories to lose one pound of weight.
    a. Write an equation that represents the number of calories a person would have to burn a day to lose 4 pounds in two weeks.
    b. How many calories would the person have to burn each day?

34. Biology Insects are the largest class of animals in the world with at least 800,000 species. The fastest-moving insect is the large tropical cockroach. It scurries at speeds of up to 3.36 miles per hour, which is about 2.3 feet per second.

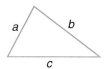

    a. Write an equation that represents how many miles the roach can travel in 1.5 hours.

    b. How many miles can the roach travel in 1.5 hours?

    c. Can the roach travel 100 feet in less than a minute? Explain.

35. Write the stems that would be used to form a stem-and-leaf plot for the following set of data. (Lesson 1–4)

    2.7, 5.9, 2.0, 7.7, 5.2, 6.0, 5.4, 9.9, 5.4

36. Hockey The stem-and leaf plot at the right shows the greatest number of goals scored by a National Hockey League player during any one season. Mario Lemieux is credited with the fourth greatest number of goals in a season. How many goals did Lemieux score during that season? (Lesson 1–4)

Stem	Leaf
7	1 2 3 6 6 6
8	5 6 7
9	2

7 | 2 = 72

37. Evaluate $5(13 - 7) - 22$. (Lesson 1–3)

38. Write the next two expressions for the pattern $2a + 1, 4a + 3, 6a + 5, 8a + 7, \ldots$ (Lesson 1–2)

39. Write a verbal expression for $x^5 - 5$. (Lesson 1–1)

40. Geometry Write an algebraic expression for the perimeter of the triangle at the right. (Lesson 1–1)

SELF TEST

Write an algebraic expression for each verbal expression. (Lesson 1–1)

1. the sum of three times a and the square of b **2.** w to the fifth power minus 37

3. Patterns Each Tuesday morning at Central High School, the first seven bells ring at 8:00, 8:43, 8:47, 9:30, 9:34, 10:17, and 10:21 A.M. Give the times of the other morning bells. (Lesson 1–2)

Evaluate each expression. (Lesson 1–3)

4. $5(8 - 3) + 7 \cdot 2$ **5.** $6(4^3 + 2^2)$ **6.** $(9 - 2 \cdot 3)^3 - 27 + 9 \cdot 2$

Make a stem-and-leaf plot for each set of data. (Lesson 1–4)

7. 67, 85, 54, 48, 89, 77   **8.** 236, 450, 748, 254, 755, 347, 97, 386

Find the solution set for each open sentence if the replacement set is {4, 5, 6, 7, 8}. (Lesson 1–5)

9. $x + 2 > 7$ **10.** $9x - 20 = x^2$

1-6

Identity and Equality Properties

Super Bowl Ticket Prices 1967–1995

What YOU'LL LEARN

- To recognize and use the properties of identity and equality, and
- to determine the multiplicative inverse of a number.

Why IT'S IMPORTANT

You can use the identity and equality properties to evaluate expressions and solve equations.

APPLICATION
Football

According to the graph at the right, the price of Super Bowl tickets increased by $0 from 1975 to 1977. Also, from 1981 to 1983, ticket prices stayed at $40. The open sentences below represent each situation.

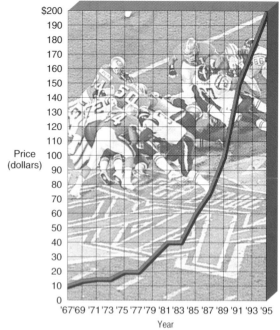

Price (dollars)

The solution for each equation is 0. Super Bowl ticket prices increased $0 during each time period.

Equations such as these can be summarized in algebraic terms. The sum of any number and 0 is equal to that number. Zero is called the **additive identity.**

Additive Identity Property	For any number *a*, *a* + 0 = 0 + *a* = *a*.

The equations below also represent the ticket prices from 1975 to 1977 and from 1981 to 1983. The variable *m* represents the number of times of increase.

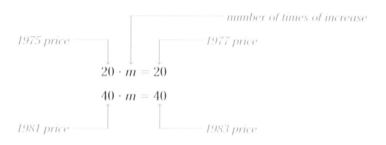

The solution for each equation is 1. Since the product of any number and 1 is equal to the number, 1 is called the **multiplicative identity.**

Multiplicative Identity Property	For any number a, $a \cdot 1 = 1 \cdot a = a$.

Suppose you purchase a number of Super Bowl tickets for $200 each. If you sell three of them at face value ($200), your profit is $0. The following equation describes the situation.

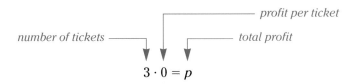

$$3 \cdot 0 = p$$

In this equation, one of the factors is 0 and the value of p is 0. This equation suggests the following property.

Multiplicative Property of Zero	For any number a, $a \cdot 0 = 0 \cdot a = 0$.

Two numbers whose product is 1 are called **multiplicative inverses** or **reciprocals.** Zero has no reciprocal because any number times 0 is 0.

Multiplicative Inverse Property	For every nonzero number $\frac{a}{b}$, where $a, b \neq 0$, there is exactly one number $\frac{b}{a}$ such that $\frac{a}{b} \cdot \frac{b}{a} = 1$.

Example ① **Name the multiplicative inverse of each number or variable. Assume that no variable equals zero.**

a. 5

Since $5 \cdot \frac{1}{5} = 1$, $\frac{1}{5}$ is the multiplicative inverse of 5.

b. x

Since $x \cdot \frac{1}{x} = 1$, the multiplicative inverse is $\frac{1}{x}$.

x ≠ 0; why?

c. $\frac{2}{3}$

Using the property, the multiplicative inverse is $\frac{3}{2}$.

At the beginning of algebra class, Ms. Escalante gave each of her students a single strip of paper 8 inches long. She instructed the students to divide their paper strips any way they wished.

Staci left her strip as one 8-inch strip.

Amad cut his strip to form a 6-inch strip and a 2-inch strip.

Liam cut his strip to form a 5-inch strip and a 3-inch strip.

Using the strips of paper, we know the following to be true.

$$8 = 8 \qquad 6 + 2 = 6 + 2 \qquad 3 + 5 = 3 + 5$$

The **reflexive property of equality** says that any quantity is equal to itself.

Reflexive Property of Equality	For any number a, $a = a$.

Using the paper strips, we can show the following statements are true.

If $8 = 6 + 2$, then $6 + 2 = 8$.

If $3 + 5 = 6 + 2$, then $6 + 2 = 3 + 5$.

The **symmetric property of equality** says that if one quantity equals a second quantity, then the second quantity also equals the first.

Symmetric Property of Equality	For any numbers a and b, if $a = b$, then $b = a$.

A third property can also be shown using the paper strips.

If $3 + 5 = 8$ and $8 = 6 + 2$, then $3 + 5 = 6 + 2$.

If $8 = 3 + 5$ and $3 + 5 = 6 + 2$, then $8 = 6 + 2$.

The **transitive property of equality** says that if one quantity equals a second quantity and the second quantity equals a third quantity, then the first and third quantities are equal.

Transitive Property of Equality	For any numbers a, b, and c, if $a = b$ and $b = c$, then $a = c$.

We know that $5 + 3 = 6 + 2$. Since $5 + 3$ is equal to 8, we can substitute 8 for $5 + 3$ to get $8 = 6 + 2$. The **substitution property of equality** says that a quantity may be substituted for its equal in any expression.

Substitution Property of Equality	If $a = b$, then a may be replaced by b in any expression.

You can use the properties of identity and equality to justify each step when evaluating an expression.

Example **2**

Sales

The pep club at Roosevelt High School is selling submarine sandwiches, lemonade, and apples at the district swim meet. Each sandwich costs $2.00 to make and sells for $3.00. Each glass of lemonade costs $0.25 to make and sells for $1.00. Each apple costs the club $0.25, and the members have decided to sell apples for $0.25 each. Write an expression that represents the profit for 80 sandwiches, 150 glasses of lemonade, and 40 apples. Evaluate the expression, indicating the property used in each step.

$$80(3.00 - 2.00) + 150(1.00 - 0.25) + 40(0.25 - 0.25)$$

$= 80(1) + 150(0.75) + 40(0)$	*Substitution* $(=)$
$= 80 + 150(0.75) + 40(0)$	*Identity* $(\times)$
$= 80 + 112.50 + 40(0)$	*Substitution* $(=)$
$= 80 + 112.50 + 0$	*Multiplicative prop. of 0*
$= 192.50 + 0$	*Substitution* $(=)$
$= 192.50$	*Identity* $(+)$

The club would make a profit of $192.50.

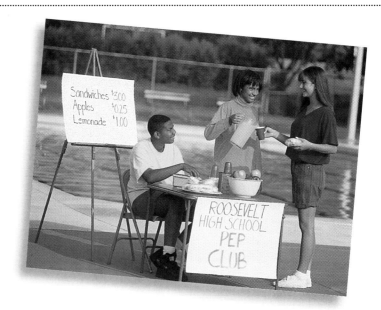

CHECK FOR UNDERSTANDING

Communicating Mathematics

Study the lesson. Then complete the following.

1. **Define** the term *identity* in your own words.

2. **Explain** whether or not 1 can be the additive identity.

3. **Explain** why 0 does *not* have a multiplicative inverse.

4. **Name** the multiplicative inverse of 1.

5. Write a paragraph explaining how to determine the multiplicative inverse of a number.

Guided Practice

Name the multiplicative inverse of each number or variable. Assume that no variable represents zero.

6. 7

7. $\frac{9}{2}$

8. c

Match the expressions in the left-hand column with the properties in the right-hand column.

9. $0 \cdot 36 = 0$

10. $1(68) = 68$

11. $14 + 16 = 14 + 16$

12. $(9 - 7)(5) = 2(5)$

13. $\frac{3}{4} \times \frac{4}{3} = 1$

14. $0 + g = g$

15. If $8 + 1 = 9$, then $9 = 8 + 1$.

a. Additive identity property

b. Multiplicative identity property

c. Multiplicative property of 0

d. Multiplicative inverse property

e. Reflexive property (=)

f. Symmetric property (=)

g. Substitution property (=)

Name the property used in each step.

16.
$$(14 \cdot \tfrac{1}{14} + 8 \cdot 0) \cdot 12 = (1 + 8 \cdot 0) \cdot 12$$
$$= (1 + 0) \cdot 12$$
$$= 1 \cdot 12$$
$$= 12$$

Evaluate each expression. Name the property used in each step.

17. $6(12 - 48 \div 4) + 9 \cdot 1$

18. $3 + 5(4 - 2^2) - 1$

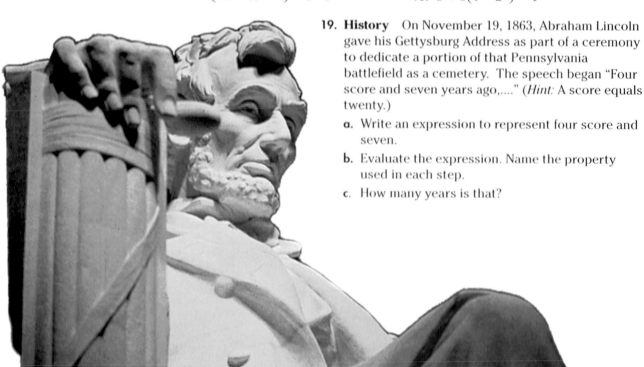

19. History On November 19, 1863, Abraham Lincoln gave his Gettysburg Address as part of a ceremony to dedicate a portion of that Pennsylvania battlefield as a cemetery. The speech began "Four score and seven years ago,....." (*Hint:* A score equals twenty.)

a. Write an expression to represent four score and seven.

b. Evaluate the expression. Name the property used in each step.

c. How many years is that?

Practice   **Name the multiplicative inverse of each number or variable. Assume that no variable represents zero.**

20. 9

21. $\frac{1}{9}$

22. $\frac{1}{4}$

23. p

24. $\frac{2}{a}$

25. $1\frac{1}{2}$

Name the property or properties illustrated by each statement.

26. If $7 \cdot 2 = 14$, then $14 = 7 \cdot 2$.

27. $8 + (3 + 9) = 8 + 12$

28. $(10 - 8)(5) = 2(5)$

29. $mnp = 1mnp$

30. $\left(\frac{3}{4}\right)\left(\frac{4}{3}\right) = 1$

31. $3\left(5^2 \cdot \frac{1}{25}\right) = 3$

32. $0 + 23 = 23$

33. If $6 = 9 - 3$, then $9 - 3 = 6$.

34. $5(0) = 0$

35. $32 + 21 = 32 + 21$

36. If $4 \cdot 2 = 8$ and $8 = 6 + 2$, then $4 \cdot 2 = 6 + 2$.

Name the property used in each step.

37. $2(3 \cdot 2 - 5) + 3 \cdot \frac{1}{3} = 2(6 - 5) + 3 \cdot \frac{1}{3}$

$$= 2(1) + 3 \cdot \frac{1}{3}$$

$$= 2 + 3 \cdot \frac{1}{3}$$

$$= 2 + 1$$

$$= 3$$

38. $26 \cdot 1 - 6 + 5(12 \div 4 - 3) = 26 \cdot 1 - 6 + 5(3 - 3)$

$$= 26 \cdot 1 - 6 + 5(0)$$

$$= 26 - 6 + 5(0)$$

$$= 26 - 6 + 0$$

$$= 20 + 0$$

$$= 20$$

39. $7(5 \cdot 3^2 - 11 \cdot 4) = 7(5 \cdot 9 - 11 \cdot 4)$

$$= 7(45 - 44)$$

$$= 7 \cdot 1$$

$$= 7$$

Evaluate each expression. Name the property used in each step.

40. $4(16 \div 4^2)$

41. $(15 - 8) \div 7 \cdot 25$

42. $(8 \cdot 3 - 19 + 5) + (3^2 + 8 \cdot 4)$

43. $(2^5 - 5^2) + (4^2 - 2^4)$

44. $8[6^2 - 3(11)] \div 8 \cdot \frac{1}{3}$

45. $5^3 + 9\left(\frac{1}{3}\right)^2$

46. Think about the relationship "is less than," represented by the symbol $<$. Does $<$ work with each of the following? Explain why or why not. Give examples that support your reasoning.

   a. reflexive property

   b. symmetric property

   c. the transitive property

47. Entertainment The students in Mr. Toshio's class are planning an ice cream party. They have selected the ice cream flavors shown in the table below. The prices listed are for $\frac{1}{2}$- cup servings. A survey of the students shows that 12 students want vanilla, 15 want chocolate, and 10 want chocolate chip cookie dough.

Flavor	Brand	Price
Vanilla	Breyers	21¢
Chocolate	Edy's/Dreyer's Grand	23¢
Chocolate Chip Cookie Dough	Ben & Jerry's	67¢

   a. Write an expression that represents how much it will cost to purchase ice cream for the class if each student gets a 1-cup serving.

   b. Evaluate the expression. Name the property used in each step.

   c. What is the total cost?

48. Postal Rates Patricia wants to mail a package to her cousin in Los Angeles. The cost of first class mail is $0.32 for the first ounce and $0.23 for each additional ounce or fraction of an ounce. Patricia's package weighs 14.4 ounces.

   a. Write an expression that represents the cost to mail the package.

   b. Evaluate the expression. Name the property used in each step.

   c. How much will Patricia need to pay in postage?

49. *True or False*: $15 \div 3 + 7 < 13$. (Lesson 1–5)

50. Solve $m = (18 - 3) \div (3^2 - 2^2)$. (Lesson 1–5)

51. *True or False*: $(2n^2 + 6) \div 4 < 5$, when $n = 3$. (Lesson 1–5)

52. When is a back-to-back stem-and-leaf plot used? (Lesson 1–4)

53. Write the first and last names of ten of your classmates. Make a stem-and-leaf plot showing the number of letters in these ten names. (Lesson 1–4)

54. Evaluate $5(7 - 2) - 3^2$. (Lesson 1–3)

55. Evaluate $xy - 2y$ when $x = 6$ and $y = 9$. (Lesson 1–3)

56. Find the next two terms in the sequence 1, 4, 7, 10, ... (Lesson 1–2)

57. Write an algebraic expression for the number of months in y years. (Lesson 1–1)

1-7A The Distributive Property

Materials: algebra tiles   product mat

A Preview of Lesson 1–7

Throughout your study of mathematics, you have used rectangles to model multiplication. For example, the figure below shows the multiplication $2(3 + 1)$ as a rectangle that is 2 units wide and $3 + 1$ units long. The model shows that the expression $2(3 + 1)$ is equal to $2 \cdot 3 + 2 \cdot 1$. The sentence $2(3 + 1) = 2 \cdot 3 + 2 \cdot 1$ illustrates the distributive property.

You can use special tiles called **algebra tiles** to form rectangles that model multiplication. A 1-tile is a square that is 1 unit long and 1 unit wide. Its area is 1 square unit. An x-tile is a rectangle that is 1 unit wide and x units long. Its area is x square units.

Activity   **Find the product $3(x + 2)$ by using algebra tiles. First, model $3(x + 2)$ as the area of a rectangle.**

Step 1   The rectangle has a width of 3 units and a length of $x + 2$ units. Use your algebra tiles to mark off the dimensions on a product mat.

Step 2   Using the marks as a guide, make the rectangle with algebra tiles.

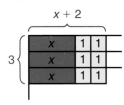

The rectangle has 3 x-tiles and 6 1-tiles. The area of the rectangle is $x + 1 + 1 + x + 1 + 1 + x + 1 + 1$, or $3x + 6$. Thus, $3(x + 2) = 3x + 6$.

Model   **Find each product by using algebra tiles.**

1. $2(x + 1)$ **2.** $5(x + 2)$ **3.** $2(2x + 1)$ **4.** $2(3x + 3)$

Draw   **Tell whether each statement is true or false. Justify your answer with algebra tiles and a drawing.**

5. $3(x + 3) = 3x + 3$ **6.** $x(3 + 2) = 3x + 2x$

Write   **7. You Decide**   Helen says that $3(x + 4) = 3x + 4$, but Adita says that $3(x + 4) = 3x + 12$.

     a. Which classmate is correct?

     b. Use words and/or models to give an explanation that demonstrates the correct equation.

The Distributive Property

What YOU'LL LEARN

- To use the distributive property to simplify expressions.

APPLICATION
Food Service

In the school cafeteria, each student can pick one hot or cold entree and one salad from the following in order to get a "Meal Deal" lunch at a special low price.

Hot Lunch	Cold Lunch	Salads
• Tacos • Meat loaf and mashed potatoes • Steamed veggies and rice • Spaghetti	• Submarine sandwich • Turkey club sandwich • Tuna salad with tomato	• Tossed salad • Fruit salad

Why IT'S IMPORTANT

You can use the distributive property to evaluate expressions and solve equations.

Jenine wants to know how many different lunches are possible. She can make a table to represent all the possible combinations.

Hot Lunch				
	Entree			
Salads	Tacos	Meat Loaf	Veggies	Spaghetti
Tossed	x	x	x	x
Fruit	x	x	x	x

Cold Lunch			
	Entree		
Salads	Submarine	Club Sandwich	Tuna Salad
Tossed	x	x	x
Fruit	x	x	x

There are $2 \cdot 4$ or 8 possible hot lunches.
There are $2 \cdot 3$ or 6 possible cold lunches.
There are $(2 \cdot 4) + (2 \cdot 3)$ possible lunches from which students can choose.
Note that $(2 \cdot 4) + (2 \cdot 3) = 8 + 6$ or 14.

Jenine can also represent this problem by using the following table.

Salads	Entree						
	Hot Lunch				Cold Lunch		
	Taco	Meat Loaf	Veggie	Spaghetti	Sub	Club Sand.	Tuna
Tossed	x	x	x	x	x	x	x
Fruit	x	x	x	x	x	x	x

According to this table, there are 2 types of salads times the total number of entrees, $4 + 3$, or 7.

$$2(4 + 3) = 2 \cdot 7 \text{ or } 14$$

Either way you look at it, there are 14 possible lunches. That's because the following is true.

$$2(4 + 3) = 2 \cdot 4 + 2 \cdot 3$$

This is an example of the **distributive property.**

Distributive Property	**For any numbers _a_, _b_, and _c_,** $a(b + c) = ab + ac$ and $(b + c)a = ba + ca;$ $a(b - c) = ab - ac$ and $(b - c)a = ba - ca.$

Notice that it doesn't matter whether _a_ is placed on the right or the left of the expression in parentheses.

The symmetric property of equality allows the distributive property to be written as follows.

$$\text{If } a(b + c) = ab + ac, \text{ then } ab + ac = a(b + c).$$

Example ❶

INTEGRATION

Geometry

The Oak Grove High School Spirit Club wants to make a banner for the championship football game. The students have two large pieces of paper to use for the banner. One piece is 5 feet by 13 feet, and the other is 5 feet by 10 feet. They plan to use both pieces for the banner. Find the total area of the banner.

The total area of the banner can be found in two ways.

13 ft   10 ft

5 ft [   ]   5 ft [   ]

Method 1: Add the areas of the smaller rectangles.

$A = w\ell_1 + w\ell_2$

$\quad = 5(13) + 5(10)$

$\quad = 65 + 50$

$\quad = 115$

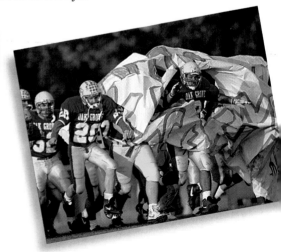

Method 2: Multiply the width by the length.

$A = w\ell$

$\quad = 5(13 + 10) \quad w = 5, l = 13 + 10$

$\quad = 5(23)$

$\quad = 115$

13 ft   10 ft

5 ft

The area is 115 square feet.

You can use the distributive property to multiply in your head.

Example 2 Use the distributive property to find each product.

a. $7 \cdot 98$

$7 \cdot 98 = 7(100 - 2)$

$\quad = 700 - 14$

$\quad = 686$

b. $8(6.5)$

$8(6.5) = 8(6 + 0.5)$

$\quad = 48 + 4$

$\quad = 52$

A **term** is a number, a variable, or a product or quotient of numbers and variables. Some examples of terms are x^3, $\frac{1}{4}a$, and $4y$. The expression $9y^2 + 13y^2 + 3$ has three terms.

Like terms are terms that contain the same variables, with corresponding variables having the same power. In the expression $8x^2 + 2x^2 + 5a + a$, $8x^2$ and $2x^2$ are like terms, and $5a$ and a are also like terms.

We can use the distributive property and properties of equality to show that $3x + 8x = 11x$. In this expression, $3x$ and $8x$ are like terms.

$$3x + 8x = (3 + 8)x \qquad \textit{Distributive property}$$

$$= 11x \qquad \textit{Substitution (=)}$$

The expressions $3x + 8x$ and $11x$ are called **equivalent expressions** because they denote the same number. An expression is in **simplest form** when it is replaced by an equivalent expression having no like terms and no parentheses.

Example 3 Simplify $\frac{1}{4}x^2 + 2x^2 + \frac{11}{4}x^2$.

In this expression, $\frac{1}{4}x^2$, $2x^2$, and $\frac{11}{4}x^2$ are like terms.

$$\frac{1}{4}x^2 + 2x^2 + \frac{11}{4}x^2 = \left(\frac{1}{4} + 2 + \frac{11}{4}\right)x^2 \qquad \textit{Distributive property}$$

$$= 5x^2 \qquad \textit{Substitution (=)}$$

The **coefficient** of a term is the numerical factor. For example, in $23ab$, the coefficient is 23. In xy, the coefficient is 1 since, by the multiplicative identity property, $1 \cdot xy = xy$. Like terms may also be defined as terms that are the same or that differ only in their coefficients.

Example **Name the coefficient in each term.**

    a. $145x^2y$ The coefficient is 145.

    b. ab^2 The coefficient is 1 since $ab^2 = 1ab^2$.

    c. $\dfrac{4a^2}{5}$ The coefficient is $\dfrac{4}{5}$ because $\dfrac{4a^2}{5}$ can be written as $\dfrac{4}{5} \cdot a^2$.

Example ❺ **Simplify each expression.**

    a. $4w^4 + w^4 + 3w^2 - 2w^2$ **b.** $\dfrac{a^3}{4} + 2a^3$

       Remember that $w^4 = 1w^4$.   Remember that $\dfrac{a^3}{4} = \dfrac{1}{4}a^3$.

       $4w^4 + w^4 + 3w^2 - 2w^2$ $\dfrac{a^3}{4} + 2a^3 = \dfrac{1}{4}a^3 + 2a^3$

          $= (4 + 1)w^4 + (3 - 2)w^2$ $= \left(\dfrac{1}{4} + 2\right)a^2$

          $= 5w^4 + 1w^2$ $= 2\dfrac{1}{4}a^3$

          $= 5w^4 + w^2$

CHECK FOR UNDERSTANDING

Communicating Mathematics

Study the lesson. Then complete the following.

1. **Explain** why the equation $2(a - 3) = 2a - 3$ is *not* a true sentence.

2. **Write** an expression that meets the following conditions.

    a. It has four terms,

    b. three terms are like terms, and

    c. one has a coefficient of 1.

3. **Describe** how you would simplify $3(2x - 4)$.

4. **Draw** a rectangular model or use tiles to represent $4(x + 1)$.

Guided Practice

Match an expression in the left-hand column with the equivalent expression in the right-hand column.

    5. $8(10 + 4)$ **a.** $(4 + x)2$

    6. $(12 - 3)6$ **b.** $8 \cdot 10 + 8 \cdot 4$

    7. $(4 \cdot 2) + (x \cdot 2)$ **c.** $2x - 2$

    8. $2(x + 5)$ **d.** $2x + 10$

    9. $2(x - 1)$ **e.** $(12)(6) - (3)(6)$

Use the distributive property to rewrite each expression without parentheses.

    10. $3(2x + 6)$ **11.** $2(a - b)$

Use the distributive property to find each product.

    12. $15 \cdot 99$ **13.** $28\left(2\dfrac{1}{7}\right)$

Name the coefficient of each term.

14. $2.5cd$ **15.** $7a^2b$ **16.** $\frac{3b}{5}$

Name the like terms in each expression.

17. $4y^4 + 3y^3 + y^4$ **18.** $3a^2 + 4c + a + 3b + c + 9a^2$

Simplify each expression, if possible. If not possible, write in simplest form.

19. $t^2 + 2t^2 + 4t$ **20.** $25x^2 + 5x$

21. $16a^2b + 7a^2b + 3ab^2$ **22.** $7p + q - p + \frac{2q}{3}$

23. Employment Maria and Mark are salesclerks at a local department store. Each earns \$5.35 per hour. Maria works 24 hours per week, and Mark works 32 hours per week. Write two expressions to represent the amount the two of them earn per week.

EXERCISES

Practice

Use the distributive property to rewrite each expression without parentheses.

24. $2(4 + t)$ **25.** $(g - 9)5$

26. $5(x + 3)$ **27.** $8(3m + 6)$

28. $28\left(y - \frac{1}{7}\right)$ **29.** $a(5 - b)$

Use the distributive property to find each product.

30. $5 \cdot 97$ **31.** $\left(3\frac{1}{17}\right) \times 17$

32. $16(102)$ **33.** $24(2.5)$

34. $999 \cdot 6$ **35.** 3×215

Simplify each expression, if possible. If not possible, write in simplest form.

36. $15x + 18x$ **37.** $14a^2 + 13b^2 + 27$

38. $10n + 3n^2 + 9n^2$ **39.** $5a + 7a + 10b + 5b$

40. $7(3x^2y - 4xy^2 + xy)$ **41.** $13p^2 + p$

42. $5(6a + 4b - 3b)$ **43.** $3(x + 2y) - 2y$

44. $\frac{2}{3}\left(c - \frac{3}{4}\right) + c(1 + b)$ **45.** $a + \frac{a}{5} + \frac{2}{5}a$

46. $4(3g + 2) + 2(g + 3)$ **47.** $3(x + y) + 2(x + y) + 4x$

Programming

48. The program at the right tests values of A, B, and C to determine if the distributive property holds for division. That is, does $\frac{A + B}{C} = \frac{A}{C} + \frac{B}{C}$?

```
Program: DISTPROP
: Prompt A,B,C
: If(A + B)/C = A/C + B/C
: Then
: Disp "YES, IT HOLDS."
: Else
: Disp "TRY AGAIN."
: END
```

  a. Run the program for 10 different sets of values for A, B, and C. Does the property hold?

  b. How could you change the program to test if $(A + B)^C = A^C + B^C$? Does the distributive property hold for powers?

49. If $2(b + c) = 2b + 2c$, does $2 + (b \cdot c) = (2 + b)(2 + c)$? Choose values for b and c to show that these may be true or find *counterexamples* to show that they are not.

50. **Economics** According to *American Demographics,* the average 13- to 15-year-old male received \$16.15 per week for allowance in 1993. The average 16- to 19-year-old female received \$32.45 per week. Toni was 17 years old, and her brother Carlos was 14 in 1993. Assume that they got the average amount for their weekly allowance.

a. Write two expressions to represent the amount their parents paid in allowance during the month of February.

b. What was the amount their parents paid in allowance during that month?

51. **Geometry** The largest permanently installed theater screen has an area of 6768 square feet. It is at the Keong Emas Imax Theater in Jakarta, Indonesia. The movie screen below represents a typical movie screen in U.S. movie theaters.

a. Write an expression for the perimeter of the movie screen below and then simplify the expression.

Keong Emas Imax Theater

x feet

$(x + 14)$ feet

b. Evaluate the expression if $x = 17$.

c. Find the area of the movie screen if $x = 17$.

52. Name the property illustrated by the following statement: *If $19 - 3 = 16$, then $16 = 19 - 3$.* (Lesson 1–6)

53. Name the property illustrated by $9 \times 0 = 0$. (Lesson 1–6)

54. Find the solution set for inequality $3x - 5 > 7$ if the replacement set is $\{2, 3, 4, 5, 6\}$. (Lesson 1–5)

55. **Physics** Sound travels 1129 feet per second through air. (Lesson 1–5)

a. Write an equation that represents how many feet sound can travel in 2 seconds when it is traveling through air.

b. How many feet can sound travel in 2 seconds when traveling through air?

56. Make a stem-and-leaf plot for the following set of data. (Lesson 1–4)
37, 45, 36, 51, 55, 29, 45, 58, 36

57. Evaluate $\dfrac{4^2 - 2^3}{24 - 2(10)}$. (Lesson 1–3)

58. What are the next two expressions for the pattern $5a$, $10a^2$, $15a^3$, $20a^4$, ...? (Lesson 1–2)

59. **Culture** Each year in the Chinese calendar is named for one of 12 animals. Every 12 years, the same animal is named. If 1992 was the Year of the Monkey, how many years in the 20th century were Years of the Monkey? (Lesson 1–2)

60. Write an algebraic expression for *37 less than 2 times a number k.* (Lesson 1–1)

Commutative and Associative Properties

- To recognize and use the commutative and associative properties to simplify expressions.

Why IT'S IMPORTANT

You can use the communitative and associative properties to evaluate expressions and solve equations.

APPLICATION

Networks

The map below shows the location of Leticia's house and school. It also shows the time in minutes that it takes Leticia to walk around her neighborhood.

Yesterday, Leticia walked to school by way of Lincoln Street and Washington Avenue. She took $3 + 3 + 3 + 2 + 2 + 2$ or 15 minutes to walk to school. Today, Leticia walked to school by way of Madison Avenue and Wilson Street. She took $2 + 2 + 2 + 3 + 3 + 3$ or 15 minutes. The time it took Leticia to walk to school is the same for both days.

$$3 + 3 + 3 + 2 + 2 + 2 = 2 + 2 + 2 + 3 + 3 + 3$$

In this equation, the addends are the same, but their order is different. The **commutative property** says that the order in which you add or multiply two numbers does not change their sum or product.

Commutative Property	For any numbers a and b, $a + b = b + a$ and $a \cdot b = b \cdot a$.

One easy way to find the sum or product of numbers is to group or *associate* the numbers. The **associative property** says that the way you group three numbers when adding or multiplying does not change their sum or product.

Associative Property	For any numbers a, b, and c, $(a + b) + c = a + (b + c)$ and $(ab)c = a(bc)$.

You can group $11 + 12 + 7 + 9 + 7 + 6$ to make mental addition easy.

$$11 + 12 + 7 + 9 + 7 + 6 = 11 + 9 + 7 + 6 + 7 + 12 \quad \textit{Commutative } (+)$$
$$= (11 + 9) + (7 + 6 + 7) + 12 \quad \textit{Associative } (+)$$
$$= 20 + 20 + 12$$
$$= 52$$

The commutative and associative properties can be used with the other properties you have studied when evaluating and simplifying expressions.

Example ① **Juan Martinez is a product manager for a cereal company. Part of his job is to determine what size box should be used to package the company's Toasty Oatsies cereal. Juan can choose from among the following sizes: $8''$ by $11''$ by $2\frac{1}{2}''$, $8\frac{1}{2}''$ by $10''$ by $2''$, or $7\frac{7}{8}''$ by $11''$ by $3''$. Juan wants to know which package will hold the most cereal.**

Why has the order of the factors changed?

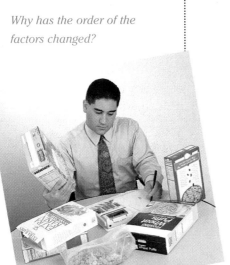

The box with the greatest volume will hold the most cereal. To find the volume of each box, multiply the length times the width times the height.

$$8 \times 11 \times 2\frac{1}{2} = 8 \times 2\frac{1}{2} \times 11 \quad \textit{Commutative } (\times)$$
$$= 20 \times 11 \quad \textit{Substitution } (=)$$
$$= 220$$
$$8\frac{1}{2} \times 10 \times 2 = 8\frac{1}{2} \times 2 \times 10 \quad \textit{Commutative } (\times)$$
$$= 17 \times 10 \quad \textit{Substitution } (=)$$
$$= 170$$
$$7\frac{7}{8} \times 11 \times 3 = 7\frac{7}{8} \times (11 \times 3) \quad \textit{Associative } (\times)$$
$$= 7\frac{7}{8} \times 33 \quad \textit{Substitution } (=)$$
$$= \frac{63}{8} \times 33$$
$$= \frac{2079}{8} \text{ or } 259\frac{7}{8}$$

The box that is $7\frac{7}{8}''$ by $11''$ by $3''$ has the greatest volume.

The chart below summarizes the properties you can use to simplify expressions.

The following properties are true for any numbers a, b, and c.		
	Addition	**Multiplication**
Commutative	$a + b = b + a$	$ab = ba$
Associative	$(a + b) + c = a + (b + c)$	$(ab)c = a(bc)$
Identity	0 is the identity. $a + 0 = 0 + a = a$	1 is the identity. $a \cdot 1 = 1 \cdot a = a$
Zero		$a \cdot 0 = 0 \cdot a = 0$
Distributive	$a(b + c) = ab + ac$ and $(b + c)a = ba + ca$	
Substitution	If $a = b$, then a may be substituted for b.	

Example **a. Write** an algebraic expression for the verbal expression *the sum of two and the square of t increased by the sum of t squared and 3.*

b. Then simplify the algebraic expression, indicating all of the properties used.

a.

the sum of two and the square of t	increased by	the sum of t squared and 3
$2 + t^2$	$+$	$t^2 + 3$

b. $2 + t^2 + t^2 + 3 = t^2 + t^2 + 2 + 3$ *Commutative (+)*

$= (t^2 + t^2) + (2 + 3)$ *Associative (+)*

$= (1 \cdot t^2 + 1 \cdot t^2) + (2 + 3)$ *Multiplicative identity*

$= (1 + 1)t^2 + (2 + 3)$ *Distributive property*

$= 2t^2 + 5$ *Substitution (=)*

CHECK FOR UNDERSTANDING

Communicating Mathematics

Study the lesson. Then complete the following.

1. **Illustrate** the associative property of multiplication with a numerical equation.

2. **Explain** how a property or properties can be used to find the product of $5(6.5)(2)$ without using a calculator or paper and pencil.

3. **Explain** the difference between the commutative and associative properties.

4. **Write** a short explanation as to whether or *not* there is a commutative property of division. Consider these examples.

$$5 \div 3 = 1\frac{2}{3} \qquad\qquad 3 \div 5 = \frac{3}{5}$$

 M<small>ATH</small> J<small>OURNAL</small>

5. **Assess Yourself** Do you believe that you understand the properties described in this chapter? Which properties were easiest for you to understand? List any properties that give you difficulty and explain why.

Guided Practice

Name the property illustrated by each statement.

6. $(6 + 4) + 2 = 6 + (4 + 2)$ 7. $7 + 6 = 6 + 7$

8. $(2 + 5) + x = 7 + x$ 9. $7(ab) = (7a)b$

10. Name the property used in each step.

    a. $ab(a + b) = (ab)a + (ab)b$

    b. $\qquad\qquad = a(ab) + (ab)b$

    c. $\qquad\qquad = (a \cdot a)b + a(b \cdot b)$

    d. $\qquad\qquad = a^2b + ab^2$

Simplify.

11. $4a + 2b + a$ 12. $3p + 2q + 2p + 8q$

13. $3(4x + y) + 2x$ 14. $6(0.4x + 0.2y) + 0.5x$

15. Write an algebraic expression for the verbal expression *the product of six and the square of z, increased by the sum of seven, z^2, and 6.* Then simplify, indicating the properties used.

Practice

Name the property illustrated by each statement.

16. $67 + 3 = 3 + 67$

17. $1 \cdot b^2 = b^2$

18. $10x + 10y = 10(x + y)$

19. $(5 \cdot m) \cdot n = 5 \cdot (m \cdot n)$

20. $(3x^2) \cdot 0 = 0$

21. $4(a + 5b) = 4a + 20b$

22. $(2 + 3)a + 7 = 5a + 7$

23. $fh + 2g = hf + 2g$

24. $7a + \left(\frac{1}{2}b + c\right) = \left(7a + \frac{1}{2}b\right) + c$

25. $(5x^2 + x + 3) + 15 = (5x^2 + x) + (3 + 15)$

26. Name the property used in each step.

a. $3c + 5(2 + c) = 3c + 5(2) + 5c$

b. $\qquad = 3c + 5c + 5(2)$

c. $\qquad = (3c + 5c) + 5(2)$

d. $\qquad = (3 + 5)c + 5(2)$

e. $\qquad = 8c + 10$

Simplify.

27. $4x + 5y + 6x$

28. $8a + 3b + a$

29. $5x + 3y + 2x + 7y$

30. $4 + 6(ac + 2b) + 2ac$

31. $2(3x + y) + 4x$

32. $4y^4 + 3y^2 + y^4$

33. $16a^2 + 16 + 16a^2$

34. $3.2(x + y) + 2.3(x + y) + 4x$

35. $\frac{1}{4}x + 2x + 2\frac{3}{4}x$

36. $0.5[3x + 4(3 + 2x)]$

37. $\frac{3}{4} + \frac{2}{3}(m + 2n) + m$

38. $\frac{3}{5}\left(\frac{1}{2}p + 2q\right) + 2p$

Write an algebraic expression for each verbal expression. Then simplify, indicating the properties used.

39. twice the sum of s and t decreased by s

40. half of the sum of p and $2q$ increased by three-fourths q

41. five times the product of x and y increased by $3xy$

42. four times the sum of a and b increased by twice the sum of a and $2b$

43. Find the following product.
$$\frac{1}{2} \cdot \frac{2}{3} \cdot \frac{3}{4} \cdot \ldots \cdot \frac{98}{99} \cdot \frac{99}{100}$$
Recall that an ellipsis "..." means "to continue the pattern." Write a sentence or two explaining how you found the product.

Critical Thinking

44. Is subtraction commutative? Write a short explanation that includes examples to support your answer.

45. Suppose the operation ✳ is defined for all numbers a and b as $a ✳ b = a + 2b$. Is the operation ✳ commutative? Give examples to support your answer.

Applications and Problem Solving

46. **Chemistry** Chemists use water to dilute acid. One important rule of chemistry is to always pour the acid into the water. Pouring water into acid could produce spattering and burns.

a. Would you say that combining acid and water is commutative?

b. Give an example from your life that is commutative.

c. Give an example from your life that is not commutative.

47. Physics The pressure you experience as you ride a roller coaster is produced by the fast changes in direction. This pressure is what pins you down to the back or bottom of your seat. This pressure is known as *g-force*. The formula used to evaluate the g-force is, $G = \dfrac{(speed)^2}{32.2 \times radius\ of\ the\ curve}$, where the speed is in feet per second, and the radius of the curve is in feet. The acceleration of gravity in $\dfrac{ft}{sec^2}$ is 32.2.

a. If you are riding the Steel Phantom at 60 feet per second (about 40 mph) and the car is turning a curve with a radius of 30 feet, what g-force will you experience?

b. The g-force has an effect on your weight. For example, if you weigh 100 pounds and experience a g-force of 2 g's, your weight at that moment feels like 2×100 or 200 pounds. What will your weight feel like while taking the curve at 60 feet per second?

Mixed Review

48. Name the coefficient of $\dfrac{4m^2}{5}$.   (Lesson 1–7)

49. Environment According to the U.S. Environmental Protection Agency, a typical family of four uses 100 gallons of water flushing the toilet, 80 gallons of water showering and bathing, and 8 gallons of water using the bathroom sink each day. Write two expressions that represent the amount of water a typical family of four uses for these purposes in *d* days.   (Lesson 1–7)

50. Name the property illustrated by the following statement.
If $a + b = c$ and $a = b$, then $a + a = c$.   (Lesson 1–6)

51. Name the multiplicative inverse of $\dfrac{2}{3}$.   (Lesson 1–6)

52. *True or False:* $\dfrac{3k-3}{7} + 13 < 17$, when $k = 8$.   (Lesson 1–5)

53. Weather The following data represents the average daily temperature in Fahrenheit for Santiago, Chile, for two weeks during the month of February. Organize the data using a stem-and-leaf plot.   (Lesson 1–4)

101, 95, 100, 100, 99, 101, 97,
99, 97, 100, 103, 101, 101, 99

54. Evaluate $(25 - 4) \div (2^2 - 1^3)$.   (Lesson 1–3)

55. Find the next two items in the sequence.   (Lesson 1–2)

56. Write an algebraic expression for *five times p to the sixth power*.   (Lesson 1–1)

A Preview of Graphs and Functions

What YOU'LL LEARN

- To interpret graphs in real-world settings, and
- to sketch graphs for given functions.

Why IT'S IMPORTANT

You need to know how to sketch and interpret graphs in order to study trends and display data.

APPLICATION
Economics

The dollar value of a car begins to decrease immediately after it is sold. This is called *depreciation*. Mathematically, depreciation can be defined by the following open sentence.

depreciation = original cost of car − value of car when sold

The table below shows how a typical $15,000 car depreciates over a period of five years. From the table, you can see that as the age of the car increases, the value decreases. Why do you think this happens?

Age of Car (year)	Approximate Value of Car
0	$15,000
1	11,000
2	8000
3	6000
4	5000
5	4000

This information can also be presented in a graph. The graph below shows the relationship between the current value of the car and the age of the car.

The value of the car is said to be a **function** of the age of the car. A function is a relationship between input and output. In a function, the output depends on the input. There is exactly one output for each input.

There are two number lines on the graph. The line that represents the current value of the car is the **vertical axis.** The line that represents the age of the car is the **horizontal axis.**

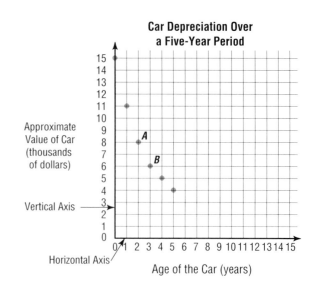

Car Depreciation Over a Five-Year Period

Approximate Value of Car (thousands of dollars)

Vertical Axis

Horizontal Axis

Age of the Car (years)

Notice that there are six points shown on the graph. Each point represents the dollar value of the car (in thousands) at the end of each year. For example, after year 1, the value is $11,000. You can express this relationship as (1, 11). What do points *A* and *B* represent?

A(2, 8) means that at the end of year 2 the value of the car is about $8000.

B(3, 6) means that at the end of year 3 the value of the car is about $6000.

(1, 11), (2, 8), and (3, 6) are called **ordered pairs.** The order in which the pair of numbers is written is important. Ordered pairs are used to locate points on the graph. The first number in an ordered pair corresponds to the horizontal axis, and the second number in an ordered pair corresponds to the vertical axis. The ordered pair (0, 0) corresponds to the **origin**.

Many real-world situations can be modeled using functions.

Example ① **Shim owns a farm market. The amount a customer pays for sweet corn depends on the number of ears that are purchased. Shim sells a dozen ears of corn for $3.00.**

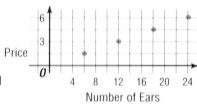

APPLICATION

Sales

a. **Make a table showing the price of various purchases of sweet corn.**

Number of Dozen	Ears of Corn	Price
$\frac{1}{2}$	6	$1.50
1	12	$3.00
$1\frac{1}{2}$	18	$4.50
2	24	$6.00

b. **Write four ordered pairs that represent the number of ears of corn and the price of the corn.**

Four ordered pairs can be determined from the chart. These ordered pairs are (6, 1.50), (12, 3.00), (18, 4.50), and (24, 6.00).

c. **Describe a set of axes that could be used to graph the number of ears of corn and the price of the corn.**

Let the horizontal axis represent the number of ears of corn, and let the vertical axis represent the price of the corn.

d. **Draw a graph that shows the relationship between the number of ears of corn and the price.**

Graph the four points described by the ordered pairs.

e. **As you read the graph from left to right, describe the trend you see. Explain.**

The graph goes upward, because the price increases as the number of ears increases. The price is a function of the number of ears purchased.

In Example 1, the price a customer pays depends on the number of ears purchased. Therefore, the number of ears purchased is called the **independent variable** or **quantity,** and the price is called the **dependent variable** or **quantity.** Usually the independent variable is graphed on the horizontal axis and the dependent variable is graphed on the vertical axis.

You can use a graph without scales on either axis to show the general shape of the graph that models a situation.

Example ②

For a certain time, Lucinda jogs up a hill at a steady speed. Then she runs down the hill and picks up her speed.

a. **What happens to her speed when Lucinda jogs at a steady pace? when Lucinda runs down the hill?**

The speed remains the same when her jogging pace is steady. The speed increases when she runs downhill.

b. **Identify the independent and dependent quantities.**

Time is the independent quantity, and speed is the dependent quantity.

c. **Match the graph to the situation.**

Graph 1 shows a steady speed (a horizontal line), followed by an increase in speed (a line that goes up) the longer she exercises.

A **relation** is a set of ordered pairs. The set of first numbers of the ordered pairs is the **domain** of the relation. The domain contains values of the independent variable. The set of second numbers of the ordered pairs is the **range** of the relation. The range contains values of the dependent variable.

Example ③
Ryan is not feeling well and has missed a day of school. The graph shows Ryan's temperature as a function of the time of day.

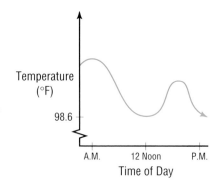

a. **Write a story about Ryan's illness.**

When Ryan got up in the morning, he had a high fever. As the morning went on, his temperature went down. By noon it was almost normal, but his temperature went up again after lunch. Then his mom gave him some medication. By evening his fever was gone, and his temperature was back to normal.

b. **Identify a reasonable domain and range for this problem.**

The domain includes the 24 hours in a day. A reasonable range would include all of the temperatures between the highest and lowest temperature a person can have and still be alive.

Communicating Mathematics

Study the lesson. Then complete the following.

1. **Explain** why the order of the numbers in an ordered pair is important.

2. Refer to the application at the beginning of the lesson. Extend the graph to predict the value of the car after 6 years and after 10 years.

3. Use the graph at the right to answer the following questions.

   a. **Describe** the information that the graph shows.

   b. **Identify** the variable represented along each axis.

   c. **Identify** two points on the graph and write the ordered pair for each point. Explain what each point represents.

   d. **Explain** what the ordered pair (2000, 6.2) represents.

   e. **Write** a statement that explains the function in the graph.

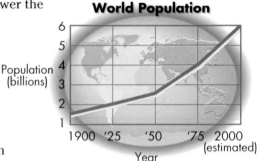

4. **Write** a story that explains what the graph in the cartoon at the left might represent. Identify the independent and dependent variables in your story.

HERMAN

"I really look forward to your cheery little visits."

Guided Practice

5. Match each description with the most appropriate graph.

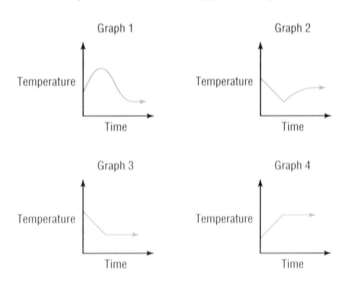

a. In August, you enter a hot house and turn on the air conditioner.

b. In January, you enter a cold house and turn up the thermostat to 68° F.

c. You put ice cubes in your fruit punch and then drink it slowly.

d. You put a cup of water in the microwave and heat it for one minute.

6. Sketch a reasonable graph for the following statement. *How hard you hit your thumb with a hammer impacts on how much it hurts.* Name the independent and dependent variables.

7. The basketball coach at City College is recruiting basketball players. The graphs below describe two of the players being recruited.

Determine whether each statement is *true* or *false*. Explain your reasoning.

a. The younger player is faster than the older player.

b. One player made more 2-point and more 3-point shots than the other.

c. The older player made more 2-point shots.

d. The younger player tried and made more free-throw shots.

EXERCISES

Practice

8. Identify the graph that matches the following statement. Explain your answer.

A bus frequently stops to pick up passengers.

9. Identify the graph that represents annual income as a function of age. Explain your answer.

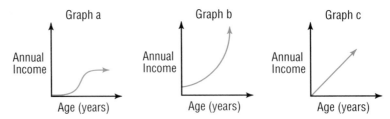

10. The graph at the right shows the amount of money in Jorge's savings account as a function of time.

a. Describe what is happening to Jorge's money. Explain why the graph rises and falls at particular points.

b. Describe the elements in the domain and range.

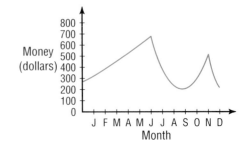

11. Match each table of data to the most appropriate graph.

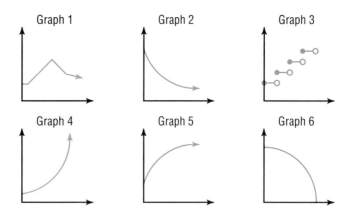

Graph 1   Graph 2   Graph 3

Graph 4   Graph 5   Graph 6

a. Primary Seismic Wave

Distance from Epicenter (km)	1000	2000	3000	4000	5000	6000	7000
Time (minutes)	2	4	5.8	7.2	8.5	9.7	10.8

b. Cost of First-Class Postage Stamps

Year	1983	1985	1987	1989	1991	1993	1995
Cost of Stamp	22¢	25¢	25¢	25¢	29¢	29¢	32¢

c. Immigration into the United States

Year	1987	1988	1989	1990	1991	1992	1993
Legal Admissions (millions)	0.6	0.6	1.1	1.5	1.8	1.0	0.9

d. Development of Human Fetus

Time (weeks)	4	8	12	16	20	24	28
Mass (grams)	0.5	1	28	110	300	650	1200

e. Height of Falling Object

Time (seconds)	0	0.5	1.0	1.5	2.0	2.5	3.0
Height (feet)	200	196	184	164	136	100	56

f. Cooling of Boiling Water

Time (minutes)	0	1	2	3	4	5	6
Temperature (°F)	100	50	25	12.5	6.25	3.125	1.5625

Sketch a reasonable graph for each situation.

12. Rashaad likes to trade basketball cards. During the first two weeks of the month, he added lots of cards to his collection. Then, he lost some cards and sold others.

13. A radio-controlled car moves along and then crashes against a wall.

14. Phyllis is riding her bike at a steady pace. Then she has a flat tire. She walks to a gas station to have the tire repaired. When the tire is fixed, she continues her ride.

15. The height of a football that was thrown into the end zone for a touchdown compared to the distance the football is from the end zone.

16. Which of the sports listed will produce a graph like the one below? Explain why you chose the sport you did and give reasons why you did not choose the other sports.

Archery   High Jumping
Biking   Javelin Throwing
Fishing   Pole Vaulting
Golf   Skateboarding
High Diving Sky Diving

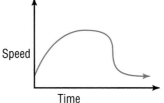

Speed / Time

**Applications and
Problem Solving**

17. Charities The graph at the right shows the amount of money raised by the Muscular Dystrophy Association over a 14-year period.

  a. Name the independent and dependent variables.

  b. What does the ordered pair (1991, 45) represent?

  c. Write a statement about the amount of money raised from 1981 to 1986.

Money Raised by MDA Telethon

Millions of Dollars

Source: Muscular Dystrophy Association

  d. Write a statement about the amount of money raised from 1987 to 1994.

  e. Use the information to predict the amount of money raised each year since 1994. Research to see if your prediction is correct.

18. Music Business The graph at the right shows the sales in millions of dollars for different music categories. Write two conclusions you can make from the information presented in the graph.

Music Sales by Categories

Sales (millions of dollars)

= Rock
= Rap
= Country

Year

Source: The Recording Industry Association of America

Mixed Review

19. Simplify $5p + 7q + 9 + 4q + 2p$. (Lesson 1–8)

20. Simplify $9a + 14(a + 3)$. (Lesson 1–7)

21. Evaluate $14(1) - 27(0)$. (Lesson 1–6)

22. *True or False:* $5m + 6^2 = 56$, when $m = 4$. (Lesson 1–5)

23. Geometry The volume of a cone equals one-third the area of the base B times the height h. Write an algebraic expression for the volume of a cone.
(Lesson 1–1)

VOCABULARY

After completing this chapter, you should be able to define each term, property, or phrase and give an example or two of each.

Algebra
additive identity (p. 37)
algebraic expression (p. 6)
algebra tiles (p. 44)
associative property
 (p. 51)
base (p. 7)
coefficient (p. 47)
commutative property
 (p. 51)
dependent variable (p. 58)
distributive property
 (p. 46)
domain (p. 58)
equation (p. 33)
equivalent expressions
 (p. 47)
evaluate (p. 8)
exponent (p. 7)
factors (p. 6)
function (p. 56)
horizontal axis (p. 56)
independent variable
 (p. 58)

inequality (p. 33)
like terms (p. 47)
multiplicative identity
 (p. 38)
multiplicative inverse
 (p. 38)
multiplicative property
 of zero (p. 38)
open sentence (p. 32)
order of operations (p. 19)
ordered pairs (p. 57)
power (p. 7)
product (p. 6)
range (p. 58)
reciprocal (p. 38)
reflexive property of
 equality (p. 39)
relation (p. 58)
replacement set (p. 33)
simplest form (p. 47)
solution (p. 32)
solution set (p. 33)
solving the open sentence
 (p. 32)

substitution property of
 equality (p. 39)
symmetric property of
 equality (p. 39)
term (pp. 13, 47)
transitive property of
 equality (p. 39)
variable (p. 6)
vertical axis (p. 56)

Discrete Mathematics
element (p. 33)
set (p. 33)
sequence (p. 13)

Statistics
data (p. 25)
statistics (p. 25)
stem-and-leaf plots (p. 25)

Problem Solving
look for a pattern
 (p. 13)

UNDERSTANDING AND USING THE VOCABULARY

Choose the letter of the term that best matches each statement or phrase.

1. For any number a, $a + 0 = 0 + a = a$.

2. For any number a, $a \cdot 1 = 1 \cdot a = a$.

3. For any number a, $a \cdot 0 = 0 \cdot a = 0$.

4. For any number a, there is exactly one number $\frac{1}{a}$ such that $a \cdot \frac{1}{a} = \frac{1}{a} \cdot a = 1$.

5. For any number a, $a = a$.

6. For any numbers a and b, if $a = b$, then $b = a$.

7. For any numbers a and b, if $a = b$, then b may be substituted for a.

8. For any numbers a, b, and c, if $a = b$ and $b = c$, then $a = c$.

9. For any numbers a, b, and c, $a(b + c) = ab + ac$.

a. additive identity property

b. distributive property

c. multiplicative identity property

d. multiplicative inverse property

e. multiplicative property of zero

f. reflexive property

g. substitution property

h. symmetric property

i. transitive property

STUDY GUIDE AND ASSESSMENT

SKILLS AND CONCEPTS

OBJECTIVES AND EXAMPLES

Upon completing this chapter, you should be able to:

• translate verbal expressions into mathematical expressions and vice versa (Lesson 1–1)

Write an algebraic expression for the sum of twice a number x and fifteen.

$$2x + 15$$

Write a verbal expression for $4x^2 - 13$.
four times a number x squared minus thirteen

• solve problems by extending sequences
(Lesson 1–2)

Give the next two items for each sequence.

$3, 7, 15, 31, \ldots$

$3 \quad 7 \quad 15 \quad 31 \quad 63 \quad 127$

$+4 \; +8 \; +16 \; +32 \; +64$

The next two terms are 63 and 127.

• use the order of operations to evaluate real number expressions (Lesson 1–3)

$5 \times 2^2 - 15 \div 3 + 4$	$[3(6) - 4^2]^3 \times 15 \div 5$
$= 5 \times 4 - 15 \div 3 + 4$	$= [3(6) - 16]^3 \times 15 \div 5$
$= 20 - 15 \div 3 + 4$	$= [18 - 16]^3 \times 15 \div 5$
$= 20 - 5 + 4$	$= [2]^3 \times 15 \div 5$
$= 15 + 4$	$= 8 \times 15 \div 5$
$= 19$	$= 120 \div 5$
	$= 24$

REVIEW EXERCISES

Use these exercises to review and prepare for the chapter test.

Write an algebraic expression for each verbal expression.

10. the sum of a number x and twenty-one

11. a number x to the fifth power

12. the difference of three times a number x and 8

13. five times the square of a number x

Write a verbal expression for each algebraic expression.

14. $2p^2$ 15. $3m^5$

16. $\frac{1}{2}x + 2$ 17. $4m^2 - 2m$

Give the next two items for each pattern.

18.

19. $2, 4, 8, 16, \ldots$ 20. $1, 1, 2, 3, 5, \ldots$

21. $x + y, 2x + y, 3x + y, \ldots$

Consider the following sequence.
$$10, 100, 1000, 10{,}000, \ldots$$

22. Give the next two terms of the sequence.

23. Express the first four terms in the sequence as a power of 10.

24. What are the 9th and 13th terms?

Evaluate each expression.

25. $3 + 2 \times 4$ 26. $(10 - 6) \div 8$

27. $18 - 4^2 + 7$ 28. $0.8(0.02 + 0.05) - 0.006$

29. $(3 \times 1)^3 - (4 + 6) \div (9.1 + 0.9)$

Evaluate each expression when x = 0.1, t = 3, and y = 2.

30. $t^2 + 3y$ 31. xty^3 32. $ty \div x$

33. the product of y and x plus t raised to the second power

OBJECTIVES AND EXAMPLES	REVIEW EXERCISES

OBJECTIVES AND EXAMPLES

• display and interpret data on a stem-and-leaf plot   (Lesson 1–4)

Stem	Leaf
2	6 8
3	3 5 6 8 9
4	0 2 3 4 4 4 5 6 7 7 8 8 9
5	0 1 2 2 2 3 3 4 6 7
6	1 3 5 9
7	0 3 $2 \mid 6 = 26$

The highest entry is 73. The lowest entry is 26. The total number of entries is 36. Most of the data is in the interval 40–49.

• solve open sentences by performing arithmetic operations   (Lesson 1–5)

State whether each sentence is true or false.

$x + 3 \geq 5$, $x = 1$

$1 + 3 \geq 5$

$4 \geq 5$

false

$5^2 - 3y \leq 1$, $y = 8$

$5^2 - 3(8) \leq 1$

$25 - 24 \leq 1$

$1 \leq 1$

true

• recognize and use the properties of identity and equality   (Lesson 1–6)

$36 + 7 \times 1 + 5(2 - 2)$

$= 36 + 7 \times 1 + 5(0)$ *Substitution* $(=)$

$= 36 + 7 + 5(0)$ *Multiplicative identity*

$= 36 + 7$ *Multiplicative prop. of zero*

$= 42$ *Substitution* $(=)$

• use the distributive property to simplify expressions   (Lesson 1–7)

$5(t + 3)$

$= 5(t) + 5(3)$

$= 5t + 15$

$2x^2 + 4x^2 + 7x$

$= (2 + 4)x^2 + 7x$

$= 6x^2 + 7x$

REVIEW EXERCISES

The ages (in years) at the time of death for the presidents of the United States are listed below.

67, 90, 83, 85, 73, 80, 78, 79, 68, 71, 53, 65, 74, 64, 77, 56, 66, 63, 70, 49, 56, 67, 71, 58, 60, 72, 67, 57, 60, 90, 63, 88, 78, 46, 64, 81

34. Make a stem-and-leaf plot of the data.

35. How old was the oldest president at death?

36. How old was the youngest president at death?

37. Does the total number of entries match the number of presidents? Explain.

38. In which interval did most of the ages fall?

State whether each sentence is *true* or *false* for the value of the variable given.

39. $x + 13 = 22$, $x = 8$

40. $2b + 2 < b^3$, $b = 2$

41. $(y + 4) \div (y + 2) \leq 2$, $y = 2$

42. Find the solution set for $3x + 1 \leq 13$ if the replacement set is $\{2, 4, 6, 8\}$.

Solve each equation.

43. $y = 4\frac{1}{2} + 3^2$

44. $y = 5[2(4) - 1^3]$

Name the multiplicative inverse of each number or variable.

45. 3   **46.** y **47.** 0.2

Name the property illustrated by each statement.

48. $xy = 1xy$ **49.** $0 + 22x = 22x$

50. Evaluate $2[3 \div (19 - 4^2)]$ and name the property used in each step.

Use the distributive property to rewrite each expression without parentheses.

51. $2(4 + 7)$ **52.** $4(x + 1)$ **53.** $3\left(\frac{1}{3} - p\right)$

Use the distributive property to find each product.

54. 6×103 **55.** 3×98 **56.** $12(1.5)$

Simplify each expression.

57. $3m + 5m + 12n - 4n$ **58.** $2p(1 + 16r)$

OBJECTIVES AND EXAMPLES

- when simplifying expressions recognize and use the commutative and associative properties
  (Lesson 1-8)

$3x + 7xy + 9x$

$= 3x + 9x + 7xy$	*Commutative (+)*
$= (3 + 9)x + 7xy$	*Distributive*
$= 12x + 7xy$	*Substitution (=)*

- interpret graphs in real-world settings
  (Lesson 1-9)

Sketch a graph for the following statement.

As a band gets more popular, their prices increase to a point and then level off.

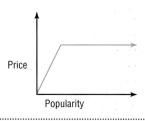

REVIEW EXERCISES

Name the property illustrated by each statement.

59. $7(p + q) = 7(q + p)$

60. $3 + (x + y) = (3 + x) + y$

61. $(b + a)c = c(b + a)$

62. $2(mn) = (2m)n$

63. Simplify $2x + 2y + 3x + 3y$.

Write an algebraic expression for each verbal expression. Then simplify, indicating the properties used.

64. five times the sum of x and y decreased by $2x$

65. twice the product of p and q increased by pq

Match each description with the most appropriate graph.

66. the speed of light

67. an airplane taking off, then landing

68. a car approaching a stop sign

APPLICATIONS AND PROBLEM SOLVING

69. Food and Nutrition   There are 80 Calories in one serving of skim milk and 8 servings in a half gallon.   (Lesson 1-1)

   a. Write an expression that describes how many Calories you get if you drink s servings of skim milk.

   b. How many Calories do you consume if you drink 4 servings of skim milk a day?

   c. How many Calories are in a half gallon of skim milk?

70. Investments   The equation $I = prt$ describes simple interest on a savings account, where I is the amount of interest, p is the amount deposited, r is the annual interest rate, and t is the time in years.   (Lesson 1-3)

   a. Solve for I if $p = 100$, $r = 0.05$, and $t = 2$.

   b. If you deposit $200 in a savings account that earns 6% interest (6% = 0.06), how much money will be in your account after the first year?

71. Geometry   The triangle inequality states that if you have a triangle, then the sum of any two of its sides must be greater than its third side. (Lesson 1-5)

   a. Verify that a triangle with sides of length 3, 4, and 5 units satisfies the triangle inequality.

   b. Suppose you want to construct a triangle such that two of the sides are 3 feet and 6 feet long. What is the minimum length of the third side? (Use x for the length of the third side.)

A practice test for Chapter 1 is provided on page 787.

ALTERNATIVE ASSESSMENT

COOPERATIVE LEARNING PROJECT

Operating a Music Store   In this chapter, you learned how to model verbal sentences with algebraic sentences. The basic principles of modeling verbal sentences involve being able to determine whether or not you have an equality or an inequality and determining what your variables are.

In this project, imagine that you and your friends plan to open a new music store that will sell only compact discs. After some research, you find that the monthly rent for your store will be $500 and that you can expect to spend about $200 a month for utilities. Assume that your inventory of compact discs must be at least 400 different titles with 4 discs per title and that the wholesale cost for compact discs is $8.00 per disc.

Outline the costs for starting your business and plan how much in CD sales you would need to show a profit each month.

Follow these steps to organize your business.

• Determine the cost to open your new business.

• Determine your monthly overhead costs.

• Find a price for which you will sell your CDs.

• Write an algebraic model that describes the number of CDs you have to sell in order to break even each month.

• Draw a graph to show that if sales increase over time, the profits will also increase.

• Write several paragraphs describing your plan and incorporate your algebraic models and graphs to help support it.

THINKING CRITICALLY

• Does the distributive property also apply to multiplication? In other words, does the product $a(bc) = (ab)(ac)$? Why or why not?

• Create a sequence of numbers in which there is more than one pattern. Explain why this is possible.

PORTFOLIO

Do you organize your work so that you and anyone else reading it can tell what you were thinking? Find an example of your work that you have done that is well organized and list the qualities that make it so. Then find an example of your work that isn't as well organized and list what you could have done to make it more so. Place both of these in your portfolio.

SELF EVALUATION

Are you confident with your knowledge of mathematics? Most math students would rather not try to solve a math problem than try to and get the wrong answer.

Assess yourself. How confident are you? Would you rather not attempt a problem, or are you willing to give your best effort and learn from the outcome? Describe how you can become more confident in your problem solving both in mathematics and in your daily life.

the Greenhouse Effect

MATERIALS NEEDED

lamp with a 100-watt bulb

stopwatch

thermometer

plastic zipper-style sandwich bag

ruler

The *greenhouse effect* refers to the warming of Earth by the sun. The main culprit in the greenhouse effect is carbon dioxide (CO_2). Carbon dioxide in the atmosphere works like the panes of glass on a greenhouse. Glass is transparent to visible light, allowing the sun's rays to warm Earth's surface. But when the surface gives off excess heat, the hot air stays in the greenhouse, which continues to keep the air warm. Similarly, CO_2 in the atmosphere absorbs the sun's infrared rays, allowing some of the excess heat to stay in the atmosphere rather than escaping into space. How much heat is retained depends on how much CO_2 is in the air.

Over the last 200 years, the amount of CO_2 in our atmosphere has increased, raising Earth's average temperature. Some scientists predict that if global warming continues and Earth's average temperature goes up an additional 3° to 8°F,

we could see a marked increase in the number of weather-related disasters like heat waves, droughts, floods, and hurricanes.

In this Investigation, you will use mathematics to analyze aspects of the greenhouse effect and report your findings. As a scientist commissioned to explore the greenhouse effect, you will conduct two experiments. The first experiment is a *control experiment* that will be used to compare data with the second experiment, called the *greenhouse experiment*. Both experiments will involve temperature as a *function* of distance.

Make an Investigation Folder in which you can store all of your work on this Investigation for future use.

CONTROL EXPERIMENT				
Test	Distances Between Thermometer and Bulb	First Temperature	Second Temperature	Difference in Temperatures
1				
2				
3				
4				

CONTROL EXPERIMENT

1 Begin by copying the chart above onto a sheet of paper.

2 Set your lamp about 5 inches away from the thermometer. Make sure the lamp is turned off.

3 Measure and record the temperature and the distance between the thermometer and the bulb.

4 Turn the lamp on and leave it on for five minutes. Use the stopwatch to keep accurate time.

5 Record the new temperature and find the difference of the temperatures.

6 Let the thermometer and bulb cool to room temperature and repeat the process, moving the lamp a little farther from the thermometer. Conduct the test a total of four times at four different distances. Make sure you let the thermometer cool to room temperature between each test.

GREENHOUSE EXPERIMENT

7 Begin by making a chart like the one you made for the Control Experiment.

8 Make sure that the thermometer has cooled to room temperature. Then record the temperature.

9 Put the thermometer in the plastic bag and close the bag. Make sure that the bulb of the thermometer does not touch the bag. Then place the thermometer the same distance from the lamp as you did in the control experiment (about 5 inches). Record the distance between the thermometer and the lamp bulb. Then turn the lamp on and leave it on for five minutes.

10 Record the new temperature. Then find the difference of the temperatures.

11 Conduct four tests, using the same distances you used in the control experiment. Make sure you take the thermometer out of the bag and let it cool to room temperature between each test. Be sure to take accurate readings and measurements.

You will continue working on this Investigation throughout Chapters 2 and 3.

Be sure to keep your chart and materials in your Investigation Folder.

The Greenhouse Effect Investigation

Working on the Investigation
Lesson 2–2, p. 83

Working on the Investigation
Lesson 2–6, p. 111

Working on the Investigation
Lesson 3–2, p. 154

Working on the Investigation
Lesson 3–6, p. 177

Closing the Investigation

End of Chapter 3, 184

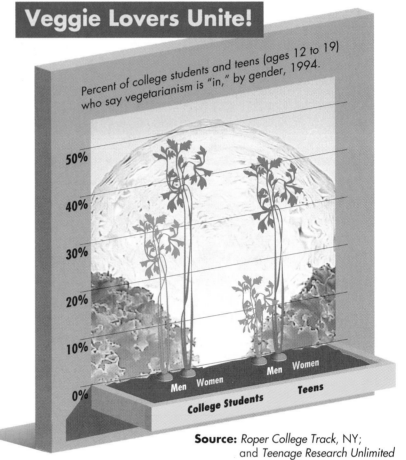

CHAPTER 2

Exploring Rational Numbers

Objectives

In this chapter, you will:

- display and interpret statistical data on line plots,
- add, subtract, multiply, and divide rational numbers,
- find square roots, and
- write equations and formulas.

Veggie Lovers Unite!

Percent of college students and teens (ages 12 to 19) who say vegetarianism is "in," by gender, 1994.

50%
40%
30%
20%
10%
0%

Men Women   Men Women

College Students   Teens

Source: *Roper College Track*, NY; and *Teenage Research Unlimited*

TIME *Line*

Hamburgers and pepperoni pizzas have been teen diet staples for decades. But lately there is a growing trend toward meatless meals. Why do teens switch to "veggie values" and go for tofu and eggplant? Are there benefits to a diet without meat? Are there dangers to a diet without meat?

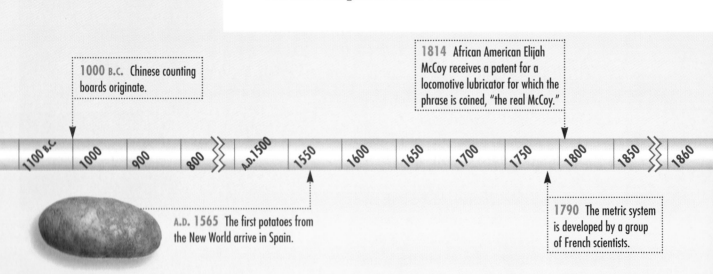

1000 B.C. Chinese counting boards originate.

1814 African American Elijah McCoy receives a patent for a locomotive lubricator for which the phrase is coined, "the real McCoy."

1100 B.C. 1000 900 800 A.D.1500 1550 1600 1650 1700 1750 1800 1850 1860

A.D. 1565 The first potatoes from the New World arrive in Spain.

1790 The metric system is developed by a group of French scientists.

Chapter Project

What would you eat if you didn't eat meat? Research nutrition and dietary resource materials at the library to help you devise a healthy vegetarian diet and a healthy meat diet for an average teenager.

- Plan two diets. One should be for a no animal meat vegetarian, and one should be for a meat-eating person.

- Devise a seven-day menu for each diet, listing all nutrition facts such as calories, fat,

cholesterol, sodium, carbohydrates, proteins, and vitamins.

- Can a healthy diet be maintained without animal products? Is an increase in fruit and vegetable consumption healthy? Explain your reasoning.

- Draw some conclusions about individual diets and food production from your research. Share your findings with the class.

The concern for animals is one reason teenagers give up eating meat. Groups like Earth 2000, a teens-only advocacy group founded by **Danny Seo,** want the focus of their generation to be "no animal cruelty, no meat." Danny and members of Earth 2000 cite teen celebrities like Michael Stipe of R.E.M., Eddie Vedder of Pearl Jam, and Jennie Garth of *Beverly Hills 90210* as vegetarian role models.

Danny also cites health reasons for maintaining a meatless diet, such as cutting back on the intake of fat. In a balanced vegetarian diet, only about one third of the calories come from fat.

1950 Luis Buñuel, Spanish filmmaker, directs his Mexican film *Los Olvidados,* "The Forgotten."

1980 Gae Veit founds Shingobee Construction in Minnesota, a successful business that hires mainly other American Indian workers.

1890 1900 1910 1920 1930 1940 1950 1960 1970 1980 1990 2000

1930 Ruth Wakefield invents the Toll House cookie.

1991 Carolyn Napoli co-invented the technology that can be applied to block genes in plants to reduce sugar and oil content.

- To state the coordinate of a point on a number line,
- to graph integers on a number line, and
- to add integers by using a number line.

Why IT'S IMPORTANT

You can use number lines to add and subtract integers and to display data.

APPLICATION
Entertainment

Have you ever played Monopoly®? Most likely you have at one time or another. Parker Brothers has sold over 100 million sets of the game of Monopoly since it was invented in 1933 by Charles Darrow.

In order to play Monopoly, you must know how to move playing pieces around the board. To do this, you add the numbers rolled on the dice and then move your playing piece the corresponding number of spaces forward. You also need to know how to move your playing piece *backward* when directed to do so. Understanding how to add and subtract integers can help you to play.

Nidi, Lynn, Michael, and Domingo are playing a game of Monopoly. The chart below shows a turn for each player. How can you represent Nidi's moves using numerals?

Most Landed-on Squares in Monopoly®

1. Illinois Ave.
2. Go
3. B&O Railroad
4. Free Parking
5. Tennessee Ave.

Player	Turn 1
Nidi	Rolls 2; lands on Chance — go back 3 spaces.
Lynn	Rolls double 3s; rolls 7.
Michael	Rolls 6 — go directly to jail (back 20).
Domingo	Rolls 9; lands on Community Chest — advance to next utility (forward 11).

You can solve problems like the one above by using a **number line.** A number line is drawn by choosing a starting position, usually 0, and marking off equal distances from that point. The set of **whole numbers** is often represented on a number line. This set can be written {0, 1, 2, 3, ...}, where "..." means that the set continues indefinitely.

Although only a portion of the number line is shown, the arrowhead indicates that the line and the set of numbers continue.

We can use the expression $2 - 3$ to represent Nidi's moves.

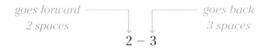

goes forward 2 spaces   *goes back 3 spaces*

$$2 - 3$$

The number line below shows that the value of $2 - 3$ should be 1 less than 0. However, there is no whole number that corresponds to 1 less than 0. You can write the number 1 *less than* 0 as -1. This is an example of a **negative number.**

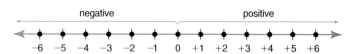

Any nonzero number written without a sign is understood to be positive.

To include negative numbers on the number line, extend the number line to the left of zero and mark off equal distances. The points to the right of zero are named using the *positive sign* $(+)$. The points to the left of zero are named using the *negative sign* $(-)$. Zero is neither positive nor negative.

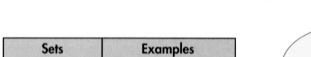

Read "-5" as *negative* 5.   Read "$+5$" as *positive* 5.

The set of numbers used on the number line above is called the set of **integers.** This set can be written $\{..., -6, -5, -4, -3, -2, -1, 0, 1, 2, 3, 4, 5, 6, ...\}$.

Venn diagrams are figures often used to represent sets of numbers.

Sets	Examples
Natural numbers	1, 2, 3, 4, 5, ...
Whole numbers	0, 1, 2, 3, 4, ...
Integers	..., -2, -1, 0, 1, 2, ...

Natural Numbers

Notice that the natural numbers are a subset of the whole numbers and the whole numbers are a subset of the integers.

To **graph** a set of numbers means to draw, or plot, the points named by those numbers on a number line. The number that corresponds to a point on a number line is called the **coordinate** of that point.

Example ① **Name the set of numbers graphed.**

a.

The bold arrow means that the graph continues indefinitely in that direction.

The set is $\{-5, -4, -3, -2, -1, 0, 1, 2, ...\}$.

b.

The set is $\{-4, -2, -1, 1, 3\}$.

You can use a number line to add integers.

MODELING MATHEMATICS

Adding Integers

To find the sum of −7 and −5, follow the steps below.

Step 1
Draw an arrow starting at 0 to −7, the first addend.

Step 2
Starting at −7, draw an arrow and go to the left 5 units long. This represents the second addend, −5.

Step 3
The arrow ends at the sum, −12.

So, $-7 + (-5) = -12$.

Notice that parentheses are used in the equation so that the sign of the number is not confused with a subtraction sign.

Your Turn

a. What addition sentence is modeled on the number line shown at the right?

b. Draw a number line to show the addition sentence $-3 + (-2) = -5$.

c. Use a number line to find $-4 + 5$.

Example ② At 4:08 A.M., the temperature in Casper, Wyoming, was −10°F. By 1:30 P.M., the temperature had risen 17° to the daytime high. What was the high temperature?

APPLICATION
Meteorology

Draw and label a number line. Draw an arrow from 0 to −10. Then draw an arrow to the right 17 units long.

The high temperature of the day was 7°F.

CHECK FOR UNDERSTANDING

Communicating Mathematics

Study the lesson. Then complete the following.

1. **Explain** how you would find the sum $-4 + 6$ on a number line.

2. **Draw** a number line to show that $-17 < 0$, $-3 < 1$, and $-5 > -10$.

3. Refer to the application at the beginning of the lesson. Draw number lines to show the first Monopoly® moves for Lynn, Michael, and Domingo.

4. **Research** instances where integers are used in real-life situations by collecting newspaper clippings that show integers.

5. Use a number line to find $5 + (-6)$.

Guided Practice

Name the set of numbers graphed.

6.
$-6\ -5\ -4\ -3\ -2\ -1\ 0\ 1\ 2\ 3$

7.
$-7\ -6\ -5\ -4\ -3\ -2\ -1\ 0\ 1\ 2$

Graph each set of numbers on a number line.

8. {0, 2, 4, 6}

9. {−1, 0, 1, 2, 3,...}

10. {integers greater than −3}

Write a corresponding addition sentence for each diagram.

11.
$-8\ -7\ -6\ -5\ -4\ -3\ -2\ -1\ 0\ 1$

12.
$-3\ -2\ -1\ 0\ 1\ 2\ 3\ 4\ 5\ 6$

Find each sum. If necessary, use a number line.

13. $-8 + 3$

14. $-7 + (-15)$

15. Football The Barnesville Bruins' offense lined up on their 20-yard line. They gained 5 yards on the first down, lost 3 yards on the second down, gained 7 yards on the third down, and punted on the fourth down. What was the net gain or loss after the third down?

EXERCISES

Practice

Name the set of numbers graphed.

16.
$-2\ -1\ 0\ 1\ 2\ 3\ 4\ 5\ 6\ 7$

17.
$-6\ -5\ -4\ -3\ -2\ -1\ 0\ 1\ 2\ 3$

18.
$-1\ 0\ 1\ 2\ 3\ 4\ 5\ 6\ 7\ 8$

19.
$-8\ -7\ -6\ -5\ -4\ -3\ -2\ -1\ 0\ 1$

20.
$-4\ -3\ -2\ -1\ 0\ 1\ 2\ 3\ 4\ 5$

21.
$-5\ -4\ -3\ -2\ -1\ 0\ 1\ 2\ 3\ 4$

Graph each set of numbers on a number line.

22. {−4, −3, −1, 3}

23. {..., −2, −1, 0, 1}

24. {integers between −6 and 10}

25. {integers less than 0}

26. {integers between −2 and 5}

27. {integers less than −3 +(−1)}

28. {whole numbers greater than −4 + 4}

29. {integers less than 0 but greater than −6}

30. {integers less than or equal to −3 and greater than or equal to −10}

Find each sum. If necessary, use a number line.

31. $8 + 5$

32. $-6 + (-7)$

33. $7 + (-12)$

34. $-4 + (-9)$

35. $0 + (-12)$

36. $5 + (-3)$

37. $-3 + 9$

38. $-13 + 13$

39. $-14 + (-9)$

40. Write three different equations using integers that have a sum of -14.

For Exercises 41–42, write an open sentence using addition. Then solve each problem.

41. Meteorology On February 10, the low temperature in Houlton, Maine, and Fort Yukon, Alaska, was $-17°$. On the same date, the high temperature in Lajitas, Texas, was $82°$ higher. What was the temperature in Lajitas, Texas, on February 10?

42. Geography Use the map below to determine what time it is in each city if it is 10:00 A.M. in New York City.
   a. Hong Kong
   b. Los Angeles
   c. Rio de Janeiro
   d. Bombay

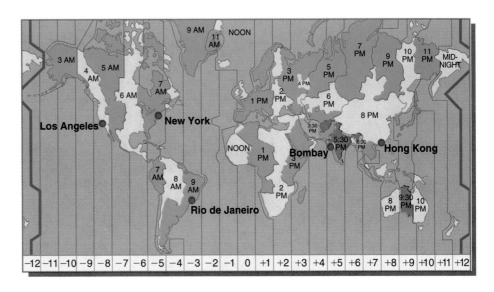

43. Meteorology *Windchill factor* is an estimate of the cooling effect the wind has on a person in cold weather. Meteorologists use a chart like the one below to predict the windchill factor. Use the chart below to find each windchill factor.
   a. 20°F, 10 mph
   b. 0°F, 15 mph
   c. -10°F, 5 mph
   d. 30°F, 15 mph

Wind speed (mph)	Actual temperature (°Fahrenheit)						
	30	20	10	0	−10	−20	−30
	Equivalent temperature (°Fahrenheit)						
0	30	20	10	0	−10	−20	−30
5	27	16	6	−5	−15	−26	−36
10	16	4	−9	−21	−33	−46	−58
15	9	−4	−18	−31	−45	−58	−72

44. Baseball Sometimes baseball team owners develop incentive clauses for hitters in an effort to win more games. Suppose an owner decided to pay each hitter $500 for each run scored, $400 for each hit, and $500 for each run batted in. Look in the sports section of a newspaper. Choose a team's box score and calculate the value of the incentive clause for the first five hitters in a game.

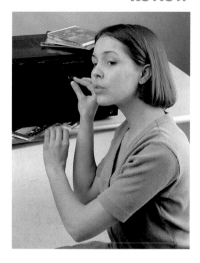

45. Entertainment Juanita has the volume on her stereo turned up because her parents are not home. Suddenly, she sees her father's car pull into the driveway. She runs to her room to turn down the volume. Juanita's father comes in, grabs an umbrella, and leaves, so Juanita returns the volume to its previous level. Sketch a reasonable graph to show the volume of Juanita's stereo during this time. (Lesson 1–9)

46. Physical Fitness Mitchell likes to exercise regularly. On Mondays, he likes to walk two miles, run three miles, sprint one-half of a mile, and then walk for another mile. Identify the graph that best represents Mitchell's heart rate as a function of time. (Lesson 1–9)

a.

Heart Rate

Time

b.

Heart Rate

Time

c.

Heart Rate

Time

47. Name the property illustrated by $8(2 \cdot 6) = (2 \cdot 6)8$. (Lesson 1–8)

48. Simplify $16a + 21a + 30b - 7b$. (Lesson 1–7)

49. Name the property illustrated by $(12 - 9)(4) = 3(4)$. (Lesson 1–6)

50. Solve $5(7) + 6 = x$. (Lesson 1–5)

51. Evaluate $9(0.4 + 1.2) - 0.5$. (Lesson 1–3)

52. Patterns Find the next three numbers in the pattern 5, 6.5, 8, 9.5, ... (Lesson 1–2)

Upper Level

= 1
= 5
= 5
= 5

Lower Level $16 \times 20 = 320$

= 1
= 5

$6 \times 1 = 6$

Mathematics and SOCIETY

Mayan Mathematics

The excerpt below is from an article that appeared in *The Columbus Dispatch* on January 15, 1995. It describes the number system developed by the Mayan Indians, an ancient civilization that lived in southern Mexico and Central America until the 16th century.

THE MAYAS USED A COUNTING SYSTEM based on 20. And while we add digits to the left of numbers to show their increased size, the Mayas piled them upwards. In our decimal system, as you look at a number from right to left, each digit tells us how much value it has, based on the number 10. The number 326, for instance, is really 3 hundreds, 2 tens and 6 ones....The Mayas used the same principle, but used the vigesimal system (based on 20). Units were represented by dots. Five dots were represented by a horizontal line. ■

1. What combinations of dots and horizontal lines (bars) are needed to write the number 252 using the Mayan system? (*Hint:* Two levels are required.)

2. What uses would the Mayas and other ancient civilizations have had for the number systems they developed?

3. Would a number system based on a number other than 10 be as accurate as the decimal system? Would it be as easy to use? Explain your answers.

Integration: Statistics
Line Plots

What YOU'LL LEARN

- To interpret numerical data from a table, and
- to display and interpret statistical data on a line plot.

Why IT'S IMPORTANT

Line plots are a useful way to display data.

APPLICATION
Television

Are you a fan of Will on the television show *Fresh Prince of Bel-Air*? The Nielsen ratings are used to determine which shows are popular and which shows are not. They are also used to determine which shows should be continued or canceled. The chart at the right shows the Nielsen ratings for Monday, February 6, 1995.

Nielsen Ratings

Monday, February 6, 1995

TIME	PROGRAM	VIEWERS (rounded to nearest million)
8:00	Fresh Prince of Bel-Air (NBC)	22
	The Nanny (CBS)	19
	Melrose Place (Fox)	14
	Coach (ABC)	15
	Star Trek: Voyager (UPN)	14
8:30	Dave's World (CBS)	21
	Blossom (NBC)	19
	Sneakers (ABC)	16
9:00	Murphy Brown (CBS)	22
	Serving in Silence (NBC)	19
	Models, Inc. (Fox)	10
	Platypus Man (UPN)	5
9:30	Cybill (CBS)	19
	Pig Sty (UPN)	4
10:00	Chicago Hope (CBS)	19

In some cases, data can be presented on a number line. Numerical data displayed on a number line is called a **line plot.** The data in the table above can be presented in a line plot as follows.

Step 1 Draw and label a number line. You can see that the data in the table ranges from 4 million to 22 million viewers. In order to represent the data on a number line, a *scale* must be used that includes this range of values. You can use a scale from 0 to 25 with *intervals* of five.

Step 2 Draw the line plot. Write a "v" for each TV show above its share of viewers. A completed line plot for the Nielsen ratings is shown below.

Notice that some data values are located between marked intervals on the number line.

Example

APPLICATION
Basketball

In 1994 and 1995, the Houston Rockets won back-to-back NBA championships. The table at the right shows the standings for each of the 27 NBA teams for the 1994–1995 season on April 28, 1995.

a. Make a line plot to show the number of wins by each playoff team.

Final Regular-Season Standings

EASTERN CONFERENCE	W	L	WESTERN CONFERENCE	W	L
ATLANTIC			**MIDWEST**		
y-Orlando	57	25	y-San Antonio	62	20
x-New York	55	27	x-Utah	60	22
x-Boston	35	47	x-Houston	47	35
Miami	32	50	x-Denver	41	41
New Jersey	30	52	Dallas	36	46
Philadelphia	24	58	Minnesota	21	61
Washington	21	61			
CENTRAL			**PACIFIC**		
z-Indiana	52	30	z-Phoenix	59	23
x-Charlotte	50	32	x-Seattle	57	25
x-Chicago	47	35	x-L.A. Lakers	48	34
x-Cleveland	43	39	x-Portland	44	38
x-Atlanta	42	40	Sacramento	39	43
Milwaukee	34	48	Golden State	26	56
Detroit	28	54	L.A. Clippers	17	65

x–clinched playoff berth; y–clinched conference; z–clinched division

Source: *USA Today*, 4-28-95

The scale ranges from 35 to 65 with intervals of 5.

The teams with an x, y, or z next to their names made the playoffs. The number of wins by the playoff teams ranges from 35 to 62. You can use a "w" to represent the number of each team's wins.

b. How many playoff teams won fewer than 50 games?
From the line plot, you can see that 8 teams won fewer than 50 games. These teams were Boston, Chicago, Cleveland, Atlanta, Houston, Denver, L.A. Lakers, and Portland.

c. Which playoff team had the best record? the worst record?
San Antonio had the best record and Boston had the worst record of the playoff teams.

d. How many teams made the playoffs?
Count the number of w's on the line plot. Sixteen teams made the playoffs.

The data in the application at the beginning of the lesson was the result of an actual survey. The data in the example was collected by checking NBA records. Data can be collected by taking actual measurements, by conducting surveys or polls, by using questionnaires, by simulation, or by consulting reference materials.

It is important that you know how the data were obtained. For example, would you want to draw conclusions about changing the name of your school mascot based on a result of a survey of seniors only? Why or why not?

MODELING MATHEMATICS

Line Plots

Materials: small packages of plain M&M's®

In 1994 and 1995, the Mars Company conducted surveys to determine whether a new color should be added to M&M's®. The choices were blue, pink, purple, or leave them as they are. The color blue was chosen.

Your Turn

a. Open a package of M&M's. Separate the candies by color. Find the total number of each color.

b. Make a line plot to show the number of each color. Use b for brown, r for red, y for yellow, o for orange, g for green, and bl for blue.

c. Do the colors cluster around any number?

d. Make a class line plot of your data. Are the data in the class line plot different or the same as yours? Explain.

e. Make a class line plot showing the total number of M&M's in each of your packages. Do the packages have the same number in them?

Communicating Mathematics

Study the lesson. Then complete the following.

1. **Describe** a situation in which it would be valuable to obtain data in order to make a decision.

2. **Compare and contrast** line plots and tables as a means of reporting data. What are the advantages and disadvantages of each?

MODELING MATHEMATICS

3. **Develop** a survey and collect data on the Monday night television viewing habits of your classmates. Make a line plot of the data. Compare your class findings to those of another class and to a Nielsen ratings chart from a magazine or newspaper. What conclusions can you draw?

Guided Practice

State the scale you would use to make a line plot for the following data. Then draw the line plot.

4. 50, 50, 30, 30, 20, 10, 60, 40

5. 52, 43, 67, 69, 37, 76

6. **Retail Sales** The advertisement below lists the cars for sale at Texas Motors.

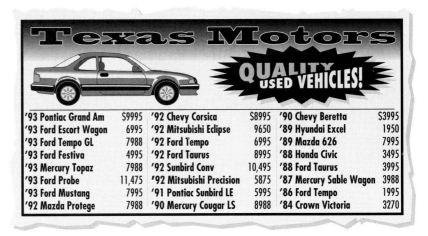

'93 Pontiac Grand Am	$9995	'92 Chevy Corsica	$8995	'90 Chevy Beretta	$3995
'93 Ford Escort Wagon	6995	'92 Mitsubishi Eclipse	9650	'89 Hyundai Excel	1950
'93 Ford Tempo GL	7988	'92 Ford Tempo	6995	'89 Mazda 626	7995
'93 Ford Festiva	4995	'92 Ford Taurus	8995	'88 Honda Civic	3495
'93 Mercury Topaz	7988	'92 Sunbird Conv	10,495	'88 Ford Taurus	3995
'93 Ford Probe	11,475	'92 Mitsubishi Precision	5875	'87 Mercury Sable Wagon	3988
'93 Ford Mustang	7995	'91 Pontiac Sunbird LE	5995	'86 Ford Tempo	1995
'92 Mazda Protege	7988	'90 Mercury Cougar LS	8988	'84 Crown Victoria	3270

a. Make a line plot to show how many cars from each year are in stock.

b. What is the price range for the used cars?

c. Make a line plot to show the costs of the cars. Round each price to the nearest thousand dollars. Let each number on the number line represent thousands of dollars.

Applications and Problem Solving

7. **Animals** The speeds of 20 of the fastest animals in miles per hour according the *The World Almanac*, 1995, are listed below.

40	61	50	50	32	70	35	30	50	45
43	40	30	30	35	45	42	32	40	30

a. Make a line plot of the data.

b. What is the fastest speed of any animal? Research to find the name of the fastest animal.

c. What is the slowest speed of the 20 animals?

d. Which speed occurred most frequently?

e. How many animals had a speed of at least 40 mph?

f. How many animals had speeds greater than 30 and less than 40 mph?

8. **History** The table below lists the fifty United States and the years they entered the Union.

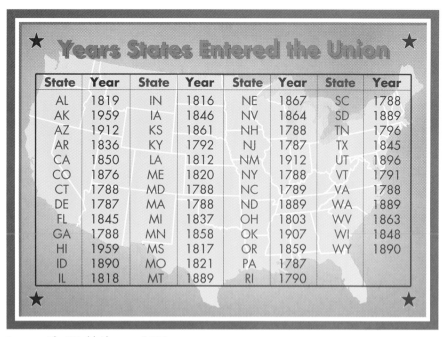

State	Year	State	Year	State	Year	State	Year
AL	1819	IN	1816	NE	1867	SC	1788
AK	1959	IA	1846	NV	1864	SD	1889
AZ	1912	KS	1861	NH	1788	TN	1796
AR	1836	KY	1792	NJ	1787	TX	1845
CA	1850	LA	1812	NM	1912	UT	1896
CO	1876	ME	1820	NY	1788	VT	1791
CT	1788	MD	1788	NC	1789	VA	1788
DE	1787	MA	1788	ND	1889	WA	1889
FL	1845	MI	1837	OH	1803	WV	1863
GA	1788	MN	1858	OK	1907	WI	1848
HI	1959	MS	1817	OR	1859	WY	1890
ID	1890	MO	1821	PA	1787		
IL	1818	MT	1889	RI	1790		

Source: *The World Almanac, 1995*

a. What is the range of years for states entering the Union?

b. Make a line plot that shows the year each state joined the Union.

c. Do the years cluster around a certain number? If so, which years are those? What would explain this?

9. **Weather** The average rainfall per year, rounded to the nearest inch, for ten cities around the world is shown on the map below.

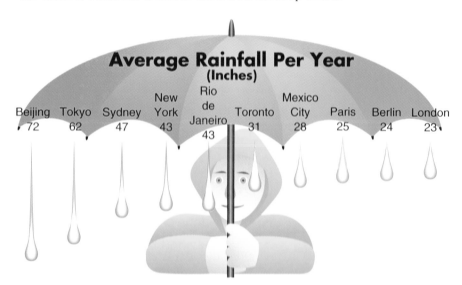

Average Rainfall Per Year
(Inches)

Beijing	Tokyo	Sydney	New York	Rio de Janeiro	Toronto	Mexico City	Paris	Berlin	London
72	62	47	43	43	31	28	25	24	23

a. Make a line plot of the data. What scale did you use?

b. Do the average amounts of rain cluster about a certain number?

c. Do you think that the average rainfall for your community is close to that of any of the cities above? Why or why not?

d. Look up the average rainfall for your community. How does it compare to the average rainfall for London? for Tokyo?

10. **Tornadoes** A tornado is a powerful, twisting windstorm. The winds of a tornado are the most dangerous winds that occur on Earth, with speeds of more than 200 miles per hour. The table below lists the average number of tornadoes that occurred per year in each state for a 30-year period.

TORNADO OCCURRENCE BY STATE, 1962–1991							
STATE	AVG.	STATE	AVG.	STATE	AVG.	STATE	AVG.
AL	22	IN	20	NE	37	SC	10
AK	0	IA	36	NV	1	SD	29
AZ	4	KS	40	NH	2	TN	12
AR	20	KY	10	NJ	3	TX	139
CA	5	LA	28	NM	9	UT	2
CO	26	ME	2	NY	6	VT	1
CT	1	MD	3	NC	15	VA	6
DE	1	MA	3	ND	21	WA	2
FL	53	MI	19	OH	15	WV	2
GA	21	MN	20	OK	47	WI	21
HI	1	MS	26	OR	1	WY	12
ID	3	MO	26	PA	10		
IL	27	MT	6	RI	0		

Source: National Severe Storm Forecast Center

a. Make a line plot of the average number of tornado occurrences.

b. Do any of the numbers appear clustered? If so, which ones?

c. Which state had the highest average? Why do you think it had the most tornadoes?

11. **School** Mr. Thomas and Ms. Martinez each asked 10 students from their algebra classes how many hours they spent talking on the telephone last week. The results are shown in the line plot below. Use this plot to answer each question.

Number of Hours on the Telephone

```
                X
        X   ●   X   X
X       ●   ●   ●   X
X   X   ●   ●   ●   ●   X   X
+---+---+---+---+---+---+---+---+
0   1   2   3   4   5   6   7   8
```

x – Mr. Thomas' class
● – Ms. Martinez's class

a. Which group of 10 students talked on the telephone the most?

b. Does the pattern for the number of hours spent talking on the telephone appear to be the same for both groups? Explain.

12. **Winter Olympics** The table below lists the medal standings for the 1994 Winter Olympics.

Winter Olympics, Medal Standings

Nation	Gold	Silver	Bronze	Total
Norway	10	11	5	26
Germany	9	7	8	24
Russia	11	8	4	23
Italy	7	5	8	20
United States	6	5	2	13
Canada	3	6	4	13
Switzerland	3	4	2	9
Austria	2	3	4	9
South Korea	4	1	1	6
Finland	0	1	5	6

a. Make a line plot that shows the number of gold medals won by the countries.

b. Write three questions that can be answered using the line plot.

Critical Thinking

13. Use the line plot in Exercise 11. Find the average number of hours that the students in Mr. Thomas and Ms. Martinez's classes spent on the telephone. Do the values support your answer to Exercise 11b? Explain.

Mixed Review

14. Graph $\{...,-5, -4, -3\}$ on a number line. (Lesson 2-1)

15. Name the property illustrated by $6(jk) = (6j)k$. (Lesson 1-8)

16. Use the distributive property to find $15(124)$. (Lesson 1-7)

17. Solve $m = \dfrac{22 - 8}{7}$. (Lesson 1-5)

18. Evaluate $np + st$ when $n = 7$, $p = 6$, $s = 4$, and $t = 5$. (Lesson 1-3)

WORKING ON THE

In·ves·ti·ga·tion

Refer to the Investigation on pages 68−69.

the Greenhouse Effect

1 Examine the data from the control experiment. Look for patterns and relationships in the data. Analyze how the change in temperature is a function of the distance. Describe your analysis in writing. Be sure to explain any patterns you see in the data.

2 Examine the data from the greenhouse experiment. Look for patterns and relationships in the data. Analyze how the change in temperature is a function of the distance. Describe your analysis in writing. Use the terms *function*, *independent*, *dependent*, *domain*, and *range*.

3 Compare the two experiments. List the similarities and differences in the data and explain the factors that might cause them.

4 Think about the design of the two experiments. How were they different? Why do you think they were designed the way they were? How do they relate to the greenhouse effect?

Add the results of your work to your Investigation Folder.

2-3A Adding and Subtracting Integers

A Preview of Lesson 2–3

Materials: counters ☐ integer mat

You can use counters to help you understand addition and subtraction of integers. In these activities, yellow counters represent positive integers, and red counters represent negative integers.

Rules for Integer Models	
A zero pair is formed by pairing one positive counter with one negative counter.	$\boxed{+ \ -}$
You can remove or add zero pairs to a set because removing or adding zero does not change the value of the set.	

Activity 1   Use counters to find the sum −2 + 3.

Step 1 Place 2 negative counters and 3 positive counters on the mat.

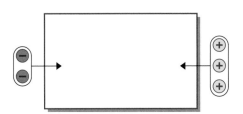

Step 2 Remove the 2 zero pairs. Since 1 positive counter remains, the sum is 1. Therefore, −2 + 3 = 1.

Activity 2   Use counters to find the difference 3 − (−2).

Step 1 Place 3 positive counters on the mat. There are no negative counters, so you cannot remove 2 negatives. Add 2 zero pairs to the mat. Adding zero pairs does not change the value of the set.

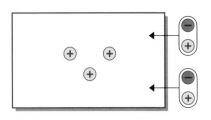

Step 2 Now remove 2 negative counters. Since 5 positive counters remain, the difference is 5. Therefore, 3 − (−2) = 5.

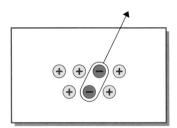

Model   **Find each sum or difference by using counters.**

1. $4 + 2$ 2. $4 + (-2)$ 3. $-4 + 2$ 4. $-4 + (-2)$

5. $4 - 2$ 6. $-4 - (-2)$ 7. $4 - (-2)$ 8. $-4 - 2$

Draw   **Tell whether each statement is *true* or *false*. Justify your answer with a drawing.**

9. $5 - (-2) = 3$ 10. $-5 + 7 = 2$ 11. $2 - 3 = -1$ 12. $-1 - 1 = 0$

Write   13. Write a paragraph explaining how to find the sum of two integers without using counters. Be sure to include all possibilities.

2-3

Adding and Subtracting Integers

What YOU'LL LEARN

- To find the absolute value of a number, and
- to add and subtract integers.

Why IT'S IMPORTANT

You can add and subtract integers to help you solve problems involving weather, business, and golf.

APPLICATION
Space Science

On February 10, 1995, Dr. Bernard A. Harris, Jr. became the first African-American to walk in space. During the walk, he and Dr. Michael Foale, a British-born astronaut who is a U.S. citizen, were exposed to temperatures as cold as $-125°$F. The scheduled five-hour spacewalk to test thermal improvements to NASA spacesuits was cut short by 30 minutes because the astronauts experienced icy fingers.

Drs. Michael Foale and Bernard Harris in space

Usually, spacewalkers spend some of their time basking in the sun's rays where temperatures can be as high as 200°F. But throughout the spacewalk, Drs. Harris and Foale remained in the shadow of the shuttle Discovery and Earth, the coldest possible spot.

As you can see, spacewalkers must be able to survive in extreme temperatures. You can use a number line to determine the range of the temperature extremes. -125 and 200 are graphed on the number line below. Notice that $-125°$F is 125 units from 0 and 200°F is 200 units from 0. So, the total number of units from -125 to 200 is $125 + 200$ or 325.

The temperature range from $-125°$F to 200°F is 325°F.

You used the idea of **absolute value** to find the range from $-125°$F to 200°F.

Definition of Absolute Value	**The absolute value of a number is its distance from zero on a number line.**

Since distance cannot be less than zero, absolute values are always greater than or equal to zero.

The symbol for the absolute value of a number is two vertical bars around the number.

Note that the absolute value bars can serve as grouping symbols.

$|-125| = 125$ is read *The absolute value of -125 equals 125.*

$|200| = 200$ is read *The absolute value of 200 equals 200.*

$|-7 + 6| = 1$ is read *The absolute value of the quantity $-7 + 6$ equals 1.*

You can evaluate expressions involving absolute value.

Example **1** Evaluate $-|x + 6|$ if $x = -10$.

$$-|x + 6| = -|-10 + 6| \qquad \textit{Substitution property of equality}$$
$$= -|-4|$$
$$= -(4) \qquad\qquad \textit{The absolute value of } -4 \textit{ is } 4.$$
$$= -4$$

You can use absolute value to add integers.

Same Signs

a. $4 + 3 = 7$ *Notice that the sign of each addend is positive. The sum is positive.*

b. $-4 + (-3) = -7$ *Notice that the sign of each addend is negative. The sum is negative.*

Different Signs

c. $6 + (-8) = -2$ *Notice that $8 - 6 = 2$. Since the integer -8 has the greater absolute value, the sign of the sum is also negative.*

d. $-6 + 8 = 2$ *Notice that $8 - 6 = 2$. Since the integer 8 has the greater absolute value, the sign of the sum is also positive.*

These examples suggest the following rules.

Adding Integers	**To add integers with the *same sign,* add their absolute values. Give the result the same sign as the integers.** **To add integers with *different signs,* subtract the lesser absolute value from the greater absolute value. Give the result the same sign as the integer with the greater absolute value.**

Example **2** Find each sum.

a. $-10 + (-17)$

$$-10 + (-17) = -(|-10| + |-17|) \qquad \textit{Both numbers are negative,}$$
$$= -(10 + 17) \qquad\qquad \textit{so the sum is negative.}$$
$$= -27$$

b. $39 + (-22)$

$$39 + (-22) = +(|39| - |-22|) \qquad \textit{Subtract the absolute values. Since the}$$
$$= +(39 - 22) \qquad\qquad \textit{number with the greater absolute value,}$$
$$= 17 \qquad\qquad\qquad \textit{39, is positive, the sum is positive.}$$

c. $-28 + 16$

$$-28 + 16 = -(|-28| - |16|) \qquad \textit{Subtract the absolute values. Since the}$$
$$= -(28 - 16) \qquad\qquad \textit{number with the greater absolute value,}$$
$$= -12 \qquad\qquad\qquad \textit{28, is negative, the sum is negative.}$$

Notice that both +2 and −2 are 2 units away from 0. That is, −2 and 2 have the same absolute value.

Every positive integer can be paired with a negative integer. These pairs are called **opposites.** For example, the opposite of +2 is −2.

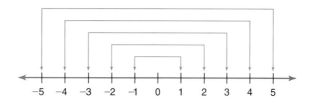

A number and its opposite are called **additive inverses** of each other. Look at the sum of each number and its additive inverse below. What pattern do you notice?

$$2 + (-2) = 0 \qquad -67 + 67 = 0 \qquad 409 + (-409) = 0$$

These examples suggest the following rule.

Additive Inverse Property	**For every number a, $a + (-a) = 0$.**

Additive inverses can be used when you subtract numbers.

Subtraction	**Addition**
additive inverses	
$6 - 2 = 4$	$6 + (-2) = 4$

same result

It appears that subtracting a number is equivalent to adding its additive inverse.

Subtracting Integers	**To subtract a number, add its additive inverse.** **For any numbers a and b, $a - b = a + (-b)$.**

Example ❸ Find each difference.

a. $6 - 14$

Method 1
Use the rule.

To subtract 14, add −14.
$$6 - 14 = 6 + (-14)$$
$$= -8$$

Method 2
Use models.

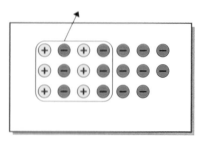

Place 6 positive counters on the mat. Add 14 negative counters. Form and remove six zero pairs.

$$6 - 14 = -8$$

b. $-12 - (-8)$

Method 1	**Method 2**
Use the rule.	Use a scientific calculator.

Method 1

Use the rule.

To subtract −8, add its inverse, +8.

$$-12 - (-8) = -12 + (+8)$$
$$= -4$$

Method 2

Use a scientific calculator.

The [+/−] key on a scientific calculator, called the *change-sign key*, changes the sign of the number on the display. You can find $-12 - (-8)$ by using this key.

Enter: 12 [+/−] [−] 8 [+/−] [=] −4

c. $43 - (-26)$

$$43 - (-26) = 43 + 26$$
$$= 69$$

Check by using a calculator.

Enter: 43 [−] 26 [+/−] [=] 69 ✔

The plural of matrix is matrices.

A **matrix** is a rectangular arrangement of elements in rows and columns. Although matrices are sometimes used as a problem-solving tool, their importance extends to another branch of mathematics called **discrete mathematics.** Discrete mathematics deals with finite or discontinuous quantities. The distinction between continuous and discrete quantities is one that you have encountered throughout your life. Think of a staircase. You can slide your hand up the banister, but you have to climb the stairs one by one. The banister represents a continuous quantity, like the graph of {all numbers greater than 3}. However, each step represents a discrete quantity, like an element in a matrix, or a point in the graph of {2, 4, 7, 12}.

When entries in corresponding positions of two matrices are equal, the matrices are equal.

$$\begin{bmatrix} 4 & 2 \\ -2 & 5 \end{bmatrix} = \begin{bmatrix} 4 & 2 \\ -2 & 5 \end{bmatrix} \qquad \begin{bmatrix} 4 & 2 \\ -2 & 5 \end{bmatrix} \neq \begin{bmatrix} 2 & 4 \\ -2 & 1 \end{bmatrix}$$

You can add or subtract matrices only if they have the same number of rows and columns. You add and subtract matrices by adding or subtracting corresponding entries.

Example ❹

APPLICATION

Business

The Paw Print at Gahanna Lincoln High School in Gahanna, Ohio, is a school store run by seniors in a yearlong marketing education class. The store is stocked with school items such as sweatshirts, T-shirts, school supplies, and snacks. The sales, cost of goods, and expenses for each semester in 1994 and 1995, rounded to the nearest dollar, are shown below.

Semester 1	Sales	Cost of Goods	Expenses
1994	$47,981	$32,627	$12,999
1995	$70,018	$49,013	$12,705

Semester 2	Sales	Cost of Goods	Expenses
1994	$31,988	$21,752	$8666
1995	$30,008	$21,005	$5445

a. Find the total sales, cost of goods, and expenses for each year at The Paw Print.

To find the total sales, cost of goods, and expenses for each year, add the matrices for semester 1 and semester 2.

$$\begin{bmatrix} \$47{,}981 + \$31{,}988 & \$32{,}627 + \$21{,}752 & \$12{,}999 + \$8666 \\ \$70{,}018 + \$30{,}008 & \$49{,}013 + \$21{,}005 & \$12{,}705 + \$5445 \end{bmatrix}$$

$$= \begin{bmatrix} \$79{,}969 & \$54{,}379 & \$21{,}665 \\ \$100{,}026 & \$70{,}018 & \$18{,}150 \end{bmatrix}$$

b. Find the store's gross income, or profit, by year. The gross income is the difference between the sales and the sum of the cost of goods and expenses.

Profit for 1994 $= \$79{,}969 - (54{,}379 + 21{,}665)$
$\qquad\qquad\qquad = \$3925$

Profit for 1995 $= \$100{,}026 - (70{,}018 + 18{,}150)$
$\qquad\qquad\qquad = \$11{,}858$

The rule for subtracting integers and the distributive property can be used to combine like terms.

Example 5 **Simplify $2x - 5x$.**

$2x - 5x = 2x + (-5x)$ *To subtract 5x, add its additive inverse, $-5x$.*

$\qquad\quad = [2 + (-5)]x$ *Use the distributive property.*

$\qquad\quad = -3x$

CHECK FOR UNDERSTANDING

Communicating Mathematics

Study the lesson. Then complete the following.

1. **Explain** how a zero pair of counters illustrates the additive inverse property.

2. **Write** an integer on a piece of paper. Have a friend explain how to find its absolute value without a number line and without knowing the integer.

3. **Explain** to a friend how you would find each sum or difference below.

   a. $-3 + (-1)$ **b.** $5 + (-7)$ **c.** $4 - 9$ **d.** $6 - (-5)$

4. **a.** Find a value for x if $x = -x$.

   b. Find a value for x if $|x| = -|x|$.

MODELING MATHEMATICS

5. Write a statement that represents the model shown below.

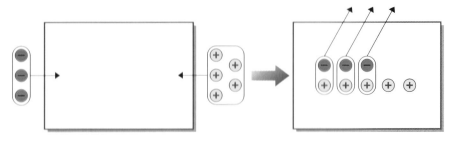

6. Find $-9 - (-4)$ by using counters.

Guided Practice

State the additive inverse and absolute value of each integer.

7. $+7$ **8.** -20 **9.** 0

Find each sum or difference.

10. $-12 + (-8)$ **11.** $-11 - (-7)$ **12.** $-36 + 15$

13. $18 - 29$ **14.** $|-6 - 4|$ **15.** $-|5 - 5|$

Simplify each expression.

16. $-16y - 5y$ **17.** $32c - (-8c)$ **18.** $-9d + (-6d)$

Evaluate each expression if $x = -5$, $y = 3$, and $z = -6$.

19. $y - 7$ **20.** $16 + z$ **21.** $|8 + x|$

Let $A = \begin{bmatrix} 3 & 2 \\ 5 & 0 \end{bmatrix}$ and $B = \begin{bmatrix} 4 & -2 \\ -1 & 6 \end{bmatrix}$.

22. Find $A + B$.   **23.** Find $A - B$.

24. Space Science Refer to the application at the beginning of the lesson. The Discovery astronauts were working in temperatures ranging from $-125°F$ to $200°F$.

   a. Which is colder, $-125°F$ or $200°F$?

   b. Write an equation that uses addition or subtraction to show how much colder one temperature is than the other.

EXERCISES

Practice

State the additive inverse and absolute value of each integer.

25. $+12$ **26.** -45 **27.** -302

Find each sum or difference.

28. $18 + (-16)$ **29.** $47 + (-47)$ **30.** $-9 + 32$

31. $0 - 32$ **32.** $16 - (-23)$ **33.** $-5 + (-11)$

34. $-104 + 16$ **35.** $-9 + (-61)$ **36.** $-18 - 4$

37. $9 - 24$ **38.** $-21 - (-24)$ **39.** $-13 - (-8)$

Simplify each expression.

40. $29t - 17t$ **41.** $17b - (-23b)$ **42.** $-6w + (-13w)$

43. $6p + (-35p)$ **44.** $54y - 47y$ **45.** $-5d + 31d$

Evaluate each expression if $b = -4$, $d = 2$, and $p = -5$.

46. $b + 14$ **47.** $d - 6$ **48.** $15 + p$

49. $d + (-17)$ **50.** $d + 7$ **51.** $|p|$

52. $|6 + b|$ **53.** $|p - 3|$ **54.** $-|-22 + p|$

Find each sum or difference.

55. $|-58 + (-41)|$ **56.** $|-93| - (-43)$ **57.** $-|-345 - (-286)|$

Discrete Mathematics

58. $\begin{bmatrix} 3 & 2 \\ 4 & -1 \end{bmatrix} + \begin{bmatrix} 0 & 5 \\ -3 & -4 \end{bmatrix}$ **59.** $\begin{bmatrix} 2 & 4 \\ -1 & 0 \end{bmatrix} + \begin{bmatrix} -2 & -1 \\ 3 & 2 \end{bmatrix}$

60. $\begin{bmatrix} 1 & 6 \\ -4 & -5 \end{bmatrix} - \begin{bmatrix} -3 & -6 \\ 2 & -4 \end{bmatrix}$ **61.** $\begin{bmatrix} 2 & 3 \\ 4 & 2 \\ 5 & -5 \end{bmatrix} - \begin{bmatrix} 6 & 8 \\ 8 & 7 \\ -5 & -2 \end{bmatrix}$

62. Geometry Use the number line below to find the length of segment *QS* if segment *RT* has a length of 15 units.

Q   R   S   T

4 units   9 units

Yokohama Landmark Tower

63. Elevators The 70-floor Yokohama Landmark Tower houses the world's fastest passenger elevators. Passengers travel from the second floor to the 69th floor in 40 seconds at an average speed of 28 miles per hour. If an employee rode the elevator up to his office on the 67th floor, then came down 43 floors to deliver a report, which floor is he on now?

64. Puzzles The sum of the numbers in each row, column, and diagonal of a magic square is the same. In the magic square at the right, the sum is 0. New magic squares can be formed by adding the same number to each entry. Complete each magic square.

1	2	−3
−4	0	4
3	−2	−1

a.

3	4	−1

b.

−2		
	−3	
		−4

c.

	−7	

65. Business The manager of The Best Bagel Shop keeps a record of each type of bagel sold each day at their two stores. Two days of sales are shown below.

Day	Store	Type of Bagel			
		Sesame	Poppy	Blueberry	Plain
Monday	East	120	80	64	75
	West	65	105	77	53
Tuesday	East	112	76	56	74
	West	69	95	82	50

  a. Write a matrix for each day's sales.
  b. Find the sum of the two days' sales using matrix addition.
  c. Subtract Tuesday's sales from Monday's sales by using matrix subtraction. What does this matrix represent?

66. Golf In golf, scores are based on *par.* Par 72 means that a golfer should hit the ball 72 times to complete 18 holes of golf. A score of 68, or 4 under par, is written as −4. A score of 2 over par is written as +2. At the Oldsmobile Classic LPGA tournament in June 1995, at a par-72 course in Michigan, Helen Alfredsson shot 65, 74, 69, and 71 during four rounds of golf.
  a. Use integers to write Ms. Alfredsson's score for each round as over or under par.
  b. Add the integers to find Ms. Alfredsson's overall score.
  c. Was Ms. Alfredsson's score under or over par? Would you want to have her score? Explain.

67. Trees The table below lists twenty different species of American trees and their highest recorded height in feet. (Lesson 2–2)

American Trees					
Type	**Height (feet)**	**Type**	**Height (feet)**	**Type**	**Height (feet)**
American Beech	130	Bitter Cherry	104	Mockernut Hickory	125
American Chestnut	110	Black Maple	118	Pecan	143
American Elm	100	Black Walnut	105	Shellback Hickory	105
American Sweetgum	136	Cherrybark Oak	124	Siberian Elm	146
Arizona Sycamore	114	Common Hackberry	111	Virginia Pine	120
Balsam Poplar	138	Digger Pine	114	Weeping Willow	114
Bishop Pine	112	Honeylocust	115		

Source: *The World Almanac, 1995*

a. What is the range of heights for the trees?
b. Make a line plot showing the height of the trees.
c. Do the heights cluster around a certain number?
d. Which tree is the tallest? the shortest?

68. Write a corresponding addition sentence for the diagram at the right. (Lesson 2–1)

69. Sketch a reasonable graph for the following situation.
The water level in Craig Pond is extremely high at the beginning of the summer. Throughout the summer, the water level lowers steadily. By the end of the summer, the pond is almost dry. (Lesson 1–9)

70. Name the property illustrated by $3 + 4 = 4 + 3$. (Lesson 1–8)

71. Simplify $23y^2 + 32y^2$. (Lesson 1–7)

Name the property illustrated by each statement. (Lesson 1–6)

72. $3abc \cdot 0 = 0$ **73.** If $12xy - 3 = 4$, then $4 = 12xy - 3$.

74. Solve $p = 6\frac{1}{4} + \frac{1}{2}$. (Lesson 1–5)

75. Statistics Write the stems that would be used for a plot of the following data: 29.1, 34.7, 20.05, 74.9, 64, 14.2, 84.9, 38.26, 20.9, 34, 59.42, 37.107, and 43.676. (Lesson 1–4)

76. Evaluate $19 + 5 \cdot 4$. (Lesson 1–3)

77. Patterns Find the sixth term in the sequence 4, 8, 12, 16, ... (Lesson 1–2)

78. Write an algebraic expression for the sum of *n* and 33. (Lesson 1–1)

Rational Numbers

What YOU'LL LEARN

- To compare and order rational numbers, and
- to find a number between two rational numbers.

Why IT'S IMPORTANT

You need to know the values of rational numbers to solve problems involving consumerism and archaeology.

APPLICATION
Recycling

Many people all over the world recycle their aluminum cans in order to help our environment. The graph below shows the fraction of aluminum cans that have been recycled over the years.

Source: The Aluminum Association

The numbers shown in the graph are examples of **rational numbers.**

Definition of a Rational Number	A rational number is a number that can be expressed in the form $\frac{a}{b}$, where a and b are integers and b is not equal to 0.

Examples of rational numbers expressed in the form $\frac{a}{b}$ are shown below.

Notice that all integers are rational numbers.

Rational Numbers	4	$-3\frac{3}{4}$	0.250	0	$0.33\overline{3}$
Form $\frac{a}{b}$	$\frac{4}{1}$	$-\frac{15}{4}$	$\frac{1}{4}$	$\frac{0}{1}$	$\frac{1}{3}$

Rational numbers can be graphed on a number line in the same manner as integers. The number line below is separated into fourths to show the graphs of some common fractions and decimals.

You can compare rational numbers by graphing them on a number line. Recall that a mathematical sentence that uses < and > to compare two expressions is called an *inequality*.

The following statements can be made about the graphs of -5, -2, $1\frac{1}{2}$ and 3.5 shown on the number line below.

a. The graph of -2 is to the left of the graph of 3.5. $-2 < 3.5$

b. The graph of $1\frac{1}{2}$ is to the right of the graph of -5. $1\frac{1}{2} > -5$

These examples suggest the following rule.

Comparing Numbers on the Number Line	If a and b represent any numbers and the graph of a is to the left of the graph of b, then $a < b$. If the graph of a is to the right of the graph of b, then $a > b$.

If $<$, $>$, and $=$ are used to compare two numbers, then the following property applies.

Comparison Property	For any two numbers a and b, exactly one of the following sentences is true.
	$a < b$ $\qquad$ $a = b$ $\qquad$ $a > b$

The symbols $\neq$, $\leq$, and $\geq$ can also be used to compare numbers. The chart below shows several inequality symbols and their meanings.

Symbol	Meaning
$<$	is less than
$>$	is greater than
$\neq$	is not equal to
$\leq$	is less than or equal to
$\geq$	is greater than or equal to

Example ❶ **Replace each _?_ with $<$, $>$, or $=$ to make each sentence true.**

a. -75 _?_ 13

Since any negative number is less than any positive number, the true sentence is $-75 < 13$.

Note that the closed ends of $<$ and $>$ always point to the lesser number.

b. -14 _?_ $-22 + 9$

-14 _?_ $-22 + 9$

-14 _?_ -13 $\qquad$ *Simplify.*

Since -14 is less than -13, the true sentence is $-14 < -22 + 9$.

c. $\dfrac{3}{8}$ _?_ $-\dfrac{7}{8}$

Since 3 is greater than -7, the true sentence is $\dfrac{3}{8} > -\dfrac{7}{8}$.

You can use **cross products** to compare two fractions with different denominators. When two fractions are compared, the cross products are the products of the terms on the diagonals.

Comparison Property for Rational Numbers	For any rational numbers $\frac{a}{b}$ and $\frac{c}{d}$, with $b > 0$ and $d > 0$: 1. if $\frac{a}{b} < \frac{c}{d}$, then $ad < bc$, and 2. if $ad < bc$, then $\frac{a}{b} < \frac{c}{d}$.

This property also holds if $<$ is replaced by $>$, $\leq$, $\geq$, or $=$.

Example ② **Replace each __?__ with $<$, $>$, or $=$ to make each sentence true.**

a. $\frac{7}{13}$ __?__ $\frac{4}{15}$

$7(15)$ __?__ $13(4)$

$105 > 52$

The true sentence is $\frac{7}{13} > \frac{4}{15}$.

b. $\frac{7}{8}$ __?__ $\frac{8}{9}$

$7(9)$ __?__ $8(8)$

$63 < 64$

The true sentence is $\frac{7}{8} < \frac{8}{9}$.

Every rational number can be expressed as a terminating or repeating decimal. You can use a calculator to write rational numbers as decimals.

Example ③ **Use a calculator to write the fractions $\frac{3}{8}$, $\frac{4}{5}$, and $\frac{1}{6}$ as decimals. Then write the fractions in order from least to greatest.**

It is helpful to know these commonly used fraction-decimal equivalencies by memory.

$\frac{1}{2} = 0.5$, $\frac{1}{3} = 0.\overline{3}$,

$\frac{1}{4} = 0.25$, $\frac{1}{5} = 0.2$,

$\frac{1}{8} = 0.125$

$\frac{3}{8} = 0.375$ This is a terminating decimal.

$\frac{4}{5} = 0.8$ This is also a terminating decimal.

$\frac{1}{6} = 0.16666\ldots$ or $0.1\overline{6}$ This is a repeating decimal.

In order from least to greatest, the decimals are $0.1\overline{6}$, 0.375, 0.8. So, the fractions in order from least to greatest are $\frac{1}{6}$, $\frac{3}{8}$, $\frac{4}{5}$.

You can use a calculator to compare the cost per unit, or **unit cost,** of two similar items. Many people comparison shop to find the best buys at the grocery store. The item that has the lesser unit cost is the better buy.

$$\text{unit cost} = \text{total cost} \div \text{number of units}$$

Example ④

APPLICATION
Consumerism

Erica and Gabriella are running in a 5-kilometer race to raise money for The Multiple Sclerosis Foundation. They want to buy a sports beverage to drink after they finish the race. Which is the better buy?

Use a calculator to find the unit cost of each brand. In each case, the unit cost is expressed in cents per ounce.

unit cost of 20-ounce bottle: 1.09 $+$ 20 $=$ *0.0545*

unit cost of 32-ounce bottle: 1.69 $+$ 32 $=$ *0.0528*

Since $0.053 < 0.055$, the 32-ounce All Sport® drink is the better buy.

Can you always find another rational number between two rational numbers? One point that lies between any two points is their midpoint. Consider $\frac{1}{3}$ and $\frac{1}{2}$.

To find the average of two numbers, add the numbers and then divide by 2.

Multiplying by $\frac{1}{2}$ is the same as dividing by 2.

To find the coordinate of the midpoint for $\frac{1}{3}$ and $\frac{1}{2}$, find the average, or mean, of the two numbers.

average: $\quad \frac{1}{2}\left(\frac{1}{3} + \frac{1}{2}\right) = \frac{1}{2}\left(\frac{5}{6}\right)$ or $\frac{5}{12}$

This process can be continued indefinitely. The pattern suggests the **density property**.

Density Property for Rational Numbers	**Between every pair of distinct rational numbers, there are infinitely many rational numbers.**

Example 5

Find a rational number between $-\frac{2}{5}$ and $-\frac{3}{8}$.

Find the average of the two rational numbers.

Method 1
Use a pencil and paper.

$$\frac{1}{2}\left[-\frac{2}{5} + \left(-\frac{3}{8}\right)\right] = \frac{1}{2}\left[-\frac{16}{40} + \left(-\frac{15}{40}\right)\right]$$
$$= \frac{1}{2}\left(-\frac{31}{40}\right)$$
$$= -\frac{31}{80}$$

Method 2
Use a scientific calculator.

Enter: (2 ÷ 5 +/−
+ 3 ÷ 8 +/−
) ÷ 2 = −0.3875

A rational number between $-\frac{2}{5}$ and $-\frac{3}{8}$ is $-\frac{31}{80}$ or -0.3875.

CHECK FOR UNDERSTANDING

Communicating Mathematics

Study the lesson. Then complete the following.

1. **Write** three examples of rational numbers.

2. **State** any assumptions you can make about y and z if $x < y$ and $x < z$.

3. **Show** two ways to find three rational numbers between $\frac{1}{5}$ and $\frac{1}{4}$.

4. **Explain** why the definition of a rational number states that b cannot be zero.

5. **You Decide**   To find a number between $\frac{3}{4}$ and $\frac{6}{7}$, John added as follows.
$$\frac{3+6}{4+7} = \frac{9}{11}; \quad \frac{9}{11} \text{ is between } \frac{3}{4} \text{ and } \frac{6}{7}.$$
He claimed that this method will always work. Do you agree? Explain.

6. Assess Yourself Describe two situations in which you would prefer to use decimals instead of fractions.

Guided Practice

Replace each __?__ with <, >, or = to make each sentence true.

7. -4 __?__ 8

8. -5 __?__ $0 - 3$

9. $\frac{5}{14}$ __?__ $\frac{25}{70}$

Write the numbers in each set in order from least to greatest.

10. $\frac{2}{3}, \frac{1}{6}, \frac{1}{2}$

11. $2.5, \frac{3}{4}, -0.5, \frac{7}{8}$

12. Which is a better buy: a 6-ounce can of tuna for \$1.59 or a package of 3 three-ounce cans for \$2.19?

13. Find a number between $\frac{1}{2}$ and $\frac{5}{7}$.

EXERCISES

Practice

Replace each __?__ with <, >, or = to make each sentence true.

14. -3 __?__ 5

15. -1 __?__ -4

16. $-6 - 3$ __?__ -9

17. 5 __?__ $8.4 - 1.5$

18. 4 __?__ $\frac{16}{3}$

19. $\frac{8}{15}$ __?__ $\frac{9}{16}$

20. $\frac{14}{5}$ __?__ $\frac{25}{13}$

21. $\frac{4}{3}(6)$ __?__ $4\left(\frac{3}{2}\right)$

22. $\frac{0.4}{3}$ __?__ $\frac{1.2}{8}$

Write the numbers in each set in order from least to greatest.

23. $\frac{6}{7}, \frac{2}{3}, \frac{3}{8}$

24. $-\frac{4}{15}, -\frac{6}{17}, -\frac{3}{16}$

25. $\frac{4}{14}, \frac{3}{23}, \frac{8}{42}$

26. $6.7, -\frac{5}{7}, \frac{6}{13}$

27. $0.2, -\frac{2}{5}, -0.2$

28. $\frac{4}{5}, \frac{9}{10}, 0.7$

Which is the better buy?

29. a 16-ounce drink for \$0.59 or a 20-ounce drink for \$0.89

30. a 32-ounce bottle of shampoo for \$3.59 or a 64-ounce bottle for \$6.99

31. a package of 48 paper plates for \$2.39 or a package of 75 paper plates for \$3.29

Find a number between the given numbers.

32. $\frac{2}{5}$ and $\frac{7}{2}$

33. $\frac{19}{30}$ and $\frac{31}{45}$

34. $-\frac{2}{15}$ and $-\frac{6}{3}$

35. Name a fraction between $\frac{1}{4}$ and $\frac{1}{2}$ whose denominator is 20.

36. Name a fraction between $\frac{1}{3}$ and $\frac{5}{6}$ whose denominator is 12.

37. Name a fraction between -0.5 and $\frac{1}{3}$ whose denominator is 6.

38. Three numbers a, b, and c satisfy the following conditions:
$b - c < 0$, $a - b > 0$, and $c - a < 0$. Which one is the greatest?

39. Find the coordinates of E, G, and H if the distances between the points are equal.

D   E   F   G   H

$\frac{1}{14}$ $\frac{1}{2}$

Applications and Problem Solving

40. Personal Computers Study the graph at the right. Write an inequality that compares the total number of PCs shipped for home use in 1995 to those shipped in 1996.

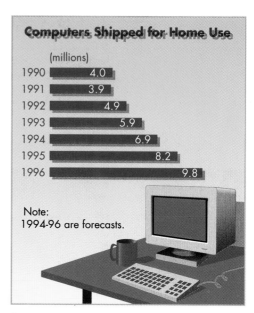

Computers Shipped for Home Use

(millions)

1990	4.0
1991	3.9
1992	4.9
1993	5.9
1994	6.9
1995	8.2
1996	9.8

Note:
1994-96 are forecasts.

Sources: Dataquest, LINK

41. Aviation When building or servicing airplanes, an airplane mechanic sometimes needs to drill holes. Drill bit sizes are given in fractions, yet many blueprints list hole diameters in decimals. One blueprint calls for a hole of 0.391 inches. Find the fractional drill size required if the hole must be drilled $\frac{1}{64}$-inch undersize.

42. Archaeology To calculate a person's height, archaeologists measure the lengths of certain bones. The bones measured are the femur or thigh bone F, the tibia or leg bone T, the humerus or upper arm bone H, and the radius or lower arm bone R. When the length of one of these bones is known, scientists can use one of the formulas below to determine the person's height h in centimeters.

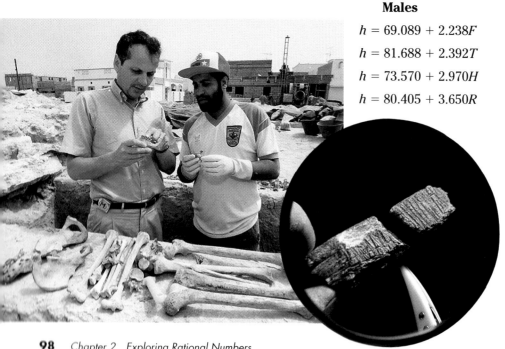

Males	Females
$h = 69.089 + 2.238F$	$h = 61.412 + 2.317F$
$h = 81.688 + 2.392T$	$h = 72.572 + 2.533T$
$h = 73.570 + 2.970H$	$h = 64.977 + 3.144H$
$h = 80.405 + 3.650R$	$h = 73.502 + 3.876R$

a. The femur of a 37-year old woman measured 47.9 cm. Use a calculator to find the height of the woman to the nearest tenth of a centimeter.

b. The humerus of a 49-year old man measured 35.7 cm. Use a calculator to find the height of the man to the nearest tenth of a centimeter.

43. State the additive inverse and absolute value of +9.   (Lesson 2–3)

44. Statistics   State the scale you would use to make a line plot for the following data. Then draw the line plot.   (Lesson 2–2)

145, 130, 135, 150, 145, 145, 140, 130, 145, 150, 130

45. Graph $\{-3, -2, 2, 3\}$ on a number line.   (Lesson 2–1)

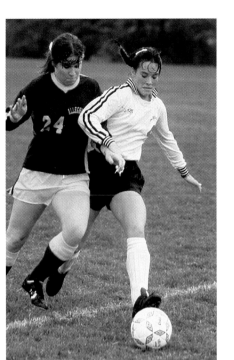

46. Soccer   Bryce, Maria, and Holly are three offensive players for the Spazmatics soccer team. The graph at the right describes the players' offensive skills in their last season.   (Lesson 1–9)

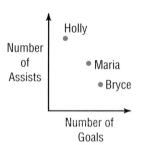

  a. Who scored the most goals?
  b. Who helped the team most with goals and assists overall?
  c. Another player, Ian, had the same number of goals as Maria and the same number of assists as Holly. Describe where on the graph a point should go that would show his offensive skills.

47. Simplify $\frac{2}{5}m + \frac{1}{5}(6n + 3m) + \frac{1}{10}(8n + 15)$.   (Lesson 1–6)

48. Patterns   Find the next three terms in the sequence 1, 3, 9, 27, ...
(Lesson 1–2)

49. Write a verbal expression for $2x^2 + 6$.   (Lesson 1–1)

SELF TEST

Graph each set of numbers on a number line.   (Lesson 2–1)

1. $\{-3, 0, 2\}$

2. {numbers greater than -2}

Find each sum or difference.   (Lesson 2–3)

3. $-9 + (-8)$

4. $6 + (-15)$

5. $23 - (-32)$

6. $-12 - 4$

7. Mountain Bikes   The table at the right shows prices of several mountain bikes.   (Lesson 2–2)

  a. Make a line plot of the data.
  b. Do the numbers cluster around any certain price?

Bike	Price
Trek 820	$325
Mongoose Threshold	270
Giant Yukon	370
Giant Rincon	300
Bianchi Timber Wolf	270
Schwinn Clear Creek	320
Schwinn Sidewinder	260
Specialized Hardrock	270
Raleigh M40GS	290
Trek 800	240

Replace each ? with <, >, or = to make each sentence true.   (Lesson 2–4)

8. $-7 - 5 \underline{\ ?\ } -11$

9. $\frac{7}{16} \underline{\ ?\ } \frac{8}{15}$

10. $\frac{5}{4}(4) \underline{\ ?\ } 8\left(\frac{1}{2}\right)$

Adding and Subtracting Rational Numbers

What YOU'LL LEARN

- To add and subtract rational numbers, and
- to simplify expressions that contain rational numbers.

Why IT'S IMPORTANT

You can add and subtract rational numbers to help you solve problems involving the stock market and track and field.

APPLICATION

Employment

In 1982, the average number of weekly overtime hours worked by a U.S. worker was $2\frac{1}{4}$ hours. This average increased $1\frac{1}{4}$ hours by 1989 and then decreased $\frac{1}{2}$ hour by 1991.

What was the average number of weekly overtime hours worked in 1991?

To answer this question, you need to find the following sum.

$$2\frac{1}{4} + 1\frac{1}{4} + \left(-\frac{1}{2}\right)$$

To add rational numbers such as these, use the same rules you used to add integers. When adding three or more rational numbers, you can use the commutative and associative properties to rearrange the addends.

LOOK BACK

You can refer to Lesson 1-8 for information on the commutative and associative properties.

Method 1:   Use the rules.

$$2\frac{1}{4} + 1\frac{1}{4} + \left(-\frac{1}{2}\right) = 3\frac{2}{4} + \left(-\frac{1}{2}\right) \qquad \textit{Add the first two addends.}$$

$$= 3\frac{1}{2} + \left(-\frac{1}{2}\right) \qquad \textit{Simplify.}$$

$$= +\left(\left|3\frac{1}{2}\right| - \left|-\frac{1}{2}\right|\right) \quad \textit{$3\frac{1}{2}$ has the greater absolute value,}$$

$$= +\left(3\frac{1}{2} - \frac{1}{2}\right) \qquad \textit{so the sign of the sum is positive.}$$

$$= 3$$

Method 2:   Use a number line.

As with integers, you can also add rational numbers by using a number line.

The average number of weekly overtime hours worked in 1991 was 3.

Example **1** Find each sum.

a. $4\frac{1}{8} + \left(-1\frac{1}{2}\right)$ *Estimate:* $4 + (-2) = 2$

$$4\frac{1}{8} + \left(-1\frac{1}{2}\right) = 4\frac{1}{8} + \left(-1\frac{4}{8}\right)$$ *The LCD is 8. Replace $-1\frac{1}{2}$ with $-1\frac{4}{8}$.*

$$= +\left(\left|4\frac{1}{8}\right| - \left|-1\frac{4}{8}\right|\right)$$ *The sum is positive. Why?*

$$= +\left(4\frac{1}{8} - 1\frac{4}{8}\right)$$

$$= +\left(3\frac{9}{8} - 1\frac{4}{8}\right)$$ *Replace $4\frac{1}{8}$ with $3\frac{9}{8}$.*

$$= 2\frac{5}{8}$$ *Compare to the estimate.*

b. $-1.34 + (-0.458)$ *Estimate:* $-1 + (-0.5) = -1.5$

$$-1.34 + (-0.458) = -(|-1.34| + |-0.458|)$$ *The numbers have the same*

$$= -(1.34 + 0.458)$$ *sign. Their sum is negative.*

$$= -1.798$$

c. $-\frac{2}{5} + 1\frac{1}{2} + \left(-\frac{2}{3}\right)$ *Estimate:* $-\frac{1}{2} + 1\frac{1}{2} + \left(-\frac{1}{2}\right) = \frac{1}{2}$

$$-\frac{2}{5} + 1\frac{1}{2} + \left(-\frac{2}{3}\right) = \left[-\frac{2}{5} + \left(-\frac{2}{3}\right)\right] + 1\frac{1}{2}$$ *Group the negative numbers together.*

$$= \left[-\frac{12}{30} + \left(-\frac{20}{30}\right)\right] + \frac{45}{30}$$ *The LCD is 30.*

$$= -\frac{32}{30} + \frac{45}{30}$$

$$= \frac{13}{30}$$ *Compare to the estimate.*

To subtract rational numbers, use the same process you used to subtract integers.

Example **2** Find each difference.

a. $-\frac{3}{5} - \left(-\frac{4}{7}\right)$

$$-\frac{3}{5} - \left(-\frac{4}{7}\right) = -\frac{3}{5} + \frac{4}{7}$$ *To subtract $-\frac{4}{7}$, add its inverse, $+\frac{4}{7}$.*

$$= -\frac{21}{35} + \frac{20}{35}$$ *The LCD is 35.*

$$= -\frac{1}{35}$$

b. $-6.24 - 8.52$

$$-6.24 - 8.52 = -6.24 + (-8.52)$$

$$= -14.76$$

Example ③ The Intel Corporation produces the Pentium™ processor for personal computers. In November of 1994, Intel announced that there was a flaw in the processor that could cause errors in complex mathematical calculations. In one week, Intel's stock dropped $2\frac{1}{4}$ points. By December 13, the stock had dropped another $2\frac{1}{8}$ points. How many points did the stock drop in this time period?

APPLICATION

Stock Market

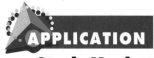

F Y I

In the 18th century, the U.S. dollar was equal to the Spanish silver dollar. This coin was often divided into eight parts, so when the stock market opened, the practice of quoting prices in eighths of a dollar began.

We can represent the drops in the price of the stock with negative numbers.

$$-2\frac{1}{4} - 2\frac{1}{8} = -2\frac{1}{4} + \left(-2\frac{1}{8}\right) \quad \textit{To subtract } 2\frac{1}{8}\textit{, add its additive inverse.}$$

$$= -2\frac{2}{8} + \left(-2\frac{1}{8}\right) \quad \textit{The LCD is 8.}$$

$$= -4\frac{3}{8}$$

Intel's stock dropped $4\frac{3}{8}$ points during this time period.

You can also evaluate expressions involving the addition and subtraction of rational numbers.

Example ④ **a.** Evaluate $p - 3.5$ if $p = 2.8$.

$$p - 3.5 = 2.8 - 3.5 \quad \textit{Replace p with 2.8.}$$

$$= -0.7$$

b. Evaluate $\frac{11}{4} - x$ if $x = \frac{27}{8}$.

$$\frac{11}{4} - x = \frac{11}{4} - \frac{27}{8} \quad \textit{Replace x with } \frac{27}{8}.$$

$$= \frac{22}{8} - \frac{27}{8} \quad \textit{The LCD is 8.}$$

$$= -\frac{5}{8}$$

CHECK FOR UNDERSTANDING

Communicating Mathematics

Study the lesson. Then complete the following.

1. **Describe** two ways to find $-2.4 + 5.87 + (-2.87) + 6.5$.

2. **Find** an example of addition or subtraction of rational numbers in a newspaper or magazine.

MODELING MATHEMATICS

3. **Draw** and use a number line to show the sum of $-\frac{4}{5}$ and $\frac{3}{5}$.

4. **Use** a number line to solve the problem in Example 3.

Guided Practice

Find each sum or difference.

5. $\frac{7}{9} - \frac{8}{9}$

6. $-\frac{7}{12} + \frac{5}{6}$

7. $-\frac{1}{8} + \left(-\frac{5}{2}\right)$

8. $-69.5 - 82.3$

9. $4.57 + (-3.69)$

10. $-\frac{2}{5} + \frac{3}{15} + \frac{3}{5}$

Evaluate each expression.

11. $a - (-5)$, if $a = 0.75$

12. $\frac{11}{2} - b$ if $b = -\frac{5}{2}$

13. Business For the week of July 22, 1995, the following day-to-day changes were recorded in the Dow Jones Industrial Average:

Monday, $+2\frac{3}{8}$; Tuesday, $-\frac{1}{4}$; Wednesday, $-\frac{3}{8}$;

Thursday, $+2\frac{1}{4}$; Friday, $-3\frac{1}{2}$.

   a. What was the net change for the week?

   b. Draw a number line that shows the change each day.

EXERCISES

Practice

Find each sum or difference.

14. $\frac{5}{8} + \left(-\frac{3}{8}\right)$

15. $-\frac{7}{8} - \left(-\frac{3}{16}\right)$

16. $-1.6 - 3.8$

17. $3.2 + (-4.5)$

18. $\frac{1}{2} + \left(-\frac{8}{16}\right)$

19. $\frac{2}{3} + \left(-\frac{2}{9}\right)$

20. $-38.9 + 24.2$

21. $-0.0007 + (-0.2)$

22. $-\frac{3}{7} + \frac{1}{4}$

23. $-5\frac{7}{8} - 2\frac{3}{4}$

24. $79.3 - (-14.1)$

25. $-0.0015 + 0.05$

26. $-9.16 - (-10.17)$

27. $-5.6 + (-9.45) + (-7.89)$

28. $\frac{1}{4} + 2 + \left(-\frac{3}{4}\right)$

29. $-0.87 + 3.5 + (-7.6) + 2.8$

30. $\frac{3}{4} + \left(-\frac{4}{5}\right) + \frac{2}{5}$

31. $-4\frac{1}{4} + 6\frac{1}{3} + 2\frac{2}{3} + \left(-3\frac{3}{4}\right)$

Evaluate each expression.

32. $-3\frac{1}{3} - b$, if $b = \frac{1}{2}$

33. $r - 2.7$, if $r = -0.8$

34. $n - 0.5$, if $n = -0.8$

35. $t - (-1.3)$, if $t = -18$

36. $\frac{4}{3} - p$, if $p = \frac{4}{5}$

37. $-\frac{12}{7} - s$, if $s = \frac{16}{21}$

Discrete Mathematics

Find each sum or difference.

38. $\begin{vmatrix} 3 & -2 \\ 5 & 6.2 \end{vmatrix} + \begin{vmatrix} 2.4 & 5 \\ -4 & 1 \end{vmatrix}$

39. $\begin{vmatrix} -1.3 & 4.2 \\ 0 & -3.4 \end{vmatrix} - \begin{vmatrix} 2.5 & 4.3 \\ -1.7 & -6.3 \end{vmatrix}$

40. $\begin{bmatrix} 1 & -2 \\ 2 & 1 \\ 7 & 4 \end{bmatrix} + \begin{bmatrix} 3 & -5 \\ 2 & 6 \\ 3 & \end{bmatrix}$

41. $\begin{bmatrix} \frac{1}{2} & 6 \\ -1 & 5 \\ 4 & -7 \end{bmatrix} + \begin{bmatrix} \frac{1}{4} & 5 \\ -4 & 3 \\ -3\frac{1}{2} & -5 \end{bmatrix}$

42. $\begin{vmatrix} 3.2 & 6.4 & -4.3 \\ 5 & -4 & 0.4 \end{vmatrix} - \begin{vmatrix} 1.4 & 3.7 & 5 \\ -1 & 5 & 0.4 \end{vmatrix}$

Critical Thinking

43. Discrete Mathematics Create four 2×2 matrices. Use the matrices to support or disprove the following statement: *Matrix addition is associative and commutative.*

44. Book Binding A textbook like this one is made up of sets of 32 pages called *signatures*. A signature is made from one piece of large paper that is folded several times. Signatures are bound together to make a book. Suppose when this book was printed, there were $1\frac{1}{2}$ rolls of paper available in the warehouse. The printer had three jobs to do that utilized this type of paper. This textbook required $\frac{1}{4}$ roll, another book required $\frac{2}{5}$ roll, and the third book needed $\frac{1}{2}$ roll. Is there enough paper available for all three jobs? Explain.

45. Discrete Mathematics An arithmetic sequence is a sequence in which the difference between any two consecutive terms is the same.
   a. Write the next three terms in the arithmetic sequence 4.53, 5.65, 6.77, 7.89, ...
   b. What is the common difference?
   c. Write the first five terms of an arithmetic sequence in which the first term is -2 and the common difference is $\frac{3}{4}$. Find the sum of the first five terms.

46. Track and Field In 1995, Sheila Hudson-Strudwick set the U.S. women's triple jump record. Suppose her first jump was 46 feet 9 inches and her next jump was $1\frac{3}{4}$ inches longer. Her record-setting jump was 1 foot $2\frac{1}{2}$ inches longer than her second jump. What was the length of her record-setting jump?

47. Replace _?_ with $<$, $>$, or $=$ to make -3.833 _?_ -3.115 true. (Lesson 2–4)

48. Evaluate $-|-9 + 2x|$ when $x = 7$. (Lesson 2–3)

49. Statistics The points scored by the winning teams in the first 29 Super Bowls are listed below. (Lesson 2–2)

35	33	16	23	16	24	14	24
16	21	32	27	35	31	27	26
27	38	38	46	39	42	20	55
20	37	52	30	49			

   a. Make a line plot of the data.
   b. Do the numbers cluster around any number?

50. Restaurant Sales Rosa's Deli made 3 cheesecakes, each having 12 slices, on Sunday. On Monday, they sold 3 slices. On Tuesday, they sold one entire cheesecake and 2 additional slices. The baker made another cheesecake on Wednesday and sold nine slices. If Rosa's Deli sold 11 slices of cheesecake on Thursday, how many slices were left on Friday? (Lesson 2–1)

51. Name the property illustrated by $7(0) = 0$. (Lesson 1–6)

52. Evaluate $7[4^3 - 2(4 + 3)] \div 7 + 2$. (Lesson 1–3)

53. Patterns Find the next term in the sequence $\frac{1}{3}, \frac{3}{6}, \frac{5}{12}, ...$ (Lesson 1–2)

54. Evaluate 8^4. (Lesson 1–1)

2-6A Multiplying Integers

Materials: counters   integer mat

A Preview of Lesson 2–6

You can use counters to model multiplication of integers. Remember that yellow counters represent positive integers and red counters represent negative integers. When multiplying whole numbers, 2×4 means *two sets of four items.* When multiplying integers with counters, $(+2)(+4)$ means *to put in two sets of four positive counters.* $(-2)(+4)$ means to take out two sets of four positive counters.

Activity 1   Use counters to find the product $(+2)(-4)$.

Step 1 Place, or *put in,* two sets of four negative counters on the mat.

Step 2 Since there are eight negative counters on the mat, the product is -8. Thus, $(+2)(-4) = -8$.

Activity 2   Use counters to find the product $(-2)(+4)$.

Step 1 Add enough zero pairs so that you can *take out* two sets of four positive counters.

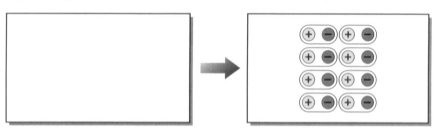

Step 2 Now take out two sets of four positive counters from the mat.

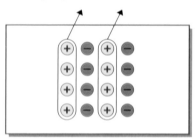

Step 3 Since there are eight negative counters on the mat, the product is -8. Thus, $-2(+4) = -8$.

Write   1. Model $-2(-4)$. Write a short paragraph explaining what $(-2)(-4)$ means.

Model   **Use counters to find each product.**

2. $2(-5)$ 3. $-2(5)$ 4. $-2(-5)$

5. $5(-2)$ 6. $-5(2)$ 7. $-5(-2)$

Write   8. How are the operations $-2(5)$ and $5(-2)$ the same? How do they differ?

Multiplying Rational Numbers

What YOU'LL LEARN

- To multiply rational numbers.

Why IT'S IMPORTANT

You can multiply rational numbers to help you solve problems involving geometry and business.

INTEGRATION
Geometry

Once called "the world's largest office building," the Pentagon, the five-story, five-sided defense building in Washington, D.C., was designed by G.E. Bergstrom to maximize space and efficiency. No two offices are more than a seven-minute walk from one another.

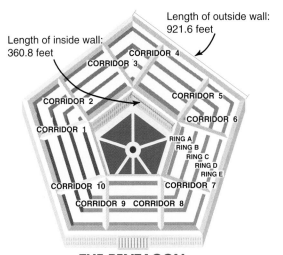

Length of outside wall: 921.6 feet

Length of inside wall: 360.8 feet

CORRIDOR 4
CORRIDOR 3
CORRIDOR 2
CORRIDOR 5
CORRIDOR 6
CORRIDOR 1
RING A
RING B
RING C
RING D
RING E
CORRIDOR 10
CORRIDOR 7
CORRIDOR 9   CORRIDOR 8

THE PENTAGON

The building is made up of 10 corridors and five rings. There are 230 restrooms, 150 stairways, and 7748 windows. Each outside wall of the building is 921.6 feet in length, slightly longer than three football fields. The inner walls, also in the shape of a pentagon, are each 360.8 feet in length. What is the outer perimeter of the Pentagon?

One way to solve this problem is to use repeated addition.

$$921.6 + 921.6 + 921.6 + 921.6 + 921.6 = 4608$$

An easier method would be to multiply 921.6 by 5.

$$5(921.6) = 4608$$

The perimeter of the Pentagon is 4608 feet.

Since this method would not work if you wanted to find the product of $\frac{2}{3}$ and $-\frac{2}{5}$ or the product of -5 and -0.3, you can use the following patterns to discover a rule for multiplying rational numbers.

$$\frac{2}{3} \times \frac{2}{5} = \frac{4}{15}$$

$$\frac{2}{3} \times \frac{1}{5} = \frac{2}{15}$$

$$\frac{2}{3} \times \frac{0}{5} = \frac{0}{15}$$

$$\frac{2}{3} \times \left(-\frac{1}{5}\right) = -\frac{2}{15}$$

$$\frac{2}{3} \times \left(-\frac{2}{5}\right) = -\frac{4}{15}$$

$$-5 \cdot 0.3 = -1.5$$

$$-5 \cdot 0.2 = -1.0$$

$$-5 \cdot 0.1 = -0.5$$

$$-5 \cdot 0 = 0$$

$$-5 \cdot (-0.1) = 0.5$$

$$-5 \cdot (-0.2) = 1.0$$

$$-5 \cdot (-0.3) = 1.5$$

The examples on the previous page suggest the following rules.

Multiplying Two Rational Numbers	The product of two numbers having the *same sign* is positive. The product of two numbers having *different signs* is negative.

Example **Find each product.**

a. $(-9.8)4$ *Estimate:* $(-10)4 = -40$

$(-9.8)4 = -39.2$ *Since the factors have different signs, the product is negative.*

b. $\left(-\dfrac{3}{4}\right)\left(-\dfrac{2}{3}\right)$ *Estimate:* $(-1)\left(-\dfrac{1}{2}\right) = \dfrac{1}{2}$

$\left(-\dfrac{3}{4}\right)\left(-\dfrac{2}{3}\right) = \dfrac{6}{12}$ or $\dfrac{1}{2}$ *Since the factors have the same sign, the product is positive.*

Sometimes you need to evaluate expressions that contain rational numbers.

Example **Evaluate** $a\left(\dfrac{5}{6}\right)^2$ **if** $a = 2$.

$a\left(\dfrac{5}{6}\right)^2 = 2\left(\dfrac{5}{6} \cdot \dfrac{5}{6}\right)$ *Replace a with 2.*

$= 2\left(\dfrac{25}{36}\right)$ *Multiply.*

$= \dfrac{\overset{1}{2}}{1}\left(\dfrac{25}{\underset{18}{36}}\right)$ or $\dfrac{25}{18}$

You may need to simplify expressions by multiplying rational numbers.

Example **3** **Simplify each expression.**

a. $(2b)(-3a)$

$(2b)(-3a) = 2(-3)ab$ *Commutative and associative properties of multiplication*

$= -6ab$

b. $3x(-3y) + (-6x)(-2y)$

$3x(-3y) + (-6x)(-2y) = -9xy + 12xy$ *Multiply.*

$= 3xy$ *Combine like terms.*

Notice that multiplying a number or expression by -1 results in the opposite of the number or expression.

$$-1(4) = -4 \qquad (1.5)(-1) = -1.5 \qquad (-1)(-3m) = 3m$$

Multiplicative Property of -1	The product of any number and -1 is its additive inverse. $-1(a) = -a$ and $a(-1) = -a$

To find the product of three or more numbers, you may want to first group the numbers in pairs.

Example **4** Find $\left(-\frac{3}{4}\right)\left(-4\frac{1}{3}\right)\left(3\frac{2}{5}\right)(4)(-1)$.

$$\left(-\frac{3}{4}\right)\left(-4\frac{1}{3}\right)\left(3\frac{2}{5}\right)(4)(-1) = \left[\left(-\frac{3}{4}\right)(4)\right]\left[\left(-4\frac{1}{3}\right)(-1)\right]\left(3\frac{2}{5}\right)$$ *Commutative and associative properties*

$$= (-3)\left(4\frac{1}{3}\right)\left(3\frac{2}{5}\right)$$

$$= \left(-\frac{\overset{1}{\cancel{3}}}{1}\right)\left(\frac{13}{\underset{1}{\cancel{3}}}\right)\left(\frac{17}{5}\right)$$

$$= -\frac{221}{5} \text{ or } -44\frac{1}{5}$$

You can multiply any matrix by a constant. This is called **scalar multiplication.** When scalar multiplication is performed, each element is multiplied by that constant and a new matrix is formed.

Scalar Multiplication of a Matrix	$m\begin{bmatrix} a & b & c \\ d & e & f \end{bmatrix} = \begin{bmatrix} ma & mb & mc \\ md & me & mf \end{bmatrix}$

Example **5**

Airlines

From July 3 through July 6, 1995, American Airlines quoted round-trip coach fares, in dollars, to and from the selected cities below.

	Chicago	Dallas	Las Vegas
Atlanta	$198.00	$198.00	$1214.00
New York	$246.00	$1224.00	$1342.00

Suppose the major airlines lower fares and begin an airfare war. To increase their air travel, American Airlines decides to lower their fares by 30% or $\frac{3}{10}$. Find the new fares for travel to the cities above.

To find the new fares, you can use scalar multiplication. If fares are reduced by $\frac{3}{10}$, travelers will pay $1 - \frac{3}{10}$ or $\frac{7}{10}$ of the original price. Multiply to find the discount prices.

$$\frac{7}{10}\begin{bmatrix} 198.00 & 198.00 & 1214.00 \\ 246.00 & 1224.00 & 1342.00 \end{bmatrix} = \begin{bmatrix} 138.60 & 138.60 & 849.80 \\ 172.20 & 856.80 & 939.40 \end{bmatrix}$$

The new fares are shown below.

	Chicago	Dallas	Las Vegas
Atlanta	$138.60	$138.60	$849.80
New York	$172.20	$856.80	$939.40

CHECK FOR UNDERSTANDING

Communicating Mathematics

Study the lesson. Then complete the following.

1. **List** the conditions under which each statement is true. Then give an example that verifies your conditions.
   a. ab is positive.   **b.** ab is negative.   **c.** ab is equal to 0.

2. **Explain** what you can conclude about a if each statement is true.
   a. a^2 is positive.   **b.** a^3 is positive.   **c.** a^3 is negative.

3. If $a = -2$, which is greater, $a + a^2$ or -4?

4. **Write** a multiplication sentence for the model at the right.

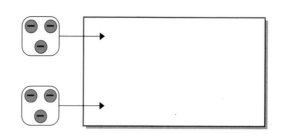

Guided Practice

Find each product.

5. $6(-3)$

6. $(-4)(-8)$

7. $(-4)(2)(-3)$

8. $\left(\dfrac{7}{3}\right)\left(\dfrac{7}{3}\right)$

9. $\left(-\dfrac{4}{5}\right)\left(-\dfrac{1}{5}\right)(-5)$

10. $\left(\dfrac{3}{5}\right)\left(-\dfrac{4}{7}\right)$

Evaluate each expression if $x = \dfrac{1}{2}$ and $y = -\dfrac{2}{3}$.

11. $3y - 4x$

12. $x^2 y$

Simplify.

13. $5s(-6t) + 2s(-8t)$

14. $6x(-7y) + (-3x)(-5y)$

Find each product.

15. $3\begin{bmatrix} -2 & 4 \\ -1 & 5 \end{bmatrix}$

16. $-5\begin{bmatrix} -1 & 0 \\ 4.5 & 8 \\ 3.2 & -4 \end{bmatrix}$

17. **Business** Employees of Glencoe/McGraw-Hill are reimbursed for the "wear and tear" that occurs to their car while on company business. One employee's odometer read 19,438.6 at the beginning of the day and 19,534.1 at the end of the day. How much should the employee be reimbursed if the rate is $0.30 per mile?

EXERCISES

Practice

Find each product.

18. $6(13)$

19. $(-5)(12)$

20. $(-7)(-6)$

21. $\left(-\dfrac{8}{9}\right)\left(\dfrac{9}{8}\right)$

22. $-\dfrac{5}{6}\left(-\dfrac{2}{5}\right)$

23. $(-5)\left(-\dfrac{2}{5}\right)$

24. $(-100)(-3.6)$

25. $(-2.93)(-0.003)$

26. $(-5)(3)(-4)$

27. $\left(\dfrac{2}{3}\right)\left(\dfrac{3}{5}\right)(-3)$

28. $\left(-\dfrac{7}{12}\right)\left(\dfrac{6}{7}\right)\left(-\dfrac{3}{4}\right)$

29. $\dfrac{6}{11}\left(-\dfrac{33}{34}\right)$

30. $(-0.075)(-5.5)$

31. $(-5.8)(-6.425)(2.3)$

32. $(-4)(0)(-2)(-3)$

33. $\dfrac{3}{5}(5)(-2)\left(-\dfrac{1}{2}\right)$

34. $(3)(-4)(-1)(-2)$

35. $\dfrac{2}{11}(-11)(-4)\left(-\dfrac{3}{4}\right)$

Evaluate each expression if $m = -\dfrac{2}{3}$, $n = \dfrac{1}{2}$, $p = -3\dfrac{3}{4}$, and $q = 2\dfrac{1}{6}$.

36. $6m$

37. nq

38. $2m - 3n$

39. $pq - m$

40. $m^2\left(-\dfrac{1}{4}\right)$

41. $n^2(q + 2)$

Simplify.

42. $-2a(-3c) + (-6y)(6r)$

43. $(5t)(-6r) - (-4s)$

44. $7m(-3n) + 3m(-4n)$

45. $5(2x - x) + 4(x + 3x)$

46. $(-6b)(-3c) - (-9a)(7b)$

47. $3.2(5x - y) - 0.3(-1.6x + 7y)$

INTEGRATION
Discrete Mathematics

Find each product.

48. $-7\begin{bmatrix} -1 & 8.2 & 0 \\ 4 & 5.6 & -1 \\ 3.2 & 7 & 7 \end{bmatrix}$

49. $\frac{1}{2}\begin{bmatrix} 4 & 12 & 6 \\ 5 & 10 & 2 \end{bmatrix}$

50. $4\begin{bmatrix} 1.3 & -2 & -4 \\ 0.5 & -0.3 & 5 \\ 6.6 & 2.1 & -8 \end{bmatrix}$

51. $-4\begin{bmatrix} 2.25 & -5.67 \\ 5.6 & 2.5 \\ -7.2 & -2.78 \end{bmatrix}$

52. $-8\begin{bmatrix} 0.2 & 4.5 \\ -1.4 & -3 \\ 3 & 2.4 \\ -7 & -3.2 \end{bmatrix}$

53. $\frac{2}{3}\begin{bmatrix} 9 & 27 & 6 \\ 0 & 3 & 4 \end{bmatrix}$

Critical Thinking

54. If a product has an even number of negative factors, what must be true of the product?

55. If a product has an odd number of negative factors, what must be true of the product?

Applications and Problem Solving

56. Discrete Mathematics A *geometric sequence* is a sequence in which the ratio of any term divided by the term before it is the same for any two terms.
  a. Write the next three terms in the geometric sequence $9, 3, 1, \frac{1}{3}, \ldots$.
  b. What is the common ratio?
  c. Write the first five terms in a geometric sequence in which the first term is -6 and the common ratio is 0.5. Then find the sum of the five terms.

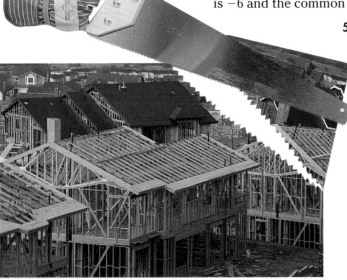

57. Construction Ryduff Builders is building homes in a new residential development. The building code for the development states that lots must have a minimum of 1250 square feet and no dimension can be less than 20 feet.
  a. Determine whether a plot plan would be approved if it measured 32 feet by 38 feet. Explain your reasoning.
  b. Determine whether a plot plan would be approved if it measured 19 feet by 70 feet. Explain your reasoning.
  c. If one dimension of a plot plan was 42 feet, find another dimension that would satisfy the building code conditions.

58. Civics The length of a flag is called the *fly,* and the width is called the *hoist.* The blue rectangle in the United States flag is called the *union.* The length of the union is $\frac{2}{5}$ of the fly, and the width is $\frac{7}{13}$ of the hoist. If the fly of a United States flag is 3 feet, how long is the union?

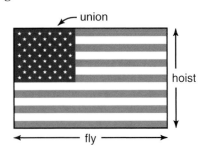

union

hoist

fly

Mixed Review

59. Find $5.7 + (-7.9)$. (Lesson 2–5)

60. Replace _?_ with $<, >,$ or $=$ to make 12 _?_ $15 - 7$ true. (Lesson 2–4)

61. Evaluate $a - 12$ if $a = -8$. (Lesson 2–3)

62. Statistics State the scale you would use to make a line plot for the following data. Then draw the line plot. (Lesson 2–2)

$$4, 9, 2, 2, 15, 7, 6, 6, 9, 12, 1, 3, 2, 11, 10, 2, 6, 4, 12, 13$$

63. Find $6 + (-13)$. If necessary, use a number line. (Lesson 2–1)

64. Write an algebraic expression for *five times the sum of x and y decreased by z*. (Lesson 1–8)

65. Find the solution set for $y + 3 > 8$ if the replacement set is $\{3, 4, 5, 6, 7\}$. (Lesson 1–5)

66. Statistics Suppose the number 27,878 is rounded to 27,900 and plotted using stem 27 and leaf 9. Write the stem and leaf for each number below if the numbers are part of the same set of data. (Lesson 1–4)
  a. 13,245　　　　**b.** 35,684　　　　**c.** 153,436

67. Evaluate $6(4^3 + 2^2)$. (Lesson 1–3)

68. Geometry Write an expression that represents the total number of small cubes in the large cube shown at the right. Then evaluate the expression. (Lesson 1–1)

WORKING ON THE

In·ves·ti·ga·tion

Refer to the Investigation on pages 68–69.

the Greenhouse Effect

Without a heat-trapping blanket of naturally occurring CO_2, it is believed that Earth would have an average surface temperature of $-17°C$, instead of its current average, $15°C$. There is evidence that Mars has little CO_2 in its atmosphere, and its temperature never exceeds $-31°C$. At the other extreme, Venus, with lots of CO_2, has an average temperature of $454°C$.

1 What would the difference in temperature be between Earth with CO_2 and Earth without CO_2? Do you think life could exist on Earth without CO_2? Explain.

2 What is the difference in temperature between Mars and Earth without CO_2? Explain what other factors contribute to the difference in temperature.

3 What is the difference in temperature between Venus and Mars? Explain what other factors contribute to the difference in temperature.

4 Make a chart that shows the three planets and their temperatures with CO_2 in their atmospheres and with little or no CO_2. Assume that the differences in the temperatures for the other planets would be the same as the difference in Earth's temperatures with CO_2 and without CO_2. You will use this information later on as you work on the Investigation.

Add the results of your work to your Investigation Folder.

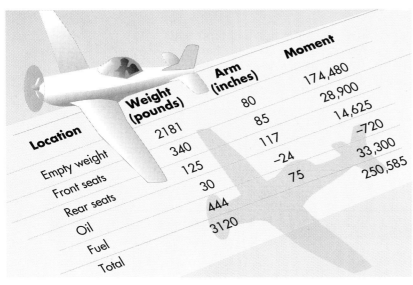

2-7

Dividing Rational Numbers

What YOU'LL LEARN

• To divide rational numbers.

Why IT'S IMPORTANT

You can divide rational numbers to help you solve problems involving drafting, nursing, and economics.

APPLICATION
Aviation

Vicki Van Meter had a lot to write in her report about her summer vacation of 1993. At 12 years old, inspired by Amelia Earhart, she became the youngest female to pilot a plane across the Atlantic Ocean. Vicki did all of the flying, navigation, and communication for her 3200-kilometer flight aboard a single-engine Cessna 210. She also had to be concerned about how supplies and fuel were loaded onto the plane.

Location	Weight (pounds)	Arm (inches)	Moment
Empty weight	2181	80	174,480
Front seats	340	85	28,900
Rear seats	125	117	14,625
Oil	30	−24	−720
Fuel	444	75	33,300
Total	3120		250,585

Before any small aircraft can take off, the pilot must be sure that the aircraft is loaded so that the center of gravity is within certain safe limits. If it is, the pilot is ready for takeoff. If not, the weight must be rebalanced.

You can use the table at the left to find the center of gravity for a particular aircraft. The safe limit for this aircraft is 82.1. For each location, the weight and the arm are multiplied to find the moment.

Now add the moments, divide by the total weight, and round to the nearest tenth to find the center of gravity.

Total Moments		*Total Weight*		*Center of Gravity*
250,585	÷	3120	≈	80.3

Since 80.3 is less than 82.1, the aircraft is safe and ready for takeoff.

You already know that the quotient of two positive numbers is positive. But how do you determine the sign of the quotient when negative numbers are involved? Since multiplication and division are inverse operations, the rule for finding the sign of the quotient of two numbers is similar to the rule for finding the sign of the product. Study these patterns.

$$-5 \cdot 8 = -40 \qquad -40 \div 8 = -5 \qquad\qquad -10 \cdot \left(-\frac{1}{2}\right) = 5 \qquad 5 \div \left(-\frac{1}{2}\right) = -10$$

These examples suggest the following rules.

Dividing Two Rational Numbers	The quotient of two numbers having the *same sign* is positive. The quotient of two numbers having *different signs* is negative.

Example **Find each quotient.**

a. $-75 \div (-15)$

This division problem may also be written as $\frac{-75}{-15}$.

$\frac{-75}{-15} = 5$ *The fraction bar indicates division.*

Since the signs are the same, the quotient is positive.

b. $\frac{72}{-8}$

$\frac{72}{-8} = -9$ *Since the signs are different, the quotient is negative.*

Recall that you can change any division expression to an equivalent multiplication expression. To divide by any nonzero number, multiply by the reciprocal of that number.

Example **2** **Find each quotient.**

a. $\frac{1}{2} \div 5$

$\frac{1}{2} \div 5 = \frac{1}{2} \cdot \frac{1}{5}$ *Multiply by $\frac{1}{5}$, the reciprocal of 5.*

$= \frac{1}{10}$ *The signs are the same, so the product is positive.*

b. $-\frac{6}{7} \div 3$

$-\frac{6}{7} \div 3 = -\frac{6}{7} \cdot \frac{1}{3}$ *Multiply by $\frac{1}{3}$, the reciprocal of 3.*

$= -\frac{6}{21}$ or $-\frac{2}{7}$ *The signs are different, so the product is negative.*

Example **3** **Three of the measurements nurses commonly use are cubic centimeters (cc), drops, and grains. Use this information to solve the following.**

Nursing

a. A doctor orders $\frac{1}{400}$ of a grain of medicine to be given to a patient. The nurse has a vial labeled $\frac{1}{200}$ grain per cc. How many cc of the medicine should the nurse give the patient?

b. The doctor also prescribes a 1000-cc intravenous (IV) pouch of fluid to be given to the patient over an 8-hour period. If there are 15 drops in 1 cc, for how many drops per minute should the nurse set the IV?

a. How many cc are in $\frac{1}{400}$ of a grain? Divide $\frac{1}{400}$ grain by $\frac{1}{200}$ grain/cc.

$\frac{1}{400}$ grain $\div \frac{1}{200}$ grain/cc $= \frac{1 \text{ grain}}{400} \cdot \frac{200 \text{ cc}}{1 \text{ grain}}$

$= \frac{1 \text{ grain}}{400} \cdot \frac{200 \text{ cc}}{1 \text{ grain}}$

$= \frac{1}{2}$ cc

The nurse should give the patient $\frac{1}{2}$ cc of the medicine.

(continued on the next page)

b. $\dfrac{1000 \text{ cc}}{8 \text{ h}} = \left(\dfrac{\overset{125}{\cancel{1000 \text{ cc}}}}{\underset{1}{\cancel{8 \text{ h}}}}\right)\left(\dfrac{\overset{1}{\cancel{15 \text{ drops}}}}{1 \text{ cc}}\right)\left(\dfrac{1 \text{ h}}{\underset{4}{\cancel{60 \text{ min}}}}\right)$ *There are 15 drops in 1 cc and*

60 minutes in 1 hour.

$= \dfrac{125 \text{ drops}}{4 \text{ min}}$

$= 31\dfrac{1}{4} \text{ drops/min}$

The nurse should set the IV for $31\dfrac{1}{4}$ drops per minute.

If a fraction has one or more fractions in the numerator or denominator, it is called a **complex fraction.** To simplify a complex fraction, rewrite it as a division sentence.

Example **Write each fraction in simplest form.**

a. $\dfrac{\frac{2}{3}}{8}$

Rewrite the fraction as $\dfrac{2}{3} \div 8$, since fractions indicate division.

$\dfrac{2}{3} \div 8 = \dfrac{2}{3} \cdot \dfrac{1}{8}$ *Multiply by $\frac{1}{8}$, the reciprocal of 8.*

$= \dfrac{2}{24}$ or $\dfrac{1}{12}$

b. $\dfrac{-5}{\frac{3}{7}}$

Rewrite the fraction as $-5 \div \dfrac{3}{7}$.

$-5 \div \dfrac{3}{7} = -5 \times \dfrac{7}{3}$ *Multiply by $\frac{7}{3}$, the reciprocal of $\frac{3}{7}$.*

$= -\dfrac{35}{3}$ or $-11\dfrac{2}{3}$ *The signs are different, so the product is negative.*

You can use the distributive property to simplify fractional expressions.

Example **Simplify $\dfrac{-3a + 16}{4}$.**

Method 1

$\dfrac{-3a + 16}{4} = (-3a + 16) \div 4$

$= (-3a + 16)\left(\dfrac{1}{4}\right)$ *To divide by 4, multiply by $\frac{1}{4}$.*

$= -3a\left(\dfrac{1}{4}\right) + 16\left(\dfrac{1}{4}\right)$ *Distributive property*

$= -\dfrac{3}{4}a + 4$

Method 2

$\dfrac{-3a + 16}{4} = -\dfrac{3a}{4} + \dfrac{16}{4}$

$= -\dfrac{3}{4}a + 4$

Communicating Mathematics

Study the lesson. Then complete the following.

1. **Compare** multiplying rational numbers to dividing rational numbers. How are the two operations similar?

2. **Complete** the sentence: Dividing by any number, except zero, is the same as ___?___ .

3. **Find** a value for x if $\frac{1}{x} > x$.

4. **You Decide** Simone says that $-\frac{4}{5}$ is equal to $\frac{-4}{-5}$. Miguel says that $-\frac{4}{5}$ is equal to $\frac{-4}{5}$ or $\frac{4}{-5}$. Which one is correct, and why?

Guided Practice

Simplify.

5. $\frac{32}{-8}$

6. $\frac{-77}{11}$

7. $-\frac{3}{4} \div 8$

8. $\frac{2}{3} \div 9$

9. $\frac{\frac{-5}{6}}{8}$

10. $\frac{54s}{6}$

11. $\frac{-300x}{50}$

12. $\frac{6b + 12}{6}$

13. **Money** On October 10, 1994, *USA Today* reported that the United States government spends about $168 million per hour. Stanley Newberg, who came to the United States from Austria in 1906, died in 1994 at age 81 and left the government $5.6 million. How long did it take the government to spend Mr. Newberg's money?

Practice

Simplify.

14. $\frac{-36}{4}$

15. $\frac{-96}{-16}$

16. $-9 \div \left(-\frac{10}{17}\right)$

17. $-\frac{2}{3} \div 12$

18. $-64 \div (-8)$

19. $-\frac{3}{4} \div 12$

20. $-18 \div 9$

21. $78 \div (-13)$

22. $-108 \div (-9)$

23. $-\frac{2}{3} \div 8$

24. $\frac{-1}{3} \div (-4)$

25. $-9 \div \left(-\frac{10}{27}\right)$

26. $\frac{\frac{5}{6}}{-10}$

27. $-\frac{\frac{7}{3}}{5}$

28. $\frac{\frac{-5}{2}}{7}$

29. $\frac{-650m}{10}$

30. $\frac{81c}{-9}$

31. $\frac{8r + 24}{8}$

32. $\frac{6a + 24}{6}$

33. $\frac{40a + 50b}{-2}$

34. $\frac{-5x + (-10y)}{-5}$

35. $\frac{42c - 18d}{-3}$

36. $\frac{-8f + (-16g)}{8}$

37. $\frac{-4a + (-16b)}{4}$

Evaluate if $a = 5$, $b = -6$, and $c = -1.5$.

38. $\frac{b}{c}$

39. $b \div a$

40. $(a + b) \div c$

41. $(a + b + c) \div 3$

42. $\frac{c}{a}$

43. $\frac{ab}{ac}$

44. The $\boxed{1/x}$ key on a scientific calculator is called the *reciprocal key*. When this key is pressed, the calculator replaces the number on the display with its reciprocal.

a. Enter 0 and then press the reciprocal key. What happened? Explain.

b. Enter a number and then press the reciprocal key twice. What happened? Predict what will happen if you press the key *n* times.

c. Enter 6.435 $\boxed{1/x}$ $\boxed{\times}$ 6.435 $\boxed{=}$. What is the result? Why?

Applications and Problem Solving

45. Drafting   Sofia Fernandez is making a dresser for her little sister's nursery. The dresser will be 30 inches high and will have a 4-inch-thick base and a $1\frac{1}{2}$-inch-thick top. Four equal-sized drawers are to fit in the remaining space, with $\frac{3}{4}$ inch between each drawer.

a. Sketch the dresser, labeling each part.

b. What is the height of each drawer?

46. Economics   Kim Lee's shoe store has a buy one, get one half-price sale. You must pay full price for the more expensive pair of shoes in order to receive half off the second pair. If you buy a single pair of shoes, you get $\frac{1}{5}$ off. Together, Sharon and LaShondra find five pair of shoes that cost $45.99, $23.88, $36.99, $19.99, and $14.99. How should they pay for the shoes in order to get the best buy?

47. Science   *Precision* is the degree of exactness to which a measurement can be reproduced. It is determined by subtracting the least measurement from the greatest measurement, and dividing by 2. Suppose you conducted an experiment to determine the length of a piece of lumber. After measuring several times, you have recorded measurements ranging between 17.239 cm and 17.561 cm, inclusive.

a. Use the plus-or-minus symbol ($\pm$) to describe your measurements.

b. What was the precision of your measurement?

48. Find $\left(-\frac{1}{5}\right)\left(\frac{3}{2}\right)(-2)$. (Lesson 2–6)

49. Evaluate $\frac{9}{4} + \frac{x}{6}$ if $x = -7$. (Lesson 2–5)

50. Name a fraction between $\frac{2}{3}$ and $\frac{7}{8}$ whose denominator is 24. (Lesson 2–4)

51. Let $A = \begin{bmatrix} 1 & 4 \\ 5 & 7 \end{bmatrix}$ and $B = \begin{bmatrix} -3 & 0 \\ -2 & 5 \end{bmatrix}$. (Lesson 2–3)

    a. Find $A + B$.   **b.** Find $A - B$.

52. Statistics The table below shows the mean scores for the mathematics portion of the Scholastic Assessment Test (SAT) by state for 1994. (Lesson 2–2)

SAT Mean Mathematics Scores, 1994									
AL	529	HI	480	MA	475	NM	528	SD	548
AK	477	ID	508	MI	537	NY	472	TN	535
AZ	496	IL	546	MN	562	NC	455	TX	474
AR	518	IN	466	MS	528	ND	559	UT	558
CA	482	IA	574	MO	532	OH	510	VT	472
CO	513	KS	550	MT	523	OK	537	VA	469
CT	472	KY	523	NE	543	OR	491	WA	488
DE	464	LA	530	NV	484	PA	462	WV	482
FL	466	ME	463	NH	486	RI	462	WI	557
GA	446	MD	479	NJ	475	SC	443	WY	521

Source: College Entrance Examination Board

    a. What is the range of scores for the fifty states?
    b. Make a line plot of the data.
    c. Do the scores cluster around a certain number? If so, which ones?

53. Simplify $8b + 12(b + 2)$. (Lesson 1–7)

54. Name the multiplicative inverse of $\frac{4}{5}$. (Lesson 1–6)

55. Statistics The list below shows the prices of several 1995 models of sport utility vehicles, rounded to the nearest hundred dollars. (Lesson 1–4)

$22,000	$19,400	$29,000	$13,000	$22,200	$17,300
$25,400	$25,100	$33,000	$20,000	$30,700	$15,400
$27,900	$34,600	$30,400	$24,500	$52,500	$17,200

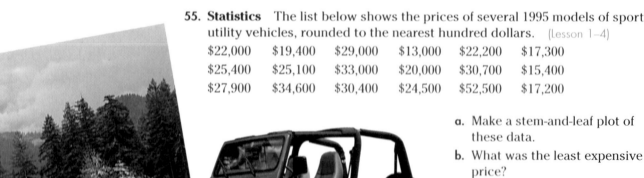

    a. Make a stem-and-leaf plot of these data.
    b. What was the least expensive price?
    c. What was the most expensive price?
    d. How many prices were over $25,000?

2–8A Estimating Square Roots

Materials: base-ten tiles

You can use base-ten tiles to model square roots. A **square root** is one of two identical factors of a number. For example, a square root of 144 is 12 since $12^2 = 144$.

Activity 1   Use base-ten tiles to find the square root of 121.

Step 1   Model 121 with base-ten tiles.

Step 2   Arrange the tiles into a square. The square root of 121 is 11 because $11^2 = 121$.

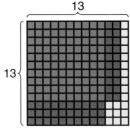

Activity 2   Use base-ten tiles to estimate the square root of 151.

Step 1   Model 151 with base-ten tiles.

Step 2   Arrange the tiles into a square. The largest square possible has 144 tiles, with 7 tiles left over.

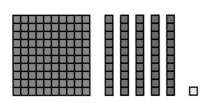

Trade a 10-tile for 10 1-tiles when necessary.

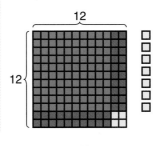

Step 3   Add tiles until you have the next larger square. You need to add 18 tiles. Since 151 is between 144 and 169, the square root of 151 is between 12 and 13.

Model   **Use base-ten tiles to estimate the square root of each number.**

    1. 20   **2.** 450   **3.** 180   **4.** 200   **5.** 2

Write   **6.** If you list all of the factors of a number in numerical order, the square root of the number either is the middle number or lies between the two middle numbers. How can you use tiles to show this?

2-8

Square Roots and Real Numbers

What YOU'LL LEARN

- To find square roots,
- to classify numbers, and
- to graph solutions of inequalities on number lines.

Why IT'S IMPORTANT

You need to know the values of irrational numbers to solve problems involving traffic safety and flying.

CONNECTION

Biology

Swedish botanist Carolus Linnaeus (1707–1778) developed the system we use today to classify every kind of living thing according to common characteristics. For example, an African elephant is from the Mammalia Class and also from the Animalia Kingdom.

In mathematics, we classify numbers that have common characteristics. So far in this text, we have classified numbers as natural numbers, whole numbers, integers, and rational numbers.

Finding a square root of 81 is the same as finding a number whose square is 81.

The square roots of perfect squares are classified as rational numbers. A **square root** is one of two equal factors of a number. For example, one square root of 81 is 9 since 9 · 9 or 9^2 is 81. A rational number, like 81, whose square root is a rational number, is called a **perfect square**.

Kingdom: Animalia

Phylum: Chordata

Class: Mammalia

Order: Proboscidea

Family: Elephantidae

Genus: Loxodonta

Species: Loxodonta africana

African elephant

It is also true that $-9 \cdot (-9) = 81$. Therefore, -9 is another square root of 81.

$$9^2 = 9 \cdot 9$$
$$= 81$$

9^2 is read "nine squared" and means that 9 is used as a factor two times.

$$(-9)^2 = (-9)(-9)$$
$$= 81$$

-9 is used as a factor two times.

| Definition of Square Root | If $x^2 = y$, then x is a square root of y. |

The symbol $\sqrt{}$, called a **radical sign**, is used to indicate a nonnegative or **principal square root** of the expression under the radical sign.

$\sqrt{81} = 9$ $\sqrt{81}$ indicates the *principal* square root of 81.

$-\sqrt{81} = -9$ -81 indicates the *negative* square root of 81.

$\pm\sqrt{81} = \pm9$ $\pm\sqrt{81}$ indicates *both* square roots of 81.

$\pm\sqrt{81}$ is read "plus or minus the square root of 81."

Example 1 Find each square root.

a. $\sqrt{25}$

The symbol $\sqrt{25}$ represents the principal square root of 25.

Since $5^2 = 25$, you know that $\sqrt{25} = 5$.

b. $-\sqrt{144}$

The symbol $-\sqrt{144}$ represents the negative square root of 144.

Since $12^2 = 144$, you know that $-\sqrt{144} = -12$.

c. $\pm\sqrt{0.16}$

The symbol $\pm\sqrt{0.16}$ represents both square roots of 0.16.

Since $(0.4)^2 = 0.16$, you know that $\pm\sqrt{0.16} = \pm0.4$.

Some square roots can be found mentally.

Most scientific calculators have a *square root* key labeled $\boxed{\sqrt{}}$ or $\boxed{\sqrt{x}}$. When you press this key, the number in the display is replaced by its principal square root.

Example 2 Use a scientific calculator to evaluate each expression if $x = 2401$, $a = 147$, and $b = 78$.

a. $\sqrt{x}$

$\sqrt{x} = \sqrt{2401}$ *Replace x with 2401.*

Enter: 2401 $\boxed{\text{2nd}}$ $\boxed{\sqrt{x}}$ *49*

Therefore, $\sqrt{2401} = 49$.

b. $\pm\sqrt{a + b}$

$\pm\sqrt{a + b} = \pm\sqrt{147 + 78}$ *Replace a with 147 and b with 78.*

Enter: $\boxed{(}$ 147 $\boxed{+}$ 78 $\boxed{)}$ $\boxed{\text{2nd}}$ $\boxed{\sqrt{x}}$ *15*

Therefore, $\pm\sqrt{a + b}$ is ± 15.

Numbers such as $\sqrt{5}$ and $\sqrt{13}$ are the square roots of numbers that are *not* perfect squares. Notice what happens when you find these square roots with your calculator.

Enter: 5 $\boxed{\text{2nd}}$ $\boxed{\sqrt{x}}$ *2.236067978...*

Enter: 13 $\boxed{\text{2nd}}$ $\boxed{\sqrt{x}}$ *3.605551275...*

These numbers continue indefinitely without any pattern of repeating digits. These numbers are not rational numbers since they are not repeating or terminating decimals. Numbers like $\sqrt{5}$ and $\sqrt{13}$ are called **irrational numbers.**

| **Definition of an Irrational Number** | An irrational number is a number that *cannot* be expressed in the form $\frac{a}{b}$, where *a* and *b* are integers and $b \neq 0$. |

The set of rational numbers and the set of irrational numbers together form the set of **real numbers**. The Venn diagram at the right shows the relationships among natural numbers, whole numbers, integers, rational numbers, irrational numbers, and real numbers.

Real Numbers

Example **Name the set or sets of numbers to which each real number belongs.**

a. 0.8333333...   This repeating decimal is a rational number since it is equivalent to $\frac{5}{6}$. *This number can also be expressed as $0.8\overline{3}$.*

b. $-\sqrt{16}$ Since $-\sqrt{16} = -4$, this number is an integer and a rational number.

c. $\frac{14}{2}$ Since $\frac{14}{2} = 7$, this number is an integer, a natural number, a whole number, and a rational number.

d. $\sqrt{120}$ Since $\sqrt{120} = 10.95445115...$, which is not a repeating or terminating decimal, this number is irrational.

The solutions to many real-world problems are irrational numbers.

Example **The area of a square is 325 square inches. Find its perimeter to the nearest hundredth.**

Geometry

First find the length of each side. Since the area of a square is the length of the side squared, find the square root of 325.

325 square inches

Enter: 325 [2nd] [√x] *18.02775638*

The length of each side is about 18.02775638 inches. Use the formula $P = 4s$ to find the perimeter.

$P = 4s$

$= 4 \cdot 18.02775638$ *Replace s with 18.02775638.*

Enter: 18.02775638 [×] 4 [=] *72.11102655l*

The perimeter is about 72.11 inches.

You have graphed rational numbers on number lines. Yet, if you graphed all of the rational numbers, the number line would still not be complete. The irrational numbers complete the number line. The graph of all real numbers is the entire number line. This is illustrated by the **completeness property**.

Completeness Property for Points on the Number Line   **Each real number corresponds to exactly one point on the number line. Each point on the number line corresponds to exactly one real number.**

Recall that equations like $x - 5 = 11$ are open sentences. Inequalities like $x < 6$ are also considered to be open sentences. To solve $x < 6$, determine what replacements for x make $x < 6$ true. All numbers less than 6 make the inequality true. This can be shown by the solution set {real numbers less than 6}. Not only does this include integers like 3, 0, and -4, but it also includes all rational numbers less than 6 such as $\frac{1}{2}$, $-5\frac{3}{8}$, and -3 and all irrational numbers less than 6 such as $\sqrt{5}$, $\sqrt{3}$, and π.

Example 5 Graph each solution set.

a. $y \geq -7$

The heavy arrow indicates that all numbers to the right of -7 are included. The *dot* indicates that the point corresponding to -7 is included in the graph of the solution set.

b. $p \neq \frac{3}{4}$

The heavy arrows indicate that all numbers to the left and to the right of $\frac{3}{4}$ are included in the graph of the solution set. The *circle* indicates that the point corresponding to $\frac{3}{4}$ is not included in the graph.

CHECK FOR UNDERSTANDING

Communicating Mathematics

Study the lesson. Then complete the following.

1. You have studied the following sets of numbers: integers, irrational, natural, rational, real, and whole numbers. Draw a number line and label at least one number from each set of numbers. Indicate which number is from each set.

2. Study the comic at the left. Explain why it is humorous.

3. **Determine** whether 36 is a perfect square. If it is, to what set of numbers does $\sqrt{36}$ belong?

4. **Explain** why 3 and -3 are both square roots of 9.

5. **Write** an inequality for the graph below.

CLOSE TO HOME   JOHN McPHERSON

© 1994 John McPherson/Dist. by Universal Press Syndicate

COUNT OFF BY THE SQUARE ROOT OF 7!

UHH....

Deep down inside, Coach Knott had always wanted to be a math teacher.

Math Journal

6. **Write** a paragraph explaining the difference between rational and irrational numbers to a classmate.

Find each square root. Use a calculator if necessary. Round to the nearest hundredth if the result is not a whole number.

7. $\sqrt{64}$ **8.** $-\sqrt{36}$ **9.** $\sqrt{122}$ **10.** $\pm\sqrt{0.08}$

Evaluate each expression. Use a calculator if necessary. Round to the nearest hundredth if the result is not a whole number.

11. $\sqrt{x}$, if $x = 256$ **12.** $\sqrt{y}$, if $y = 151$

Name the set or sets of numbers to which each real number belongs. Use N for natural numbers, W for whole numbers, Z for integers, Q for rational numbers, and I for irrational numbers.

13. $-\dfrac{3}{4}$ **14.** $\dfrac{8}{4}$ **15.** $0.6666...$ **16.** $\sqrt{13}$

Graph the solution set of each inequality on a number line.

17. $p < 7$ **18.** $r \geq -3$ **19.** $x \neq 2$

20. Discrete Mathematics In the geometric sequence 5, 15, _?_, 135, 405, the missing number is called the *geometric mean* of 15 and 135. It can be found by evaluating $\sqrt{ab}$, where a and b are the numbers on either side of the geometric mean. Find the missing number.

EXERCISES

Find each square root. Use a calculator if necessary. Round to the nearest hundredth if the result is not a whole number or simple fraction.

21. $\sqrt{169}$ **22.** $\sqrt{0.0049}$ **23.** $\sqrt{\dfrac{4}{9}}$

24. $-\sqrt{289}$ **25.** $\sqrt{420}$ **26.** $\sqrt{\dfrac{25}{64}}$

27. $\sqrt{225}$ **28.** $\pm\sqrt{1158}$ **29.** $-\sqrt{625}$

30. $\sqrt{1.96}$ **31.** $\sqrt{\dfrac{9}{25}}$ **32.** $-\sqrt{5.80}$

Evaluate each expression. Use a calculator if necessary. Round to the nearest hundredth if the result is not a whole number.

33. $\sqrt{x}$, if $x = 87$ **34.** $\pm\sqrt{t}$, if $t = 529$

35. $-\sqrt{m}$, if $m = 2209$ **36.** $\sqrt{c + d}$, if $c = 23$ and $d = 56$

37. $-\sqrt{np}$, if $n = 16$ and $p = 25$ **38.** $\pm\sqrt{\dfrac{a}{b}}$, if $a = 64$ and $b = 4$

Name the set or sets of numbers to which each real number belongs. Use N for natural numbers, W for whole numbers, Z for integers, Q for rational numbers, and I for irrational numbers.

39. $-\sqrt{49}$ **40.** 0 **41.** 0.4583

42. $0.\overline{3}$ **43.** $-\dfrac{1}{2}$ **44.** $\sqrt{49}$

45. 0.6666 **46.** $\dfrac{10}{5}$ **47.** $\sqrt{37}$

48. 3.14 **49.** $\dfrac{3}{5}$ **50.** 5

Graph the solution set of each inequality on a number line.

51. $y > -2$ **52.** $x < 1$ **53.** $p \geq -4$

54. $n \neq 6$ **55.** $c > -12$ **56.** $r \leq 4.5$

57. $b \geq -5.2$ **58.** $y \neq \frac{3}{4}$ **59.** $s \leq 5\frac{1}{2}$

60. Geometry   The volume of a rectangular solid is 100 cm³. Its height is the product of its length and width. The base of the solid is a square.

   a. Draw the solid from two different perspectives.

   b. Find the dimensions of the solid. Use a calculator as needed and round decimal answers to the nearest hundredth.

Critical Thinking

61. Determine whether $\sqrt{733}$ is an irrational number. If it is, name two consecutive integers between which its graph lies on the number line.

62. Find all numbers of the form $\sqrt{n}$ such that n is a natural number and the graph of $\sqrt{n}$ lies between each pair of numbers on the number line.

   a. 3 and 4   **b.** 5.25 and 5.5

Applications and Problem Solving

63. Military   The formula to determine the distance d in miles that an object can be seen on a clear day on the surface of the ocean is $d = 1.4\sqrt{h}$, where h is the height in feet the viewer's eyes are above the surface of the water. About how many miles can the pilot of a U.S. Coast Guard plane see if he is flying at 1275 feet?

   a. Write an equation to represent the distance. Then approximate the distance mentally.

   b. Use a calculator to find the exact distance.

64. Traffic Safety   When investigating a traffic accident, police officers often need to estimate how fast a car was traveling by measuring the length of its skid marks. The formula $s = \sqrt{24d}$ can be used to estimate the speed on a dry, concrete road. In the formula, s is the speed in miles per hour, and d is the distance in feet the car skidded after its brakes were applied. What was the approximate speed of a car that left skid marks 40 feet long?

   a. Write an equation to represent the speed. Then approximate the speed mentally.

   b. Use a calculator to find the exact speed.

65. Math History   One of the great mathematical challenges in early mathematics was to find an exact value for π. Today, 3.14 and $\frac{22}{7}$ are two frequently used approximations for π. In the third century, Chinese mathematician Liu Hui used the number $3\frac{7}{50}$ as an approximation of π. The Italian mathematician Leonardo Fibonacci used $\frac{864}{275}$ as an approximation of π. Which mathematician's approximation was closer to the generally-accepted approximation of π, 3.141592654...?

66. Business   The Spanish Club at Alexander Manor High School is selling chocolate bars to raise money for a trip to Mexico. The bars are sold only by the case, with 24 bars per case. Each bar costs $1.00. The club's goal is to raise at least $2450 in gross sales.

   a. How many cases of chocolate bars must the club sell in order to reach or exceed their goal?

   b. Graph the solution set.

67. Evaluate $(-3a)(4)\left(\frac{2}{3}\right)\left(\frac{1}{6}a\right)$ if $a = 3$. (Lesson 2–6)

68. Find $-\frac{5}{12} + \frac{3}{8}$. (Lesson 2–5)

69. Which is the better buy, a 1-pound package of lunch meat for $1.95 or a 12-ounce package for $1.80? (Lesson 2–4)

70. Find $-5 - (-6)$. (Lesson 2-3)

71. Statistics State the scale you would use to make a line plot for the following data: 504, 509, 520, 500, 513, 517, 517, 508, 502, 518, 509, 520, 504, 509, 517, 511, 520. Then draw the line plot. (Lesson 2–2)

72. Graph {0, 2, 6} on a number line. (Lesson 2–1)

73. Suppose a basketball was dropped from a tall building onto a sidewalk. Identify the graph below that best represents this situation. (Lesson 1–9)

a.   b.   c.

74. Name the property illustrated by $a + (2b + 5c) = (a + 2b) + 5c$.
(Lesson 1–8)

Laura Davies studies her putt.

75. Golf The stem-and-leaf plot below shows the earnings of the top 25 U.S. women professional golfers for 1994 according to the *Sports Almanac*. (Lesson 1–4)

Stem	Leaf
6	6 9
5	0
4	0 1 2 3 7
3	2 3 4 5 9
2	0 0 1 1 3 4 4 5 6 7 7 8 $4|3 = \$430,000$

a. How have these numbers been rounded?

b. How many of these golfers made between $300,000 and $490,000?

c. Laura Davies had the highest earnings of all women golfers in 1994. How much money did she earn?

76. Evaluate $12(19 - 15) - 3 \cdot 8$. (Lesson 1–3)

77. Find the next two terms in the sequence 2, 8, 14, 20,... (Lesson 1–2)

78. Write an algebraic expression for p to the sixth power.
(Lesson 1–1)

Problem Solving
Write Equations and Formulas

CONNECTION
Biology

Bugs, bugs, everywhere! That could be the slogan for James Fujita, an avid bug collector. Fujita has been fascinated with bugs since he was bitten by a cricket at the age of three. His collection includes hundreds of bugs—beetles, spiders, millipedes, scorpions, just to name a few—both living and dead.

Fujita has studied bugs as far away as Japan and Costa Rica, as well as in his own backyard in Oxnard, California. He has given bug presentations and workshops, donated some of his bugs to museums and zoos, and even appeared on a television talk show with some of his larger bugs. Fujita is a big believer in reading and exploring all you can about bug collecting—if you get the itch!

You can explore problems by asking and answering questions. In this text, we will use a four-step plan to solve problems. All of the steps are listed below.

Problem-Solving Plan >

1. **Explore the problem.**
2. **Plan the solution.**
3. **Solve the problem.**
4. **Examine the solution.**

The first step in solving a problem is to read and explore it until you completely understand the relationships in the given information.

Step 1: Explore the Problem

To solve a verbal problem, first read the problem carefully and explore what the problem is about.

• Identify what information is given.

• Identify what you are asked to find.

Questions	Answers
a. How old was James when he became interested in bugs?	**a.** 3 years old
b. In what countries other than the U.S. has James studied bugs?	**b.** Japan and Costa Rica
c. If James is n years old now, for how many years has he been a collector of bugs?	**c.** $n - 3$
d. If James has b bugs and finds 14 new bugs, how many bugs are in his collection now?	**d.** $b + 14$

Other strategies are:
- *look for a pattern*
- *solve a simpler problem*
- *act it out*
- *guess and check*
- *draw a diagram*
- *make a table or chart*
- *work backward*

Step 2: Plan the Solution

One strategy you can use to solve a problem is to write an equation. Choose a variable to represent one of the unspecified numbers in the problem. This is called **defining the variable.** Then use the variable to write expressions for the other unspecified numbers in the problem.

Step 3: Solve the Problem

Use the strategy you chose in Step 2 to solve the problem.

Step 4: Examine the Solution

Check your answer within the context of the original problem. Does your answer make sense? Does it fit the information in the problem?

Example ❶

Computers

Gregory Arakelian of Herndon, Virginia, set a speed record for typing the most words per minute, with no errors, on a personal computer in the KeyTronic World Invitational Type-Off on September 24, 1991. Suppose his closest competitor typed 10 fewer words per minute than Arakelian. If his closest competitor typed 148 words per minute, how many words did Arakelian type per minute?

Explore
- How many words can Arakelian type in relation to his closest competitor?   10 more words

- How many words per minute did his competitor type? 148 words

Plan
Write an equation to represent the situation. Let w represent the words per minute that Arakelian typed.

$$\underbrace{w}_{\substack{words \\ Arakelian\ typed}} \quad -10 = \quad \underbrace{148}_{\substack{words \\ competitor\ typed}}$$

Solve
$w - 10 = 148$ *Solve mentally by asking,*
$w = 158$ *"What number minus 10 is 148?"*

Arakelian typed 158 words per minute.

Examine
The problem asks how many words Arakelian typed per minute. His competitor typed 10 fewer words per minute. Since $158 - 10 = 148$, the answer makes sense.

Gregory Arakelian

In Lesson 2–4, you studied meanings for the symbols ...., *and* .....

When solving problems, many sentences can be written as equations. Use variables to represent the unspecified numbers or measures referred to in the sentence or problem. Then write the verbal expressions as algebraic expressions. Some verbal expressions that suggest the *equals sign* are listed below.

- is
- equals
- is equal to
- is the same as
- is as much as
- is identical to

Example **2** **Translate each sentence into an algebraic sentence.**

a. *Six times a number x is equal to 7 times the sum of z and y.*

$$\underbrace{Six\ times}_{6\ \times}\ \underbrace{a\ number\ x}_{x}\ \underbrace{is\ equal\ to}_{=}\ \underbrace{7\ times}_{7\ \times}\ \underbrace{the\ sum\ of\ z\ and\ y.}_{(z+y)}$$

The algebraic sentence is $6x = 7(z + y)$.

b. *A number is less than or equal to 5.*

$$\underbrace{A\ number}_{n}\ \underbrace{is\ less\ than\ or\ equal\ to}_{\leq}\ \underbrace{5.}_{5}$$

The algebraic sentence is $n \leq 5$.

Another strategy you can use to solve a problem is to write a formula. A **formula** is an equation that states a rule for the relationship between certain quantities. Sometimes you can develop a formula by making a model.

MODELING MATHEMATICS

Surface Area

Materials: ⬜ rectangular box ✂ scissors

You need to find the surface area of a rectangular box.

- Cut up a box like the one shown at the right.
- Find the area of each face.

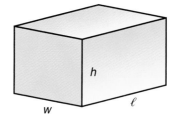

- Write the formula by adding the areas.

$S = wh + wh + \ell h + \ell h + \ell w + \ell w$

$\quad = 2wh + 2\ell h + 2\ell w$

$\quad = 2(wh + \ell h + \ell w)$

Your Turn

a. Find a box that is the shape of a cube. Take the box apart. Draw and label each side. Then write a formula for finding the surface area of the cube.

b. How are the box (rectangular prism) and cube the same? How are they different?

c. Find a formula for the surface area of the triangular prism below. Use S, h, w, and ℓ in your formula.

Example **Translate the sentence into a formula.**

The area of a circle equals the product of π and the square of the radius r.

The area of a circle equals π times the square of the radius r.

$$A \qquad = \quad \pi \quad \times \qquad\qquad r^2$$

The formula is $A = \pi r^2$.

You can also translate equations into verbal sentences or make up your own verbal problem if you are given an equation or two.

Example **Translate $x^2 + 6 = 39$ into a verbal sentence.**

$$x^2 \qquad\qquad + \quad 6 \quad = \quad 39$$

The sum of the square of a number and six equals 39.

Example **5** **Write a problem based on the given information.**

ℓ = Lawana's height in inches
$\ell + 5$ = Tatewin's height in inches
$2\ell + (\ell + 5) = 194$

Here's a sample problem.

Tatewin is 5 inches taller than Lawana. The sum of Tatewin's height and twice Lawana's is 194 inches. How tall is Lawana?

CHECK FOR UNDERSTANDING

Communicating Mathematics

Study the lesson. Then complete the following.

1. **Write** three questions you would ask yourself in order to understand the problem below.
   Consuelo's target heart rate is 140 beats per minute. Her normal heart rate is 60 beats per minute. After running for 10 minutes, her heart rate rises by 65 beats. It drops by 32 beats after a 5-minute rest period and then rises by 50 beats after another 15-minute run. Did she reach her target heart rate?

2. **Explain** whether an equation is a type of formula or a formula is a type of equation.

3. **Write** a problem that uses the formula for the area of a rectangle, $A = \ell w$.

MODELING MATHEMATICS

4. **Find** any gift box. Measure each side. Use the formula $S = 2(wh + \ell h + \ell w)$ to find its surface area, including the lid.

5. Answer the related questions for the verbal problem below.
Each question in the computation section of a mathematics test is worth 4 points. Each question in the problem-solving section is worth 6 points. Madison needs to answer 15 questions correctly to have a total of 86 points and get a B on the test. How many problem-solving questions must she answer correctly?

   a. How many points are the computation questions worth?
   b. How many questions does Madison need to answer correctly?
   c. What score does Madison need to make on the test?
   d. How many points are the problem-solving questions worth?
   e. If n represents the number of correct computation questions, how many problem-solving questions does she need to answer correctly?
   f. If n represents the number of correct answers in the computation section of the test, how does Madison compute the number of points in this section of the test?

Translate each sentence into an equation, inequality, or formula.

   6. The sum of twice x and three times y is equal to thirteen.
   7. The sum of a number and 5 is at least 48.
   8. The perimeter P of a parallelogram is twice the sum of the lengths of two adjacent sides, a and b.

Define a variable, and then write an equation for each problem. Do *not* try to solve.

   9. Olivia drove 189 miles from her home in Fort Worth, Texas, to Austin in three hours. What was her average speed?
   10. Shane has 4 more dimes than quarters and 7 fewer nickels than dimes. He has 28 coins in all. How many quarters does Shane have?
   11. Translate $a(y + 1) = b$ into a verbal sentence.
   12. Write a problem based on the given information.
   Let h = the number of hours you can work during the summer.
   $5h - 8$

EXERCISES

13. Answer the related questions for the verbal problem below.
The Johnson Sisters' Construction Company is planning to build a playground for a housing development. The blueprints allow an area of 20,000 square feet for the playground. The width of the rectangular playground is 80 feet less than its length, and its perimeter is 640 feet. Is there enough space to build the playground as designed?

   a. What is the perimeter of the playground?
   b. If y represents the length of the playground, what is its width?
   c. What is the formula for the perimeter of a rectangular shape?
   d. What is the length of the playground?
   e. What is the area of the playground?
   f. What does the problem ask?

Translate each sentence into an equation, inequality, or formula.

14. The volume V of a pyramid is equal to one third the product of the area of the base b and its height h.

15. The sum of the square of a and the cube of b is equal to twenty-five.

16. The quotient of x and y is at least 18 more than five times the sum of x and y.

17. The quantity x is less than the square of the difference of a and 4.

18. The circumference C of a circle is the product of 2, π, and the radius r.

19. Seven-eighths of the sum of a, b, and the square of c is the same as 48.

Define a variable, and then write an equation for each problem. Do _not_ try to solve.

20. Nicole has 25 CDs and tapes altogether. When she tries to pair them up, she has 4 CDs left over. How many tapes does Nicole have?

21. One number is 25 more than a second number. The sum of these two numbers is 106. Find the numbers.

22. Anna is twice as old as her brother Dennis, who is 4 years older than their brother Curtis. The sum of their ages is 32. How old is Anna?

23. Hector works for a lawn care service. He gets paid twice a month and saves $50 of each pay. After 7 months, how much will he have saved?

Translate each equation or formula below into a verbal sentence.

24. $(mn)^2 = p$ **25.** $V = \dfrac{ah}{3}$ **26.** $\dfrac{(a + 2)}{5} > 10$

Write a problem based on the given information.

27. $x =$ the number of miles from home to school
$3x = 36$

28. $p =$ the cost of a suit
$p - 25 = 150$

29. $d =$ the number of appointments a doctor has scheduled
$6d + 5 = 47$

Write a formula for each shaded area.

30.

31.

32.

Critical Thinking

33. Geometry   The two line segments below are the same length.

Write an equation that describes this situation. Then, use guess and check or any other strategy to find x.

34. Food Aaron and Ellie went out for pizza. After looking at the menu, they couldn't decide whether to order two 7-inch round pizzas or one 12-inch round pizza because the cost was the same. Which is the better buy?

a. What questions do you need to answer?

b. Draw and label diagrams that describe the situation.

c. Write the formulas or equations needed to solve the problem, then solve.

d. Ellie's favorite part of a pizza is the crust. Explain how knowing this might affect your answer.

35. Geometry The area of a trapezoid equals one half the product of its height and the sum of the lengths of its two bases.

a. Write a formula for the area of a trapezoid. Let A represent the area, let a and b represent the bases, and let h represent the height.

b. Find the area for a trapezoid with a base of 20 feet, another base of 11 feet, and a height of 7 feet.

36. Geometry Study the diagram at the right.

a. Based on the dimensions given, label the remaining sides. Assume that all angles are right angles.

b. Write a formula to represent the perimeter.

c. Write a formula to represent the area.

d. If x is 5.5 cm and y is 4 cm, find the perimeter and area.

37. Find $-\sqrt{64}$. (Lesson 2–8)

38. Simplify $-72 \div (-6)$. (Lesson 2–7)

39. Find $-4\begin{bmatrix} -2 & 0.4 \\ -3 & 5 \end{bmatrix}$. (Lesson 2–6)

40. Find a fraction between $\frac{1}{3}$ and $\frac{4}{7}$ whose denominator is 42. (Lesson 2–4)

41. State the additive inverse and absolute value of $+25$. (Lesson 2–3)

42. Identify the graph below that best represents the following situation. Brandon has a deflated balloon. He slowly fills the balloon up with air. Without tying the balloon, he lets it go. (Lesson 1–9)

43. Find the solution set for the inequality $4y + 2 > 9$ if the replacement set is $\{1, 2, 3, 4, 5\}$. (Lesson 1–5)

VOCABULARY

After completing this chapter, you should be able to define each term, property, or phrase and give an example or two of each.

Algebra
absolute value   (p. 85)
adding integers   (p. 86)
additive inverses   (p. 87)
additive inverse property
   (p. 87)
comparison property
   (p. 94)
completeness property
   (p. 121)
complex fraction   (p. 114)
cross products   (p. 94)
defining the variable
   (p. 127)
density property   (p. 96)
dividing rational numbers
   (p. 112)
formula   (p. 128)
integer   (p. 73)

irrational number
   (p. 120)
multiplicative property
   of −1   (p. 107)
multiplying rational
   numbers   (p. 107)
negative number   (p. 73)
number line   (p. 72)
opposites   (p. 87)
perfect square   (p. 119)
principal square root
   (p. 119)
radical sign   (p. 119)
rational number   (p. 93)
real number   (p. 121)
square root   (pp. 118, 119)
subtracting integers
   (p. 87)
unit cost   (p. 95)

Venn diagram   (p. 73)
whole number   (p. 72)

Geometry
coordinate   (p. 73)
graph   (p. 73)

Statistics
line plot   (p. 78)

Problem Solving
problem-solving plan
   (p. 126)

Discrete Mathematics
(p. 88)
matrix   (p. 88)
scalar multiplication
   (p. 108)

UNDERSTANDING AND USING THE VOCABULARY

State whether each sentence is *true* or *false*. If false, replace the underlined word or number to make a true sentence.

1. The absolute value of -26 is $\underline{26}$.

2. The <u>multiplicative inverse</u> of 2 is -2.

3. Terminating decimals are <u>rational</u> numbers.

4. The square root of 144 is $\underline{12}$.

5. $\underline{2\frac{1}{2}}$ is a complex fraction.

6. $-\sqrt{576}$ is an <u>irrational number</u>.

7. $\underline{225}$ is a <u>perfect square</u>.

8. $\underline{-3.1}$ is an integer.

9. $\underline{0.66\overline{6}}$ is a repeating decimal.

10. $\frac{10}{5}$ is a whole number.

SKILLS AND CONCEPTS

| OBJECTIVES AND EXAMPLES | REVIEW EXERCISES |

Upon completing this chapter, you should be able to:

Use these exercises to review and prepare for the chapter test.

• graph integers on a number line (Lesson 2–1)

Graph { . . . , −5, −4, −3}.

-7 -6 -5 -4 -3 -2 -1 0 1

Graph each set of numbers on a number line.

11. {5, 3, −1, −3}

12. {integers greater than −3}

13. {integers less than 4 and greater than or equal to −2}

• add integers by using a number line
(Lesson 2–1)

4 + (−3) = 1

-2 -1 0 1 2 3 4 5 6

Find each sum. If necessary, use a number line.

14. 4 + (−4)   **15.** 2 + (−7)

16. −8 + (−12)   **17.** −9 + 5

18. −14 + (−8)   **19.** 6 + (−11)

• display and interpret statistical data on a line plot (Lesson 2–2)

Make a line plot for the set of data.

78, 74, 86, 88, 99, 63, 85, 85, 85

```
                              X
                              X
      X            X   X     XX  X              X
  +--+--+--+--+--+--+--+--+--+--+--+--+--+--+--+--+--+--+
  60    65    70    75    80    85    90    95    100
```

The following table lists the percent of 18-year-olds with high school diplomas for each of 9 years.

Year	'50	'55	'60	'65	'70	'75	'80	'85	'90
%	56	65	72	71	77	74	72	74	72

20. Make a line plot of the data.

21. What was the lowest percent of 18-year-olds with high school diplomas?

22. In how many years was the percent with diplomas between 70 and 75%?

• add and subtract integers (Lesson 2–3)

$$-14 + (-9) = -(|-14| + |-9|)$$
$$= -(14 + 9)$$
$$= -23$$

$7 - 9 = 7 + (-9)$ *To subtract 9, add −9.*
$$= -2$$

Find each sum and difference.

23. 17 + (−9)   **24.** 14 − 36

25. −10 + 8   **26.** 18 − (−5)

27. −7 − (−11)   **28.** −17 + (−31)

29. −12 + 7   **30.** −54 − (−34)

OBJECTIVES AND EXAMPLES	REVIEW EXERCISES

compare and order rational numbers (Lesson 2–4)

$$\frac{5}{9} \ \underline{\ ?\ } \ \frac{3}{5}$$

$5 \cdot 5 \ \underline{\ ?\ } \ 9 \cdot 3$ *Find the cross products.*

$$25 < 27$$

$$\frac{5}{9} < \frac{3}{5}$$

Replace each __?__ with <, >, or = to make each sentence true.

31. $-8 \ \underline{\ ?\ } \ -14$
32. $\frac{3}{8} \ \underline{\ ?\ } \ \frac{4}{11}$

33. $-5.6 \ \underline{\ ?\ } \ -4.5$
34. $\frac{-3.6}{0.6} \ \underline{\ ?\ } \ -7$

Find a number between the given numbers.

35. $-\frac{3}{5}$ and $\frac{7}{12}$
36. $-\frac{2}{9}$ and $-\frac{5}{8}$

add and subtract rational numbers (Lesson 2–5)

$$-0.37 + 0.812 = + (|0.812| - |0.37|)$$

$$= 0.442$$

$$7\frac{3}{10} - \left(-4\frac{1}{5}\right) = 7\frac{3}{10} + 4\frac{1}{5}$$

$$= 7\frac{3}{10} + 4\frac{2}{10}$$

$$= 11\frac{5}{10} \text{ or } 11\frac{1}{2}$$

Find each sum or difference.

37. $\frac{6}{7} + \left(-\frac{13}{7}\right)$

38. $-0.0045 + 0.034$

39. $3.72 - (-8.65)$

40. $-\frac{4}{3} + \frac{5}{6} + \left(-\frac{7}{3}\right)$

41. $-4.57 - 8.69$

42. $-4.5 - 8.1$

multiply rational numbers (Lesson 2–6)

$$-4(-2) + 6(-3) = 8 + (-18)$$

$$= -10$$

$$-2\frac{1}{7}\left(3\frac{2}{3}\right) + \left(-5\frac{5}{7}\right) = \left(-\frac{15}{7}\right)\left(\frac{11}{3}\right) + \left(-\frac{40}{7}\right)$$

$$= -\frac{55}{7} + \left(-\frac{40}{7}\right)$$

$$= -\frac{95}{7}$$

Find each product.

43. $(-11)(9)$

44. $(-7)(12)(-3)$

45. $\left(\frac{3}{5}\right)\left(-\frac{5}{7}\right)$

46. $(-5.733)(-2.43)(-3.6)$

47. $-2(45)$

48. $-4\left(\frac{7}{12}\right)$

divide rational numbers (Lesson 2–7)

$$\frac{-12}{\frac{2}{3}} = -12 \div \left(-\frac{2}{3}\right)$$

$$= -12\left(-\frac{3}{2}\right)$$

$$= 18$$

Simplify.

49. $\frac{-54}{6}$

50. $-15 \div \left(\frac{3}{4}\right)$

51. $\frac{\frac{4}{5}}{-7}$

52. $\frac{-575x}{5}$

53. $218 \div (-2)$

54. $-78 \div (-6)$

OBJECTIVES AND EXAMPLES

• find square roots   (Lesson 2–8)

Evaluate $\sqrt{a+b}$ if $a = 489$ and $b = 295$.

$$\sqrt{489 + 295} = \sqrt{784}$$
$$= 28$$

REVIEW EXERCISES

Evaluate each expression. Use a calculator if necessary. Round to the nearest hundredth if the result is not a whole number.

55. $\sqrt{y}$, if $y = 196$
56. $\pm\sqrt{t}$, if $t = 112$
57. $-\sqrt{ab}$, if $a = 36$ and $b = 25$
58. $\pm\sqrt{\dfrac{c}{d}}$, if $c = 169$ and $d = 16$

APPLICATIONS AND PROBLEM SOLVING

59. **Consumerism**   The cost per cup, in cents, of 28 different liquid laundry detergents are listed below.   (Lesson 2–2)

28	17	16	18
19	21	26	15
19	19	16	14
21	12	26	17
30	17	13	18
14	22	20	12
19	9	15	12

  a. Make a line plot of the data.
  b. How many detergents cost at most 17¢ per cup?

60. **Aquatics**   A submarine descended to a depth of 432 meters and then rose 189 meters. How far below the surface of the water is the submarine now?   (Lesson 2–3)

61. **Consumerism**   Which is the better buy: 0.75 liter of soda for 89¢ or 1.25 liters of soda for $1.31?   (Lesson 2–4)

62. **Electricity**   A circuit is designed with two resistance settings R, 4.6 ohms and 5.2 ohms, and two power settings P, 1200 watts and 1500 watts. Which settings can be used so that the voltage of the circuit is between 75 volts and 85 volts? Use $V = \sqrt{PR}$.   (Lesson 2–8)

63. **Stock**   The changes in two separate stocks for one week are listed below.   (Lesson 2–5)

	M	T	W	TH	F
CompNet	$+\frac{3}{8}$	$+\frac{1}{8}$	0	$-\frac{3}{4}$	$+\frac{1}{2}$
AccuFirm	$-\frac{1}{4}$	$+\frac{1}{8}$	$+\frac{1}{8}$	$-\frac{1}{4}$	$+\frac{1}{2}$

  a. Which stock saw the largest gain for the week?
  b. Which stock had the greatest change from one day to the next?
  c. What was the change and on what days?

Define a variable, and then write an equation for each problem. Do not try to solve.   (Lesson 2–9)

64. Minal weighs 8 pounds less than Claudia. Together they weigh 182 pounds. How much does Minal weigh?

65. Three times a number decreased by 21 is 57. Find the number.

66. Four years ago, three times Cecile's age was 42, her father's age now. How old is Cecile now?

A practice test for Chapter 2 is provided on page 788.

ALTERNATIVE ASSESSMENT

COOPERATIVE LEARNING PROJECT

Cooking   In this project, you are hosting a party for the cast members of the school play. You must prepare enough chili and peanut butter brownies to serve 100 individuals. (Recipes are given below.) You will also need bowls, napkins, and spoons. You start out with $50 in your checking account. You write a check to a party goods store for the bowls, napkins, and spoons. You then go to the grocery and buy everything except the ground beef because you decide to buy the ground beef at the local meat market.

Follow these steps to prepare a grocery list and a checkbook register.

- Determine the amount of each ingredient needed for the chili and the peanut butter brownies.
- Write a grocery list including the total amount of ingredients needed.
- Go to the grocery and record the size of the package you would need to buy for each item, the quantity you would need to buy, and the price of the package.
- Keep a running total with entries for your checkbook register.
- Write several paragraphs incorporating your data and your money situation for a report to be given to the school play director.

Every time you run a negative balance in your checking account your bank deducts $10. Did you ever run a negative balance? If so, when would have been a good time to deposit more money? How much more money should you have deposited in order to cover all your checks?

THINKING CRITICALLY

- Write a sentence explaining why terminating and repeating decimals are rational numbers.
- A number is both multiplied and divided by the same rational number n, where $0 < n < 1$. Which is greater, the product or the quotient? Explain your reasoning.

PORTFOLIO

Make a grocery list containing ten items. Go to a grocery store and record the price of two different size containers for each of these items. Be sure to also include the brand names and the different sizes in your record. Make a chart of this data.

When you have completed the chart, calculate the unit cost and include it in your chart. Write a new grocery list of the ten items containing the exact brand name and size of container that should be bought in order to insure the best buy. Place these items in your portfolio.

SELF EVALUATION

Comparing numbers is often useful when solving a problem. It can be beneficial when estimating and in decision-making. When comparing numbers or items, be sure to consider other factors.

Assess yourself. When do you use comparison? Do you evaluate new situations based on comparing that situation to another situation that was similar? Describe two ways in which you have used or could use comparison to solve a problem or situation in mathematics and in your daily life.

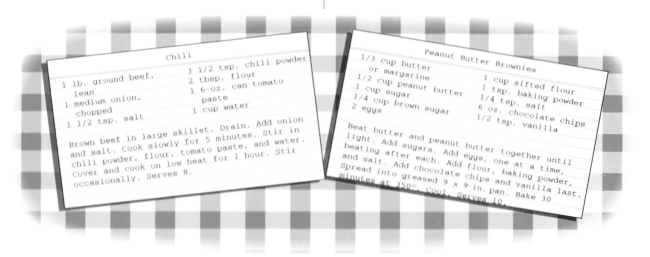

Chili

1 lb. ground beef, lean
1 medium onion, chopped
1 1/2 tsp. salt
3 1/2 tsp. chili powder
2 tbsp. flour
1 6-oz. can tomato paste
1 cup water

Brown beef in large skillet. Drain. Add onion and salt. Cook slowly for 5 minutes. Stir in chili powder, flour, tomato paste, and water. Cover and cook on low heat for 1 hour. Stir occasionally. Serves 8.

Peanut Butter Brownies

1/3 cup butter or margarine
1/2 cup peanut butter
1 cup sugar
1/4 cup brown sugar
2 eggs
1 cup sifted flour
1 tsp. baking powder
1/4 tsp. salt
6 oz. chocolate chips
1/2 tsp. vanilla

Beat butter and peanut butter together until light. Add sugars. Add eggs, one at a time, beating after each. Add flour, baking powder, and salt. Add chocolate chips and vanilla last. Spread into greased 9 x 9-in. pan. Bake 30 minutes at 350°. Cool. Serves 10.

SECTION ONE: MULTIPLE CHOICE

There are eight multiple-choice questions in this section. After working each problem, write the letter of the correct answer on your paper.

1. **Patterns**   Find the sixth term in the sequence $-6, -3\frac{1}{2}, -1, \dots$.

   A. 0

   B. $2\frac{1}{2}$

   C. 4

   D. $6\frac{1}{2}$

2. **Travel**   Travel agencies often give information on a state's temperature during the year in order to prepare their customers for their vacation. The stem-and-leaf plot shows daily high temperatures recorded for April in Ohio. Which statement best describes the data in order to prepare travelers for their visit to Ohio in April?

Stem	Leaf
8	0 2 2 5
7	0 3 3 4 5 7 7 7 8 9 9
6	1 2 3 5 5 6 7 8 8 9
5	4 7 8 8
4	9 $5\,|\,7 = 57$

   A. April temperatures can range from 49 to 85 degrees.

   B. April is mild, with temperatures usually between 60 and 80 degrees.

   C. April's temperatures can soar as high as 85 degrees.

   D. April is a cold month with temperatures as low as 49 degrees.

3. Evaluate $3t^2 - g(w - t)$ if $t = 3$, $g = 5$, and $w = -2$.

   A. -110

   B. 20

   C. 43

   D. 52

4. Tyrone is 3 years older than Bill. In 5 years, twice the sum of their ages will equal 5 times Bill's age now. Choose the equation that should be used to solve this problem.

   A. $b + (b + 3) = 5b$

   B. $2b + (b + 3) + 5 = 5b$

   C. $2[(b + 5) + (b + 3 + 5)] = 5b$

   D. $2[(b + 3) + (b - 3 + 5)] = 5b$

5. Quan is working as a mailroom clerk in a busy downtown office building for the summer. His morning route takes him from the mailroom up two floors to the payroll office, up four more floors to the attorney's office, down three floors to the cafeteria, up seven floors to the executive offices, and down twelve floors to the security desk on the first floor. On what floor is the mailroom?

   A. second floor

   B. third floor

   C. sixth floor

   D. top floor

6. Stephen bought a 5-foot-long submarine sandwich to serve at his Super Bowl party. How many pieces would he have to serve if he cut the sub into $1\frac{1}{2}$-inch pieces?

   A. 3 pieces

   B. 8 pieces

   C. 30 pieces

   D. 40 pieces

7. Simplify $10a^3 + 7 - 2(2a^3 + a) + 3$.

   A. $-2a^4 + 10a^3 + 10$

   B. $8a^3 + 2a + 10$

   C. $6a^3 - 2a + 10$

   D. $6a^3 + a + 10$

8. Choose the expression that has a value of 28.

   A. $4 + 3 \cdot 4$

   B. $\dfrac{75}{3} - 2$

   C. $8 \cdot 4 - 8$

   D. $(5 + 3) \cdot 7 \div 2$

SECTION TWO: FREE RESPONSE

This section contains seven questions for which you will provide short answers. Write your answer on your paper.

9. Rewrite the expression $7 + 15 \div 2 + 4 - 2$ using grouping symbols so that the value of the expression is $6\frac{1}{2}$.

10. Three integers x, y, and z satisfy the following conditions:

    $y - z < 0$, $x - y > 0$, and $z - x < 0$.

    Arrange the integers in order from least to greatest.

11. Graph the solution set of $k \neq 1$ on a number line.

12. **Geometry** Find the perimeter of the figure below if $x = 7$, $y = 3$, and $z = 1\frac{1}{2}$.

13. **School** The students in Miss Stickler's algebra class scored the following points on their last test: 88, 89, 85, 92, 91, 86, 90, 95, 91, 86, 90, 92, 91, 89, 91, and 90. Make a line plot of the scores and write a sentence to describe the data shown.

14. The average hourly rate of a teenage babysitter on a Saturday night is $5 in Los Angeles and $2 in Pittsburgh. Corey, who lives in Pittsburgh, baby-sits his little brother every Saturday night for 5 hours. Ella lives in Los Angeles and baby-sits her cousin for 3 hours every other Saturday night. How many Saturday nights will each have to baby-sit to earn the same amount of money?

15. What is the absolute value of a if $-a > a$?

16. Write an algebraic expression to represent *nine decreased by the product of y and 3.*

17. **Patterns** Find the next five terms of the sequence 3, 4.5, 6, ...

18. Replace _?_ with >, <, or = to make a true statement.

    $-2 - 1 \underline{\ ?\ } -2(-1)$

SECTION THREE: OPEN-ENDED

This section contains two open-ended problems. Demonstrate your knowledge by giving a clear, concise solution to each problem. Your score on these problems will depend on how well you do the following.

- Explain your reasoning.

- Show your understanding of the mathematics in an organized manner.

- Use charts, graphs, and diagrams in your explanation.

- Show the solution in more than one way or relate it to other situations.

- Investigate beyond the requirements of the problem.

19. Latisha and Joia are planning to purchase a CD player for their dorm room. Latisha works at the campus book store and earns $5.35 per hour. Joia has a job off campus as a waitress and earns $2.00 per hour, plus tips. Each student works 20 hours per week.

    A. How long will they have to work to earn enough money to buy a CD player worth $350?

    B. What other factors may affect when they can buy the CD player?

20. If $2(b + c) = 2b + 2c$, does $2 + (b \cdot c) = (2 + b)(2 + c)$? Use values for b and c to justify your answer.

Solving Linear Equations

Objectives

In this chapter, you will:

- solve equations using one or more operations,
- solve problems that can be represented by equations,
- work backward to solve problems,
- define and study angles and triangles, and
- find measures of central tendency.

The Rising Cost of College

Annual Cost (in 1992–93 dollars)

16000
14000
12000
10000
8000
6000
4000
2000
0

Private 4-year colleges

Public 4-year colleges

School Year Ending

1960 1965 1970 1975 1980 1985 1990 1995 2000 (projected)

Source: U.S. Department of Education

TIME *Line*

Over the last thirty years, tuition at public 4-year colleges has increased by about 50%, and tuition at private 4-year colleges has increased by a whopping 110%! How do you and your family intend to fund your college experience?

A.D. **240** In *Arithmetica*, Diophantus obtains negative numbers as solutions to equations, but dismisses them as "absurd."

700 B.C. | 600 | 500 | 400 | 300 | 200 | 100 | 0 | A.D. 100 | 200 | 300 | 400 | 500

575 B.C. Early standard coins, Lydia, Turkey

PEOPLE IN THE NEWS

Chapter Project

J uan Ramirez has been offered two college scholarships. Under the terms of the first scholarship, University A will pick up one-half of all his college expenses. Under the terms of the second scholarship, University B will pick up $30,000 of his college expenses. There is $27,500 in the college fund his parents set up for him when he was born.

- Suppose college fees for one year at University A, a state university, are $9500 per year and fees for one year at University B, a private university, are $19,000.

- Write an equation to represent the cost for Juan to attend University A for four years. Write another equation to represent the cost for Juan to attend University B for four years.

- Which school should Juan attend, and why? Be sure to completely explain your reasoning.

W hen **Marianne Ragins** was a high-school student in her hometown of Macon, Georgia, she applied for scholarship funds from several sources so she could go to college. By the time Marianne was a freshman at Florida A & M University, she had won more than $400,000 in scholarship offers.

When she graduates from college, Marianne will have used $120,000 of those scholarships for her college expenses. Marianne, who is now 21, is the author of *Winning Scholarships for College,* and she now conducts The Scholarship Workshop across the country.

780 Muhammed ibn Musa Al-Khowarizmi, Arabian mathematician, born; from his name we get the word *algebra.*

1848 Maria Mitchell is the first woman elected to the American Academy of Science.

800 900 1000 1100 1200 1300 1400 1500 1600 1700 1800 1900

1158 The University of Bologna, Italy, is founded.

1989 Ileana Ros-Lehtinen becomes the first Latina ever elected to Congress.

MODELING MATHEMATICS

A Preview of Lesson 3–1

3–1A Solving One-Step Equations

Materials: ▭• cups & counters   ▭=▭ equation mat

You can use cups and counters as a model for solving equations. After you model the equation, the goal is to get the cup by itself on one side of the mat by using the rules stated below.

Rules for Equation Models	
You can remove or add the same number of identical counters to each side of the equation mat without changing the equation.	← (+ + +) = (+ + +) →
You can remove or add zero pairs to either side of the equation mat without changing the equation.	← (+ –) = (–) →

Activity 1   Use an equation model to solve $x + (-3) = -5$.

Step 1 Model the equation $x + (-3) = -5$ by placing 1 cup and 3 negative counters on one side of the mat. Place 5 negative counters on the other side of the mat. The two sides of the mat represent equal quantities.

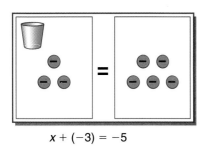

$$x + (-3) = -5$$

Step 2 Remove 3 negative counters from each side to get the cup by itself.

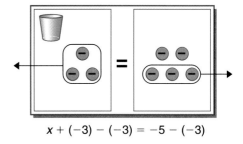

$$x + (-3) - (-3) = -5 - (-3)$$

Step 3 The cup on the left side of the mat is matched with 2 negative counters. Therefore, $x = -2$.

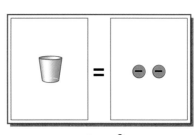

$$x = -2$$

Activity 2   Use an equation model to solve 2p = −6.

Step 1 Model the equation $2p = -6$ by placing 2 cups on one side of the mat. Place 6 negative counters on the other side of the mat.

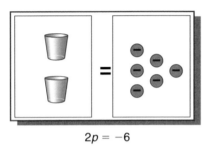

$$2p = -6$$

Step 2 Separate the counters into 2 equal groups to correspond to the 2 cups. Each cup on the left is matched with 3 negative counters. Therefore, $p = -3$.

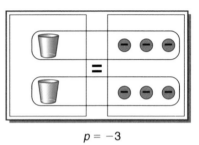

$$p = -3$$

Activity 3   Use an equation model to solve r − 2 = 3.

Step 1 Write the equation in the form $r + (-2) = 3$. Place 1 cup and 2 negative counters on one side of the mat. Place 3 positive counters on the other side of the mat. Notice that it is not possible to remove the same kind of counters from each side. Add 2 positive counters to each side.

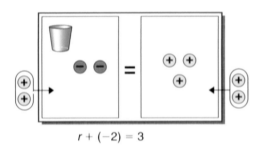

$$r + (-2) = 3$$

Step 2 Group the counters to form zero pairs. Then remove all the zero pairs. The cup on the left is matched with 5 positive counters. Therefore, $r = 5$.

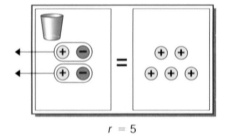

$$r = 5$$

Model   Use equation models to solve each equation.

1. $x + 4 = 5$ 2. $y + (-3) = -1$ 3. $y + 7 = -4$ 4. $3z = -9$

5. $m - 6 = 2$ 6. $-2 = x + 6$ 7. $8 = 2a$ 8. $w - (-2) = 2$

Draw   **Tell whether each number is a solution of the given equation. Justify your answer with a drawing.**

9. $-3; x + 5 = -2$ 10. $-1; 5b = -5$ 11. $-4; y - 4 = -8$

Write   12. Write a paragraph explaining how you use zero pairs to solve an equation like $m + 5 = -8$.

Solving Equations with Addition and Subtraction

APPLICATION
Football

According to the graph below, the Oakland Raiders and the Dallas Cowboys have each appeared on *Monday Night Football* a total of 41 times through the 1992–1993 season.

Most *Monday Night Football* Appearances

Source: Capital Cities/ABC

If during the next five seasons, each team appears an average of two times per season, the two teams would still have an equal number of *Monday Night Football* appearances.

$$41 = 41$$

$$41 + 5(2) = 41 + 5(2)$$ *Two appearances per season for five seasons can be represented as 5(2).*

This example illustrates the **addition property of equality.**

Addition Property of Equality	**For any numbers *a*, *b*, and *c*, if *a* = *b*, then *a* + *c* = *b* + *c*.**

Note that *c* can be positive, negative, or 0. In the equation on the left below, *c* = 5. In the equation on the right, *c* = −5.

$$18 + 5 = 18 + 5 \qquad\qquad 18 + (-5) = 18 + (-5)$$

If the same number is added to each side of an equation, then the result is an **equivalent equation.** Equivalent equations are equations that have the same solution.

$$y + 7 = 13$$ *The solution to this equation is 6.*

$$y + 7 + 3 = 13 + 3$$ *Using the addition property of equality, add 3 to each side.*

$$y + 10 = 16$$ *The solution to this equation is also 6.*

Recall that an equation mat is a model of an equation. The sides of the mat represent the sides of an equation. When you add counters to each side, the result is an equivalent equation, as shown below.

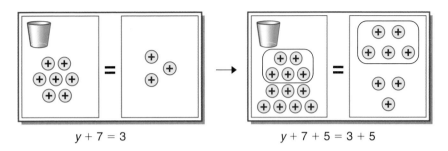

$$y + 7 = 3 \qquad\qquad y + 7 + 5 = 3 + 5$$

Remember, x means $1 \cdot x$.
The coefficient of x is 1.

To **solve an equation** means to isolate the variable having a coefficient of 1 on one side of the equation. You can do this by using the addition property of equality.

Example ❶ **Solve $23 + t = -16$.**

$$23 + t = -16$$
$$23 + t + (-23) = -16 + (-23) \quad \text{\textit{Add} -23 \textit{to each side.}}$$
$$t + 0 = -39 \qquad\qquad \text{\textit{The sum of 23 and} -23 \textit{is 0.}}$$
$$t = -39$$

To check that -39 is the solution, substitute -39 for t in the original equation.

Check:
$$23 + t = -16$$
$$23 + (-39) \overset{?}{=} -16$$
$$-16 = -16 \quad ✔$$

The solution is -39.

You can and will use equations to solve many real-world problems.

Example ❷ **Refer to the application at the beginning of the lesson. The San Francisco 49ers appeared fewer times than the Washington Redskins through the 1992–1993 season. How many more times did the Redskins appear?**

Explore Read the problem to find out what is asked. Then define a variable.

The problem asks how many more times the Washington Redskins appeared on *Monday Night Football* than the San Francisco 49ers through the 1992–1993 season. From the chart, we can see that the Redskins appeared 43 times.

Let n = the difference in the number of times each team appeared. Then $43 - n$ = the number of times the 49ers appeared.

Plan Write an equation.

You know that $43 - n$ represents the number of times the 49ers appeared. From the chart, we can see that the 49ers appeared 35 times. Thus, $43 - n = 35$.

 APPLICATION

Football

 LOOK BACK

You can refer to Lesson 2-10 for information on solving problems by writing and solving equations.

(continued on the next page)

Solve   Solve the equation and answer the problem.

$$43 - n = 35$$

$$43 + (-43) - n = 35 + (-43) \quad \textit{Add } -43 \textit{ to each side.}$$

$$-n = -8 \qquad \textit{The opposite of n is negative 8.}$$

$$n = 8 \qquad \textit{Therefore, n is positive 8.}$$

Thus, the Washington Redskins appeared 8 times more than the San Francisco 49ers.

Examine   Check to see if the answer makes sense.

If the Redskins appeared 43 times and the 49ers appeared 35 times, then the difference is $43 - 35$ or 8 times.

In addition to the addition property of equality, there is a **subtraction property of equality** that may also be used to solve equations.

Subtraction Property of Equality	For any numbers *a*, *b*, and *c*, if *a = b*, then *a − c = b − c*.

Example **Solve $190 - x = 215$.**

$$190 - x = 215$$

$$190 - x - 190 = 215 - 190 \quad \textit{Subtract 190 from each side.}$$

$$-x = 25 \qquad \textit{The opposite of x is 25.}$$

$$x = -25$$

Check:
$$190 - x = 215$$
$$190 - (-25) \stackrel{?}{=} 215$$
$$190 + 25 \stackrel{?}{=} 215$$
$$215 = 215 \quad ✔$$

The solution is -25.

Most equations can be solved in two ways. Remember that subtracting a number is the same as adding its inverse.

Example **Solve $a + 3 = -9$ in two ways.**

Method 1:   Use the subtraction property of equality.
$$a + 3 = -9$$
$$a + 3 - 3 = -9 - 3 \quad \textit{Subtract 3 from each side.}$$
$$a = -12$$

Check: $a + 3 = -9$
$$-12 + 3 \stackrel{?}{=} -9$$
$$-9 = -9 \quad ✔$$

Method 2: Use the addition property of equality.

$$a + 3 = -9$$

$$a + 3 + (-3) = -9 + (-3) \quad \textit{Add } -3 \textit{ to each side.}$$

$$a = -12 \qquad \textit{The answers are the same.}$$

The solution is -12.

Sometimes equations can be solved more easily if they are first rewritten in a different form.

Example **Solve** $y - \left(-\dfrac{3}{7}\right) = -\dfrac{4}{7}$.

This equation is equivalent to $y + \dfrac{3}{7} = -\dfrac{4}{7}$. *Why?*

$$y + \frac{3}{7} = -\frac{4}{7}$$

$$y + \frac{3}{7} - \frac{3}{7} = -\frac{4}{7} - \frac{3}{7} \quad \textit{Subtract } \tfrac{3}{7} \textit{ from each side.}$$

$$y = -\frac{7}{7} \ or \ -1$$

Check: $\quad y - \left(-\dfrac{3}{7}\right) = -\dfrac{4}{7}$

$$-\frac{7}{7} - \left(-\frac{3}{7}\right) \stackrel{?}{=} -\frac{4}{7}$$

$$-\frac{7}{7} + \frac{3}{7} \stackrel{?}{=} -\frac{4}{7}$$

$$-\frac{4}{7} = -\frac{4}{7} \quad ✔$$

The solution is -1.

EXPLORATION　　　　　CALCULATORS

You can use a scientific or graphing calculator to solve equations that involve decimals.

　If you're using a scientific calculator like the TI-34, you can use the *plus/minus* key to input negative numbers or to change the sign of any number.

　If you're using a graphing calculator like the TI-82 or TI-83, you can use the *negative* key to input negative numbers.

　On all calculators, use the *subtraction* key to subtract two numbers.

Your Turn

a. Take some time to work through the examples in this lesson using your calculator.

b. Describe the difference between the key you use to indicate a negative number on your calculator and the subtraction key.

c. Does the order in which you use these keys matter? Why or why not?

d. How would you use your calculator to solve $x + (-9.016) = 5.14$?

Example ⑥ **Solve** $b + (-7.2) = -12.5$.

This equation is equivalent to $b - 7.2 = -12.5$. *Why?*

$$b - 7.2 = -12.5$$
$$b - 7.2 + 7.2 = -12.5 + 7.2 \quad \text{\textit{Add 7.2 to each side.}}$$
$$b = -5.3$$

Method 1	**Method 2**
Check: $b + (-7.2) = -12.5$	Use a calculator.
$-5.3 + (-7.2) \stackrel{?}{=} -12.5$	5.3 $\boxed{+/-}$ $\boxed{+}$ 7.2
$-12.5 = -12.5$ ✔	$\boxed{+/-}$ $\boxed{=}$ -12.5 ✔

The solution is -5.3.

CHECK FOR UNDERSTANDING

Communicating Mathematics

Study the lesson. Then complete the following.

1. Choose the equation that is equivalent to $2.3 - w = 7.8$.
   a. $w - 2.3 = 7.8$ **b.** $w - 2.3 = -7.8$ **c.** $-w = -5.5$

2. **Write** three equivalent equations.

3. **Explain** how you can prove that 4.5 is the solution of $x - 1.2 = 3.3$.

4. **Complete:** If $x - 6 = 21$, then $x + 7 = \underline{\ ?\ }$. Explain your reasoning.

5. Write an equation for the model at the right. Then use cups and counters to solve the equation.

Guided Practice

Solve each equation. Then check your solution.

6. $m + 10 = 7$ 7. $a - 15 = -32$ 8. $5.7 + a = -14.2$

9. $y + (-7) = -19$ 10. $\frac{1}{6} - n = \frac{2}{3}$ 11. $d - (-27) = 13$

Define a variable, write an equation, and solve each problem. Then check your solution.

12. Thirteen subtracted from a number is -5. Find the number.

13. A number increased by -56 is -82. Find the number.

EXERCISES

Practice

Solve each equation. Then check your solution.

14. $k + 11 = -21$ 15. $41 = 32 - r$ 16. $-12 + z = -36$

17. $2.4 = m + 3.7$ 18. $-7 = -16 - k$ 19. $0 = t + (-1.4)$

20. $r + (-8) = 7$ 21. $h - 26 = -29$ 22. $-23 = -19 + n$

23. $-11 = k + (-5)$ 24. $r - 6.5 = -9.3$ 25. $t - (-16) = 9$

26. $-1.43 + w = 0.89$ 27. $m - (-13) = 37$ 28. $-4.1 = m + (-0.5)$

29. $-\frac{5}{8} + w = \frac{5}{8}$ 30. $x - \left(-\frac{5}{6}\right) = \frac{2}{3}$ 31. $g + \left(-\frac{1}{5}\right) = -\frac{3}{10}$

Define a variable, write an equation, and solve each problem. Then check your solution.

32. Twenty-three minus a number is 42. Find the number.

33. A number increased by 5 is equal to 34. Find the number.

34. What number decreased by 45 is -78?

35. The difference of a number and -23 is 35. Find the number.

36. A number increased by -45 is 77. Find the number.

37. The sum of a number and -35 is 98. Find the number.

Critical Thinking

38. Suppose a solution of an equation is a number n. Is it possible for $-n$ to also be a solution of this equation? If not, explain why not. If so, give an example of such an equation and its solutions n and $-n$.

Applications and Problem Solving

Write two different equations to represent each situation. Then solve the problem.

39. **Telecommunications** The graph at the right shows the growth in the number of people who own and use cellular telephones.

    a. How many more cellular phone subscribers were there in 1994 than there were in 1985?

    b. Predict how many cellular phone subscribers there will be in the year 2000. Write a justification for your prediction.

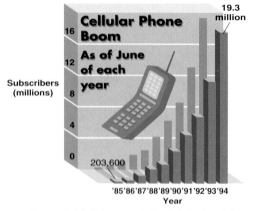

Cellular Phone Boom

As of June of each year

Subscribers (millions)

19.3 million

203,600

'85 '86 '87 '88 '89 '90 '91 '92 '93 '94
Year

Source: Cellular Telecommunications Industry Association

40. **Consumerism** According to Runzheimer International, in 1994, a tube of lipstick cost $5.94 in Los Angeles, California. The same tube cost $4.89 more than that in London, England. In Sao Paulo, Brazil, the lipstick cost $26.54.

    a. How much did the lipstick cost in London?

    b. How much more did the lipstick cost in Sao Paulo than in Los Angeles?

Mixed Review

41. Alejandra Salazar said, "I am 24 years younger than my mom and the sum of our ages is 68 years." (Lesson 2–9)

    a. How old was Mrs. Salazar when Alejandra was born?

    b. How much older than Alejandra is Mrs. Salazar now?

    c. What will be the sum of their ages in 5 years?

    d. How old was Mrs. Salazar when Alejandra was 10 years old?

    e. In ten years, how much younger than Mrs. Salazar will Alejandra be?

42. Find the principal square root of 256. (Lesson 2–8)

43. Simplify $65 \div (-13)$. (Lesson 2–7)

44. **Statistics** Odina's algebra average dropped three fourths of a point for each of eight consecutive months. If her average was originally 82, what was her average at the end of eight months? (Lesson 2–6)

45. Find the sum. $-0.23x + (-0.5 x)$ (Lesson 2–5)

46. Simplify $6(5a + 3b - 2b)$. (Lesson 1–7)

47. Evaluate $12 \div 4 + 15 \cdot 3$. (Lesson 1–3)

Solving Equations with Multiplication and Division

3-2

What YOU'LL LEARN

- To solve equations by using multiplication and division.

Why IT'S IMPORTANT

You can use equations to solve problems involving construction, sociology, and demographics.

APPLICATION
Construction

In 1990, Congress passed into law the Americans with Disabilities Act. One of the provisions of that law has to do with ramps installed on buildings to give people with disabilities access to those buildings. The law states that for a *rise* of 1″, there should be at least 12″ of *run*. The maximum rise is 30″.

Rick Hansen, wheelchair athlete

If a contractor wants to build a ramp that has a 180-inch run, what is the greatest rise that the ramp can have?

In the table at the right, the pattern suggests that the run is always 12 times the rise. Let x represent the number of inches in the rise. Then $12x$ represents the number of inches in the run. Write an equation to represent the situation.

Rise	Run
1	12
2	24
3	36
4	48
x	$12x$

Since run is dependent on rise, run is a function of rise. Rise is the independent quantity, and run is the dependent quantity. Only positive values for rise and run make sense.

$12x = 180$ *This equation will be solved in Example 3.*

To solve equations with multiplication and division, you will need new tools. Equations of the form $ax = b$, where a and/or b are fractions, are generally solved by using the **multiplication property of equality**.

Multiplication Property of Equality	**For any numbers a, b, and c, if $a = b$, then $ac = bc$.**

Example ① **Solve $\frac{g}{24} = \frac{5}{12}$.**

$$\frac{g}{24} = \frac{5}{12}$$

$$24\left(\frac{g}{24}\right) = 24\left(\frac{5}{12}\right) \quad \textit{Multiply each side by 24.}$$

$$g = 2(5) \textit{ or } 10$$

Check: $\quad \frac{g}{24} = \frac{5}{12}$

$$\frac{10}{24} \overset{?}{=} \frac{5}{12} \quad \textit{Replace g with 10.}$$

$$\frac{5}{12} = \frac{5}{12} \quad ✓ \qquad \text{The solution is 10.}$$

Example ② Solve each equation.

a. $\left(3\frac{1}{4}\right)p = 2\frac{1}{2}$

$$\left(3\frac{1}{4}\right)p = 2\frac{1}{2}$$

$$\frac{13}{4}p = \frac{5}{2}$$ *Rewrite the mixed numbers as improper fractions.*

$$\frac{4}{13}\left(\frac{13}{4}\right)p = \frac{4}{13}\left(\frac{5}{2}\right)$$ *Multiply each side by $\frac{4}{13}$, the reciprocal of $\frac{13}{4}$.*

$$p = \frac{20}{26} \text{ or } \frac{10}{13}$$ *Check this result.*

The solution is $\frac{10}{13}$.

b. $40 = -5d$

$$40 = -5d$$

$$-\frac{1}{5}(40) = -\frac{1}{5}(-5d)$$ *Multiply each side by $-\frac{1}{5}$.*

$$-8 = d$$ *Check this result.*

The solution is -8.

The equation in Example 2b, $40 = -5d$, was solved by multiplying each side by $-\frac{1}{5}$. The same result could have been obtained by dividing each side by -5. This method uses the **division property of equality**. It is often easier to use than the multiplication property of equality.

Division Property of Equality	For any numbers a, b, and c, with $c \neq 0$, if $a = b$, then $\frac{a}{c} = \frac{b}{c}$.

Example ③ Refer to the application at the beginning of the lesson. Solve $12x = 180$.

APPLICATION
Construction

$$12x = 180$$

$$\frac{12x}{12} = \frac{180}{12}$$ *Divide each side by 12.*

$$x = 15$$

x
$12x$ or $180"$

Check: $12x = 180$

$12(15) \stackrel{?}{=} 180$

$180 = 180$ ✔

The rise of the ramp could be at most 15 inches.

Example **4** **Solve $3x = -9$.**

$$3x = -9$$

$$\frac{3x}{3} = \frac{-9}{3} \quad \textit{Divide each side by 3.}$$

$$x = -3$$

Recall that an equation mat is a model of an equation. You can group each cup on the left side of the mat with the same number of counters on the right side of the mat.

$3x = -9$ $x = -3$

The solution is -3.

CHECK FOR UNDERSTANDING

Communicating Mathematics

Study the lesson. Then complete the following.

1. **You Decide** Kezia says that the equation $0x = 8$ has a solution. Doralina says it doesn't. Which one is correct, and why?

2. **Define** the terms *rise* and *run* in your own words.

3. **Refer** to the application at the beginning of the lesson. If the maximum rise is 30″, what is the maximum run?

4. **Draw** a model and explain how to find the width of a rectangle if its area is 51 m^2 and its length is 17 m.

5. Use cups and counters to model $4x = 16$. Then solve the equation.

Guided Practice

Solve each equation. Then check your solution.

6. $-8t = 56$ 7. $-5s = -85$ 8. $42.51x = 8$

9. $\frac{k}{8} = 6$ 10. $-10 = \frac{b}{-7}$ 11. $-5x = -3\frac{2}{3}$

Define a variable, write an equation, and solve the problem. Then check your solution.

12. Eight times a number is 216. What is the number?

13. The product of -7 and a number is 1.477. What is the number?

Practice

Solve each equation. Then check your solution.

14. $-4r = -28$ **15.** $5x = -45$ **16.** $9x = 40$

17. $-3y = 52$ **18.** $3w = -11$ **19.** $434 = -31y$

20. $1.7b = -39.1$ **21.** $0.49x = 6.277$ **22.** $-5.73c = 97.41$

23. $-0.63y = -378$ **24.** $11 = \dfrac{x}{5}$ **25.** $\dfrac{h}{11} = -25$

26. $\dfrac{c}{-8} = -14$ **27.** $\dfrac{2}{5}t = -10$ **28.** $-\dfrac{11}{8}x = 42$

29. $-\dfrac{13}{5}y = -22$ **30.** $3x = 4\dfrac{2}{3}$ **31.** $\left(-4\dfrac{1}{2}\right)x = 36$

Define a variable, write an equation, and solve each problem. Then check your solution.

32. Six times a number is -96. Find the number.

33. Negative twelve times a number is -156. What is the number?

34. One fourth of a number is -16.325. What is the number?

35. Four thirds of a number is 4.82. What is the number?

36. Seven eighths of a number is 14. What is the number?

Complete.

37. If $3x = 15$, then $9x = \underline{\ ?\ }$.   **38.** If $10y = 46$, then $5y = \underline{\ ?\ }$.

39. If $2a = -10$, then $-6a = \underline{\ ?\ }$.   **40.** If $12b = -1$, then $4b = \underline{\ ?\ }$.

41. If $7k - 5 = 4$, then $21k - 15 = \underline{\ ?\ }$.

Critical Thinking

42. If $-x$ and $-1 \cdot x$ always represent the same number, does $-x$ always stand for a negative number? Explain your reasoning.

Applications and Problem Solving

Write an equation to represent each situation. Then solve the problem.

43. Sociology   Research conducted by *USA Today* has shown that about 1 in 7 people in the world are left-handed.
  a. About how many left-handed people are there in a group of 350 people? 583 people?
  b. If there are 65 left-handed people in a group, about how many people are in that group?

44. Demographics   In the year 2000, there will be about $1\dfrac{1}{2}$ times the number of 15- to 19-year-olds in the U.S than there were in 1960. If the U.S Census Bureau projects that there will be 19,819,000 15- to 19-year-olds in the year 2000, about how many were there in 1960?

U.S. Population Growth of 15- to 19-year-olds

Population (millions)

25
20
15
10
5
0

1960 '70 '80 90 '94 '95 '00 '15
Year   (est.)

Source: U.S.Census Bureau

45. Telecommunications   In 1995, the long-distance company Sprint introduced Sprint Sense, a plan in which long-distance calls placed on weekends cost only $0.10 per minute.
  a. How long could you talk for $2.30?
  b. What would be the cost of an 18-minute call?

46. Solve $-11 = a + 8$. Then check your solution. (Lesson 3–1)

47. Define a variable, then write an equation for the problem below. (Lesson 2–9)

Ponderosa pines grow about $1\frac{1}{2}$ feet each year. If a pine tree is now 17 feet tall, about how long will it take for the tree to become $33\frac{1}{2}$ feet tall?

48. World Records The longest loaf of bread ever baked was 2132 feet $2\frac{1}{2}$ inches. If this loaf were cut into $\frac{1}{2}$-inch slices, how many slices of bread would there have been? (Lesson 2–7)

49. Find a number between $\frac{1}{2}$ and $\frac{6}{7}$. (Lesson 2–4)

50. Graph $\{-1, 1, 3, 5\}$ on a number line. (Lesson 2–1)

51. The graph at the right shows the number of bags of potato chips in the snack machine at Lee High School two times during an average day. (Lesson 1–9)
  a. What can you learn from looking at the graph?
  b. What do you think happened between the times indicated by the dots?
  c. If lunch at Lee is between 11:45 A.M. and 1:15 P.M., estimate the number of bags you think would be in the machine at 1:30 P.M. Justify your estimate.

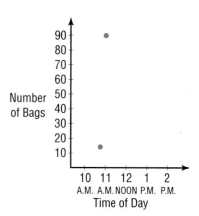

Number of Bags — Time of Day
(10 A.M. 11 A.M. 12 NOON 1 P.M. 2 P.M.)

52. Simplify $5(3x + 2y - 4y)$. (Lesson 1–7)

53. Evaluate $a + b^2 + c^2$ if $a = 6$, $b = 4$, and $c = 3$. (Lesson 1–1)

WORKING ON THE
In·ves·ti·ga·tion

Refer to the Investigation on pages 68–69.

the Greenhouse Effect

The changes in the levels of carbon dioxide (CO_2) on Earth happened naturally until nearly 200 years ago. During the Industrial Revolution of the early 1800s, a new factor was thrown into the equation. When wood and fossil fuels like coal, oil, and natural gas are burned, CO_2 is released in large quantities. Oceans and vegetation absorb the gas. But due to the widespread cutting of trees to produce goods in the 1900s, few remained to soak up excess CO_2. In 1920, atmospheric CO_2 stood at about 280 parts per million (ppm). By 1996, it had risen to 356 ppm.

1 Make a chart that shows the changes in the levels of CO_2 in our atmosphere. Assume that the levels of CO_2 grow at a constant rate over the years. Use the headings *Year, Amount of CO_2 (ppm)*, and *Change*. Start the chart with 1920, and end with the present year.

2 Write a paragraph about the patterns you see in your chart. Use the terms *function, independent, dependent, domain*, and *range*.

3 Based on the relationships you found in the previous question, determine how much CO_2 was in the atmosphere in 1800 and 1950. Explain how you found those amounts.

4 Predict the amounts of CO_2 that will be in the atmosphere in the years 2000, 2020, 2050, and 3000. Explain how you made your predictions.

Add the results of your work to your Investigation Folder.

3–3A Solving Multi-Step Equations

Materials: cups & counters   equation mat

You can use an equation model to solve equations with more than one operation or equations with a variable on each side.

A Preview of Lesson 3–3

Activity 1   Use an equation model to solve 2x + 2 = −4.

Step 1 Model the equation by placing 2 cups and 2 positive counters on one side of the mat. Place 4 negative counters on the other side of the mat. Add 2 negative counters to each side to form zero pairs on the left.

Step 2 Group the counters to form zero pairs and remove the zero pairs. Separate the remaining counters into 2 equal groups to match the 2 cups. Each cup is paired with 3 negative counters. Therefore, $x = -3$.

$$2x + 2 = -4$$

$$x = -3$$

Activity 2   Use an equation model to solve w − 3 = 2w − 1.

Step 1 Model the equation by placing 1 cup and 3 negative counters on one side of the mat. Place 2 cups and 1 negative counter on the other side of the mat. Remove 1 negative counter from each side of the mat.

Step 2 You can remove the same number of cups from each side of the mat. In this case, remove 1 cup from each side. The cup on the right is matched with 2 negative counters. Therefore, $w = -2$.

$$w - 3 = 2w - 1$$

$$w = -2$$

Model   Use equation models to solve each equation.

1. $2x + 3 = 13$
2. $2y - 2 = -4$
3. $-4 = 3a + 2$
4. $3m - 2 = 4$
5. $3x + 2 = x + 6$
6. $3x + 7 = x + 1$
7. $3x - 2 = x + 6$
8. $y + 1 = 3y - 7$
9. $2b + 3 = b + 1$

Solving Multi-Step Equations

APPLICATION
World Cultures

On June 9, 1993, Japan's Crown Prince Naruhito wed former diplomat Masako Owada. It took Ms. Owada about $2\frac{1}{2}$ hours to put on her wedding kimono, a 12-layered silk garment that weighed about 30 pounds. The prince wore a kimono of bright orange—the "sunrise color"—that can only be worn by the heir to the throne.

After the wedding, the crown prince changed into a tuxedo and the princess into a bridal gown to formally announce their marriage to the Emperor and Empress. To change into the bridal gown, Princess Masako and her attendants had to remove the wedding kimono, a process that was similar to putting on the garment, only in reverse order. You will solve certain kinds of problems by working in reverse order, or **working backward.** Work backward is one of many *problem-solving strategies* that you can use to solve problems. Here are some other problem-solving strategies.

Problem-Solving Strategies	
draw a diagram	solve a simpler (or a similar) problem
make a table or chart	eliminate the possibilities
make a model	look for a pattern
guess and check	act it out
check for hidden assumptions	list the possibilities
use a graph	identify subgoals

What YOU'LL LEARN
- To solve equations involving more than one operation, and
- to solve problems by working backward.

Why IT'S IMPORTANT
You can use equations to solve problems involving home repair, health, and world cultures.

F Y I
The *kimono* has been worn by Japanese men and women since the 7th Century A.D. It is an ankle-length gown with long wide sleeves and is secured by a sash called an *obi*.

Example **1** Due to melting, an ice sculpture loses one-half its weight every hour. After 8 hours, it weighs $\frac{5}{16}$ of a pound. How much did it weigh in the beginning?

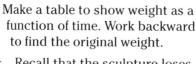

Make a table to show weight as a function of time. Work backward to find the original weight.

Recall that the sculpture loses one-half its weight every hour. Multiply the current weight of the sculpture by 2 to find its weight an hour before. Continue multiplying until you reach the original weight.

The original weight of the sculpture was 80 pounds.

Hour	Weight
8	$\frac{5}{16}$
7	$2\left(\frac{5}{16}\right) = \frac{5}{8}$
6	$2\left(\frac{5}{8}\right) = \frac{5}{4}$
5	$2\left(\frac{5}{4}\right) = \frac{5}{2}$
4	$2\left(\frac{5}{2}\right) = 5$
3	$2(5) = 10$
2	$2(10) = 20$
1	$2(20) = 40$
0	$2(40) = 80$

To solve equations with more than one operation, often called **multi-step equations,** you will undo the operations by working backward.

Example ❷ **Mrs. Guzman needs a repairperson to fix her washing machine. Since her machine is pretty old, she doesn't want to spend more than $100 for repairs. The owner of Albie's Appliances told her that a service call would cost $35 and the labor would be an additional $20 per hour. What is the maximum number of hours that the repairperson can work and keep the total cost at $100?**

APPLICATION

Home Repair

Explore   Read the problem and define the variable.
Let h represent the maximum number of hours the repair can take.

Plan   Write an equation.

$35 plus $20 per hour for h hours equals $100.$

$$35 \; + \; 20h \; = \; 100$$

Solve   Work backward to solve the equation.

$$35 + 20h = 100$$ *Undo the addition first. Use the subtraction*
$$35 - 35 + 20h = 100 - 35$$ *property of equality.*
$$20h = 65$$

$$\frac{20h}{20} = \frac{65}{20}$$ *Then undo the multiplication. Use the division property of equality.*

$$h = \frac{13}{4} \text{ or } 3\frac{1}{4}$$

The repairperson has up to $3\frac{1}{4}$ hours to fix the washing machine.

Examine   Check to see if the answer makes sense.

$$3\frac{1}{4} \times \$20 = \$65 \qquad\qquad \$35 + \$65 = \$100$$

EXPLORATION

GRAPHING CALCULATORS

You can use a graphing calculator, like the TI-83, to solve multi-step equations. The *solve(* function will solve an equation if it is rewritten as an expression that equals zero, or most commonly, in the form $ax + b - c = 0$. This function also requires that you include a guess about the solution to the equation. The format is as follows.

solve(*expression,variable,guess*)

To solve the equation $3x - 4 = 2$, access the math function by pressing the ⌑MATH⌑ key. Then choose 0, for solve. Enter the expression $3x - 4 - 2$, the variable x, and a guess about its solution. If we guess -1, then this appears on the calculator screen: solve(3X-4-2,X,-1). When you press ⌑ENTER⌑, the solution, 2, appears on the screen.

Your Turn

a. Work through the examples in this lesson by using a graphing calculator.

b. Describe the process of using a graphing calculator to solve equations in your own words.

c. Why did we input the equation $3x - 4 = 2$ as $3x - 4 - 2$?

d. How would you use a graphing calculator to solve the equation $3x - 4 = -2x + 6$?

You have seen a multi-step equation in which the first, or *leading,* coefficient is an integer. You can use the same steps if the leading coefficient is a fraction.

Example **3** **Solve each equation.**

a. $\frac{y}{5} + 9 = 6$

$$\frac{y}{5} + 9 = 6 \qquad \textit{The leading coefficient is } \frac{1}{5}.$$

$$\frac{y}{5} + 9 - 9 = 6 - 9 \qquad \textit{First, subtract 9 from each side. Why?}$$

$$\frac{y}{5} = -3$$

$$5\left(\frac{y}{5}\right) = 5(-3) \qquad \textit{Then, multiply each side by 5.}$$

$$y = -15$$

Check: $\qquad \frac{y}{5} + 9 = 6$

$$\frac{-15}{5} + 9 \stackrel{?}{=} 6$$

$$-3 + 9 \stackrel{?}{=} 6$$

$$6 = 6 \quad ✔$$

The solution is -15.

b. $\frac{d - 2}{3} = 7$

$$\frac{d - 2}{3} = 7$$

$$3\left(\frac{d - 2}{3}\right) = 3(7) \qquad \textit{Multiply each side by 3. Why?}$$

$$d - 2 = 21$$

$$d - 2 + 2 = 21 + 2 \qquad \textit{Add 2 to each side.}$$

$$d = 23$$

Check: $\qquad \frac{d - 2}{3} = 7$

$$\frac{23 - 2}{3} \stackrel{?}{=} 7$$

$$\frac{21}{3} \stackrel{?}{=} 7$$

$$7 = 7 \quad ✔$$

The solution is 23.

Consecutive integers are integers in counting order, such as 3, 4, 5. Beginning with an even integer and counting by two will result in *consecutive even integers.* For example, $-6, -4, -2, 0,$ and 2 are consecutive even integers. Beginning with an odd integer and counting by two will result in *consecutive odd integers.* For example, $-1, 1, 3,$ and 5 are consecutive odd integers.

The study of numbers and the relationships between them is called **number theory.** Number theory involves the study of odd and even numbers.

Example **4** **Find three consecutive odd integers whose sum is −15.**

Number Theory

Let n = the least odd integer.
Then $n + 2$ = the next greater odd integer,
and $n + 4$ = the greatest of the three odd integers.

$$n + (n + 2) + (n + 4) = -15$$

$$3n + 6 = -15 \qquad \textit{Simplify.}$$

$$3n + 6 - 6 = -15 - 6 \quad \textit{Subtract 6 from each side.}$$

$$3n = -21$$

$$\frac{3n}{3} = \frac{-21}{3} \qquad \textit{Divide each side by 3.}$$

$$n = -7$$

$n + 2 = -7 + 2 \qquad\qquad n + 4 = -7 + 4$

$n + 2 = -5 \qquad\qquad\quad n + 4 = -3$

The consecutive odd integers are -7, -5, and -3.

Explain why this answer makes sense.

CHECK FOR UNDERSTANDING

Communicating Mathematics

Study the lesson. Then complete the following.

1. a. **Explain** how you would solve $2p + 10 = 42$ if you had to undo the multiplication first.
   b. **Explain** why undoing the multiplication first would be inconvenient for solving the equation $7x - 4 = 24$.

2. **Determine** whether the sum of two consecutive even numbers can ever equal the sum of two consecutive odd numbers. Justify your answer.

3. If n is an even integer, explain how to find the even integer just before it.

4. **Explain** why -2 is not the solution of the equation $4x + 3x - 5 = 27$.

5. **List** three consecutive even integers if the greatest one is -18.

6. **Complete:** If $2x + 1 = 5$, then $3x - 4 = \underline{\ ?\ }$. Explain your reasoning.

MODELING MATHEMATICS

7. Write an equation for the model at the right. Then use cups and counters to solve the equation.

Guided Practice

Explain how to solve each equation. Then solve.

8. $4a - 5 = 15$
9. $7 + 3c = -11$
10. $\frac{2}{9}v - 6 = 14$

11. $3 - \frac{a}{7} = -2$
12. $\frac{x - 3}{7} = -2$
13. $\frac{p - (-5)}{-2} = 6$

Define a variable, write an equation, and solve each problem. Then check your solution.

14. Find two consecutive integers whose sum is -31.

15. Find three consecutive odd integers whose sum is 21.

16. Twenty-nine is 13 added to 4 times a number. Find the number.

Practice

Solve each equation. Then check your solution.

17. $3x - 2 = -5$ 18. $-4 = 5n + 6$ 19. $5 - 9w = 23$

20. $17 = 7 + 8y$ 21. $-2.5d - 32.7 = 74.1$ 22. $0.2n + 3 = 8.6$

23. $\frac{3}{2}a - 8 = 11$ 24. $5 = -9 - \frac{p}{4}$ 25. $7 = \frac{c}{-3} + 5$

26. $\frac{m}{-5} + 6 = 31$ 27. $\frac{g}{8} - 6 = -12$ 28. $6 = -12 + \frac{h}{-7}$

29. $\frac{b + 4}{-2} = -17$ 30. $\frac{z - 7}{5} = -3$ 31. $-10 = \frac{17 - s}{4}$

32. $\frac{4t - 5}{-9} = 7$ 33. $\frac{7n + (-1)}{6} = 5$ 34. $\frac{-3j - (-4)}{-6} = 12$

Number Theory

Define a variable, write an equation, and solve each problem. Then check your solution.

35. Twelve decreased by twice a number is -7.

36. Find three consecutive integers whose sum is -33.

37. Find four consecutive integers whose sum is 86.

38. Find two consecutive odd integers whose sum is 196.

Geometry

Make a diagram to represent each situation. Then define a variable, write an equation, and solve each problem. Check your solution.

39. The lengths of the sides of a triangle are consecutive odd integers. The perimeter is 39 meters. What are the lengths of the sides?

40. The lengths of the sides of a quadrilateral are consecutive even integers. Twice the length of the shortest side plus the length of the longest side is 120 inches. Find the lengths of the four sides.

Graphing Calculator

Solve each equation. Then use a graphing calculator to verify your solution.

41. $0.2x + 3 = 8.6$ 42. $4.91 + 7.2x = 39.4201$

43. $\frac{3 + x}{7} = -5$ 44. $\frac{-3n - (-4)}{-6} = -9$

Critical Thinking

45. Write an expression for the sum of three consecutive even integers if $3n - 1$ is the least integer.

Applications and Problem Solving

For Exercises 46–47, write an equation to represent each situation. Then solve the problem.

46. **Health** According to the American Medical Association, the average birth weight of a baby in the United States is 7.5 pounds. After birth, most babies lose an average of 1 ounce a day for the first 5 days and then gain an average of 1 ounce a day for the next 13 weeks.

    a. Suppose a baby weighs 7.5 pounds at birth. After how many days will the baby weigh 12 pounds?

    b. Draw a graph that shows the expected weight of a 7.5-pound baby for each of its first 13 weeks of life.

47. **World Cultures** Hawaii became our 50th state in 1959. It's the only state that is not on the North American continent. The English alphabet contains 2 more than twice as many letters as the Hawaiian alphabet. How many letters are there in the Hawaiian alphabet?

48. Work Backward Four families went to a baseball game. A vendor selling bags of popcorn came by. The Wilson family bought half of the bags of popcorn plus one. The Martinez family bought half of the remaining bags of popcorn plus one. The Brightfeather family bought half of the remaining bags of popcorn plus one. The Wimberly family bought half of the remaining bags of popcorn plus one, leaving the vendor with no bags of popcorn. If the Wimberlys bought 2 bags of popcorn, how many bags did each of the four families buy?

Mixed Review

49. Health One way to calculate your suggested daily intake of fat grams is to multiply your ideal weight in pounds times 0.454. (Lesson 3–2)

   a. Write an equation to represent your suggested daily intake.

   b. What is your suggested daily intake of fat grams?

50. Complete: If $x + 4 = 15$, then $x - 2 = \underline{\ ?\ }$. (Lesson 3–1)

51. Name the set or sets of numbers to which $\sqrt{11}$ belongs. Use N for natural numbers, W for whole numbers, Z for integers, Q for rational numbers, or I for irrational numbers. (Lesson 2–8)

52. Simplify $\dfrac{-200x}{50}$. (Lesson 2–7)

53. Simplify $4(7) - 3(11)$. (Lesson 2–6)

54. Name the property illustrated by $9 + (2 + 10) = 9 + 12$. (Lesson 1–6)

55. Solve $a = \dfrac{12 + 8}{4}$. (Lesson 1–5)

56. Evaluate $\dfrac{3}{4}(6) + \dfrac{1}{3}(12)$. (Lesson 1–3)

SELF TEST

Solve each equation. (Lessons 3–1, 3–2, and 3–3)

  1. $-10 + k = 34$ **2.** $y - 13 = 45$ **3.** $20.4 = 3.4y$

  4. $-65 = \dfrac{f}{29}$ **5.** $-3x - 7 = 18$ **6.** $5 = \dfrac{m - 5}{4}$

Write an equation and solve. Then check your solution.

7. Twenty-three minus a number is 42. Find the number. (Lesson 3–1)

8. Geometry The length of one side of a rectangle is 34 cm. The area of the rectangle is 68 cm². Find the width of the rectangle. (Lesson 3–2)

9. Number Theory Find two consecutive even integers whose sum is 126. (Lesson 3–3)

10. Economics According to the *Baltimore Sun*, if inflation continues at the current rate, the price of a movie ticket in 40 years will be $3.50 less than 10 times the current average price of $5.05. What will be the price of a movie ticket in 40 years? (Lesson 3–3)

Integration: Geometry
Angles and Triangles

What YOU'LL LEARN

- To find the complement and supplement of an angle, and
- to find the measure of the third angle of a triangle given the measures of the other two angles.

Why IT'S IMPORTANT

A knowledge of angles and triangles will help you further your study of geometry and trigonometry.

APPLICATION
Football

On fourth down in a football game, the team with the ball usually *punts*. A punt is a kick in which the ball is dropped by the kicker and is kicked before it hits the ground.

The longest punt ever kicked in a college football game was kicked by Pat Brady of the University of Nevada-Reno in their 1950 game against Loyola University. The punt was 99 yards long—just 1 yard short of spanning the entire length of the football field.

How far a football will travel once it is punted depends not only on the strength of the punter, but also on the angle at which the ball is kicked. If the angle is too small, the ball will travel close to the ground, then drop; if the angle is too large, the ball will travel high, but not very far. What do you think would be the best angle at which to kick the ball?

The symbol for angle is ∠.

You can use a *protractor* to measure angles, as shown below.

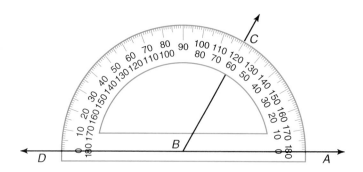

Angle *ABC* (denoted ∠*ABC*) measures 60°. However, where ray *BC* (denoted $\overrightarrow{BC}$) intersects the curve of the protractor, there are two readings, 60° and 120°. The measure of ∠*DBC* is 120°. What is the sum of the measures of ∠*ABC* and ∠*DBC*?

Supplementary Angles	Two angles are supplementary if the sum of their measures is 180°.

Example **1** The measure of an angle is three times the measure of its supplement. Find the measure of each angle.

Let x = the lesser measure. Then $3x$ = the greater measure.

$$x + 3x = 180 \quad \textit{The angles are supplementary.}$$
$$4x = 180 \quad \textit{Add like terms.}$$
$$\frac{4x}{4} = \frac{180}{4} \quad \textit{Divide each side by 4.}$$
$$x = 45$$

The measures are 45° and 3 · 45° or 135°. *Check this result.*

| **Complementary Angles** | Two angles are complementary if the sum of their measures is 90°. |

Example **2**

Top Five List

Lowest winning scores in the U.S. Masters Tournament
1. 271, Raymond Floyd, 1976
1. 271, Jack Nicklaus, 1965
3. 274, Ben Hogan, 1953
4. 275, Severiano Ballesteros, 1980
4. 275, Fred Couples, 1992

The backward slant of the face of a golf club head is called the *loft*. It is designed to drive the ball in a high arc. Assuming that the angle made by the ground and the face of the club is 79°, what is the measure of the loft?

Let x = the measure of the loft.

$$79° + x = 90° \quad \textit{The angles are complementary.}$$
$$79° - 79° + x = 90° - 79° \quad \textit{Subtract 79° from each side.}$$
$$x = 11°$$

The club has an 11° loft.

Example **3** The measure of an angle is 34° greater than its complement. Find the measure of each angle.

Let x = the lesser measure. Then $x + 34$ = the greater measure.

$$x + (x + 34) = 90 \quad \textit{The angles are complementary.}$$
$$2x + 34 = 90$$
$$2x + 34 - 34 = 90 - 34 \quad \textit{Subtract 34 from each side.}$$
$$2x = 56$$
$$\frac{2x}{2} = \frac{56}{2} \quad \textit{Divide each side by 2.}$$
$$x = 28$$

The measures are 28° and 28° + 34° or 62°. *Check this result.*

A **triangle** is a polygon with three sides and three angles. What is the sum of the measures of the three angles of a triangle?

MODELING MATHEMATICS

Angles of a Triangle

Materials: 3 sheets of construction paper   protractor   scissors

Copy each triangle below onto a sheet of construction paper. If possible, you may want to enlarge the triangles on a photocopier.

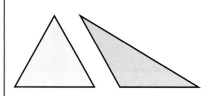

b. Cut out one of the triangles. Label the angles A, B, and C. Tear off the angles and arrange as shown.

c. When the angles were put together, what was the measure of the new angle they formed?

d. Try this activity with another triangle. What do you think is true about the measures of the angles of any triangle?

Your Turn

a. Use a protractor to measure the angles of each triangle above. What is the sum of the measures of the three angles of each triangle?

Sum of the Angles of a Triangle	**The sum of the measures of the angles in any triangle is 180°.**

In an **equilateral triangle,** each angle has the same measure. We say that the angles are **congruent.** The sides of an equilateral triangle are also congruent. What is the measure of each angle of an equilateral triangle?

Let x = the measure of each angle.

$x + x + x = 180$

$3x = 180$ *Add like terms.*

$\dfrac{3x}{3} = \dfrac{180}{3}$ *Divide each side by 3.*

$x = 60$ Each angle measures 60°.

In an **isosceles triangle,** at least two angles have the same measure. Generally, the two congruent angles are the base angles.

Example **4**   **What are the measures of the base angles of an isosceles triangle in which the vertex angle measures 45°?**

Let x = the measure of each base angle.

$x + x + 45 = 180$

$2x + 45 = 180$ *Add like terms.*

$2x + 45 - 45 = 180 - 45$ *Subtract 45 from each side.*

$2x = 135$

$\dfrac{2x}{2} = \dfrac{135}{2}$ *Divide each side by 2.*

$x = \dfrac{135}{2}$ or $67\dfrac{1}{2}$ The base angles each measure $67\dfrac{1}{2}$°.

Triangles are often classified in one of three ways. A **right triangle** has one angle that measures 90°. An **obtuse triangle** has one angle with measure greater than 90°. In an **acute triangle,** all of the angles measure less than 90°.

Example **The measures of the angles of a triangle are given as $x°$, $2x°$, and $3x°$.**

    a. What are the measures of each angle?
    b. Classify the triangle.

    a. The sum of the measures of the angles of a triangle is 180°.

$$x + 2x + 3x = 180$$
$$6x = 180 \quad \textit{Add like terms.}$$
$$\frac{6x}{6} = \frac{180}{6} \quad \textit{Divide each side by 6.}$$
$$x = 30$$

    The measures are 30°, 2(30°) or 60°, and 3(30°) or 90°.

    b. Since the triangle has one angle that measures 90°, it is a right triangle.

CHECK FOR UNDERSTANDING

Communicating Mathematics

Study the lesson. Then complete the following.

1. The words *complement* and *compliment* sound very much alike.
   a. What are their meanings?
   b. Which word has the mathematical meaning?

2. **Compare and contrast** the standard definition and the mathematical meaning of the word *supplement.*

3. **Classify** each of the three triangles in the Modeling Mathematics activity on page 164 as right, obtuse, or acute. Justify your classifications.

4. **Make up a problem** in which you must use an equation to find the measures of the angles of a triangle. Then solve the problem, explaining each step.

5. Draw a right triangle on a separate piece of paper and cut it out. Cut or tear off the two acute angles and arrange as shown. What seems to be true about the two acute angles?

Guided Practice

Find both the complement and the supplement of each angle measure.

    6. 130°   7. 11°   8. 24°

    9. $3x°$ 10. $(2x + 40)°$ 11. $(x - 20)°$

Find the measure of the third angle of each triangle in which the measures of two angles of the triangle are given.

    12. 16°, 42°   13. 50°, 45°   14. $x°$, $(x + 20)°$

Make a diagram to represent each situation. Then define a variable, write an equation, and solve each problem. Check your solution.

15. An angle measures 38° less than its complement. Find the measures of the two angles.

16. The measures of the angles of a triangle are given as $x°$, $(x + 5)°$, and $(2x + 3)°$. What are the measures of each angle?

EXERCISES

Practice

Find both the complement and the supplement of each angle measure.

17. 42°	**18.** 87°	**19.** 125°
20. 90°	**21.** 21°	**22.** 174°
23. 99°	**24.** $y°$	**25.** $3a°$
26. $(x + 30)°$	**27.** $(b - 38)°$	**28.** $(90 - z°)$

The measures of two angles of a triangle are given. Find the measure of the third angle.

29. 40°, 70°	**30.** 90°, 30°	**31.** 63°, 12°
32. 43°, 118°	**33.** 4°, 38°	**34.** $x°$, $y°$
35. $p°$, $(p - 10)°$	**36.** $c°$, $(2c + 1)°$	**37.** $y°$, $(135 - y)°$

Draw a diagram to represent each situation. Then define a variable, write an equation, and solve each problem. Check your solution.

38. One of the congruent angles of an isosceles triangle measures 37°. Find the measures of the other angles.

39. Find the measure of an angle that is 30° less than its supplement.

40. One of the angles of a triangle measures 53°. Another angle measures 37°. What is the measure of the third angle?

41. One of two complementary angles measures 30° more than three times the other. Find the measure of each angle.

42. Find the measure of an angle that is one-half the measure of its supplement.

43. The measures of the angles of a triangle are given as $x°$, $(3x)°$, and $(4x)°$. What are the measures of each angle?

Critical Thinking

44. Draw a diagram that shows the relationships among right, isosceles, equilateral, acute, scalene, and obtuse triangles.

Applications and Problem Solving

45. Carpentry A *stringer* is a triangular piece of wood on which staircases are based. If the stringer makes a 30° angle with the floor, what is the measure of the angle that the stringer makes with the vertical?

46. Aeronautics   One of the newer space shuttles is *Endeavour.* Its first flight left Earth on May 7, 1992. The oldest shuttle is *Columbia,* which flew its first flight on April 12, 1981. A space shuttle lands at an angle six times as great as that of the average commercial jet.

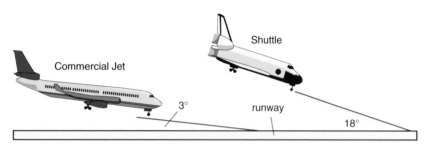

a. What is the measure of the angle the path of the jet makes with the vertical?

b. What is the measure of the angle the path of the shuttle makes with the vertical?

c. Which craft lands at a steeper angle?

47. Mountain Climbing   *Rapelling* is a technique used by climbers to make a difficult descent. Climbers back to the edge and then spring off. To avoid slipping, the climber's legs should remain at a 90° angle with the side of the ledge. The ledge at the right is 50° off the vertical. What is the measure of the climber's angle with the vertical?

Mixed Review

48. Solve $4 + 7x = 39$.   (Lesson 3–3)

49. Consumerism   When stores go out of business, there is generally a period when all of the items in the store can be purchased for a fraction of the original price. When Central Hardware went out of business, all of the items in the store began at $\frac{1}{4}$ off.   (Lesson 3–2)

a. What was the original price of a circular saw if you save $20.00 during the sale?

b. How much would you save on patio furniture that originally cost $299.00?

50. Solve $h + (-13) = -5$.   (Lesson 3–1)

51. Simplify $5(3t - 2t) + 2(4t - 3t)$.   (Lesson 2–6)

52. Find the sum $5y + (-12y) + (-21y)$.   (Lesson 2–5)

53. American History   Each number below represents the age of a U.S. president on his first inauguration.   (Lesson 2–2)

57 61 57 57 58 57 61 54 68 51 49 64 50 48

65 52 56 46 54 49 50 47 55 55 54 42 51 56

55 51 54 51 60 62 43 55 56 61 62 69 64 46

a. Make a number line plot of the ages of the presidents.

b. Do any of the ages appear to be clustered? If so, which ones?

54. Name the property illustrated by $5a + 2b = 2b + 5a$.   (Lesson 1–8)

55. Statistics   Make a stem-and-leaf plot of the data in Exercise 53.   (Lesson 1–4)

Solving Equations with the Variable on Both Sides

APPLICATION
Track and Field

What YOU'LL LEARN
- To solve equations with the variable on both sides, and
- to solve equations containing grouping symbols.

Why IT'S IMPORTANT

You can use equations to solve problems involving track and field, business, and geometry.

In the 1928 Olympics, the winner of the men's 800-meter run, Douglas Lowe of Great Britain, won the race in 1 minute 51.8 seconds, or 111.8 seconds. In that same year, the winner of the women's 800-meter run, Lina Radke of Germany, won the race in 2 minutes 16.8 seconds, or 136.8 seconds. Over the next 64 years, the men's winning times decreased an average of 0.127 seconds per year, and the women's winning times decreased an average of 0.332 seconds per year. Suppose the times continued to decrease at these rates. When will men and women have the same winning times in the 800-meter run?

Wilma Rudolph (1940–1994)

Wilma Rudolph was the first woman to win three gold medals at a single Olympic games. She won the 100-meter sprint, the 200-meter dash, and the 4 × 100-meter relay at the 1960 Olympic Games.

After x years, the men's winning times can be represented by $111.8 - 0.127x$. After x years, the women's winning times can be represented by $136.8 - 0.332x$. The times would be the same when the two expressions are equal.

$$111.8 - 0.127x = 136.8 - 0.332x$$

Many equations contain variables on each side. To solve these types of equations, first use the addition or subtraction property of equality to write an equivalent equation that has all of the variables on one side. Then solve the equation.

Method 1

$$111.8 - 0.127x = 136.8 - 0.332x \qquad \textit{Use a calculator.}$$

$$111.8 - 0.127x + 0.332x = 136.8 - 0.332x + 0.332x \qquad \textit{Add 0.332x to each side.}$$

$$111.8 + 0.205x = 136.8$$

$$111.8 - 111.8 + 0.205x = 136.8 - 111.8 \qquad \textit{Subtract 111.8 from each side.}$$

$$0.205x = 25$$

$$\frac{0.205x}{0.205} = \frac{25}{0.205} \qquad \textit{Divide each side by 0.205.}$$

$$x \approx 122$$

At these rates, the men's and women's winning times will be equal about 122 years after 1928, or in the year 2050.

Method 2

Since two of the decimals involve thousandths, another way to solve the equation would be to multiply each side by 1000 to clear the decimals.

$$111.8 - 0.127x = 136.8 - 0.332x$$

Multiply each side by 1000. $$1000(111.8) - 1000(0.127x) = 1000(136.8) - 1000(0.332x)$$

$$111,800 - 127x = 136,800 - 332x$$

Add 332x to each side. $$111,800 + 205x = 136,800$$

Subtract 111,800 from each side. $$205x = 25,000$$

Divide each side by 205. $$x \approx 122$$

Example ❶ Solve $\frac{3}{8} - \frac{1}{4}x = \frac{1}{2}x - \frac{3}{4}$.

$$\frac{3}{8} - \frac{1}{4}x = \frac{1}{2}x - \frac{3}{4}$$ *The least common denominator is 8.*

$$8\left(\frac{3}{8} - \frac{1}{4}x\right) = 8\left(\frac{1}{2}x - \frac{3}{4}\right)$$ *Multiply each side by 8.*

$$8\left(\frac{3}{8}\right) - 8\left(\frac{1}{4}x\right) = 8\left(\frac{1}{2}x\right) - 8\left(\frac{3}{4}\right)$$ *Use the distributive property.*

$$3 - 2x = 4x - 6$$ *The fractions are eliminated.*

$$3 = 6x - 6$$ *Add 2x to each side.*

$$9 = 6x$$ *Add 6 to each side.*

$$\frac{3}{2} = x$$ *Divide each side by 6.*

The solution is $\frac{3}{2}$.   *Check this result.*

TECHNOLOGY Tip

Look back to Lesson 3–3 to see how you would solve an equation with the variable on both sides by using a graphing calculator.

When solving equations that contain grouping symbols, first use the distributive property to remove the grouping symbols.

Example ❷

INTEGRATION
Geometry

One angle of a triangle measures 10° more than the second. The measure of the third angle is twice the sum of the first two angles. Find the measure of each angle.

Make a drawing. Let y = the measure of one angle. Let $y + 10$ = the measure of another angle. Then $2[y + (y + 10)]$ = the measure of the third angle. Recall that the sum of the measures of the angles in any triangle is 180°.

$$y + (y + 10) + 2[y + (y + 10)] = 180$$

$$2y + 10 + 2(2y + 10) = 180$$ *Simplify.*

$$2y + 10 + 4y + 20 = 180$$ *Use the distributive property.*

$$6y + 30 = 180$$ *Simplify.*

$$6y + 30 - 30 = 180 - 30$$ *Subtract 30 from each side.*

$$6y = 150$$

$$y = 25$$ *Divide each side by 6.*

The measures of the three angles are 25°, (25 + 10)° or 35°, and 2(25 + 35)° or 120°.   *Check this result.*

Some equations with the variable on both sides may have no solution. That is, there is no value of the variable that will result in a true equation.

Example ③ Solve $5n + 4 = 7(n + 1) - 2n$.

$5n + 4 = 7(n + 1) - 2n$

$5n + 4 = 7n + 7 - 2n$ *Distributive property*

$5n + 4 = 5n + 7$

$4 = 7$ Since $4 = 7$ is a false statement, this equation has no solution.

An equation that is true for every value of the variable is called an **identity**.

Example ④ Solve $7 + 2(x + 1) = 2x + 9$.

$7 + 2(x + 1) = 2x + 9$

$7 + 2x + 2 = 2x + 9$ *Distributive property*

$2x + 9 = 2x + 9$ *Reflexive property of equality*

Since the expressions on each side of the equation are the same, this equation is an identity. The statement $7 + 2(x + 1) = 2x + 9$ is true for all values of x.

CHECK FOR UNDERSTANDING

Communicating Mathematics

Study the lesson. Then complete the following.

1. a. **Explain** why you think winning times in the 800-meter run for men and women are growing closer.
   b. **Use a table** to show how women's winning times in the 800-meter run could catch up to men's winning times in the same race.
   c. Do you think the times will ever be the same? Why or why not?

2. **Describe** the difference between an identity and an equation with no solution.

3. **You Decide** Lauren says that to solve the equation $2[x + 3(x - 1)] = 18$, the first step should be to multiply 2 by $[x + 3(x - 1)]$. Carmen says the first step should be to multiply 3 by $(x - 1)$. Which one is correct, and why?

 MATH JOURNAL

4. **Assess Yourself** Describe your favorite sport and explain how math is used in it.

Guided Practice

Explain how to solve each equation. Then solve each equation and check your solution.

5. $3 - 4x = 10x + 10$ 6. $8y - 10 = -3y + 2$ 7. $\frac{3}{5}x + 3 = \frac{1}{5}x - 7$

8. $5.4y + 8.2 = 9.8y - 2.8$ 9. $5x - 7 = 5(x - 2) + 3$ 10. $4(2x - 1) = -10(x - 5)$

Define a variable, write an equation, and solve each problem. Then check your solution.

11. Twice the greater of two consecutive odd integers is 13 less than three times the lesser. Find the integers.

12. **Geometry** The measures of the angles of a triangle are given as $6x°$, $(x - 3)°$, and $(3x + 7)°$. What are the measures of each angle?

13. One half of a number increased by 16 is four less than two thirds of the number. Find the number.

Practice

Solve each equation. Then check your solution.

14. $6x + 7 = 8x - 13$

15. $17 + 2n = 21 + 2n$

16. $\dfrac{3n - 2}{5} = \dfrac{7}{10}$

17. $\dfrac{7 + 3t}{4} = -\dfrac{t}{8}$

18. $13.7b - 6.5 = -2.3b + 8.3$

19. $18 - 3.8x = 7.36 - 1.9x$

20. $\dfrac{3}{2}y - y = 4 + \dfrac{1}{2}y$

21. $\dfrac{3}{4}n + 16 = 2 - \dfrac{1}{8}n$

22. $-7(x - 3) = -4$

23. $4(x - 2) = 4x$

24. $28 - 2.2y = 11.6y + 262.6$

25. $1.03x - 4 = -2.15x + 8.72$

26. $7 - 3x = x - 4(2 + x)$

27. $6 = 3 + 5(y - 2)$

28. $6(y + 2) - 4 = -10$

29. $5 - \dfrac{1}{2}(b - 6) = 4$

30. $-8(4 + 9x) = 7(-2 - 11x)$

31. $2(x - 3) + 5 = 3(x - 1)$

32. $4(2a - 8) = \dfrac{1}{7}(49a + 70)$

33. $-3(2n - 5) = \dfrac{1}{2}(-12n + 30)$

Define a variable, write an equation, and solve each problem. Then check your solution.

34. Three times the greatest of three consecutive even integers exceeds twice the least by 38. Find the integers.

35. One fifth of a number plus five times that number is equal to seven times the number less 18. Find the number.

36. The difference of two numbers is 12. Two fifths of the greater number is six more than one third of the lesser number. Find both numbers.

Geometry

Make a diagram to represent each situation. Then define a variable, write an equation, and solve each problem. Check your solution.

37. The measures of the angles of a certain triangle are consecutive even integers. Find their measures.

38. One angle of a triangle measures 30° more than another angle. The measure of the third angle is three times the sum of the first two angles. Find the measure of each angle.

Programming

39. The graphing calculator program at the right can help you solve equations of the form $ax + b = cx + d$.

Write each equation in the form $ax + b = cx + d$. Then run the program to find the solution.

a. $2(2x + 3) = 4x + 6$

b. $5x - 7 = x + 3$

c. $6 - 3x = 3x - 6$

d. $6.8 + 5.4x = 4.6x + 2.8$

e. $5x - 8 - 3x = 2(x - 3)$

```
PROGRAM:SOLVE
: Disp "ENTER A,B,C,D"
: Prompt A: Prompt B: Prompt C:
  Prompt D
: If A - C ≠ 0
: Goto 2
: If D - B ≠ 0
: Goto 1
: Disp "THIS IS AN"
: Disp "IDENTITY"
: Goto 3
: Lbl 1
: Disp "NO SOLUTION"
: Goto 3
: Lbl 2
: Disp "X = ", (D - B)/(A - C)
: Lbl 3
```

40. Mathematics History   Diophantus of Alexandria was one of the greatest mathematicians of the Greek civilization. Little is known of his life except for the following problem, which was first printed in a work called the *Greek Anthology*. Although the problem was not written by Diophantus, it is believed to accurately describe his life.

Diophantus passed one sixth of his life in childhood, one twelfth in youth, and one seventh more as a bachelor. Five years after his marriage, there was born a son who died four years before his father, at half his father's (final) age. How old was Diophantus when he died?

**Applications and
Problem Solving**

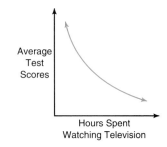

For Exercises 41–42, assume that the rates of change continue indefinitely.

41. Sales   According to the Association of Home Appliance Manufacturers, sales of room air conditioners were 4.6 million units in 1988, and sales of window fans were 0.975 million units in 1988. Since 1988, sales of room air conditioners have decreased about 0.425 million units per year, and sales of window fans have increased about 0.106 million units per year.

   a. If the trend continues, after how many years will sales of room air conditioners and window fans be equal?

   b. How would you explain these trends?

42. Pet Ownership   According to the American Veterinary Medical Association, in 1987, 34.7 million households owned a dog, and 27.7 million households owned a cat. Since 1987, dog ownership has decreased by about 0.025 million households per year, and cat ownership has increased by about 0.375 million households per year. If the trend continues, after how many years will the number of households that own dogs and cats be equal?

**Mixed
Review**

43. Find the supplement of a 32° angle.   (Lesson 3–4)

44. Work Backward   A number is decreased by 35, then multiplied by 6, then added to 87, then divided by 3. The result is 67. What is the number?
(Lesson 3–3)

45. Solve $x + 4.2 = 1.5$. Check your solution.   (Lesson 3–1)

46. Define a variable, then write an equation for the following problem.
(Lesson 2–9)

Karen is 10, and she has always admired her 15-year-old sister Kristy. Karen wants to be exactly like Kristy. She knows that she will never catch up with Kristy in age, but she has heard that when one person is 0.9 times as old as another, you can't really tell them apart. How long will it be before Karen is 0.9 times as old as Kristy?

47. Which is greater, $\frac{11}{9}$ or $\frac{12}{10}$?   (Lesson 2–4)

48. Explain what the graph at the right tells you about test scores as a function of hours spent watching television.   (Lesson 1–9)

49. Solve $14.8 - 3.75 = t$.   (Lesson 1–5)

Solving Equations and Formulas

3-6

What YOU'LL LEARN

- To solve equations and formulas for a specified variable.

Why IT'S IMPORTANT

You can use equations and formulas to solve problems involving physics and geometry.

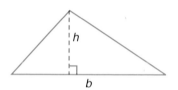

Geometry

Some equations contain more than one variable. At times you will need to solve these equations for one of the variables. For example, suppose the variables x, y, and z were all in the same equation. To *solve for x* would mean to get x by itself on one side of the equation, with no x's on the other side. In the same way, to *solve for y* would mean to get y by itself on one side of the equation, with no y's on the other side.

The formula for the area of a triangle is $A = \frac{1}{2}bh$ where b represents the length of the base and h represents the height of the triangle. Suppose you know the areas and the lengths of the bases of several triangles and you want to find the height of each triangle. Rather than solve the formula over and over for different values of A and b, it would be easier to solve the formula for h before substituting the values for the other variables.

$$A = \frac{1}{2}bh$$

$$2A = 2\left[\frac{1}{2}bh\right] \qquad \textit{Multiply each side by 2.}$$

$$2A = bh$$

$$\frac{2A}{b} = \frac{bh}{b} \qquad \textit{Divide each side by b.}$$

$$\frac{2A}{b} = h$$

When you divide by a variable in an equation, remember that division by 0 is undefined. For example, in the formula above, b cannot equal zero.

Example **1** Solve the equation $-5x + y = -56$

    a. for y, and

    b. for x.

a.
$$-5x + y = -56$$
$$-5x + 5x + y = -56 + 5x \qquad \textit{Add 5x to each side.}$$
$$y = -56 + 5x$$

b.
$$-5x + y = -56$$
$$-5x + y - y = -56 - y \qquad \textit{Subtract y from each side.}$$
$$-5x = -56 - y$$
$$\frac{5x}{5} = \frac{56 - y}{5} \qquad \textit{Divide each side by -5.}$$
$$x = \frac{56 - y}{5} \text{ or } \frac{56 + y}{5}$$

LOOK BACK

You can refer to Lesson 2-9 for information on writing equations and formulas.

Example 2 **Solve for y in $3y + z = am - 4y$.**

$$3y + z = am - 4y$$

$$3y + z - z = am - 4y - z \qquad \textit{Subtract } z \textit{ from each side.}$$

$$3y = am - 4y - z$$

$$3y + 4y = am + 4y - 4y - z \qquad \textit{Add } 4y \textit{ to each side.}$$

$$3y + 4y = am - z$$

$$(3 + 4)y = am - z \qquad \textit{Use the distributive property.}$$

$$7y = am - z$$

$$\frac{7y}{7} = \frac{am - z}{7} \qquad \textit{Divide each side by 7.}$$

$$y = \frac{am - z}{7}$$

Many real-world problems require the use of formulas. When using formulas, you may need to use **dimensional analysis.** This is the process of carrying units throughout a computation. You may also be asked to solve the formula for a specific variable. This may make it easier to use certain formulas.

Example 3 **The formula $P = \dfrac{1.2W}{H^2}$ represents the amount of pressure exerted on the floor by the heel of a shoe. In this formula, P represents the pressure in pounds per square inch (lb/in^2), W represents the weight of the person wearing the shoe in pounds, and H is the width of the heel of the shoe in inches.**

a. Find the amount of pressure exerted if a 130-pound person wore shoes with heels $\frac{1}{2}$ inch wide.

b. Solve the formula for W.

c. Find the weight of the person if the heel is 3 inches wide and the pressure exerted is 40 lb/in^2.

a. $P = \dfrac{1.2W}{H^2}$

$\quad = \dfrac{1.2(130 \text{ lb})}{\left(\frac{1}{2}\text{ in}\right)^2} \qquad \begin{array}{l} W = 130 \text{ lb,} \\ H = \frac{1}{2} \text{ in.} \end{array}$

$\quad = \dfrac{156 \text{ lb}}{\frac{1}{4}\text{ in}^2}$

$\quad = 624 \text{ lb/in}^2$

624 lb/in^2 of pressure are exerted.

b. $\qquad P = \dfrac{1.2W}{H^2}$

$\quad H^2 P = H^2\left(\dfrac{1.2W}{H^2}\right) \qquad \begin{array}{l} \textit{Multiply each} \\ \textit{side by } H^2. \end{array}$

$\quad H^2 P = 1.2W$

$\quad \dfrac{H^2 P}{1.2} = \dfrac{1.2W}{1.2} \qquad \begin{array}{l} \textit{Divide each} \\ \textit{side by 1.2.} \end{array}$

$\quad \dfrac{H^2 P}{1.2} = W$

c.

$$\frac{H^2P}{1.2} = W$$

$$\frac{(3 \text{ in.})^2(40 \text{ lb/in}^2)}{1.2} = W$$ *Estimate:* $[(3)^2 - 40] \div 1 = 360$
Will the actual answer be greater or less? Why?

$$\frac{360 \text{ lb}}{1.2} = W$$ $(9 \text{ in}^2)(40 \text{ lb/in}^2) = 360 \text{ lb}$

$$300 \text{ lb} = W$$ The person weighs 300 pounds.

CHECK FOR UNDERSTANDING

Communicating Mathematics

Study the lesson. Then complete the following.

1. **Show** how you would solve the equation in Example 2 for m.

2. **Compare and contrast** the amount of pressure being exerted in the case of the two people described in Example 3.

 MATH JOURNAL

3. Find a formula in a newspaper or magazine. Define all of the variables and explain what the purpose of the formula is.

Guided Practice

Solve each equation or formula for the variable specified.

4. $3x - 4y = 7$, for x 5. $3x - 4y = 7$, for y

6. $a(y + 1) = b$, for y 7. $4x + b = 2x + c$, for x

8. $F = G\left(\dfrac{Mm}{d^2}\right)$, for M 9. $S = \dfrac{n}{2}(A + t)$, for A

Write an equation and solve for the variable specified.

10. Twice a number x and 12 is 31 less than three times another number y. Solve for y.

EXERCISES

Practice

Solve each equation or formula for the variable specified.

11. $-3x + b = 6x$, for x 12. $ex - 2y = 3z$, for x

13. $\dfrac{y + a}{3} = c$, for y 14. $\dfrac{3}{5}y + a = b$, for y

15. $v = r + at$, for a 16. $y = mx + b$, for m

17. $I = prt$, for r 18. $\dfrac{by + 2}{3} = c$, for y

19. $H = (0.24)I^2Rt$, for R 20. $P = \dfrac{E^2}{R}$, for R

21. $4b - 5 = -t$, for b 22. $km + 5x = 6y$, for m

23. $c = \dfrac{3}{4}y + b$, for y 24. $p(t + 1) = -2$, for t

25. $\dfrac{5x + y}{a} = 2$, for a 26. $\dfrac{3ax - n}{5} = -4$, for x

Write an equation and solve for the variable specified.

27. Twice a number x increased by 12 is 31 less than three times another number y. Solve for x.

28. Five eighths of a number x is three more than one half of another number y. Solve for y.

29. Five more than two thirds of a number x is the same as three less than one half of another number y. Solve for x.

Critical Thinking

30. Weather Here's a conversation between a mother and her teenage daughter, adapted from the pages of *Games* magazine.

Mrs. Weatherby: "This thermometer's no good. It's in Celsius, and I want to know the temperature in Fahrenheit."

Mercuria: "You're so old-fashioned, Mother. Just use the formula $\frac{9}{5}C + 32 = F$, where C is the temperature in Celsius and F is the temperature in Fahrenheit. If that's too hard for you, then just double the number on the thermometer and add 30. Of course, you won't get a precise answer."

The two women each computed the temperature in degrees Fahrenheit, Mercuria with her formula and Mrs. Weatherby by the approximation. Surprisingly, they got exactly the same answer. Just how hot was it? Justify your reasoning.

Applications and Problem Solving

31. Business The formula in Exercise 17 is the formula for computing simple interest, where I is the interest, p is the principal or amount invested, r is the interest rate, and t is the time in years. Find the amount of interest earned if you were to invest $5000 at 6% interest (use 0.06), for 3 years.

32. Physics The formula $d = vt + \frac{1}{2}at^2$ is the formula for distance d, given the initial velocity v, the time t, and the acceleration a. Suppose a biker going 4.5 meters per second (m/s) passes a lamppost at the top of a hill and then accelerates down the hill at a constant rate of 0.4 m/s^2 for 12 seconds. How far does she go down the hill during this time?

33. Work The formula $s = \frac{w - 10e}{m}$ is often used by placement services to find keyboarding speeds. In the formula, w represents the number of words typed, e represents the number of errors, m represents the number of minutes typed, and s represents the speed in words per minute.

a. On a 10-minute keyboarding test, Sally typed 420 words with 6 errors. Find Sally's speed in words per minute.

b. Who has the greater speed on a 5-minute test: Isabel, who types 500 words with 14 errors, or Clarence, who types 410 words with 4 errors?

Mixed Review

34. Solve $2(x - 2) = 3x - (4x - 5)$. (Lesson 3–5)

35. Find the complement of an 85° angle. (Lesson 3–4)

36. **Work Backward** Greta Moore's plane is scheduled to leave for Dallas at 9:30 A.M. She is to pick up her tickets at the boarding gate thirty minutes before departure. It takes Ms. Moore fifty minutes to drive to the airport, ten minutes to park her car in the overnight lot, and ten more minutes to get to the gate. On her way to the airport, Ms. Moore needs to pick up some materials from her office. This detour will take about fifteen minutes. If it takes her an hour to get dressed and out the door, for what time should Ms. Moore set her alarm? (Lesson 3–3)

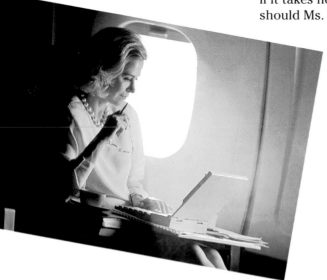

37. Write an inequality for the graph shown below. (Lesson 2–4)

$$\xleftarrow{\qquad} \underset{-2 \ -1 \ \ 0 \ \ 1 \ \ 2 \ \ 3 \ \ 4}{\oplus} \xrightarrow{\qquad}$$

38. Use a number line to find the sum of 9 and -5. (Lesson 2–1)

39. Find the solution set for $x - 3 > \dfrac{x + 1}{2}$ if the replacement set is {4, 5, 6, 7, 8}. (Lesson 1–5)

40. **Health** Your optimum exercise heart rate per minute is given by the expression $0.7(220 - a)$, where a represents your age. Find your optimum exercise heart rate. (Lesson 1–1)

WORKING ON THE

In·ves·ti·ga·tion

Refer to the Investigation on pages 68–69.

the Greenhouse Effect

Two scales are used to measure temperature. The Celsius scale, part of the metric system, was invented in 1741 by Swedish astronomer Anders Celsius (1701–1744). The Celsius scale has the freezing point of water at 0°C and the boiling point of water at 100° C. The scale most frequently used in the United States is the Fahrenheit scale, invented in 1714 by German physicist Gabriel Daniel Fahrenheit (1686–1736). On the Fahrenheit scale, the boiling point of water is 212°F, and the freezing point is 32°F.

To convert Celsius temperatures to Fahrenheit, use the formula $F = \dfrac{9}{5}C + 32$.

1 Using the formula, convert the average temperatures of Mars, Earth, and Venus from Celsius to Fahrenheit. How do the temperatures compare?

2 The greenhouse effect may raise the average temperature of Earth 8°F by the year 2050. What will be the average temperature of Earth in degrees Celsius at that time?

3 Change the formula for converting Celsius to Fahrenheit to a formula for converting Fahrenheit to Celsius.

Add the results of your work to your Investigation Folder.

Integration: Statistics
Measures of Central Tendency

What YOU'LL LEARN

- To find and interpret the mean, median, and mode of a set of data.

Why IT'S IMPORTANT

Measures of central tendency can help you easily describe a set of data.

APPLICATION
Charities

The table at the right shows the top 10 charities in amount of contributions for 1993.

When analyzing data, it is helpful to have one number that describes the entire set of data. Numbers known as **measures of central tendency** are often used to describe sets of data because they represent a centralized, or *middle,* value. Three of the most commonly used measures of central tendency are the **mean, median,** and **mode.**

Charity	Millions of Dollars
Salvation Army	$726
Catholic Charities USA	411
United Jewish Appeal	407
Second Harvest	407
American Red Cross	395
American Cancer Society	355
YMCA	317
American Heart Association	235
YWCA	218
Boy Scouts of America	211

Source: *The Chronicle of Philanthropy, Nov., 1993*

Definition of Mean	**The mean of a set of data is the sum of the numbers in the set divided by the number of numbers in the set.**

F Y I

Until 1994, the Cardwell sisters of Sweetwater, Texas, were the oldest living set of triplets. However, on October 2, 1994, the firstborn triplet, Faith, died at the age of 95. She was survived by her sisters Hope and Charity.

To find the mean of the charity data, find the sum of the dollar values and divide by 10, the number of numbers in the set. *Estimate: Is an answer of $500 reasonable?*

$$\text{mean} = \frac{726 + 411 + 407 + 407 + 395 + 355 + 317 + 235 + 218 + 211}{10}$$

$$= \frac{3682}{10}$$

$$= 368.2$$

The mean is $368.2 million. Notice that the amount collected by the Salvation Army, $726 million, is far greater than the amounts collected by the other charities. Because the mean is an average of several numbers, a single number that is so much greater than the others can affect the mean a great deal. In extreme cases, the mean becomes less representative of the values in a set of data.

The median is another measure of central tendency.

Definition of Median	The median of a set of data is the middle number when the numbers in the set are arranged in numerical order.

To find the median of the charity data, arrange the dollar values in order.

726   411   407   407   395   355   317   235   218   211

If there were an odd number of dollar values, the middle one would be the median. However, since there is an even number of dollar values, the median is the average of the two middle values, 395 and 355.

$$\text{median} = \frac{395 + 355}{2}$$

$$= \frac{750}{2}$$

$$= 375$$

The median is $375 million. Notice that the number of values that are greater than the median is the same as the number of values that are less than the median.

A third measure of central tendency is the mode.

Definition of Mode	The mode of a set of data is the number that occurs most often in the set.

To find the mode of the charity data, look for the number that occurs most often.

726   411   407   407   395   355   317   235   218   211

In this set, 407 appears twice. Thus, $407 million is the mode of the data.

It is possible for a set of data to have more than one mode. For example, the set of data {2, 3, 3, 4, 6, 6} has two modes, 3 and 6.

Based on our results, the charity data has mean $368.2 million, median $375 million, and mode $407 million. As you can see, the mean, median, and mode are rarely the same value.

Example ➊

CONNECTION

History

When the Declaration of Independence was signed in 1776, George III was king of Great Britain. After George III, Great Britain had seven monarchs before Queen Elizabeth II was crowned queen in 1952. The table at the right shows the number of years that each monarch reigned. Find the mean, median, and mode of the data.

Monarch	Reign (years)
George III	59
George IV	10
William IV	7
Victoria	63
Edward VII	9
George V	25
Edward VIII	1
George VI	15

$$\text{mean} = \frac{59 + 10 + 7 + 63 + 9 + 25 + 1 + 15}{8}$$

$$= \frac{189}{8}$$

$$= 23.625 \quad \text{The mean is 23.625 years.}$$

(continued on the next page)

median   1   7   9   10   15   25   59   63

The median is $\frac{10 + 15}{2}$ or $12\frac{1}{2}$ years. Why do you think the mean and median are so different? Which value, mean or median, do you think is more representative of the data?

mode   Since each number appears exactly once, there is no mode.

You can also determine the mean, median, and mode of a set of data by examining stem-and-leaf plots.

Example 2

APPLICATION

Demographics

Since it would be inconvenient to list every stem, we have used dots to indicate that some stems are missing.

The stem-and-leaf plot at the right shows population data for the 20 largest cities in the United States to the nearest ten thousand. Find the mean, median, and mode of the data.

Stem	Leaf
73	2
⋮	⋮
34	9
⋮	⋮
27	8
⋮	⋮
16	9 3
⋮	⋮
10	3 1
⋮	⋮
8	8
7	9 9 4 3
6	8 4 4 3 3 3 1
5	7

8 | 8 = 880,000 people

mean   Add the 20 values and then divide by 20. Since the sum of the 20 values is 2791, the mean is $\frac{2791}{20}$ or 139.55. This represents a mean population of 1,395,500.

median   Since there are 20 values, the median is the average of the 10th and 11th values. Counting from the top down, you will find that the 10th and the 11th values are 79 and 74.

The average of 79 and 74 is $\frac{79 + 74}{2}$ or 76.5. Thus, the median population is 765,000.

mode   There are three entries for 63. Thus, the mode is 630,000.

EXPLORATION

GRAPHING CALCULATORS

You can use a graphing calculator and the MEAN and MEDIAN functions to find the mean and median of a list of numbers. The format is as follows.

mean(*{a,b,c,d,...z}*)   median(*{a,b,c,d,...z}*)

To find the mean of the data in Example 2, access the LIST MATH function by pressing [2nd] [LIST] [▶]. Then choose 3, for mean or 4, for median. Enter the list of numbers, beginning with a left brace ([2nd] [{]) and ending with a right brace ([2nd] [}]) and a right parenthesis. When you press [ENTER], the mean or median appears on the screen.

Your Turn

Take time to work through the examples in this lesson using a graphing calculator.

Communicating Mathematics

Study the lesson. Then complete the following.

1. **Explain** why you think the mean, median, and mode are rarely the same value.

2. **Discuss** which of the measures of central tendency you would use to describe the data in Example 2. Explain your reasoning.

3. **Describe** the steps you would take to find the mean, median, and mode of the data represented by the line plot shown at the right.

4. Suppose you conducted a survey in which you asked your classmates, "What's your favorite rock group?" Which measure of central tendency could you use to analyze the data, and why?

MATH JOURNAL

5. Find some data that interests you in a newspaper or magazine. Find the mean, median, and mode and discuss which of these measures you think best describes the data.

Complete.

6. Measures of central tendency represent _?_ values of a set of data.

7. If the numbers in a set of data are arranged in numerical order, then the _?_ of the set is the middle number.

8. Extremely high or low values affect the _?_ of a set of data.

9. If all the numbers in a set of data occur the same number of times, then the set has no _?_ .

Guided Practice

Find the mean, median, and mode for each set of data.

10. 4, 6, 12, 5, 8

11. 8, 8, 8, 8, 9

12.

Stem	Leaf
7	3 5
8	2 2 4
9	0 4 7 9
10	5 8
11	4 6 *9|4 = 94*

List six numbers that satisfy each set of conditions.

13. The mean is 50, the median is 40, and the mode is 20.

14. The mean is 70, the median is 70, and the modes are 65 and 70.

15. **Basketball** The table at the right shows the names of players most often named most valuable player of the NBA.

   a. Find the mean, median, and mode of the data.

   b. Which measure of central tendency best represents these data, and why?

Player	MVP Wins
Kareem Abdul-Jabbar	6
Bill Russell	5
Wilt Chamberlain	4
Larry Bird	3
Magic Johnson	3
Moses Malone	3
Michael Jordan	3

Source: *World Almanac, 1995*

Practice

Find the mean, median, and mode for each set of data.

16. 2, 4, 7, 9, 12, 15

17. 300, 34, 40, 50, 60

18. 23, 23, 23, 12, 12, 12

19. 10, 3, 17, 1, 8, 6, 12, 15

20. 7, 19, 9, 4, 7, 2

21. 2.1, 7.4, 13.9, 1.6, 5.21, 3.901

22.

Stem	Leaf
5	3 6 8
6	5 8
7	0 3 7 7 9
8	1 4 8 8 9
9	9 $6\,|\,8 = 68$

23.

Stem	Leaf
19	3 5 5
20	2 2 5 8
21	5 8 8 9 9 9
22	0 1 7 8 9 $21\,|\,5 = 215$

24. List ten numbers that satisfy each set of conditions.

  a. The mean is less than all but one of the numbers.

  b. The mean is greater than all but one of the numbers.

  c. The mean is greater than all of the numbers.

  d. Would you be able to complete parts a, b, and c if the word *mean* were replaced by the word *median*? Why or why not?

25. The mean of a set of ten numbers is 5. When the greatest number in the set is eliminated, the mean of the new set of numbers is 4. What number was eliminated from the original set of numbers?

26. The first of three consecutive odd integers is $2n + 1$.

  a. What is the mean of the three integers?

  b. What is the median?

Critical Thinking

27. This paragraph appeared in the *San Diego Union-Tribune* on August 8, 1993.

> Lawyers at the top of their profession earn more than $1 million a year, while the typical attorney makes almost $67,000—far more than the average American. . . . According to census figures, lawyers and judges earned a median pay of $66,784 in 1991, which means that half of them made more than that amount and half made less. . . . However R. Wilson Montjoy II of Jackson, Mississippi, said, "We have lots more people at or below the median than above it."

  a. To which measure of central tendency do you think the word *average* refers?

  b. Has *median* been defined correctly in this passage? Why or why not?

  c. In the statement by R. Wilson Montjoy II, did he use the term *median* correctly? If not, which term should he have used? Justify your answer.

Applications and Problem Solving

28. Demographics   The table at the right shows population data for 10 South American countries, according to the 1990 Census.

  a. Find the mean, median, and mode of the data.

  b. If you were reporting these data in a newspaper article, which measure would you use, and why?

Country	Population (millions)
Argentina	32
Bolivia	7
Brazil	150
Chile	13
Colombia	33
Ecuador	10
Paraguay	5
Peru	22
Uruguay	3
Venezuela	19

29. Geography The areas of each of the South American countries listed in the table on the previous page are as follows (in millions of square miles): 1085, 139, 7778, 670, 3166, 1068, 177, 1135, 154, and 1234, respectively. Find the mean, median, and mode area of the countries.

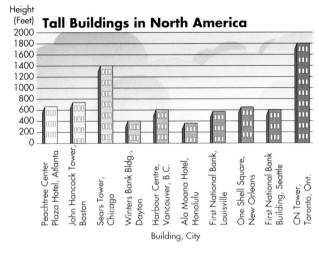

Height (Feet)

Tall Buildings in North America

2000
1800
1600
1400
1200
1000
800
600
400
200
0

Peachtree Center Plaza Hotel, Atlanta
John Hancock Tower, Boston
Sears Tower, Chicago
Winters Bank Bldg., Dayton
Harbour Centre, Vancouver, B.C.
Ala Moana Hotel, Honolulu
First National Bank, Louisville
One Shell Square, New Orleans
First National Bank Building, Seattle
CN Tower, Toronto, Ont.

Building, City

30. American Landmarks The graph at the left shows the height in feet of several tall buildings in North America. Find the mean and median height of the buildings.

31. Commuting On her way home from work, Luisa drives 10 miles through the city at about 30 mph and then 10 miles on the highway at about 50 mph. What is her average speed? Use $d = rt$. (*Hint:* The answer is *not* 40 mph.)

Mixed Review

32. Solve $\frac{a+5}{3} = 7x$ for x.   (Lesson 3–6)

33. Solve $\frac{3}{4}n - 3 = 9$. Then check your solution.   (Lesson 3–3)

34. Evaluate $|k| + |m|$ if $k = 3$ and $m = -6$.   (Lesson 2–3)

35. Use a number line to add $-6 + (-14)$.   (Lesson 2–1)

36. Evaluate $\left(13 + \frac{2}{5} \cdot 5\right)(3^2 - 2^3)$. Indicate the property used in each step.   (Lesson 1–8)

37. Simplify $5 + 7(ac + 2b) + 2ac$.   (Lesson 1–7)

Mathematics and **SOCIETY**

Teen Talk Barbies

The article below appeared in *Antique Week* on August 30, 1993.

SOME TIME AGO, MATTEL RECALLED ANY Teen Talk Barbies programmed to say "Math class is tough" because some women's groups complained the comment perpetuated the view that girls are bad at mathematics. According to a news piece in a December issue of *The Atlanta Journal and Constitution*, only five Barbie buyers in the United States turned in their dolls because of Mattel's offer. While some talking Barbies may have found math tough, the consumers/collectors who own the doll won't have trouble adding up the profits.... Savvy collectors were tearing open Teen Talk boxes trying to find a Barbie who talked about math. Mike Huen, a professional toy collector, estimates the doll, which sold for $35, will be worth several hundred dollars now. ■

1. Why do you think some women's groups objected to what the dolls say? Do you think that what the dolls say has any influence on the girls who play with them? Why or why not?

2. Why do you think only five Barbie buyers have returned their dolls? Explain.

3. Do you think it is right for people to make money on this doll? Justify your reasoning.

Investigation

the Greenhouse Effect

Refer to the Investigation on pages 68–69.

For more than a decade, scientists have warned that cars and factories spew so many gases into the atmosphere that Earth could soon be affected by disastrous climatic changes. The loss of rain forests is reducing the number of trees to offset the large increases in carbon dioxide in our atmosphere. The possible consequences are so frightening that it makes sense to slow the buildup of CO_2 through preventive measures, such as encouraging energy conservation, developing alternatives to fossil fuels, and preventing the destruction of the rain forests.

Analyze

You have conducted experiments and done research on this important problem. It is now time to analyze your findings and state your conclusions.

PORTFOLIO ASSESSMENT

You may want to keep your work on this Investigation in your portfolio.

1 Create a graph of the data from your experiments. Let the vertical axis represent the changes in temperature. Let the horizontal axis represent the distances between the lamp and the thermometer. Use different colors to indicate the control experiment and the greenhouse experiment.

2 Use the data from your experiments to draw a conclusion about the relationship between the temperature and the distance between the thermometer and the lamp. Describe this relationship as a function, and justify your reasoning.

3 Use the data from Working on the Investigation in Lesson 2–6 to make a chart.

Use the headings *Planet, Temperature Without CO_2, Temperature With CO_2,* and *Distance from the Sun.*

4 Draw a conclusion about the relationship between the distance from the sun and the temperature of Venus and Earth with CO_2. Describe this relationship as a function, and justify your reasoning.

5 Draw a conclusion about the relationship between the distance from the sun and the temperature of Mars and Earth with CO_2. Describe this relationship as a function, and justify your reasoning.

6 How does your experiment explore the temperatures of the planets? How are the relationships similar and different?

7 Predict what the temperature of Mars would be if it had the same amount of CO_2 that Earth has. Explain your calculations.

Write

Some scientists predict that by the year 2050, Earth's atmosphere will contain between 500 and 700 ppm of CO_2. Write a one-page paper explaining the situation to a group of concerned citizens. Consider some of the following questions to address in your paper.

8 What factors will cause this increase in CO_2?

9 How different is this estimate from the amount of CO_2 calculated earlier?

10 In your opinion, what will the approximate average temperature on Earth be at that time?

11 What might occur if the temperature reaches that point? What are some of the dangers involved?

VOCABULARY

After completing this chapter, you should be able to define each term, property, or phrase and give an example or two of each.

Algebra

addition property of equality (p. 144)
consecutive integers (p. 158)
dimensional analysis (p. 174)
division property of equality (p. 151)
equivalent equation (p. 144)
identity (p. 170)
multi-step equations (p. 157)
multiplication property of equality (p. 150)
number theory (p. 158)
solve an equation (p. 145)
subtraction property of equality (p. 146)

Statistics

mean (p. 178)
measures of central tendency (p. 178)
median (p. 178)
mode (p. 178)

Geometry

acute triangle (p. 165)
complementary angles (p. 163)
congruent (p. 164)
equilateral triangle (p. 164)
isosceles triangle (p. 164)
obtuse triangle (p. 165)
right triangle (p. 165)
supplementary angles (p. 162)
triangle (p. 163)

Problem Solving

work backward (p. 156)

UNDERSTANDING AND USING THE VOCABULARY

Choose the letter of the term that best matches each statement or phrase.

1. If $a = b$, then $a + c = b + c$.
2. If $a = b$, then $a - c = b - c$.
3. If $a = b$, then $ac = bc$.
4. If $a = b$ and $c \neq 0$, then $\frac{a}{c} = \frac{b}{c}$.
5. a triangle in which each angle has the same measure
6. a triangle in which two angles have the same measure
7. a triangle in which one angle measures 90°
8. a triangle in which one angle measures more than 90°
9. a triangle in which all of the angles measure less than 90°
10. the sum of the numbers divided by the number of numbers
11. the middle number when data is in numerical order
12. the number that occurs most often in a set of data

a. acute triangle
b. addition property of equality
c. division property of equality
d. equilateral triangle
e. isosceles triangle
f. mean
g. median
h. mode
i. multiplication property of equality
j. obtuse triangle
k. right triangle
l. subtraction property of equality

STUDY GUIDE AND ASSESSMENT

SKILLS AND CONCEPTS

OBJECTIVES AND EXAMPLES

Upon completing this chapter, you should be able to:

• solve equations by using addition or subtraction (Lesson 3–1)

$$x - 13 = 45$$
$$x - 13 + 13 = 45 + 13$$
$$x = 58$$

$$y - (-33) = 14$$
$$y + 33 = 14$$
$$y + 33 - 33 = 14 - 33$$
$$y = -19$$

• solve equations by using multiplication or division (Lesson 3–2)

$$\frac{4}{9}t = -72$$
$$\frac{9}{4}\left(\frac{4}{9}t\right) = \frac{9}{4}(-72)$$
$$t = -162$$

$$-5c = -8$$
$$\frac{-5c}{-5} = \frac{-8}{-5}$$
$$c = \frac{8}{5}$$

• solve equations involving more than one operation (Lesson 3–3)

$$34 = 8 - 2t$$
$$34 - 8 = 8 - 8 - 2t$$
$$26 = -2t$$
$$\frac{26}{-2} = \frac{-2t}{-2}$$
$$-13 = t$$

$$\frac{d+5}{3} = -9$$
$$3\left(\frac{d+5}{3}\right) = 3(-9)$$
$$d + 5 = -27$$
$$d = -32$$

REVIEW EXERCISES

Use these exercises to review and prepare for the chapter test.

Solve each equation. Then check your solution.

13. $r - 21 = -37$
14. $14 + c = -5$
15. $-27 = -6 - p$
16. $b + (-14) = 6$
17. $r + (-11) = -21$
18. $d - (-1.2) = -7.3$

19. A number decreased by 14 is -46. Find the number.

20. The sum of two numbers is -23. One of the numbers is 9. What is the other number?

21. Eighty-two increased by some number is -34. Find the number.

Solve each equation. Then check your solution.

22. $6x = -42$
23. $-7w = -49$
24. $\frac{3}{4}n = 30$
25. $-\frac{3}{5}y = -50$
26. $\frac{5}{2}x = -25$
27. $\frac{5}{12} = \frac{r}{24}$

28. Negative seven times a number is -56. Find the number.

29. Three fourths of a number is -12. What is the number?

30. One and two thirds of a number equals one and a half. What is the number?

Solve each equation. Then check your solution.

31. $4t - 7 = 5$
32. $6 = 4n + 2$
33. $\frac{y}{3} + 6 = -45$
34. $\frac{c}{-4} - 8 = -42$
35. $\frac{4d + 5}{7} = 7$
36. $\frac{7n + (-1)}{8} = 8$

37. Four times a number decreased by twice the number is 100. What is the number?

38. Find four consecutive integers whose sum is 130.

39. Find two consecutive integers such that twice the lesser integer increased by the greater integer is 49.

OBJECTIVES AND EXAMPLES	REVIEW EXERCISES

• find the complement and supplement of an angle
(Lesson 3–4)

Find the complement and supplement of an angle with measure 70°.

complement: $90° - 70°$ or $20°$

supplement: $180° - 70°$ or $110°$

Find both the complement and supplement of each angle measure.

40. $28°$ **41.** $69°$

42. $5x°$ **43.** $(y + 20)°$

44. Find the measure of an angle that is 10° more than its supplement.

45. Find the measure of an angle that is one-half the measure of its complement.

• find the measure of the third angle of a triangle given the measures of the other two angles
(Lesson 3–4)

The measures of two angles of a triangle are 89° and 90°. The measure of the third angle is $180° - (89° + 90°)$ or $1°$.

Find the measure of the third angle of each triangle in which the measures of two angles of the triangle are given.

46. $16°, 47°$

47. $45°, 120°$

48. $y°, x°$

49. $(z - 30)°, z°$

• solve equations with the variable on both sides and solve equations containing grouping symbols   (Lesson 3–5)

$$-3(x + 5) = 3(x - 1)$$
$$-3x - 15 = 3x - 3$$
$$-3x - 15 + 15 = 3x - 3 + 15$$
$$-3x = 3x + 12$$
$$-3x - 3x = 3x - 3x + 12$$
$$\frac{-6x}{-6} = \frac{12}{-6}$$
$$x = -2$$

Solve each equation. Then check your solution.

50. $4(3 + 5w) = -11$

51. $\frac{2}{3}n + 8 = \frac{1}{3}n - 2$

52. $3x - 2(x + 3) = x$

53. $\frac{4 - x}{5} = \frac{1}{5}x$

54. The sum of two numbers is 25. Twelve less than four times one of the numbers is 16 more than twice the other number. Find both numbers.

• solve equations and formulas for a specified variable   (Lesson 3–6)

Solve $\frac{x + y}{b} = c$ for x.

$$b\left(\frac{x + y}{b}\right) = b(c)$$
$$x + y = bc$$
$$x = bc - y$$

Solve each equation for the variable specified.

55. $5x = y$, for x

56. $ay - b = c$, for y

57. $yx - a = cx$, for x

58. $\frac{2y - a}{3} = \frac{a + 3b}{4}$, for y

OBJECTIVES AND EXAMPLES

• find and interpret the mean, median, and mode of a set of data   (Lesson 3–7)

Find the mean, median, and mode of 1.5, 3.4, 5.4, 5.6, 5.7, 6.2, 6.8, 7.1, 7.1, 8.4, and 9.9.

mean: The sum of the 11 values is 67.1.
$\frac{67.1}{11} = 6.1$

median: The 6th value is 6.2.

mode: 7.1

REVIEW EXERCISES

59. School   Marisa's scores on the 25-point quizzes in her English class are 20, 21, 18, 21, 22, 22, 24, 21, 20, 19, and 23. Find the mean, median, and mode of her scores.

60. Business   Of the 42 employees at Pirate Printing, sixteen make $4.75 an hour, four make $5.50 an hour, three make $6.85 an hour, six make $4.85 an hour, and thirteen make $5.25 an hour. Find the mean, median, and mode of the hourly wages.

APPLICATIONS AND PROBLEM SOLVING

61. Food and Nutrition   According to the National Eating Trends Service, in 1984, the average person ate 71 meals in a restaurant and purchased 20 meals to go. The number of meals eaten in a restaurant decreased 0.7 meals per year, and the number of meals purchased to go increased 1.3 meals per year.   (Lesson 3–5)

a. After how many years will people eat the same number of meals in restaurants as they order to go and eat at home?

b. How would you explain these trends?

62. Health   You can estimate your ideal weight with the following formulas.

Male	**Female**
$w = 100 + 6(h - 60)$	$w = 100 + 5(h - 60)$

In each formula, w is your ideal weight in pounds and h represents your height in inches. (Lesson 3–6)

a. Find your ideal weight.

b. Solve each formula for h.

c. How tall is a man that is at his ideal weight of 170 pounds?

63. Home Economics   According to the *Dallas Morning News,* the average refrigerator lasts $2\frac{1}{4}$ years less than twice as long as the average color television. The average color television lasts for 8 years. How long does the average refrigerator last?   (Lesson 3–3)

64. Postal Service   The graph below shows the cost of mailing a first-class letter in several countries in 1994.   (Lesson 3–7)

a. Find the mean, median, and mode cost.

b. Round each cost to the nearest cent. Then find the mean, median, and mode cost and compare your two answers.

Source: *USA TODAY* research

A practice test for Chapter 3 is provided on page 789.

ALTERNATIVE ASSESSMENT

COOPERATIVE LEARNING PROJECT

Balances and Solving Equations   In this chapter, you learned the addition, subtraction, multiplication, and division properties of equality. Each definition begins with an equation in *balance, a = b*, and states that if something is done to one side of the equation, the equation can be kept in balance by doing the same thing to the other side.

A basic principle is that two quantities in balance will remain in balance until something is done to only one of the quantities. To bring the quantities back into balance, you must do the same thing to the other quantity or undo what was done to the first quantity.

In this project, you will construct a device to demonstrate the principle of balance and how it relates to solving equations. All balances have certain features in common. A horizontal bar is balanced on a vertical support. Pans are suspended from the ends of the bar. A pointer attached to the bar moves along a scale on the support, indicating when the pans are balanced.

Follow these steps to design and build your balance.

- Outline a plan you can follow.
- Use materials you can easily find. The precision of your balance depends more on the care and accuracy with which you build it than it does on the materials you use.
- Carry out your plan.
- Determine how you might use your balance to represent equations. Include ways to solve equations by using the balance.
- Write several paragraphs describing how you can solve equations using the balance. Be sure to give examples of equations and their solutions.

THINKING CRITICALLY

- Write an equation that has no solution. Explain why there is no solution.
- Write an equation that has an infinite number of solutions. Explain why.

PORTFOLIO

Select one of the assignments from this chapter that you found especially challenging. Revise your work if necessary and place it in your portfolio. Explain why you found it to be a challenge.

SELF EVALUATION

One characteristic of a good problem solver is persistence. If you don't succeed the first time you attempt something, you should learn from your mistakes and try again.

Assess yourself. How persistent are you? How do you react when you fail at something? List two or three ways that you can make a conscious effort to be more persistent in solving problems both in your mathematics studies and in your daily life.

In·ves·ti·ga·tion

Go Fish!

MATERIALS NEEDED

paper lunch bag

2 bags of dry beans, each a different color

5-oz. paper cup

Imagine that you are asked to determine the number of fish in a nearby pond. To count the fish one by one, you could remove the fish from the pond and stack them to one side, or mark each fish so you would not count them over and over again. Counting like this could be hazardous to a fish's health!

To determine the number of animals in a population, scientists often use the *capture-recapture* method. A number of animals are captured, carefully tagged, and returned to their native habitat. Then a second group of animals is captured and counted, and the number of tagged animals is noted. Scientists then use proportions to estimate the number in the entire population.

In this Investigation, you will work in pairs to model the process used by scientists to estimate the number of fish in a large lake. The lunch bag will represent the lake, the beans will represent fish, and the paper cup will represent a net.

Make an Investigation Folder in which you can store all of your work on this Investigation for future use.

CASTING	A Total Number of Tagged Fish	B Total Number of Fish in Sample	C Number of Tagged Fish in Sample	D Estimate
1				
2				
3				
4				
5				

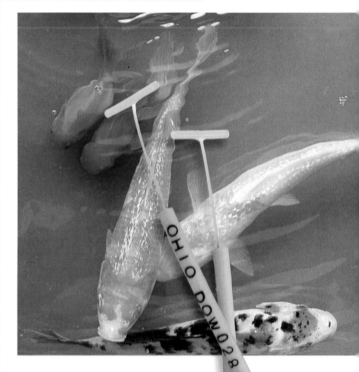

CAPTURE

1 Begin by copying the chart above onto a sheet of paper.

2 Write your name on the paper bag. Empty one bag of beans into the bag.

3 Use your net to remove a sample of fish. Count the number of fish you netted. Since this number will remain constant for all casts, you can record this number in each row of column A.

4 Replace all of the beans you counted with beans of a different color. Put these "tagged fish" back into the lake and gently shake your bag to mix the fish.

RECAPTURE

5 For your first casting, use your net to remove a sample of fish. Count the total number of fish in your sample and record this number in column B. Then count the number of tagged fish in this sample and record this number in column C of your chart. Return these fish to the lake and gently shake your bag to mix the fish.

6 Cast your net a second time and record your findings. Continue casting and recording until you have counted five samples.

You will continue working on this Investigation throughout Chapters 4 and 5.

Be sure to keep your chart and materials in your Investigation Folder.

Go Fish! Investigation

Working on the Investigation
Lesson 4–1, p. 200

Working on the Investigation
Lesson 4–5, p. 227

Working on the Investigation
Lesson 5–2, p. 269

Working on the Investigation
Lesson 5–6, p. 302

Closing the Investigation
End of Chapter 5, p.314

Using Proportional Reasoning

Objectives

In this chapter, you will:

- solve proportions,
- find the unknown measures of the sides of two similar triangles,
- use trigonometric ratios to solve right triangles,
- solve percent problems,
- find the probability and odds of a simple event, and
- solve problems involving direct and inverse variation.

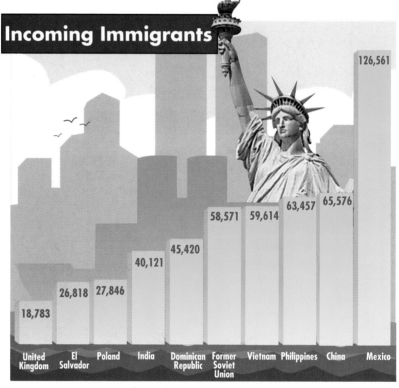

Incoming Immigrants

126,561

65,576

63,457

59,614

58,571

45,420

40,121

27,846

26,818

18,783

United Kingdom | El Salvador | Poland | India | Dominican Republic | Former Soviet Union | Vietnam | Philippines | China | Mexico

Source: U.S. Immigration and Naturalization Service

Some people blame immigrants, both legal and illegal, for many of the economic and social problems in the United States. Others believe immigrants bring vitality, diversity, and economic strength to the United States. What effect does immigration have on American society? Do you have any personal experiences with new immigrants?

TIME Line

280 B.C. The Colossus of Rhodes, standing 110 feet tall, is built on the Greek island of Rhodes.

1893 Poet Paul Laurence Dunbar writes *Oak and Ivy*.

400 B.C. 300 200 100 0 A.D. 100 1500 1600 1700 1880 1885 1890 1895

A.D. 1662 Englishman John of Gaunt publishes the first book of statistics.

Chapter Project

Ayrris **Layug Aunario** won the grand prize in *Filipinas* magazine's 1994 National Essay Contest with an essay describing what it means to be Filipino-American. The 16-year-old from Waukegan, Illinois, is a student at the Illinois Mathematics and Science Academy, a public school for gifted students in science and mathematics.

Ayrris' essay explains what it was like to be a 10-year-old immigrant to the U.S., arriving in Chicago and learning to live with people of many different cultures and beliefs. He had to master English quickly and soon learned the American virtues of independence, self-reliance, and assertiveness. Ayrris plans to attend college in the field of physics.

Work in cooperative groups to take a heritage survey of students in your school.

- Determine the geographic birthplace of each student and the birthplace of their parents or guardians. If possible, gather information about the birthplace of each student's grandparents and great-grandparents.

- Use a world map to pinpoint each geographic location. Connect the generations of each student using colored yarn.

- Draw some conclusions about the birthplaces of the students you surveyed. Share your findings with the class.

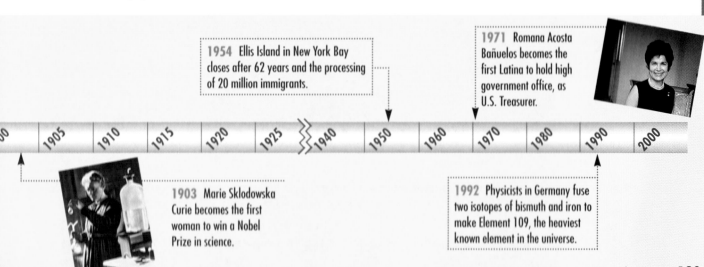

1954 Ellis Island in New York Bay closes after 62 years and the processing of 20 million immigrants.

1971 Romana Acosta Bañuelos becomes the first Latina to hold high government office, as U.S. Treasurer.

1900 | 1905 | 1910 | 1915 | 1920 | 1925 | 1940 | 1950 | 1960 | 1970 | 1980 | 1990 | 2000

1903 Marie Sklodowska Curie becomes the first woman to win a Nobel Prize in science.

1992 Physicists in Germany fuse two isotopes of bismuth and iron to make Element 109, the heaviest known element in the universe.

4-1A Ratios

A Preview of Lesson 4-1

Materials: ☐⊐ tape measure

You can collect data to determine whether there is a relationship between the length of an individual's head and an individual's body.

Activity 1 **Use a tape measure to find the length of your head.**

Step 1 Measure your head from the top of your skull to the bottom of your chin. Since the top of your head is round, hold something flat, such as a piece of cardboard, across the top of your head to get an accurate measurement.

Step 2 Make a measuring "stick" with the length of your head as one unit to find the following measurements.
- total height
- chin to waist
- waist to hip
- knee to ankle
- ankle to bottom of bare heel
- underarm to elbow
- elbow to wrist
- wrist to tip of finger
- shoulder to tip of finger

Step 3 Make a table to record the relationship, or *ratio*, of each measurement to the measurement of your head.

Body Proportions of an Average Adult

Draw 1. The average adult is seven and one-half heads tall. Use the information in your table to draw an outline of the proportions of your body. How does your drawing compare to the one above?

Write 2. Compare your ratios to those of your classmates. Find the number of heads tall of a random student in your class. Write a paragraph explaining how you compare to this average.

Ratios and Proportions

What YOU'LL LEARN

- To solve proportions.

Why IT'S IMPORTANT

You can use proportions to solve problems involving food and entertainment.

APPLICATION
Food

How would you like to eat chocolate for a living? That's exactly what Carl Wong does every day as associate director of product development for Hershey Foods. His job is to create new candy bars as well as taste test existing candy bars in order to improve them.

Hershey Foods keeps their candy recipes confidential, but the basic ingredients found in a batch of Hershey's chocolate bars are sugar, cocoa beans, milk, and flavorings. The ingredients for a batch of candy are shown in the table below.

Ingredient	Parts Per Batch
Sugar	10
Cocoa Beans	5
Milk	4
Flavorings	1

A **ratio** is a comparison of two numbers by division. The ratio of x to y can be expressed in the following ways.

$$x \text{ to } y \qquad\qquad x{:}y \qquad\qquad \frac{x}{y}$$

Ratios are often expressed as fractions in simplest form. A ratio that is equivalent to a whole number is written with a denominator of 1.

In the application at the beginning of the lesson, the table shows that for every 10 parts sugar in a batch of chocolate bars, there are 5 parts cocoa beans. The ratio of sugar to cocoa beans is $\frac{10}{5}$. Suppose Hershey uses 30 pounds of sugar and 15 pounds of cocoa beans in a batch of chocolate bars. This ratio is $\frac{30}{15}$. Is this ratio different than the first ratio of $\frac{10}{5}$? When simplified, both ratios are equivalent to $\frac{2}{1}$.

$$\left.\begin{array}{c} \overset{5}{\overbrace{}} \\ \dfrac{10}{5} = \dfrac{2}{1} \\ \underset{5}{\underbrace{}} \end{array}\right. \qquad\qquad \left.\begin{array}{c} \overset{15}{\overbrace{}} \\ \dfrac{30}{15} = \dfrac{2}{1} \\ \underset{15}{\underbrace{}} \end{array}\right.$$

An equation stating that two ratios are equal is called a **proportion**. So, $\frac{10}{5} = \frac{30}{15}$ is a proportion.

F Y I

Almost 500 new candy and chocolate products were introduced in the first four months of 1995.

To solve a proportion, cross multiply.

One way to determine if two ratios form a proportion is to check their cross products. In the proportion at the right, the cross products are $10 \cdot 15$ and $5 \cdot 30$. In this proportion, 10 and 15 are called the **extremes,** and 5 and 30 are called the **means.**

$$\frac{10}{5} = \frac{30}{15}$$

$$10(15) = 5(30)$$

extremes = means

$$150 = 150$$

The cross products of a proportion are equal.

Means-Extremes Property of Proportions	**In a proportion, the product of the extremes is equal to the product of the means.** If $\frac{a}{b} = \frac{c}{d}$, then $ad = bc$.

Example 1 Use cross products to determine whether each pair of ratios forms a proportion.

a. $\frac{2}{3}, \frac{12}{18}$

$$2 \cdot 18 \stackrel{?}{=} 3 \cdot 12$$

$$36 = 36$$

So, $\frac{2}{3} = \frac{12}{18}$.

This is a proportion.

b. $\frac{2.5}{6}, \frac{3.4}{5.2}$

Enter: 2.5 $\boxed{\times}$ 5.2 $\boxed{=}$ *13*

Enter: 6 $\boxed{\times}$ 3.4 $\boxed{=}$ *20.4*

Since $13 \neq 20.4$, $\frac{2.5}{6} \neq \frac{3.4}{5.2}$.

This is not a proportion.

You can write proportions that involve a variable and then use cross products to solve the proportion.

Example 2

APPLICATION

Food

Refer to the application at the beginning of the lesson. Suppose Hershey makes a batch of chocolate with 75 pounds of cocoa beans. How many gallons of milk will they use?

You know that the ratio of cocoa beans to milk is 5:4. Let m represent the gallons of milk.

$$\frac{5 \text{ parts cocoa beans}}{4 \text{ parts milk}} = \frac{75 \text{ pounds of cocoa beans}}{m \text{ gallons of milk}}$$

$$\frac{5}{4} = \frac{75}{m}$$

$$5m = 4(75) \quad \textit{Find the cross products.}$$

$$5m = 300$$

$$m = 60$$

In a batch of chocolate with 75 pounds of cocoa beans, Hershey needs to use 60 gallons of milk.

A ratio called a **scale** is used when making a model to represent something that is too large or too small to be conveniently drawn at actual size. The scale compares the size of the model to the actual size of the object being modeled.

Example ③

APPLICATION
Entertainment

In the movie *Jurassic Park*, the dinosaurs were scale models and so was the sport utility vehicle that the T-Rex overturned. The vehicle was made to the scale of 1 inch to 8 inches. The actual vehicle was about 14 feet long. What was the length of the model sport utility vehicle?

First, change 14 feet to 168 inches. Then let ℓ represent the length of the model vehicle.

$$\begin{array}{c} scale \rightarrow \\ actual \rightarrow \end{array} \frac{1}{8} = \frac{\ell}{168}$$

$$168 = 8\ell \quad \textit{Find the cross products.}$$

$$21 = \ell$$

The model sport utility vehicle was 21 inches long.

You can solve proportions by using a calculator.

Example ④ **Solve each proportion.**

a. $\dfrac{5}{4.25} = \dfrac{11.32}{m}$

Method 1
Use paper and a pencil.

$$\frac{5}{4.25} = \frac{11.32}{m}$$

$$5m = 4.25(11.32)$$

$$5m = 48.11$$

$$m = 9.622$$

Method 2
Use a calculator.

By the means-extremes property, $5m = 4.25(11.32)$. Multiply the means. Then divide by 5.

Enter: 4.25 ⊠ 11.32 ÷ 5 = *9.622*

Rounded to the nearest hundredth, the solution is 9.62.

b. $\dfrac{x}{3} = \dfrac{x+5}{15}$

$$\frac{x}{3} = \frac{x+5}{15}$$

$$15x = 3(x + 5) \quad \textit{Means-extremes property}$$

$$15x = 3x + 15 \quad \textit{Distributive property}$$

$$12x = 15$$

$$x = \frac{5}{4} \quad \textit{Check this result.}$$

The solution is $\dfrac{5}{4}$.

The ratio of two measurements having different units of measure is called a **rate.** For example, 30 miles per gallon is a rate. Proportions are often used to solve problems involving rates.

Example **5**

APPLICATION
State Fair

In the first 30 minutes of the opening day of the Texas State Fair, 1252 people entered the gates. If this attendance rate continued, how many people visited the fair during the operating hours of 8:00 A.M. and 12:00 midnight the first day?

Explore   Let p represent the number of people attending the fair on opening day.

Plan   Write a proportion for the problem.

$$\frac{1252}{0.5} = \frac{p}{16}$$ *Notice that both ratios compare the number of people per hour.*

Solve $$\frac{1252}{0.5} = \frac{p}{16}$$

$$1252(16) = 0.5p$$

$$40{,}064 = p$$

If the attendance rate continued, 40,064 people visited the fair on opening day.

Examine   Use estimation to check your answer. About 1250 people entered the fair every 30 minutes. This means that about 2500 people entered every hour. The fair was open for 16 hours each day. Therefore, about 16×2.5 thousand or 40,000 people entered, and the answer is reasonable.

CHECK FOR UNDERSTANDING

Communicating Mathematics

Study the lesson. Then complete the following.

1. **Explain** how to determine whether two ratios are equivalent.

2. **Explain** how to use a calculator to solve a proportion.

3. **Find** three examples of ratios in a newspaper or magazine.

MODELING MATHEMATICS

4. **Draw** an outline of an infant's body if the average infant is three heads long.

Guided Practice

Use cross products to determine whether each pair of ratios forms a proportion.

5. $\frac{3}{2}, \frac{21}{14}$

6. $\frac{2.3}{3.4}, \frac{0.3}{3.6}$

Solve each proportion.

7. $\frac{2}{3} = \frac{8}{x}$

8. $\frac{4}{w} = \frac{2}{10}$

9. $\frac{3}{15} = \frac{1}{y}$

10. $\frac{5.22}{13.92} = \frac{b}{48}$

11. $\frac{1.1}{0.6} = \frac{8.47}{n}$

12. $\frac{x}{1.5} = \frac{2.4}{1.6}$

13. **Travel**   A 96-mile trip requires 6 gallons of gasoline. At that rate, how many gallons would be required for a 152-mile trip?

Practice

Use cross products to determine whether each pair of ratios forms a proportion.

14. $\frac{6}{8}, \frac{22}{28}$

15. $\frac{4}{5}, \frac{16}{20}$

16. $\frac{4}{11}, \frac{12}{33}$

17. $\frac{8}{9}, \frac{16}{17}$

18. $\frac{2.1}{3.6}, \frac{5}{7}$

19. $\frac{0.4}{0.8}, \frac{0.7}{1.4}$

Solve each proportion.

20. $\frac{3}{4} = \frac{x}{8}$

21. $\frac{a}{45} = \frac{3}{15}$

22. $\frac{y}{9} = \frac{-7}{16}$

23. $\frac{3}{5} = \frac{x+2}{6}$

24. $\frac{w+2}{5} = \frac{7}{5}$

25. $\frac{x}{8} = \frac{0.21}{2}$

26. $\frac{5+y}{y-3} = \frac{14}{10}$

27. $\frac{m+9}{5} = \frac{m-10}{11}$

28. $\frac{r+7}{-4} = \frac{r-12}{6}$

29. $\frac{85.8}{t} = \frac{70.2}{9}$

30. $\frac{z}{33} = \frac{11.75}{35.25}$

31. $\frac{0.19}{2} = \frac{0.5x}{12}$

32. $\frac{2.405}{3.67} = \frac{g}{1.88}$

33. $\frac{x}{4.085} = \frac{5}{16.33}$

34. $\frac{3t}{9.65} = \frac{21}{1.066}$

Critical Thinking

35. Mariah is exactly eight years older than her cousin Louis.

   a. Copy and complete the table below.

Louis' age	1	2	3	6	10	20	30
Mariah's age	9						

   b. Find the ratio in decimal form of Mariah's age to Louis' age for each pair of ages listed in the table.

   c. Write a formula for the ratio of Mariah's age to Louis' age when Louis is y years old.

   d. As Mariah and Louis grow older, explain what happens to the ratio of their ages.

   e. Explain whether the ratio of their ages will ever equal 1.

Applications and Problem Solving

36. Biology A flea, usually less than an eighth of an inch long, can long jump about thirteen inches and high jump about eight inches. If humans could jump in proportion to the flea, it would take only nine leaps to go one mile! At this rate, how many leaps would it take a human to travel 693 miles from Louisville, Kentucky to Norfolk, Virginia?

37. Movies When rating movies in 1994, critic Gene Siskel gave four thumbs up to every five thumbs up given by his partner Roger Ebert. If Mr. Siskel gave thumbs up to 68 movies, how many movies did Mr. Ebert rate favorably?

38. Recycling When a pair of blue jeans is made, the leftover denim scraps can be recycled to make stationery, pencils, and more denim. One pound of denim is left after making every five pairs of jeans. How many pounds of denim would be left from 250 pairs of jeans?

39. Statistics Find the mean, median, and mode for the following set of data. (Lesson 3–7)

19, 21, 18, 22, 46, 18, 17

40. Solve $a = \frac{v}{t}$ for t. (Lesson 3–6)

41. Define a variable, write an equation, and then solve the following problem. (Lesson 3–2)
Four times a number decreased by twice the number is 100. What is the number?

42. Solve $-15 + d = 13$. (Lesson 3–1)

43. Sales Tanya is a sales representative for Incredible Universe. A CD and tape player manufacturer is having a sales competition. Anyone who sells more than 40 CD players in one day wins a trip to Hawaii. Tanya sold 93 CD and tape players in one day for a total of $9695. The CD players sold for $135 and the tape players for $80. (Lesson 2–9)

  a. How many CD and tape players were sold?

  b. What was the total amount of Tanya's CD and tape player sales?

  c. How many CD players does Tanya need to sell for her trip award?

  d. If p represents the number of CD players sold, what was the number of tape players sold?

  e. Will Tanya win the trip to Hawaii?

44. Golf In four rounds of a recent golf tournament, Heather shot 3 under par, 2 over par, 4 under par, and 1 under par. What was her score for the tournament? (Lesson 2–5)

45. Evaluate $|m - 4|$ if $m = -6$. (Lesson 2–3)

46. Simplify $x^2 + \frac{7}{8}x - \frac{x}{8}$. (Lesson 1–7)

WORKING ON THE

In·ves·ti·ga·tion

Refer to the Investigation on pages 190–191.

Since it is impossible to count all the fish in the lake, you can use your samples to estimate the population in the lake.

1 Write a proportion that relates the numbers in your chart and the estimated number of fish in the lake.

2 Imagine that the only information you had was the data you recorded after your first casting.

What would have been your estimate of the number of fish in the lake? Justify your reasoning. Record your estimate in the last column of your chart.

3 Make estimates for each of your castings and record them in the last column of your chart.

4 Taking all of your castings into consideration, if you had to give an official estimate of the number of fish in the lake, what would it be? Justify your reasoning.

Add the results of your work to your Investigation Folder.

Integration: Geometry
Similar Triangles

APPLICATION
Architecture

"Rock n' roll is here to stay," is what you might hear after visiting the Rock and Roll Hall of Fame and Museum in Cleveland, Ohio. Memorabilia from musicians such as the Beatles, U2, and Carlos Santana cover the walls inside. The automatic teller machines in the building are even shaped like jukeboxes.

The building itself, designed by architect I.M. Pei, is made up of several geometric forms—a triangle, a cylinder, a rectangle, and a trapezoid. The triangular section of the building is similar to the one Pei designed for the Louvre in Paris, France. Both are made up of similar triangles.

Two figures are **similar** if they have the same shape, but not necessarily the same size.

If **corresponding angles** of two triangles have equal measures, the triangles are similar. The two triangles above are similar. We write this as $\triangle ABC \sim \triangle DEF$. The order of the letters indicates the angles that correspond.

corresponding angles	corresponding sides
$\angle A$ and $\angle D$	$\overline{AB}$ and $\overline{DE}$
$\angle B$ and $\angle E$	$\overline{BC}$ and $\overline{EF}$
$\angle C$ and $\angle F$	$\overline{AC}$ and $\overline{DF}$

The sides opposite corresponding angles are called **corresponding sides**. Compare the measures of the corresponding sides. Note that AC means the measure of $\overline{AC}$, DF means the measure of $\overline{DF}$ and so on.

$$\frac{AB}{DE} = \frac{6}{18} = \frac{1}{3} \qquad \frac{BC}{EF} = \frac{5}{15} = \frac{1}{3} \qquad \frac{AC}{DF} = \frac{3}{9} = \frac{1}{3}$$

When the measures of the corresponding sides form equal ratios, the measures are said to be *proportional*.

Similar Triangles	If two triangles are similar, the measures of their corresponding sides are proportional, and the measures of their corresponding angles are equal.

Proportions can be used to find the measures of the sides of similar triangles when some measurements are known.

Example ❶ Surveyors normally use instruments to measure objects that are too large or too far away to measure by hand. They also can use the shadows that objects cast to find the height of the objects without measuring them directly. How can a surveyor use a telephone pole that is 25 feet tall and casts a shadow 20 feet long to find the height of a building that casts a shadow 52 feet long?

$\triangle NOP$ is similar to $\triangle MRQ$.

$\dfrac{RQ}{OP} = \dfrac{QM}{PN}$ *Corresponding sides of similar triangles are proportional.*

$\dfrac{25}{x} = \dfrac{20}{52}$

$20x = 25(52)$ *Find the cross products.*

$20x = 1300$

$x = 65$

The building is 65 feet high.

Similar triangles may be positioned so that the corresponding parts are not obvious.

Example ❷ **Find the missing measures of the sides if each pair of triangles below are similar.**

Remember that the sum of the measures of the angles in a triangle is 180°.

a. The measure of $\angle Q$ is $180° - (117° + 27°)$ or $36°$.

The measure of $\angle T$ is $180° - (117° + 36°)$ or $27°$.

Since the corresponding angles have equal measures, $\triangle QRS \sim \triangle VUT$. This means that the lengths of the corresponding sides are proportional.

$\dfrac{SQ}{TV} = \dfrac{QR}{VU}$ $\dfrac{SQ}{TV} = \dfrac{SR}{TU}$

$\dfrac{33}{12} = \dfrac{x}{6}$ $\dfrac{33}{12} = \dfrac{y}{8}$

$12x = 33(6)$ $12y = 33(8)$

$12x = 198$ $12y = 264$

$x = 16.5$ $y = 22$

The missing measures are 16.5 and 22.

b. $\triangle NRT$ is similar to $\triangle MST$.

$\dfrac{RN}{SM} = \dfrac{NT}{MT}$

$\dfrac{9}{x} = \dfrac{12}{28}$

$12x = 9(28)$ *Find the cross products.*

$12x = 252$

$x = 21$

The measure of $\overline{SM}$ is 21 meters.

CAREER CHOICES

Land Surveyors use technical expertise to determine official land, space, and water boundaries. They plan field work, research legal records, record survey results, and prepare maps and reports. High school students should take courses in algebra, geometry, trigonometry, and drafting. Many colleges offer 4-year programs leading to a B.S. degree in surveying.

For more information, contact:

American Congress on Surveying and Mapping 5410 Grosvenor Lane Bethesda, MD 20814–2122

Communicating Mathematics

Study the lesson. Then complete the following.

1. **Name** the pairs of corresponding angles in Example 2a.

2. **List** what you know about the angles and sides of two triangles that are similar.

3. **List** the corresponding angles and corresponding sides if $\triangle RED \sim \triangle SOX$. Then write three proportions that correspond to these triangles.

MODELING MATHEMATICS

4. **Model** a triangle using toothpicks or straws. Then make a similar triangle using more toothpicks or straws. Write a proportion to show that the two triangles are similar.

Guided Practice

5. For the pair of similar triangles at the right, name the triangle that is similar to $\triangle XYZ$. Make sure you have the letters in the correct order.

Determine whether each pair of triangles is similar. Justify your answer.

6.

78°

52°   52°

7.

56°   40°

84°

84°

$\triangle KLM$ and $\triangle NOP$ are similar. For each set of measures given, find the measures of the remaining sides.

8. $k = 24$, $\ell = 30$, $m = 15$, $n = 16$

9. $k = 9$, $n = 6$, $o = 8$, $p = 4$

10. $n = 6$, $p = 2.5$, $\ell = 4$, $m = 1.25$

L

m   k

K ℓ M

O

p   n

N   o   P

11. If a tree 6 feet tall casts a shadow 4 feet long, how high is a flagpole that casts a shadow 18 feet long at the same time of day?

J

6 ft

K 4 ft L

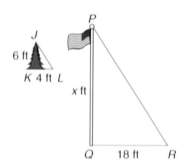

P

x ft

Q   18 ft   R

Practice

For each pair of similar triangles, name the triangle that is similar to △*WXY*.

12.

13.

14.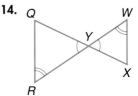

Determine whether each pair of triangles is similar. Justify your answer.

15.

16.

17.

18.

19.

20.

△*ABC* and △*DEF* are similar. For each set of measures given, find the measures of the remaining sides.

21. $c = 11, f = 6, d = 5, e = 4$

22. $a = 5, d = 7, f = 6, e = 5$

23. $a = 17, b = 15, c = 10, f = 6$

24. $a = 16, e = 7, b = 13, c = 12$

25. $d = 2.1, b = 4.5, f = 3.2, e = 3.4$

26. $f = 12, d = 18, c = 18, e = 16$

27. $c = 5, a = 12.6, e = 8.1, f = 2.5$

28. $f = 5, a = 10.5, b = 15, c = 7.5$

29. $a = 4\frac{1}{4}, b = 5\frac{1}{2}, e = 2\frac{3}{4}, f = 1\frac{3}{4}$

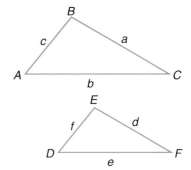

Critical Thinking

30. △*ABC* and △*DEF* are similar. If the ratio of *AC* to *DF* is 2 to 3, what is the ratio of the area of △*ABC* to the area of △*DEF*?

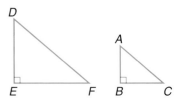

31. Draw an equilateral triangle. Try to cut it into three similar parts with no two parts that are congruent. Describe the process you used.

32. Billiards Meda is playing billiards on a table like the one show at the right. If she can make her next shot, she figures she will have no trouble winning. She wants to strike the cue ball at D, bank it at C, and hit another ball at the mouth of pocket A. Use similar triangles to find where Meda's cue ball should strike the rail.

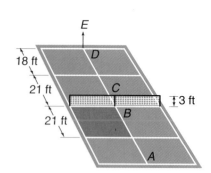

33. Tennis When serving in tennis, a player hits the ball standing at point E as shown in the figure at the right. To be a good serve, the ball must go over the net and land in the shaded area. Assume that all serves travel in a straight line and that all serves just clear the net. The figure below shows the view from the sideline.

Let $\overline{EY}$ represent the flight of the ball.

a. Suppose you hit the ball when it is 9 feet above the ground. Write a proportion and solve for d to determine whether the serve is good.

b. Repeat the calculation of the distance d if the ball is hit from 8 feet above the ground and from 10 feet above the ground. Use the results to help you determine which players have an easier time hitting a good serve, tall players or short players. Explain.

34. Bridges Haylee is making a model of the Mark Clark Expressway, a continuous truss bridge that spans the Cooper River in Charleston, South Carolina. Parts of the trusses are made up of triangles. Haylee plans to make the model bridge to the following scale: 1 inch to 12 feet. If the height of the triangle on the actual bridge is 40 feet, what will the height be on the model?

Mark Clark Expressway

Mixed Review

35. Drafting The scale on the blueprint for a house is 1 inch to 3 feet. If the living room on the blueprint is $5\frac{1}{2}$ inches by 7 inches, what are the dimensions of the actual room? *(Lesson 4–1)*

36. Solve $7 = 3 - \dfrac{n}{3}$. *(Lesson 3–3)*

37. Table Tennis Each Ping-Pong™ ball weighs about $\frac{1}{10}$ of an ounce. How many Ping-Pong balls weigh 1 pound altogether? *(Lesson 3–2)*

38. Solve $x + (-8) = -31$. *(Lesson 3–1)*

39. Ecology Americans use about 2.5 million plastic bottles every hour. How many do they use in one day? How many in one week? *(Lesson 2–6)*

40. Replace __?__ with $<$, $>$, or $=$ to make $\dfrac{8}{15}$ __?__ $\dfrac{9}{16}$ true. *(Lesson 2–4)*

41. Evaluate $(19 - 12) \div 7 \cdot 23$. Name the property used in each step. *(Lesson 1–6)*

Integration: Trigonometry
Trigonometric Ratios

What YOU'LL LEARN

- To use trigonometric ratios to solve right triangles.

Why IT'S IMPORTANT

You can use the trigonometric ratios to solve problems involving architecture and archaeology.

APPLICATION

Architecture

The Leaning Tower of Pisa, a bell tower in Pisa, Italy, currently leans 16.5 feet off center. The tower was closed in 1990 in order for workers to attempt to stabilize the tower's foundation to prevent it from eventually collapsing. By 1994, the tower's lean had decreased about $\frac{2}{5}$ of an inch. No date has been scheduled for reopening the tower.

If you were to draw a line from the top of the tower to the ground, a right triangle would be formed with the ground. If enough is known about a right triangle, certain ratios can be used to find the measures of the remaining parts of the triangle. These ratios are called **trigonometric ratios.** *Trigonometry is an area of mathematics that studies angles and triangles.*

A typical right triangle is shown at the right.

  a is the measure of the side *opposite* $\angle A$, $\overline{BC}$.

  b is the measure of the side *opposite* $\angle B$, $\overline{AC}$.

  c is the measure of the side *opposite* $\angle C$, $\overline{AB}$.

  a is *adjacent* to $\angle B$ and $\angle C$.

  b is *adjacent* to $\angle A$ and $\angle C$.

  c is *adjacent* to $\angle A$ and $\angle B$.

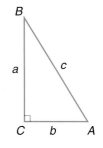

The side opposite $\angle C$, the right angle, is called the **hypotenuse.** The other two sides are called **legs.**

Three common trigonometric ratios are defined as follows.

Definition of Trigonometric Ratios	
	sine of $\angle A$ = $\dfrac{\text{measure of leg opposite } \angle A}{\text{measure of hypotenuse}}$
	$\sin A = \dfrac{a}{c}$
	cosine of $\angle A$ = $\dfrac{\text{measure of leg adjacent to } \angle A}{\text{measure of hypotenuse}}$
	$\cos A = \dfrac{b}{c}$
	tangent of $\angle A$ = $\dfrac{\text{measure of leg opposite } \angle A}{\text{measure of leg adjacent to } \angle A}$
	$\tan A = \dfrac{a}{b}$

Notice that sine, cosine, and tangent are abbreviated as sin, cos, and tan, respectively.

Example Find the sine, cosine, and tangent of each acute angle. Round your answers to the nearest thousandth.

$$\sin J = \frac{\text{opposite leg}}{\text{hypotenuse}}$$

$$= \frac{7}{25} \text{ or } 0.280$$

$$\cos J = \frac{\text{adjacent leg}}{\text{hypotenuse}}$$

$$= \frac{24}{25} \text{ or } 0.960$$

$$\tan J = \frac{\text{opposite leg}}{\text{adjacent leg}}$$

$$= \frac{7}{24} \text{ or about } 0.292$$

$$\sin L = \frac{\text{opposite leg}}{\text{hypotenuse}}$$

$$= \frac{24}{25} \text{ or } 0.960$$

$$\cos L = \frac{\text{adjacent leg}}{\text{hypotenuse}}$$

$$= \frac{7}{25} \text{ or } 0.280$$

$$\tan L = \frac{\text{opposite leg}}{\text{adjacent leg}}$$

$$= \frac{24}{7} \text{ or about } 3.429$$

Consider triangles *QTV* and *MSO*. Since corresponding angles have the same measure, $\triangle QTV \sim \triangle MSO$. Recall that in similar triangles, the corresponding sides are proportional.

$$\frac{m}{q} = \frac{s}{t}$$

$$\frac{q}{s} \cdot \frac{m}{q} = \frac{q}{s} \cdot \frac{s}{t} \quad \textit{Multiply each side by } \frac{q}{s}.$$

$$\frac{m}{s} = \frac{q}{t}$$

$$\sin M = \sin Q$$

In general, the sine of a 64° angle of a right triangle will be the same number no matter how large or small the triangle is. A similar result holds for cosine and tangent.

You can use a calculator to find the values of trigonometric functions or to find the measure of an angle.

Example Find the value of sin 64° to the nearest ten thousandth.

Use a scientific calculator.

Enter: 64 [SIN] *0.898794046*   *The calculator must be in degree mode.*

Rounded to the nearest ten thousandth, sin 64° ≈ 0.8988.

Example Find the measure of ∠*P* to the nearest degree.

Since the length of the opposite and adjacent sides are known, use the tangent ratio.

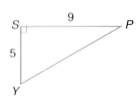

$$\tan P = \frac{\text{opposite leg}}{\text{adjacent leg}}$$

$$= \frac{5}{9}$$

Use a scientific calculator to find the angle measure with a tangent of $\frac{5}{9}$.

Enter: 5 [÷] 9 [=] [2nd] [TAN⁻¹] *29.0546041*   *The TAN⁻¹ key "undoes" the original function.*

To the nearest degree, the measure of ∠*P* is 29°.

Many real-world applications involve trigonometry.

Example **4**

APPLICATION
Architecture

Refer to the application at the beginning of the lesson. Engineers working to correct the lean of the Leaning Tower of Pisa could check the angle the Tower makes with the ground to find any increase in the lean. Use the measurements to find the angle to the nearest degree that the Tower made with the ground.

Since the length of adjacent side and hypotenuse are known, use the cosine.

$$\cos x = \frac{\text{adjacent leg}}{\text{hypotenuse}}$$

$$\cos x = \frac{16.5}{179}$$

$\cos x = 0.09217877$ *Use a calculator to find x.*

$$x \approx 85°$$

To the nearest degree, the measure of the angle is 85°.

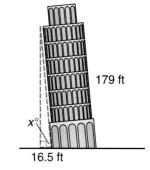

179 ft

$x°$

16.5 ft

You can find the missing measures of a right triangle if you know the measure of two sides of the triangle or the measure of one side and one acute angle. Finding all of the measures of the sides and the angles in a right triangle is called **solving the triangle**.

Example **5** **Solve ΔABC.**

The measure of ∠C is 90° − 42° or 48°.

$$\sin 42° = \frac{x}{22}$$

$$0.6691 \approx \frac{x}{22}$$

$$14.7 \approx x$$

Thus, $\overline{BC}$ is about 14.7 inches long.

A

42°   22 in.

y in.

B   *x* in.   *C*

$$\cos 42° = \frac{y}{22}$$

$$0.7431 \approx \frac{y}{22}$$

$$16.3 \approx y$$

Thus, $\overline{AB}$ is about 16.3 inches long.

In the real world, trigonometric ratios are often used to find distances or lengths that cannot be measured directly. In these applications, you will sometimes use an angle of elevation or an angle of depression. An **angle of elevation** is formed by a horizontal line of sight and a line of sight above it. An **angle of depression** is formed by a horizontal line of sight and a line of sight below it.

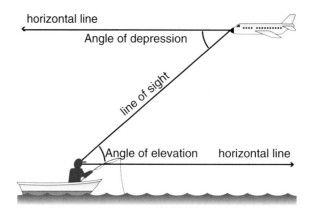

horizontal line

Angle of depression

line of sight

Angle of elevation   horizontal line

MODELING MATHEMATICS

Make a Hypsometer

Materials: protractor straw tape paper clip string

You can make a hypsometer to find the angle of elevation of an object that is too tall to measure.

Straw

180
170
160
150
140
130
120
110
100
90
10 20 30 40 50 60 70 80

String

Paper Clip

Your Turn

a. Tie one end of a piece of string to the middle of a straw. Tie the other end of string to a paper clip.

b. Tape a protractor to the side of the straw. Make sure that the string hangs freely to create a vertical or plumb line.

c. Find an object outside that is too tall to measure directly, such as a basketball hoop, a flagpole, or the school building.

d. Look through the straw to the top of the object you are measuring. Find a horizontal line of sight as shown above. Find the measurement where the string and protractor intersect. Determine the angle of elevation by subtracting this measurement from 90°.

e. Use the equation tan (angle sighted) = $\frac{\text{height of object} - x}{\text{distance of object}}$, where x represents the distance from the ground to your eye level, to find the height of the object.

f. Compare your answer with someone who measured the same object. Did your heights agree? Why or why not?

Example ⑥

APPLICATION
Balloons

What would the Macy's Thanksgiving Day Parade be without balloons? In 1994, Dr. Suess' The Cat in the Hat made its first appearance. It took 36 people to maneuver the cat along the parade route. Suppose you were watching the parade. You noticed that two of the handling lines formed a right triangle with the ground. You estimated the angle of elevation to the bottom of the balloon to be 60°, and the handlers to be about 35 feet away from each other. Find the altitude of the bottom of the balloon to the nearest foot.

Explore   You know the distance from one handler to another and the approximate angle of elevation. You need to find the height from the handlers to the bottom of the balloon.

(continued on the next page)

Plan Let *h* represent the distance to the bottom of the balloon in feet. Use the tangent ratio to solve this problem.

Solve

$$\tan 60° = \frac{\text{opposite leg}}{\text{adjacent}}$$

$$\tan 60° = \frac{h}{35}$$

$$1.7321 \approx \frac{h}{35}$$

$$1.7321(35) \approx h$$

$$60.6235 \approx h$$

The altitude of the bottom of the balloon is about 61 feet above the handlers.

Examine You can examine the solution by determining the angle of elevation in the equation $\tan x = \frac{61}{35}$. Since $x = 60.15°$, the solution is reasonable.

CHECK FOR UNDERSTANDING

Communicating Mathematics

Study the lesson. Then complete the following.

1. **Use** the figure at the right to answer the following questions.

   a. What is the measure of the leg adjacent to $\angle C$?

   b. What is the measure of the leg opposite $\angle C$?

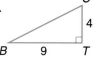

2. **List** the trigonometric ratio or ratios that involve the measure of the hypotenuse.

3. **Explain** how to determine which trigonometric ratio to use when solving for an unknown measure of a right triangle.

4. **Name** the angles of elevation and depression in the drawing at the right.

5. **State** which trigonometric ratio you could use to find *x*.

6. **Use** a hypsometer to find the height of a tree in your community. Then write a paragraph explaining how you found the height.

For each triangle, find sin Y, cos Y, and tan Y to the nearest thousandth.

7.

8.

Use a calculator to find the value of each trigonometric ratio to the nearest ten thousandth.

9. $\cos 35°$ 10. $\sin 63°$ 11. $\tan 7°$

Use a calculator to find the measure of each angle to the nearest degree.

12. $\cos C = 0.9613$ 13. $\sin X = 0.7193$ 14. $\tan W = 2.4752$

For each triangle, find the measure of the marked acute angle to the nearest degree.

15.

16.

Solve each right triangle. State the side lengths to the nearest tenth and the angle measures to the nearest degree.

17.

18.

19.

20. **Transportation** Find the angle of elevation of a train track if a train in the mountains rises 15 feet for every 250 feet it moves along the track.

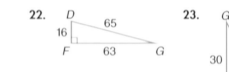
EXERCISES

For each triangle, find sin G, cos G, and tan G to the nearest thousandth.

21.

22.

23.

For each triangle, find sin _G_, cos _G_, and tan _G_ to the nearest thousandth.

24.

25.

26.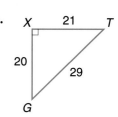

Use a calculator to find the value of each trigonometric ratio to the nearest ten thousandth.

27. sin 21°

28. cos 15°

29. sin 76°

30. tan 56°

31. cos 68°

32. tan 30°

Use a calculator to find the measure of each angle to the nearest degree.

33. tan A = 0.6473

34. cos B = 0.7658

35. sin F = 0.3823

36. cos Q = 0.2993

37. sin R = 0.8827

38. tan J = 8.8988

For each triangle, find the measure of the marked acute angle to the nearest degree.

39.

40.

41.

42.

43.

44.

Solve each right triangle. State the side lengths to the nearest tenth and the angle measures to the nearest degree.

45.

46.

47.

48.

49.

50.

51.

52.

53.

Programming

To run the program repeatedly, press ENTER *after each run is completed and the prompt will appear again for new values of a and b.*

54. Suppose ∠*ABC* has a right angle *C*, and the measures of its legs are *a* and *b*. The graphing calculator program at the right asks you to enter the lengths of the legs of the triangle. It then calculates the measure of the hypotenuse *c*, the measures of the acute angles, and the trigonometric ratios for each acute angle. The PAUSE commands in the program stop the calculator's computations and allow you to record the information shown on the screen. Press ENTER to continue the program. Make sure your calculator is in degree mode before running the program.

```
PROGRAM: LEGS
: Disp "ENTER LENGTHS OF",
  "LEGS A AND B"
: Prompt A,B
: √(A² + B²)→C
: Disp "SIDE C =",C
: Disp "ANGLE A =", tan⁻¹ (A/B)
: Disp "ANGLE B =", tan⁻¹ (B/A)
: Pause
: Disp "sin (A)=",A/C
: Disp "sin (B)=",B/C
: Disp "cos (A)=",B/C
: Pause
: Disp "cos (B)=",A/C
: Disp "tan (A)=",A/B
: Disp "tan (B)=",B/A
```

Use the program for each pair of values for *a* and *b*.

a. 3, 4 **b.** 5, 12
c. 20, 40 **d.** 125, 100

Critical Thinking

55. Use the triangle below to determine which of the following statements are true. Justify your answers.

a. $\sin C = \cos B$

b. $\cos B = \dfrac{1}{\sin B}$

c. $\tan C = \dfrac{\cos C}{\sin C}$

d. $\tan C = \dfrac{\sin C}{\cos C}$

e. $\sin B = (\tan B)(\cos B)$

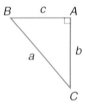

Applications and Problem Solving

56. Archaeology The largest of the pyramids of Egypt has a square base with sides 755 feet long. ∠*PQR* has a measure of 52°. The top of the pyramid is no longer there. What was the pyramid's original height ($\overline{RP}$) to the nearest foot?

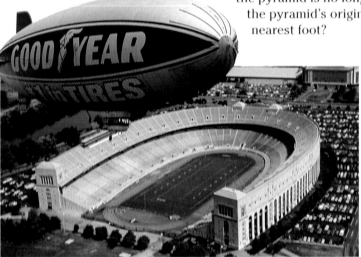

57. Make a Drawing On September 30, 1995, the Ohio State Buckeyes defeated the Notre Dame Fighting Irish in football as the Goodyear blimp relayed aerial views from above. Suppose the blimp was directly over the 50-yard line. A photographer on the ground estimated the angle of elevation to be 85° when she was 65 yards away from the point on the ground directly under the blimp. How high is the blimp? Round your answer to the nearest foot.

58. Civil Engineering The city of Mansfield plans to build a bridge across Spring Lake. Use the information in the diagram at the right to find the distance across Spring Lake. (Lesson 4–2)

59. Solve $\frac{x}{5} = \frac{x+3}{10}$. (Lesson 4–1)

60. Baseball The batting averages for 10 players on a baseball team are 0.234, 0.253, 0.312, 0.333, 0.286, 0.240, 0.183, 0.222, 0.297, and 0.275. Find the median batting average for these players. (Lesson 3–7)

61. Solve $\frac{2}{5}y + \frac{y}{2} = 9$. (Lesson 3–3)

62. Write an equation to represent the situation below. Then solve the problem. (Lesson 3–2)

Joyce Conners paid $47.50 for five football tickets. What was the cost per ticket?

63. Name the sets of numbers to which $\sqrt{36}$ belongs. Use N for natural numbers, W for whole numbers, Z for integers, Q for rational numbers, and I for irrational numbers. (Lesson 2–8)

64. Simplify $\frac{3a+9}{3}$. (Lesson 2–7)

65. Find $\left(-\frac{2}{3}\right)\left(-\frac{1}{5}\right)$. (Lesson 2–6)

66. Find $-13 + (-8)$. (Lesson 2–3)

67. Simplify $8x + 2y + x$. (Lesson 1–8)

68. Evaluate $(9 - 2 \cdot 3)^3 - 27 + 9 \cdot 2$. (Lesson 1–6)

69. Evaluate 5^3. (Lesson 1–1)

Morgan's Conjecture

The excerpt below appeared in an article in *The Columbus Dispatch* on December 21, 1994.

RYAN MORGAN WOULD HAVE GOTTEN an "A" in geometry even if he hadn't unearthed a mathematical treasure. But the persistent Baltimore high school sophomore pushed a hunch into a theory. He calls it "Morgan's Conjecture" and is hoping it will soon be "Morgan's Theorem." In geometric circles, developing a theorem is a big deal—especially if you're only 15....What did Ryan see?...When the sides of a triangle are divided by an odd number larger than one, and when lines are drawn from the division points on the sides of the triangle to the vertices, there always will be a hexagon in the interior of the triangle. And the area of that hexagon always will be a predictable fraction of the area of the larger triangle. The fraction is determined by a complex formula that Ryan worked out. ■

1. In talking about his work, the student said, "It's mostly just luck in playing around with it." What other factors besides luck might have been involved?

2. Are you surprised to learn that students can discover theorems? Why or why not?

3. Is there an area of mathematics that interests you that you would like to explore further? How would you do it?

Percents

APPLICATION
Cosmetics

According to *Good Housekeeping*, 3 out of 5 teenage girls wear mascara. What is the rate of teenage girls wearing mascara per 100 girls?

You can solve this problem by using a proportion. The ratio of teenage girls who wear mascara to the total number of teenage girls is $\frac{3}{5}$. Write a proportion that sets $\frac{3}{5}$ equal to a ratio with a denominator of 100.

$$\frac{3}{5} = \frac{n}{100}$$

$$60 = n$$

Teenage girls wear mascara at a rate of 60 per 100, or 60 percent. A **percent** is a ratio that compares a number to 100. Percent also means *per hundred*, or *hundredths*. Percents can be expressed with a percent symbol (%), as fractions, or as decimals.

$$60\% = \frac{60}{100} = 0.60$$

Example **①** Write $\frac{3}{4}$ as a percent.

Method 1
Use a proportion.

$$\frac{3}{4} = \frac{n}{100}$$

$$300 = 4n$$

$$75 = n$$

Thus, $\frac{3}{4}$ is equal to $\frac{75}{100}$ or 75%.

Method 2
Use a calculator.

$$\frac{3}{4} = \frac{n}{100}$$

$$100\left(\frac{3}{4}\right) = n$$

Enter: 100 ⊠ 3 ÷ 4 ═ 75

Proportions are often used to solve percent problems. One of the ratios in these proportions is always a comparison of two numbers called the **percentage** and the **base**. The other ratio, called the **rate**, is a fraction with a denominator of 100.

$$\begin{array}{l} percentage \rightarrow \\ base \rightarrow \end{array} \left. \begin{array}{l} 3 \\ 4 \end{array} = \begin{array}{l} 75 \\ 100 \end{array} \right\} \leftarrow Rate$$

Percent Proportion	$\dfrac{\text{Percentage}}{\text{Base}} = \text{Rate}$ or $\dfrac{\text{Percentage}}{\text{Base}} = \dfrac{r}{100}$

The $\dfrac{percentage}{base}$ represents $\dfrac{part}{whole}$

Example **a. 30 is what percent of 50?**   **b. 20 is what percent of 30?**

Use the percent proportion.

$$\frac{\text{Percentage}}{\text{Base}} = \frac{r}{100}$$

The percentage is 30. $\frac{30}{50} = \frac{r}{100}$
The base is 50.

$$3000 = 50r$$

$$60 = r$$

Thus, 30 is 60% of 50.

Use a calculator.

$$\frac{\text{Percentage}}{\text{Base}} = \frac{r}{100}$$

$$\frac{20}{30} = \frac{r}{100}$$

Enter: 20 ÷ 30 × 100 =

66.66666667

Thus, 20 is $66.\overline{6}$% of 30. This can also be written as $66\frac{2}{3}$%.

You can also write equations to solve problems with percents.

Example **a. 60% of what number is 54?**   **b. What number is 40% of 37.5?**

60% of what number is 54?

$$\frac{60}{100} \cdot x = 54$$

$$0.6x = 54$$

$$x = 90$$

Thus, 60% of 90 is 54.

What number is 40% of 37.5?

$$x = \frac{40}{100} \cdot 37.5$$

$$x = 15$$

Thus, 15 is 40% of 37.5.

TECHNOLOGY *Tip*

Most calculators have a key labeled %. To find 28% of 58.4 on a scientific calculator, enter:

28 2nd % ×

58.4 = *16.352.*

MODELING MATHEMATICS

Making Circle Graphs

Materials: compass   protractor

You can use a circle graph to compare parts of a whole. Follow the steps to display the data below in a circle graph.

Where's the Remote?	
Number of Times TV Remote Misplaced Per Week	Number of People Responding
Never	220
1–5	190
5 or more	85
Don't Know	5

Your Turn

a. Find the total number of people surveyed.

b. Find the ratio that compares the number of people that responded for each category to the total number of people that responded.

c. Since there are 360° in a circle, multiply each ratio by 360 to find the number of degrees for each section of the graph. Round to the nearest degree.

d. Use a compass to draw a circle and a radius as shown at the right.

e. Use a protractor to draw angles. Start with the least number of degrees. Repeat for the remaining sections. Label each section and give the graph a title.

f. What is the sum of the percents in your graph?

Percents are also used in simple interest problems. **Simple interest** is the amount paid or earned for the use of money. The formula $I = prt$ is used to solve problems involving simple interest. In the formula, I represents the interest, p represents the amount of money invested, called the *principal*, r represents the annual interest rate, and t represents time in years.

Example ④

Finance

a. **Luis Hernandez has some money he has saved over the summer from mowing lawns. He wants to put some of it in a 6-month certificate of deposit (CD) account at BancOne that would pay 6% annual interest. He doesn't want to put all of his money in the CD because he wants some spending money. He is hoping to earn $45 in interest to buy a new video game. How much money should Luis put in the CD?**

Explore   Let p = the amount of money Luis should deposit.

Plan $I = prt$
             $45 = p(0.06)(0.5)$ *Write 6% as 0.06 and 6 months as 0.5 years.*

Solve $45 = 0.03p$ *Multiply 0.06 by 0.5.*
             $1500 = p$ *Divide each side by 0.03.*

Examine   When p is 1500, $I = (1500)(0.06)(0.5)$ or $45. Luis should deposit $1500.

b. **Whitney Williamson has $30,000 she would like to invest. She has a choice of two interest-paying bonds, one offering a 6% annual interest and the other paying 7.5% annual interest. She would like to earn $2100 in interest in one year. If she earns any more than $2100, she would need to pay a higher rate of income tax. How much money should she invest in each bond?**

Explore   Let n = the amount of money invested at 6%.
             Then $30,000 - n$ = the amount of money invested at 7.5%.

Plan   *interest on 6% investment*
             $n(0.06)(1)$ or $0.06n$

             interest on 7.5% investment
             $(30,000 - n)(0.075)(1)$ or $2250 - 0.075n$

Solve   *interest at 6%* + *interest at 7.5%* = *total interest*

             $0.06n$ + $(2250 - 0.075n) =$ 2100
             $-0.015n + 2250 = 2100$ *Add 0.06n and*
             $-0.015n = -150$ *−0.075n.*
             $n = 10,000$

Whitney should invest $10,000 at 6% and $30,000 - $10,000 or $20,000 at 7.5%.

Examine $(10,000)(0.06)(1) + (20,000)(0.075)(1) = 2100$

The answer checks.

Communicating Mathematics

Study the lesson. Then complete the following.

1. **Compare** $\frac{r}{100}$ in the percent proportion and r in the simple interest formula. How are they the same? How are they different?

2. **Explain** how to find 57% of 42 using a scientific calculator.

3. **Explain** why an investor might invest a greater amount of money in a lower rate of interest instead of a higher rate.

4. **Read** the following excerpt from an article about the America's Cup yacht race in *Motor Boating & Sailing* magazine. Explain what is incorrect in this excerpt and why.

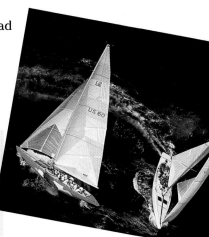

With a Ph.D. from MIT, (Dr. Koch) takes a scientific view of things, and quantifies the effort involved in winning the Cup as 55 percent boat speed, 20 percent tactics, 20 percent crew work, 5 percent luck, and less than 5 percent physical strength.

MODELING MATHEMATICS

5. **Draw** a circle graph to represent the number of students in your school by grade level.

Guided Practice

Write each ratio as a percent and then as a decimal.

6. $\frac{3}{4}$

7. $\frac{43}{100}$

8. $\frac{2}{25}$

Use a proportion to answer each question.

9. Eleven is what percent of 20?

10. What percent of 80 is 45?

Use an equation to answer each question.

11. What number is 30% of 50?

12. What percent of 75 is 16?

Use $I = prt$ to find the missing quantity.

13. Find I if $p = \$1500$, $r = 7\%$, and $t = 6$ months.

14. Find p if $I = \$196$, $r = 10\%$, and $t = 7$ years.

15. **Finance** Kaylee Richardson invested $7200 for one year, part at 10% annual interest and the rest at 14% annual interest. Her total interest for the year was $960. How much money did she invest at each rate?

EXERCISES

Practice

Write each ratio as a percent and then as a decimal.

16. $\frac{67}{100}$

17. $\frac{6}{20}$

18. $\frac{5}{8}$

19. $\frac{7}{10}$

20. $\frac{5}{6}$

21. $\frac{9}{5}$

22. $\frac{2}{3}$

23. $\frac{25}{40}$

24. $\frac{20}{8}$

Use a proportion to answer each question.

25. Thirty-five is what percent of 70?   **26.** What percent of 60 is 18?

27. Eight is what percent of 64?   **28.** Six is what percent of 15?

29. What percent of 2 is 8?   **30.** What percent of 14 is 4.34?

Use an equation to answer each question.

31. What percent of 160 is 4?   **32.** What number is 25% of 56?

33. Twelve is 16.6% of what number?   **34.** Thirty-two is what percent of 80?

35. 17.56 is 2.5% of what number?   **36.** What percent of 75 is 30?

Solve.

37. $64.93 is what percent of $231.90?

38. Find 112% of $500.

39. Find 81% of 32.

Use $I = prt$ to find the missing quantity.

40. Find r if $I = \$5920$, $p = \$4000$, and $t = 3$ years.

41. Find r if $I = \$780$, $p = \$6500$, and $t = 1$ year.

42. Find I if $p = \$3200$, $r = 9\%$, and $t = 18$ months.

43. Find I if $p = \$5000$, $r = 12\frac{1}{2}\%$, and $t = 5$ years.

44. Find t if $I = \$2160$, $p = \$6000$, and $r = 8\%$.

45. Find p if $I = \$756$, $r = 9\%$, and $t = 3\frac{1}{2}$ years.

Critical Thinking

46. If x is 225% of y, then y is what percent of x?

Applications and Problem Solving

47. Health   Physicians in the United States treated 4.1 million broken bones in 1992. The chart at the right shows the fractures by age. Find the percentage of fractures by age.

Age	Fractures
Under 18	1.6 million
18–44	1.4 million
45–64	600,000
65 and older	500,000

Source: American Academy of Orthopedic Surgeons

48. Finance   Melanie Morgan invested $5000 for one year, part at 9% annual interest and the rest at 12% annual interest. The interest from the investment at 9% was $198 more than the interest from the investment at 12%. How much money did she invest at 9%?

49. Finance   Juan and Rosita Diaz have invested $2500 at 10% annual interest. They have $6000 more to invest. At what rate must they invest the $6000 to have a total of $9440 at the end of the year?

50. Statistics   According to The Alden Group, desserts in the home are not always eaten in the kitchen or dining room. In fact, only 30% of desserts are eaten in the kitchen and 14% in the dining room. Eighteen percent are most often eaten in the living room, 10% in the den, and an amazing 28% are eaten in the bedroom. Make a circle graph showing these data.

51. **Consumerism**   Luke's father used his credit card to purchase a $330 television. The bank that issued the credit card charges an annual interest rate of 19.8%. Each month they charge interest on the unpaid balance of the account. If Luke's father does not make any more purchases and makes no payments, how much interest will he be charged next month?

52. **Nutrition**   The table at the right lists the number of calories and grams of fat in eight popular pizza toppings. The estimates are based on two slices of a 14-inch pizza. To find the percentage of calories from fat, multiply the fat grams by 9 and divide the result by the number of calories. Which of the toppings gets the highest percentage of its calories from fat?

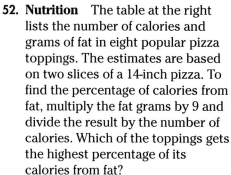

Topping	Calories	Fat Grams
Extra Cheese	168	8
Sausage	97	8
Pepperoni	80	7
Black Olives	56	5
Ham	41	2
Onion	11	<1
Green Pepper	5	0
Mushrooms	5	0

53. **Data Analysis**   A survey was taken recently to determine what Americans would serve President Clinton and his family for dinner. The survey results are shown at the right.

What Would You Serve?	
steak and potatoes	39%
lasagna	25%
chicken marsala	21%
poached salmon	18%
hamburgers or hot dogs	15%
spaghetti and meatballs	14%
pasta primavera	13%
chef salad	10%
pizza	9%
spaghetti with tomato sauce	8%
soup and sandwiches	5%

Source: The Ragu Corporation

   a. Suppose 258 people were surveyed. How many people said they would serve chicken marsala? How many said pizza?

   b. The sum of the percentages in the survey is not 100%. Why do you think this is the case?

54. **Money**   The number of coins minted in 1994 is shown in the table at the right. Make a circle graph of the data.

U.S. Coins Minted in 1994	
Pennies	14,120,000,000
Nickels	1,480,000,000
Dimes	2,600,000,000
Quarters	1,760,000,000
Half-Dollars	40,000,000

Mixed Review

General Sherman

55. **Trees**   While visiting the Sequoia National Park in California, Tyler stops near a tree called the "General Sherman." The guide tells him it is the world's largest tree according to its volume of wood. Tyler is standing about 160 feet from the tree. If he is looking at the top of the tree at a 60° angle, about how tall is the tree?   (Lesson 4–3)

56. **Geometry** $\triangle ABC$ and $\triangle RQP$ are similar. If $a = 10$, $r = 5$, $q = 3$, and $p = 4$, find the measures of the remaining sides.   (Lesson 4–2)

57. Solve $\dfrac{9}{x - 8} = \dfrac{4}{5}$.   (Lesson 4–1)

58. Statistics In a girl's basketball game between East High School and Monroe High School, the East players' individual scores were 12, 4, 5, 3, 11, 23, 4, 6, 7, and 8. Find the mean, median, and mode of the individual points. (Lesson 3–7)

59. Solve $y + (-7.5) = -12.2$. (Lesson 3-1)

60. Diving Greg Louganis of the United States has won a record of five world titles in diving. He won the highboard title in 1978, the highboard and springboard in 1982 and 1986, and four Olympic gold medals in 1984 and 1988. The total score for a dive is the sum of the scores of each dive times the degree of difficulty of the dive. His final dive in 1988 had a degree of difficulty of 3.4. If his scores were 9, 8, and 8.5, find his total score for the dive. (Lesson 2–6)

61. Find $5y + (-12y) + (-21y)$. (Lesson 2–5)

62. Statistics Mr. Crable asked his students how many hours of sleep they got on the average each night for a week. He labeled the data with G for girls and B for boys. Use Gs and Bs to make a line plot of these data. (Lesson 2–2)

| 10-G | 7.5-B | 9-B | 9-G | 8-G | 8.5-G | 9.5-B | 8.5-B | 8.5-G |
| 9.5-B | 8-G | 8-B | 10-B | 10-G | 9.5-G | 7.5-G | 9-G | 8.5-B |

63. Solve $\frac{21-3}{12-3} = x$. (Lesson 1–5)

64. Patterns Find the next three terms in the sequence 3, 8.5, 14, 19.5.... (Lesson 1–2)

SELF TEST

Solve each proportion. (Lesson 4–1)

1. $\frac{2}{10} = \frac{1}{a}$

2. $\frac{3}{5} = \frac{24}{x}$

3. $\frac{y}{4} = \frac{y+5}{8}$

4. $\triangle WYQ$ and $\triangle MVT$ shown at the right are similar. Find the measures of the missing sides. (Lesson 4–2)

Solve each right triangle. State the side lengths to the nearest tenth and the angle measures to the nearest degree. (Lesson 4–3)

5.

6.

7. Refer to the triangles in Exercises 5 and 6. Determine whether the triangles are similar. (Lesson 4–2)

8. Thirty-six is 45% of what number? (Lesson 4–4)

9. Fifty-five is what percent of 88? (Lesson 4–4)

10. Research Suppose 6% of 8000 people polled regarding an election expressed no opinion. How many people had an opinion? (Lesson 4–4)

Percent of Change

mallow plant

APPLICATION
Food

If you've ever bought a package of marshmallows, you've mostly bought air. That's right, marshmallows are 80% air! Today, marshmallows come in a variety of flavors such as vanilla, chocolate, toasted coconut, and fruit flavors like laser lime and astroberry. The marshmallow was invented in ancient Egypt and made with honey and the sap of the mallow plant that grows in swamps, or marshes. Since then, the marshmallow has gone through many changes.

Today, marshmallows are mass produced and made from cornstarch and gelatin. In 1955, the height of marshmallow making in the United States, there were 30 marshmallow companies. Today there are only a handful.

Let's say there are only six marshmallow companies today. You can write a ratio that compares the amount of decrease in marshmallow companies to the original number of companies in 1955. This ratio can be written as a percent. First, subtract to find the amount of change: $30 - 6 = 24$. Then divide by the original number of companies.

$$\frac{amount\ of\ decrease\ \rightarrow}{original\ amount\ \rightarrow} = \frac{r}{100}$$

$$24 \cdot 100 = 30 \cdot r \qquad \textit{Find the cross products.}$$

$$80 = r \qquad \textit{Solve for r.}$$

The amount of decrease is $\frac{80}{100}$ or 80% of the original number of companies. So, we can say that the **percent of decrease** is 80%.

The **percent of increase** can be found in a similar way.

Example ①
CONNECTION
Biology

F Y I

In 1982, the California Condor Recovery Program, made up of a team of specialists and zoos, was formed to assist in the capturing, breeding, and freeing of the condors into the wild.

In 1982, the California condor was on the verge of extinction. The population had dwindled to a total of 21, making the condor the rarest bird in North America. According the Refuge Department of the U.S. Fish and Wildlife Service, however, there were 64 condors in existence in 1992 and 103 as of October 1995. Find the percent of increase in the population of California condors from 1992 to 1995.

Method 1
First, subtract to find the amount of change.

$$103 - 64 = 39$$

$$\frac{amount\ of\ increase}{original\ number} = \frac{39}{64}$$

$$\frac{39}{64} = \frac{r}{100}$$

$$60.9375 = r$$

The percent of increase is about 61%.

Method 2
Divide the new amount by the original amount.

$$103 \div 64 = 1.609375$$

Subtract 1 from the result and write the decimal as a percent.

$$1.609375 - 1 = 0.609375 \text{ or } 60.9\%$$

The percent of increase is about 61%.

Sometimes an increase or decrease is given as a percent, rather than an amount. Two applications of percent of change are discounts and sales tax.

Example ❷

APPLICATION

Sales

A discount of 20% means that the price is decreased by 20%.

Ayita received a coupon in the mail offering her a 20% discount on the price of a pair of jeans at The Limited. Ayita visited the store and found a pair of jeans she wants to buy that cost $48. What will be the discounted price?

Explore   The original price is $48.00, and the discount is 20%.

Plan   You want to find the amount of discount, then subtract that amount from $48.00. The result is the discounted price.

Estimate: *20% of $48 → 2(10% of $50) = 2($5) or $10.*

Solve   20% of $48.00 = 0.20(48.00)   *Note that* $20\% = \frac{20}{100} = 0.20.$
                  = 9.60

Subtract this amount from the original price.

$48.00 − $9.60 = $38.40

The discounted price will be $38.40.

Examine   Solve the problem another way. The discount was 20%, so the discounted price will be 80% of the original price.   *100% − 20% = 80%*

Find 80% of $48.00.   0.80(48.00) = 38.40

This method produces the same discounted price, $38.40.

How does this compare to the estimate?

Example ❸

APPLICATION

Sales

Sales tax of $5\frac{3}{4}\%$ means that the price is increased by $5\frac{3}{4}\%$.

Karen Danko, a sophomore at Westerville North High School, is ordering a class ring. The ring she has chosen costs $169.99. She also needs to pay a sales tax of $5\frac{3}{4}\%$. What is the total price?

Explore   The price is $169.99 and the tax rate is $5\frac{3}{4}\%$.

Plan   First find $5\frac{3}{4}\%$ of $169.99. Then add the result to $169.99.

Estimate: $5\frac{3}{4}\%$ *of $169.99 →* $\frac{1}{2}$ *(10% of 170) =* $\frac{1}{2}$ *($17) or $8.50. The total will be about $170 + $8.50 or $178.50.*

Solve $5\frac{3}{4}\%$ of $169.99 = 0.0575(169.99)   *Note that* $5\frac{3}{4}\% = 0.0575.$
                      = 9.774425   *Round 9.774425 to 9.78, since tax is always rounded up.*

Add this amount to the original price.

$169.99 + $9.78 = $179.77

The total price is $179.77.

Examine   The tax rate is $5\frac{3}{4}\%$, so the total price was $100\% + 5\frac{3}{5}\%$ or 105.75% of the purchase price. Find 105.75% of $169.99.

(1.0575)(169.99) = 179.76442

Thus, the total price of $179.77 is correct.

How does this compare to the estimate?

Suppose a newspaper ad states that a discount store's microwave ovens are discounted 20% from the manufacturer's suggested retail price (MSRP). The store is running a special sale that says all kitchen appliances have been discounted an additional 15%. What is the final sale price of a microwave oven whose MSRP is $250?

There are two ways you might interpret the ad in the paper.

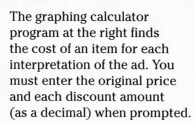

- The discounts can be successive. That is, 20% is taken off the MSRP and then 15% is taken off the resulting price.

- The discounts can be combined. So the price is 35% off the MSRP.

The graphing calculator program at the right finds the cost of an item for each interpretation of the ad. You must enter the original price and each discount amount (as a decimal) when prompted.

```
PROGRAM:DISCOUNT
: Disp "ORIG. PRICE"
: Input P
: Disp "1ST DISCOUNT?"
: Input A
: Disp "2ND DISCOUNT?"
: Input B
: P(1-A)(1-B)→C
: P(1-(A+B))→D
: Disp "SUCCESSIVE",
  "DISCOUNT",C
: Disp "COMBINED",
  "DISCOUNT",D
```

Your Turn

Copy the following table and use the program to complete it.

	Price	First Discount	Second Discount	Sale Price Successive Discount	Sale Price Combined Discount
a.	$49.00	20%	10%		
b.	$185.00	25%	10%		
c.	$12.50	30%	12.5%		
d.	$156.95	30%	15%		

e. What is the relationship between the sale price using successive discounts and the sale price using combined discounts? Which one is usually used in stores?

CHECK FOR UNDERSTANDING

Communicating Mathematics

Study the lesson. Then complete the following.

1. **Explain** how to find a percent of increase.

2. **Explain** how to find the sales tax if you know the rate of tax.

3. **Write** the equation you would use to find the percent of increase on a car that cost $17,972 in 1995 and $19,705 in 1996.

4. **Assess Yourself** Find an advertisement or article in the newspaper that shows a percent of change. Determine whether it is a percent of increase or decrease. Explain your reasoning.

State whether each percent of change is a percent of increase or a percent of decrease. Then find the percent of increase or decrease.

5. original: $50
   new: $70

6. original: $200
   new: $172

7. original: 72 ounces
   new: 36 ounces

Find the final price of each item. When there is a discount and sales tax, first compute the discount price and then compute the sales tax and final price.

8. CD player: $149
   discount: 15%

9. athletic shoes: $89.99
   discount: 10%
   sales tax: 6%

10. sweater: $45
    sales tax: 6.5%

11. **Sales** Kevin Mason paid $205.80 for his senior class photographs. This included 5% sales tax. What was the cost of the pictures before taxes?

EXERCISES

Practice

State whether each percent of change is a percent of increase or a percent of decrease. Then find the percent of increase or decrease. Round to the nearest whole percent.

12. original: $100
    new: $59

13. original: 324 people
    new: 549 people

14. original: 58 homes
    new: 152 homes

15. original: 66 dimes
    new: 30 dimes

16. original: $53
    new: $75

17. original: 15.6 liters
    new: 11.4 liters

18. original: $3.78
    new: $2.50

19. original: 231.2 mph
    new: 236.4 mph

20. original: 124 tons
    new: 137 tons

Find the final price of each item. When there is a discount and sales tax, first compute the discount price and then compute the sales tax and final price.

21. VCR: $219
    sales tax: 6.5%

22. jeans: $39.99
    discount: 15%
    sales tax: 4%

23. book: $19.95
    discount: 5%
    sales tax: 5%

24. concert tickets: $52.50
    sales tax: 7%

25. in-line skates: $99.99
    discount: 20%
    sales tax: 6.75%

26. hiking boots: $59
    discount: 10%
    sales tax: 5.5%

27. compact disc: $15.88
    sales tax: 4.5%

28. amusement park
    tickets: $37.50
    sales tax: 6%

29. software: $29.99
    discount: 6%
    sales tax: 6.75%

Graphing Calculator

30. Use a graphing calculator to find the sale price of each item using successive discounts and then one combined discount.

    a. price, $89; discounts, 25% and 10%

    b. price, $254; discounts 30% and 14.5%

Critical Thinking

31. An amount is increased by 15%. The result is decreased by 15%. Is the final result equal to the original amount? Explain your reasoning with an example.

32. **Sales** The advertisement below was used by a local optician's office in Walkersville, Maryland.

The 100% Price Pledge!
We guarantee that we will simply not be undersold. If you find the same eyewear as ours at a lower price within 30 days of your purchase, we'll pay you **110%** of the difference.

Suppose you bought a pair of glasses from this optician for $150. You then found the same glasses from a competitor for $125. What should the optician do to honor the pledge?

Applications and Problem Solving

33. **Predictions** The World Future Society predicts that by the year 2020, airplanes will be able to carry 1400 passengers. Today's biggest jets can carry 600 people. What will be the percent of increase of airplane passengers?

34. **Computers** In 1995, America Online had about 3,000,000 users. Over the next decade, users are expected to increase from a few million to the tens of millions. Suppose the number of users increases by 150% by the year 2000. How many users will there be in the year 2000?

35. **Taxes** As the saying goes, time is money. The graph at the right shows a function of how much time it takes out of each eight-hour day to earn enough money to pay a day's worth of taxes. What was the percent of increase from 1970 to 1994?

Source: *The Universal Almanac*, 1995

36. **Sales** Music Systems, Inc. allows a 10% discount if a purchase is paid for within 30 days. An additional 5% discount is given if the purchase is paid for within 15 days. Brent Goodson buys a sound system that originally cost $360. If he pays the entire amount at the time of purchase, how much does he pay for his system after the successive discounts?

37. **Health** The excerpt below is from the March 1994 issue of *Runner's World* magazine.

When you first get up in the morning, your muscles and soft tissues are tight. In fact, at that time your muscles are generally about 10 percent shorter than their normal resting length. As you move around, they stretch to their normal length. Then when you start to exercise, your muscles stretch even more, to about 10 percent longer than resting length. This means you have a 20 percent change in muscle length from the time you get out of bed until your muscles are well warmed up.

According to the basic laws of physics, muscles work more efficiently when they are longer; they can exert more force with less effort. This means, too, that longer muscles are much less prone to injury.

What is the percent change in muscle length from the time a runner gets up in the morning until after he or she has exercised for a while?

38. Consumer Awareness A windsuit that costs $75 is discounted 25%. The sales tax is 6%.

    a. Would it be better for the customer to have the discount taken off before adding the sales tax or after the sales tax has been added? Explain your reasoning.

    b. Do you think stores have a choice in which order to calculate discounts and sales tax? Explain your reasoning.

39. Automobiles As soon as a new car is purchased and driven away from the dealership, it begins to lose its value, or depreciate. Alonso bought a 1994 Plymouth Neon for $9559. One year later, the value of the car was $8500. What is the percent of decrease of the value of the car?

Mixed Review

40. Entertainment A theater was filled to 75% capacity. How many of the 720 seats were filled? (Lesson 4–4)

41. Geometry A chimney casts a shadow 75 feet long when the angle of elevation of the sun is 41°. How tall is the chimney? (Lesson 4–3)

42. Construction To paint his house, Lonnie needs to purchase an extension ladder that reaches at least 24 feet off the ground. Ladder manufacturers recommend the angle formed by the ladder and the ground be no more than 75°. What is the shortest ladder he could buy to reach 24 feet safely? (Lesson 4–3)

43. Solve $0.2x + 1.7 = 3.9$. (Lesson 3–6)

44. Geometry The measures of two angles of a triangle are 38° and 41°. Find the measure of the third angle. (Lesson 3–4)

45. Dietetics A hospitalized diabetic is on a carefully controlled 2000-calorie diet distributed over five feedings a day. There are two heavy feedings consisting each of $\frac{2}{7}$ of the total calories. There are also three light feedings consisting each of $\frac{1}{7}$ of the total calories. Determine the number of calories for a heavy feeding and for a light feeding. (Lesson 2–6)

46. Statistics State the scale you would use to make a line plot for the following data. Then draw the line plot. 4.2, 5.3, 7.6, 9.6, 7.3, 6.7 (Lesson 2–2)

WORKING ON THE

In·ves·ti·ga·tion

Refer to the Investigation on pages 190–191.

Go Fish!

1 Count all of the fish in your lake. Record the actual population and your estimates on a class chart.

2 How close were your estimates? How would you account for the difference between your estimates and the actual number of fish in the lake?

3 What percent of the fish in the lake was tagged? What percent of the fish in your samples was tagged?

4 Suppose you went fishing at that lake and caught one fish. What is the probability, or chance, that the fish you caught would be tagged? Justify your reasoning.

5 Write a paragraph or two that relates proportions, percents, and probability in this situation.
You will learn more about probability in the next lesson.

Add the results of your work to your Investigation Folder.

Integration: Probability
Probability and Odds

APPLICATION
Contests

Have you ever heard anyone say that one man's trash is another man's treasure? For garbage truck driver Craig Randall of East Bridgewater, Massachusetts, the treasure was a Wendy's soft-drink cup. By peeling off a contest sticker on the discarded cup, Randall won the grand prize of $200,000 toward the purchase of a home! Since there was only one grand prize available and 24,059,900 cups were distributed, the probability of winning the grand prize was $\frac{1}{24,059,900}$.

You can calculate the chance, or **probability,** that a certain event, such as winning a particular prize, will happen. The **probability of an event** is a ratio of the number of ways a certain event can occur to the number of possible outcomes. The numerator is the number of favorable outcomes, and the denominator is the total number of possible outcomes. *The probability of an event may be written as a percent, a fraction, or a decimal.*

For example, suppose you want to know the probability of getting a 2 on one roll of a die. When you roll a die, there are six possible outcomes. Of these outcomes, only one is favorable, a 2. Therefore, the probability is $\frac{1}{6}$. We write $P(2)$ to represent *the probability of getting a 2 on one roll of a die.*

Definition of Probability	P(event)	$\dfrac{\text{number of favorable outcomes}}{\text{total number of possible outcomes}}$

Example **1** Lauren Slocum is representing her sophomore class in a *Scrabble* tournament. She is the first player to select her seven tiles out of the 100 available tiles. The distribution of tiles is shown at the right. Find the probability of each selection below.

Letters	Number of Tiles
J, K, Q, X, Z	1
B, C, F, H, M, P, V, W, Y, blank	2
G	3
D, L, S, U	4
N, R, T	6
O	8
A, I	9
E	12

APPLICATION
Games

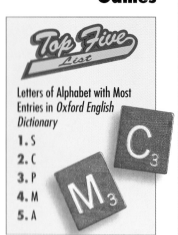

Letters of Alphabet with Most Entries in *Oxford English Dictionary*

1. S
2. C
3. P
4. M
5. A

a. an O on the very first selection

There are eight O tiles and 100 tiles in all. $P(\text{selecting an O}) = \frac{8}{100}$ or $\frac{2}{25}$

The probability that Lauren selects an O is $\frac{2}{25}$ or 8%.

b. an E if 5 tiles have been selected and 2 of them were Es

There are 12 E tiles. If 2 have been chosen, 10 remain. There are $100 - 5$ or 95 tiles from which to select.

$P(\text{choosing an E}) = \frac{10}{95}$ or $\frac{2}{19}$

The probability of selecting an E is $\frac{2}{19}$ or about 10%.

The probability that an event will occur is somewhere between 0 and 1.

- A probability of 0 means that it is impossible for an event to occur.
- A probability of 1 means that an event is certain to occur.
- A probability between 0 and 1 means that an event is neither impossible nor certain.

Based on the situations above, the probability of any event can be expressed as $0 \leq P(\text{event}) \leq 1$.

Some outcomes have an equal chance of occurring. We say that such outcomes are **equally likely.** When an outcome is chosen without any preference, we say that the outcome occurs at **random.**

MODELING MATHEMATICS

Probability

Materials: grid paper

What are the chances that two of the Oscar-winning actors or actresses from 1983 to 1993 share the same birthday? Considering that there are only 24 people in this group, you would probably say the chances are slim.

You may be surprised that in a group of only 24 people, the likelihood that two of them have the same birthday is just about 50%, or one out of two.

For example, in this group of Academy Award winners, Holly Hunter and William Hurt were both born on March 20.

Your Turn

a. Survey a group of 24 people in your school. Tally the results.

b. Does your group contain a shared birthday? Display your findings in a chart or graph.

c. Combine your results with those of your classmates. Does your class graph differ from your individual graph? Explain.

d. Use the graph below to find the likelihood of birthdays shared by the Presidents of the United States. Research to find any shared Presidential birthdays. Compare your findings. Do you agree or disagree with the percentages? Explain your reasoning.

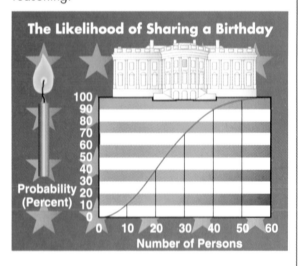

The Likelihood of Sharing a Birthday

Another way to express the chance of an event's occurring is with **odds.** The odds of an event is the ratio that compares the number of ways an event can occur to the number of ways it *cannot* occur.

Definition of Odds	The odds of an event occurring is the ratio of the number of ways the event can occur (successes) to the number of ways the event cannot occur (failures).

Odds = number of successes : number of failures

Example **2** Jersey Mike's Submarine Shop has a business card drawing for a free lunch every Tuesday. Four coworkers from Invo Accounting put their business cards in the bowl for the drawing. If 80 cards were in the bowl, what are the odds that one of the coworkers will win a free lunch?

The coworkers have 4 of the 80 business cards. Thus, there are $80 - 4$ or 76 business cards that will not be winning cards for the coworkers.

Odds of winning = $\dfrac{\text{number of chances of}}{\text{drawing winning card}}$: $\dfrac{\text{number of chances of}}{\text{drawing other cards}}$

= 4:76 or 1:19   *1:19 is read "1 to 19".*

Example **3** Refer to the application at the beginning of the lesson. The probability of winning any prize or food discount in Wendy's peel-off sticker contest was 20% or $\dfrac{1}{5}$. Find the odds of winning a prize or discount.

Do you have a better chance of winning a prize or not winning a prize?

If the probability of winning a prize or discount is $\dfrac{1}{5}$, then the number of successes (prize) is 1, while the total number of outcomes is 5. This means that the number of failures (no prize) must be $5 - 1$ or 4.

Odds of prize = number of successes : number of failures

= 1:4   *This is read "1 to 4."*

CHECK FOR UNDERSTANDING

Communicating Mathematics

Study the lesson. Then complete the following.

1. **Refer** to Example 1. Find the probability of selecting a Z if 25 tiles have been selected and one of them was a Z.

2. **Give** examples of an impossible event and a certain event when rolling a die.

3. **Tell** what the odds would be for an event not occurring if the odds for the event occurring are 3:5.

4. **Write** a problem in which the answer will be a probability of $\dfrac{3}{4}$.

MODELING MATHEMATICS

5. Choose another group of 24 people to survey. Record their birthdays. Does your group contain a shared birthday?

Guided Practice

Determine the probability of each event.

6. This is an algebra book.   7. A coin will land tails up.

Find the probability of each outcome if a die is rolled.

8. a 6   9. a number greater than 2

Find the odds of each outcome if a die is rolled.

10. a number greater than 2   11. not a 3

12. If the probability that an event will occur is $\dfrac{3}{7}$, what are the odds that the event will *not* occur?

Practice

Determine the probability of each event.

13. A coin will land heads up.

14. There is a January 1 this year.

15. A baby will be a boy.

16. Pigs will fly.

17. You will roll a multiple of 2 on a die.

18. A person's birthday is in June.

Find the probability of each outcome if a computer randomly chooses a letter in the word "probability."

19. the letter b

20. $P(not\ i)$

21. the letter e

22. $P(vowel)$

23. *not b* or *y*

24. the letters *a* or *t*

Find the odds of each outcome if you randomly select a coin from a jar containing 80 pennies, 100 nickels, 70 dimes, and 50 quarters.

25. selecting a quarter

26. selecting a nickel

27. selecting a penny

28. *not* selecting a nickel

29. *not* selecting a dime

30. selecting a penny or dime

A card is selected at random from a standard deck of 52 cards.

31. What is the probability of selecting a red card?

32. What is the probability of selecting a queen?

33. What are the odds of selecting a heart?

34. What are the odds of *not* selecting a black 6?

35. If the odds that an event will occur are 8:5, what is the probability that the event will occur?

36. If the probability that an event will occur is $\frac{2}{3}$, what are the odds that the event will occur?

Critical Thinking

37. Geometry If a point inside the figure at the right is chosen at random, what is the probability that it will be in the shaded region?

38. Geometry If a point inside the figure at the right is chosen at random, what is the probability that it will *not* be in the shaded region?

39. Games The game board on *Jeopardy!* is divided into 30 squares. There are six categories with five answers in each category. In the Double Jeopardy! round, 2 Daily Double squares are hidden among the 30 squares.

   a. What are the odds of choosing a Daily Double square on the first selection?

   b. If you are making the fourth selection of the Double Jeopardy! round, what is the probability of choosing a Daily Double square if one has been chosen?

40. School Have you ever had to make an oral presentation before your classmates? Did the teacher tell everyone to be prepared and then choose the first day's presenters at random?

   a. If there are 35 students in your class, and four students will be chosen to give a presentation each day until everyone has made a presentation, what is the probability that you will be chosen to make your presentation the first day?

   b. If you have not been chosen by the fourth day, what is the probability that you will be chosen on the fourth day?

41. Statistics The stem-and-leaf plot at the right shows the total points earned by each of the 26 teams in the National Hockey League for the 1993–1994 season.

   a. What is the probability that a team earned 97 points?

   b. What is the probability that a team earned less than 80 points?

   c. What are the odds in favor of a team scoring more than 100 points?

Stem	Leaf
11	2
10	0 1 6
9	1 5 6 7 7 7 8
8	0 2 3 4 5 7 8
7	1 1 6
6	3 4 6
5	7
4	
3	7 *9	1 = 91*

42. Chocolate Illinois is the number one candy-producing state. The Little Chocolatier shop in Sterling, Illinois, is known for its pecan dumplings. The owners used to make nine pounds of pecan dumplings at one time. Now they make 300 pounds in the same amount of time! Find the percent of change. (Lesson 4–5)

43. Geometry In a parking garage, there are 20 feet between each level. Each ramp to a level is 130 feet long. Find the measure of the angle of elevation of each ramp. (Lesson 4–3)

44. Solve $\frac{5}{8}x + \frac{3}{5} = x$. (Lesson 3–5)

45. Work Backward Kristen spent one fifth of her money for gasoline. Then she spent half of what was left for a haircut. She bought lunch for $7. When she got home, she had $13 left. How much did Kristen have originally? (Lesson 3–3)

46. Simplify $\pm\sqrt{1764}$. (Lesson 2–8)

47. Find $\frac{17}{21} + \left(-\frac{13}{21}\right)$. (Lesson 2–5)

48. School Each number below represents the age of a student in Ms. Wallace's evening calculus class at DeSantis Community College. (Lesson 1–4)

   22 17 25 24 19 27 33 16 35 26 20 18 24 33 18 19 48
   36 19 23 55 18 18 19 27 18 19 25 17 32 19 45 19 20 30

   a. Make a stem-and-leaf plot of the data.

   b. How many people attend Ms. Wallace's class?

   c. What is the difference in ages between the oldest and youngest person in class?

   d. What is the most common age for a student in the class?

   e. Which age group is most widely represented in the class?

49. Evaluate $8(a - c)^2 + 3$ if $a = 6$, $b = 4$, and $c = 3$. (Lesson 1–3)

Weighted Averages

What YOU'LL LEARN

- To solve mixture problems, and
- to solve problems involving uniform motion.

Why IT'S IMPORTANT

You can use weighted averages to solve problems involving travel, transportation, and chemistry.

APPLICATION

School

In Mr. Calloway's American History class, semester grades are based on five unit exams, one semester exam, and a long-term project. Each unit exam is worth 15% of the grade, the semester exam is worth 20% of the grade, and the project is worth 5% of the grade.

Parker's scores for the semester are given in the table at the right. If 100% is a perfect score for each item, find Parker's average for the semester.

Semester Grades	
Item	Score
Unit Exam A	79%
Unit Exam B	83
Unit Exam C	96
Unit Exam D	91
Unit Exam E	89
Semester Exam	90
Project	95

In order to find Parker's average for the semester, you may want to add the percentages and then divide by the number of items, 7. But this assumes that each has the same weight. So, you need to find the **weighted average** of the scores.

Definition of Weighted Average	**The weighted average _M_ of a set of data is the sum of the product of each number in the set and its weight divided by the sum of all the weights.**

You can find the weighted average for the application above by multiplying each score by the percentage of the semester grade that it represents and then dividing by the sum of the weights, or 100.   _5(15) + 1(20) + 1(5) = 100_

$$M = \frac{15(79 + 83 + 96 + 91 + 89) + 20(90) + 5(95)}{100}$$

$$= \frac{8845}{100} \text{ or } 88.45$$

So, Parker's average for the semester is about 88%.

Mixture problems involve weighted averages. In a mixture problem, the weight is usually a price or a percentage of something.

Example ①

PROBLEM SOLVING
Make a Chart

Suppose the Central Perk coffee shop sells a cup of espresso for $2.00 and a cup of cappuccino for $2.50. On Friday, Rachel sold 30 more cups of cappuccino than espresso, and she sold $178.50 worth of espresso and cappuccino. How many cups of each were sold?

Explore   Let _e_ represent the number of cups of espresso sold. Then _e_ + 30 represents the number of cups of cappuccino sold.

(continued on the next page)

Plan   Make a chart of the information.

	Number of Cups	Price Per Cup	Total Price
espresso	e	$2.00	$2e$
cappuccino	$e + 30$	$2.50	$2.5(e + 30)$

Solve

$$\underbrace{\text{total sales of espresso}} + \underbrace{\text{total sales of cappuccino}} = \underbrace{\text{total sales}}$$

$$2e + 2.5(e + 30) = 178.50$$
$$2e + 2.5e + 75 = 178.50$$
$$4.5e + 75 = 178.50$$
$$4.5e = 103.50$$
$$e = 23$$

There were 23 cups of espresso sold. There were $23 + 30$, or 53 cups of cappuccino sold.

Examine   If 23 cups of espresso were sold, the total sales of those cups of coffee would be 23($2.00) or $46.00. If 53 cups of cappuccino were sold, the total sales of those cups of coffee would be 53($2.50) or $132.50.

Since $46.00 + $132.50 = $178.50, the solution is correct.

Sometimes mixture problems are expressed in terms of percents.

Example ② An advertisement for an orange drink claims that the drink contains 10% orange juice. Jamel needs 6 quarts of the drink to serve at a party and he wants the drink to contain 40% orange juice. How much of the 10% drink and pure orange juice should Jamel mix to obtain 6 quarts of a mixture that contains 40% orange juice?

Food Preparation

Explore   Let p represent the amount of pure juice to be added.

Plan   Make a chart of the information.

	Quarts	Amount of Orange Juice
10% juice	$6 - p$	$0.10(6 - p)$
Pure juice	p	$1.00p$
40% juice	6	$0.40(6)$

Solve

$$\underbrace{\text{amount of orange juice in 10\% juice}} + \underbrace{\text{amount of orange juice in pure juice}} = \underbrace{\text{amount of orange juice in 40\% juice}}$$

$$0.10(6 - p) + 1.00p = 0.40(6)$$
$$0.6 - 0.1p + 1.00p = 2.4$$
$$0.9p = 1.8$$
$$p = 2$$

Jamel needs to combine 2 quarts of pure orange juice with $6 - 2$ or 4 quarts of 10% juice to obtain a 6-quart mixture that is 40% orange juice. *Examine this solution.*

Motion problems are another application of weighted averages. When an object moves at a constant speed, or rate, it is said to be in **uniform motion**. The formula $d = rt$ is used to solve uniform motion problems. In the formula, d represents distance, r represents rate, and t represents time. You can also use equations and charts when solving motion problems.

Example ③

Travel

On Friday, Shenae and her brother Rafiel went to visit their grandparents for the weekend. Luckily, traffic was light and they were able to make the 50-mile trip in exactly one hour. On Sunday, they weren't so lucky. The trip home took exactly two hours. What was their average speed for the round trip?

To find the average speed for each leg of the trip, rewrite $d = rt$ as $r = \dfrac{d}{t}$.

Going	Returning
$r = \dfrac{d}{t}$	$r = \dfrac{d}{t}$
$= \dfrac{50 \text{ miles}}{1 \text{ hour}}$ or 50 miles per hour	$= \dfrac{50 \text{ miles}}{2 \text{ hours}}$ or 25 miles per hour

You may think that the average speed of the trip would be $\dfrac{50 + 25}{2}$ or 37.5 miles per hour. However, Shenae did not drive at these speeds for equal amounts of time. You can find the weighted average for their trip.

Round Trip

$$M = \frac{50(1) + 25(2)}{3}$$

$$= \frac{100}{3} \text{ or } 33\frac{1}{3}$$

Their average speed was $33\frac{1}{3}$ miles per hour.

Example ④

Transportation

Two city buses leave their station at the same time, one heading east and the other heading west. The eastbound bus travels at 35 miles per hour, and the westbound bus travels at 45 miles per hour. In how many hours will they be 60 miles apart?

Explore   Draw a diagram to help analyze the problem.

60 miles

Westbound bus
45 mph

Bus Station

Eastbound bus
35 mph

Plan   Organize the information in a chart. Let t represent the number of hours until they are 60 miles apart. Remember that $rt = d$.

Bus	r	t	d	
Eastbound	35	t	$35t$	*The eastbound bus travels 35t miles.*
Westbound	45	t	$45t$	*The westbound bus travels 45t miles.*

eastbound distance   +   *westbound distance*   =   *total distance*

$$35t \quad + \quad 45t \quad = \quad 60$$

(continued on the next page)

Solve $35t + 45t = 60$

$80t = 60$

$t = \dfrac{3}{4}$

In $\dfrac{3}{4}$ hour, or 45 minutes, the buses will be 60 miles apart.

Examine   To check the answer, find the distance each bus could travel in

$\dfrac{3}{4}$ hour and see if it totals 60 miles.

$$\left(\dfrac{3}{4}\,h\right)(35\text{ mph}) + \left(\dfrac{3}{4}\,h\right)(45\text{ mph}) \overset{?}{=} 60\text{ mi}$$

$$26\dfrac{1}{4}\text{ mi} + 33\dfrac{3}{4}\text{ mi} \overset{?}{=} 60\text{ mi}$$

$$60\text{ mi} = 60\text{ mi} \quad\checkmark$$

CHECK FOR UNDERSTANDING

Communicating Mathematics

Study the lesson. Then complete the following.

1. **Tell** what the d, r, and t represent in the formula $d = rt$.

2. **Explain** why it is sometimes helpful to use charts and diagrams.

MATH JOURNAL

3. **Describe** your favorite mode of transportation. Then write a problem using the formula $d = rt$ and your favorite mode.

Guided Practice

4. **Chemistry**   Joshua is doing a chemistry experiment that calls for a 30% solution of copper sulfate. He has 40 mL of 25% solution. How many milliliters of 60% solution should Joshua add to obtain the required 30% solution?

	Amount of Solution (mL)	Amount of Copper Sulfate
25% solution	40	
60% solution	x	
30% solution		

5. **Sales**   The Cookie Crumbles Company sells two kinds of cookies daily: peanut butter at $6.50 per dozen and chocolate chip at $9.00 per dozen. Yesterday, Cookie Crumbles sold 85 dozen more peanut butter than chocolate chip cookies. The total sales for both were $4055.50. How many dozen of each were sold?

6. **Air Travel**   An airplane flies 1000 miles due east in 2 hours and 1000 miles due south in 3 hours. What is the average speed of the airplane?

7. **Boating**   *The Yankee Clipper* leaves the pier at 9:00 A.M. at 8 knots (nautical miles per hour). A half hour later, *The Riverboat Rover* leaves the same pier in the same direction traveling at 10 knots. At what time will *The Riverboat Rover* overtake *The Yankee Clipper*?

Applications and Problem Solving

8. Sales The Madison Local High School marching band sold gift wrap to earn money for a band trip to Orlando, Florida. The gift wrap in solid colors sold for $4.00 per roll, and the print gift wrap sold for $6.00 per roll. The total number of rolls sold was 480, and the total amount of money collected was $2340. How many rolls of each kind of gift wrap were sold?

	Number of Rolls	Price Per Roll	Total Price
Solid	r	$4	$4r$
Print	$480 - r$	$6	$6(480 - r)$

9. Money Rochelle has $2.55 in dimes and quarters. She has eight more dimes than quarters. How many quarters does she have?

10. Sales The Nut House sells walnuts for $4.00 a pound and cashews for $7.00 a pound. How many pounds of cashews should be mixed with 10 pounds of walnuts to obtain a mixture that sells for $5.50 a pound?

11. Travel Ryan and Jessica Wilson leave their home at the same time, traveling in opposite directions. Ryan travels at 57 miles per hour and Jessica travels at 65 miles per hour. In how many hours will they be 366 miles apart?

12. Travel At 7:00 A.M., Brooke leaves home to go on a business trip driving 35 miles per hour. Fifteen minutes later, Bart discovers that Brooke forgot her presentation materials. He drives 50 miles per hour to catch up with her. If Bart is delayed 30 minutes with a flat tire, when will he catch up with Brooke?

13. Cycling Two cyclists begin traveling in the same direction on the same bike path. One travels at 20 miles per hour, and the other at 14 miles per hour. After how many hours will they be 15 miles apart?

14. Entertainment A local radio station is having a drawing to win tickets to see En Vogue in concert. The winners will be driven to the concert in a limousine that holds eight passengers. Concert tickets cost $22 for adults and $19 for students. After renting the limousine, the radio station has $167 left in their giveaway budget to buy the eight tickets. How many adult tickets and student tickets can the station buy?

15. Travel Pablo is driving 40 miles per hour. After he has driven 30 miles, his brother Ricardo starts driving in the same direction. At what rate must Ricardo drive to catch up with Pablo in 5 hours?

16. Automotives A car radiator has a capacity of 16 quarts and is filled with a 25% antifreeze solution. How much must be drained off and replaced with pure antifreeze to obtain a 40% antifreeze solution?

17. Law Enforcement A state patrol officer is chasing a car he believes is speeding. The officer is traveling 70 miles per hour to try to catch up with the car. But he can't say later that the other car was going 70 mph because the driver could ask, "If we were going the same speed, how did you catch up? Obviously I was going slower than you. I wasn't speeding." The officer sees the driver pass a mile marker, one-fourth of a mile ahead. From the marker, it takes the officer five minutes $\left(\frac{1}{12}\text{ of an hour}\right)$ to catch up. How fast was the driver going?

18. **Travel**   An express train travels 80 kilometers per hour from Ironton to Wildwood. A local train, traveling at 48 kilometers per hour, takes 2 hours longer for the same trip. How far apart are Ironton and Wildwood?

19. **Colleges**   Emilio is trying to decide what college he would like to attend. He has determined his five most important factors in choosing a school. A score of 1 is the lowest and 4 is the highest. The table at the right shows his ratings for one school he is considering.

	Importance Factor (1 to 4)	Rating (1 to 4)	Total
Distance from home	4	4	16
Academic reputation	3	2	6
Social climate	3	1	3
Quality of dorms	1	4	4
Community	2	3	6

   a. Find the weighted average of his ratings for this school.

   b. What is the highest rating that one of Emilio's schools can receive?

20. **School**   Many schools base a student's grade point average, or GPA, on the grade a student receives as well as on the number of credits received for the class. Mercedes' grades for this semester are listed in the table at the right. Find Mercedes' GPA if a grade of A equals 4 and a B equals 3.

Grade Card

Class	Credits Rating	Grade
Honors Algebra	1	A
Biology	1	B
English 1	1	A
Spanish 2	1	B
Phys. Ed.	$\frac{1}{2}$	A

Critical Thinking

21. Write a mixture problem for the equation $1.00x + 0.20(40 - x) = 0.28(40)$.

22. Monica Morrison drove to work on Monday at 40 miles per hour and arrived one minute late. She left at the same time on Tuesday, drove at 45 miles per hour, and arrived one minute early. How far does Monica drive to work?

Mixed Review

23. **Probability**   Find the probability of getting a number less than 1 if a die is rolled.   (Lesson 4–6)

24. **Baseball**   In 1994, DeMarini Sports Inc. sold 7500 aluminum softball bats to softball enthusiasts. In 1995, the company expected to increase sales by 60%. How many bats did the company expect to sell in 1995?   (Lesson 4–5)

25. **Finance**   Selena Cruz wants to invest a portion of her $16,000 savings at 8% annual interest and the balance at a safer 5% annual interest. She hopes to earn $1130 in the next year to pay for a Caribbean cruise. How much money should she invest at 8%?   (Lesson 4–4)

26. **Model Trains**   Model trains are scaled-down replicas of real trains that come in a variety of sizes, called scales. One of the most popular-sized models is called the HO. Every dimension of the HO model measures $\frac{1}{87}$ that of a real engine. The HO model of a modern diesel locomotive is about 8 inches long. About how many feet long is the real locomotive?   (Lesson 4–1)

27. Solve $5.3 - 0.3x = -9.4$.   (Lesson 3–3)

28. Solve $24 = -2a$.   (Lesson 3–2)

29. Name the property illustrated in $(a + 3b) + 2c = a + (3b + 2c)$.   (Lesson 1–8)

Direct and Inverse Variation

4-8

What YOU'LL LEARN

• To solve problems involving direct and inverse variation.

Why IT'S IMPORTANT

You can use direct and inverse variation to solve problems involving work and music.

F Y I

A low-flow shower head uses only 2.5 gallons of water per minute.

APPLICATION
Water Usage

Do you find yourself singing in the shower? If you are like the average American, you have enough time to sing several songs. The national average for time spent in the shower is 12.2 minutes. A standard shower head uses about 6 gallons of water per minute. That means, you use 73.2 gallons of water for each shower you take. Taking one shower per day for one year would use 26,718 gallons of water!

The number of gallons of water used depends *directly* on the amount of time spent in the shower. The table below shows the number of gallons of water used y as a function of time in the shower x.

x (minutes)	3	6	9	12	15
y (gallons)	18	36	54	72	90

The relationship between the number of minutes in the shower and the gallons of water used is shown by the equation $y = 6x$. This type of equation is called a **direct variation.** We say that y *varies directly as* x or y *is directly proportional to* x. This means that as x increases in value, y increases in value, or as x decreases in value, y decreases in value.

Definition of Direct Variation	**A direct variation is described by an equation of the form $y = kx$, where $k \neq 0$.**

In the equation $y = kx$, k is called the **constant of variation**. To find the constant of variation, divide each side by x.

$$\frac{y}{x} = k$$

Example **①**

APPLICATION
Employment

In this situation, Julio's pay is the dependent quantity and the number of hours Julio works is the independent quantity.

Julio's wages vary directly as the number of hours that he works. If his wages for 5 hours are \$29.75, how much will they be for 30 hours?

First, find Julio's hourly pay. Let x = number of hours Julio works, and let y = Julio's pay. Find the value of k in the equation $y = kx$. The value of k is the amount of money Julio is paid per hour.

$k = \dfrac{y}{x}$

$k = \dfrac{29.75}{5}$

$k = 5.95$

Julio is paid \$5.95 per hour.

(continued on the next page)

Next find out how much Julio's wages will be for 30 hours.

$y = kx$

$y = 5.95(30)$

$y = 178.50$

Thus, Julio's wages will be $178.50 for 30 hours of work.

Using the table in the application at the beginning of the lesson, many proportions can be formed. Two examples are shown below.

number of minutes
↓   ↓

$$\frac{3}{18} = \frac{9}{54} \qquad \text{number of minutes} \; \diagdown\frac{3}{9} = \frac{18}{54}\diagup \; \text{gallons of water}$$

↑   ↑
gallons of water

x_1 is read "x sub 1."

Two general forms for proportions like these can be derived from the equation $y = kx$. Let (x_1, y_1) be a solution for $y = kx$. Let a second solution be (x_2, y_2). Then $y_1 = kx_1$ and $y_2 = kx_2$.

$y_1 = kx_1$ *This equation describes a direct variation.*

$\dfrac{y_1}{y_2} = \dfrac{kx_1}{kx_2}$ *Use the division property of equality. Since y_2 and kx_2 are equivalent, you can divide the left side by y_2 and the right side by kx_2.*

$\dfrac{y_1}{y_2} = \dfrac{x_1}{x_2}$ *Simplify.*

Another proportion can be derived from this proportion.

$x_2y_1 = x_1y_2$ *Find the cross products of the proportion above.*

$\dfrac{x_2y_1}{y_1y_2} = \dfrac{x_1y_2}{y_1y_2}$ *Divide each side by y_1y_2.*

$\dfrac{x_2}{y_2} = \dfrac{x_1}{y_1}$ *Simplify.*

You can use either of these forms to solve problems involving direct proportion.

Example   **2**   **If y varies directly as x, and $y = 28$ when $x = 7$, find x when $y = 52$.**

Use $\dfrac{y_1}{y_2} = \dfrac{x_1}{x_2}$ to solve the problem.

$\dfrac{28}{52} = \dfrac{7}{x_2}$ *Let $y_1 = 28$, $x_1 = 7$, and $y_2 = 52$.*

$28x_2 = 52(7)$ *Find the cross products.*

$28x_2 = 364$

$x_2 = 13$

Thus, $x = 13$ when $y = 52$.

The reverse of direct variation is **inverse variation**. We say that *y varies inversely as x*. This means that as *x* increases in value, *y* decreases in value, or as *y* decreases in value, *x* increases in value. For example, the more miles you drive, the less gasoline you have in the tank.

Definition of Inverse Variation	An inverse variation is described by an equation of the form $xy = k$, where $k \neq 0$.

Sometimes inverse variations are written in the form $y = \dfrac{k}{x}$.

Example ③

Music

F Y I

Midori, a classical violinist, got her first violin when she was 4. At age 10, she gave a solo performance with the New York Philharmonic. She now performs internationally.

The length of a violin string varies inversely as the frequency of its vibrations. A violin string 10 inches long vibrates at a frequency of 512 cycles per second. Find the frequency of an 8-inch string.

Let ℓ represent the length in inches and f represent the frequency in cycles per second. Find the value of k.

$$\ell f = k$$

$(10)(512) = k$ *Substitute 10 for ℓ and 512 for f.*

$5120 = k$ *The constant of variation is 5120.*

Next, find the frequency, in cycles per second, of the 8-inch string.

$\ell f = k$ *Use the same inverse variation equation.*

$8 \cdot f = 5120$ *Replace ℓ with 8 and k with 5120.*

$f = \dfrac{5120}{8}$ *Divide each side by 8.*

$f = 640$

The frequency of an 8-inch string is 640 cycles per second.

Let (x_1, y_1) be a solution of an inverse variation, $xy = k$. Let (x_2, y_2) be a second solution. Then $x_1 y_1 = k$ and $x_2 y_2 = k$.

$x_1 y_1 = k$

$x_1 y_1 = x_2 y_2$ *You can substitute $x_2 y_2$ for k because $x_2 y_2 = k$.*

The equation $x_1 y_1 = x_2 y_2$ is called the product rule for inverse variations. Study how it can be used to form a proportion.

$x_1 y_1 = x_2 y_2$

$\dfrac{x_1 y_1}{x_2 y_1} = \dfrac{x_2 y_2}{x_2 y_1}$ *Divide each side by $x_2 y_1$.*

$\dfrac{x_1}{x_2} = \dfrac{y_2}{y_1}$ *Notice that this proportion is different from the proportion for direct variation.*

You can use either the product rule or the proportion rule to solve problems involving inverse variation.

Example **If y varies inversely as x, and $y = 5$ when $x = 15$, find x when $y = 3$.**

Let $x_1 = 15$, $y_1 = 5$, and $y_2 = 3$. Solve for x_2.

Method 1

Use the product rule.

$$x_1 y_1 = x_2 y_2$$

$$15 \cdot 5 = x_2 \cdot 3$$

$$\frac{75}{3} = x_2$$

$$25 = x_2$$

Method 2

Use the proportion.

$$\frac{x_1}{x_2} = \frac{y_2}{y_1}$$

$$\frac{15}{x_2} = \frac{3}{5}$$

$$75 = 3x_2$$

$$25 = x_2$$

Thus, $x = 25$ when $y = 3$.

If you have observed people on a seesaw, you may have noticed that the heavier person must sit closer to the fulcrum (pivot point) to balance the seesaw. A seesaw is a type of *lever,* and all lever problems involve inverse variation.

Suppose weights w_1 and w_2 are placed on a lever at distances d_1 and d_2, respectively, from the fulcrum. The lever is balanced when $w_1 d_1 = w_2 d_2$. This property of levers is illustrated at the right.

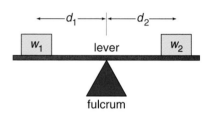

Example ⑤ **The fulcrum is placed in the middle of a 20-foot seesaw. Cholena, who weighs 120 pounds, is seated 9 feet from the fulcrum. How far from the fulcrum should Antonio sit if he weighs 135 pounds?**

Physics

Let $w_1 = 120$, $d_1 = 9$, and $w_2 = 135$. Solve for d_2.

$$w_1 d_1 = w_2 d_2$$
$$120 \cdot 9 = 135 \cdot d_2$$
$$1080 = 135 d_2$$
$$d_2 = 8 \qquad \text{Antonio should sit 8 feet from the fulcrum.}$$

CHECK FOR UNDERSTANDING

Communicating Mathematics

Study the lesson. Then complete the following.

1. **Tell** whether an equation of the form $xy = k$, where $k \neq 0$, represents a direct or inverse variation.

2. Refer to the application at the beginning of the lesson. Describe a reasonable domain and range.

3. **Explain** how you find the constant of variation in a direct variation.

4. **You Decide** Morgan says that a person's height varies directly as his or her age. Do you agree? Explain.

Determine which equations represent inverse variations and which represent direct variations. Then find the constant of variation.

   5. $mn = 5$ 6. $a = -3b$

Solve. Assume that y varies directly as x.

   7. If $y = 27$, when $x = 6$, find x when $y = 45$.

   8. If $y = -7$ when $x = -14$, find y when $x = 20$.

Solve. Assume that y varies inversely as x.

   9. If $y = 99$ when $x = 11$, find x when $y = 11$.

   10. If $y = -6$ when $x = -2$, find y when $x = 5$.

   11. **Physics** An 8-ounce weight is placed at one end of a yardstick. A 10-ounce weight is placed at the other end. Where should the fulcrum be placed to balance the yardstick?

EXERCISES

Determine which equations represent inverse variations and which represent direct variations. Then find the constant of variation.

   12. $c = 3.14d$ 13. $15 = rs$ 14. $\dfrac{35}{p} = q$

   15. $s = \dfrac{9}{t}$ 16. $\dfrac{1}{3}x = z$ 17. $4a = b$

Solve. Assume that y varies directly as x.

   18. If $y = -8$ when $x = -3$, find x when $y = 6$.

   19. If $y = 12$ when $x = 15$, find x when $y = 21$.

   20. If $y = 2.5$ when $x = 0.5$, find y when $x = 20$.

   21. If $y = 4$ when $x = 12$, find y when $x = -24$.

   22. If $y = -6$ when $x = 9$, find y when $x = 6$.

   23. If $y = 2\frac{2}{3}$ when $x = \frac{1}{4}$, find y when $x = 1\frac{1}{8}$.

Solve. Assume that y varies inversely as x.

   24. If $y = 9$ when $x = 8$, find y when $x = 6$.

   25. If $x = 2.7$ when $y = 8.1$, find y when $x = 3.6$.

   26. If $y = 24$ when $x = -8$, find y when $x = 4$.

   27. If $x = 6.1$ when $y = 4.4$, find x when $y = 3.2$

   28. If $y = 7$ when $x = \frac{2}{3}$, find y when $x = 7$.

   29. If $x = \frac{1}{2}$ when $y = 16$, find x when $y = 32$.

   30. Assume that y varies inversely as x.

      a. If the value of x is doubled, what happens to the value of y?

      b. If the value of y is tripled, what happens to the value of x?

Applications and Problem Solving

31. Space The weight of an object on the moon varies directly as its weight on Earth. With all of his gear on, Neil Armstrong weighed 360 pounds on Earth. When he became the first person to step on the moon on July 20, 1969, he weighed 60 pounds. Tara weighs 108 pounds on Earth. What would she weigh on the moon?

32. Physics Pam and Adam are seated on the same side of a seesaw. Pam is 6 feet from the fulcrum and weighs 115 pounds. Adam is 8 feet from the fulcrum and weighs 120 pounds. Kam is seated on the other side of the seesaw, 10 feet from the fulcrum. If the seesaw is balanced, how much does Kam weigh?

33. Music The pitch of a musical tone varies inversely as its wavelength. If one tone has a pitch of 440 vibrations per second and a wavelength of 2.4 feet, find the wavelength of a tone that has a pitch of 660 vibrations per second.

Mixed Review

34. Travel At 8:00 A.M., Alma drove west at 35 miles per hour. At 9:00 A.M., Reiko drove east from the same point at 42 miles per hour. At what time will they be 266 miles apart? (Lesson 4–7)

35. Probability If the probability that an event will occur is $\frac{2}{3}$, what are the odds that the event will occur? (Lesson 4–6)

Neil Armstrong on the moon

36. Retail Sales A department store buys clothing at wholesale prices and then marks the clothing up 25% to sell at retail price to customers. If the retail price of a jacket is $79, what was the wholesale price? (Lesson 4–5)

37. Finance Hiroko invested $11,700, part at 5% interest and the balance at 7% interest. If her annual earnings from both investments is $733, how much is invested at each rate? (Lesson 4–4)

38. Solve $\frac{2x}{5} + \frac{x}{4} = \frac{26}{5}$ for x. (Lesson 3–6)

39. Solve $\frac{x}{4} + 9 = 6$. (Lesson 3–3)

40. Simplify $\frac{4a + 32}{4}$. (Lesson 2–7)

VOCABULARY

After completing this chapter, you should be able to define each term, property, or phrase and give an example or two of each.

Algebra
base (p. 215)
constant of variation (p. 239)
direct variation (p. 239)
extremes (p. 196)
inverse variation (p. 241)
means (p. 196)
odds (p. 229)
percent (p. 215)
percentage (p. 215)
percent of decrease (p. 222)
percent of increase (p. 222)
percent proportion (p. 215)
proportion (p. 195)

rate (pp. 197, 215)
ratio (p. 195)
scale (p. 197)
simple interest (p. 217)
uniform motion (p. 235)
weighted average (p. 233)

Geometry
angle of depression (p. 208)
angle of elevation (p. 208)
corresponding angles (p. 201)
corresponding sides (p. 201)
hypotenuse (p. 206)
legs (p. 206)

similar (p. 201)
similar triangles (p. 201)
trigonometric ratios (p. 206)

Probability
equally likely (p. 229)
probability (p. 228)
probability of an event (p. 228)
random (p. 229)

Trigonometry
cosine (p. 206)
sine (p. 206)
solving the triangle (p. 208)
tangent (p. 206)

UNDERSTANDING AND USING THE VOCABULARY

Choose the correct term to complete each sentence.

1. The angle formed by a person in a radio tower looking up at an airplane in flight is called the (*angle of depression, angle of elevation*).

2. The equation $\frac{3}{5} = \frac{9}{15}$ is a (*proportion, ratio*).

3. If $\triangle ABC \sim \triangle DEF$, then their corresponding angles are (*equal, proportional*), and their corresponding sides are (*equal, proportional*).

4. A ratio (*can, cannot*) be expressed in the following ways: $\frac{4}{5}$, 4:5, and 4 to 5.

5. In a right triangle, the sides that form the right angle are called the (*hypotenuse, legs*) .

6. In $\frac{2}{x} = \frac{3}{9}$, 3 and x are called the (*extremes, means*).

7. The (*odds, probability*) that a certain event will occur is the ratio of the number of ways it can occur to the number of ways it cannot occur.

8. In the ratio $\frac{5}{28}$, 5 is the (*base, percentage*), and 28 is the (*base, percentage*).

9. A(n) (*direct variation, inverse variation*) is described by an equation of the form $xy = k$, where $k \neq 0$.

10. The tangent of an angle is defined as the measure of the (*adjacent, opposite*) leg divided by the measure of the (*adjacent leg, hypotenuse*).

SKILLS AND CONCEPTS

OBJECTIVES AND EXAMPLES

Upon completing this chapter, you should be able to:

• solve proportions   (Lesson 4–1)

$$\frac{x}{3} = \frac{x+1}{2}$$

$2(x) = 3(x + 1)$ *Find the cross products.*

$2x = 3x + 3$

$-3 = x$

• find the unknown measures of the sides of two similar triangles   (Lesson 4–2)

$$\frac{10}{5} = \frac{6}{a}$$

$10a = 5(6)$

$10a = 30$

$a = 3$

• use trigonometric ratios to solve right triangles
(Lesson 4–3)

$$\sin A = \frac{\text{measure of leg opposite } \angle A}{\text{measure of hypotenuse}}$$

$$\cos A = \frac{\text{measure of leg adjacent } \angle A}{\text{measure of hypotenuse}}$$

$$\tan A = \frac{\text{measure of leg opposite } \angle A}{\text{measure of leg adjacent } \angle A}$$

• use trigonometric ratios to solve right triangles
(Lesson 4–3)

$\cos M = 0.3245$

Enter: 0.3245 [2nd] [COS⁻¹]   *71.064715*

The measure of $\angle M$ is about 71°.

REVIEW EXERCISES

Use these exercises to review and prepare for the chapter test.

Use cross products to determine whether each pair of ratios forms a proportion.

11. $\dfrac{10}{3}, \dfrac{150}{45}$ 12. $\dfrac{8}{7}, \dfrac{2}{1.75}$

13. $\dfrac{30}{12}, \dfrac{5}{4}$ 14. $\dfrac{2.7}{3.1}, \dfrac{8.1}{9.3}$

Solve each proportion.

15. $\dfrac{6}{15} = \dfrac{n}{45}$ 16. $\dfrac{x}{11} = \dfrac{35}{55}$

17. $\dfrac{y+4}{y-1} = \dfrac{4}{3}$ 18. $\dfrac{z-7}{6} = \dfrac{z+3}{7}$

△ABC and △DEF are similar. For each set of measures given, find the measures of the remaining sides.

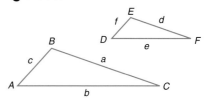

19. $c = 16, b = 12, a = 10, f = 9$
20. $a = 8, c = 10, b = 6, f = 12$
21. $c = 12, f = 9, a = 8, e = 11$
22. $b = 20, d = 7, f = 6, c = 15$

For △ABC, find each value to the nearest thousandth.

23. $\cos B$
24. $\tan A$
25. $\sin B$
26. $\cos A$
27. $\tan B$
28. $\sin A$

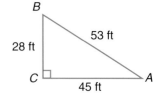

Use a calculator to find the measure of each angle to the nearest degree.

29. $\tan M = 0.8043$
30. $\sin T = 0.1212$
31. $\tan Q = 5.9080$
32. $\cos F = 0.7443$

OBJECTIVES AND EXAMPLES	REVIEW EXERCISES

use trigonometric ratios to solve right triangles
(Lesson 4–3)

Solve right $\triangle ABC$ if $m\angle B = 40°$ and $c = 6$.

$m\angle A = 180° - (90° + 40°)$ or $50°$

$$\cos 40° = \frac{a}{6} \qquad\qquad \sin 40° = \frac{b}{6}$$

$$0.7660 \approx \frac{a}{6} \qquad\qquad 0.6428 \approx \frac{b}{6}$$

$$0.7660(6) \approx a \qquad\qquad 0.6428(6) \approx b$$

$$4.596 \approx a \qquad\qquad 3.857 \approx b$$

Solve each right triangle. State the side lengths to the nearest tenth and the angle measures to the nearest degree.

33.

34.

35.

36.

---

solve percent problems   (Lesson 4–4)

Nine is what percent of 15?

$$\frac{9}{15} = \frac{r}{100} \qquad \textit{Use a proportion.}$$

$$9(100) = 15r \qquad \textit{Find the cross products.}$$

$$900 = 15r$$

$$60 = r$$

Solve.

37. What number is 60% of 80?

38. Twenty-one is 35% of what number?

39. Eighty-four is what percent of 96?

40. What percent of 17 is 34?

41. What number is 0.3% of 62.7?

42. Find 0.12% of $5200.

---

solve problems involving percent of increase or decrease   (Lesson 4–5)

original: $120   new: $114

amount of decrease: $120 − $114 = $6

$$\frac{\text{amount of decrease}}{\text{original}} = \frac{r}{100}$$

$$\frac{6}{120} = \frac{r}{100}$$

$$5 = r$$

The percent of decrease is 5%.

State whether each percent of change is a percent of increase or a percent of decrease. Then find the percent of increase or decrease. Round to the nearest whole percent.

43. original: $40
    new: $35

44. original: 97 cases
    new: 115 cases

45. original: $35
    new: $37.10

46. original: $50
    new: $88

47. original: 1500
        employees
    new: 1350 employees

48. original: 12,500
        students
    new: 11,800 students

---

solve problems involving discounts or sales tax
(Lesson 4–5)

running shoes: $74; discount: 15%; sales tax: 6%

15% of $74 = (0.15)(74) or $11.10

$74 − $11.10 = $62.90 ← *sale price*

6% of $62.90 = (0.06)(62.90) or $3.77

$62.90 + $3.77 = $66.67 ← *price with tax*

The total price is $66.67 for the shoes.

Find the final price of each item. When there is a discount and sales tax, compute the discount price first.

49. calculator: $81
    sales tax: 5.75%

50. gasoline: $21.50
    discount: 2%

51. used car: $8,690
    sales tax: 6.7%

52. dress: $89
    discount: 25%
    sales tax: 7%

OBJECTIVES AND EXAMPLES

● find the probability of a simple event and find the odds of a simple event (Lesson 4–6)

Find the probability of a computer randomly choosing the letter I in the word MISSISSIPPI.

$$\frac{\text{number of favorable outcomes}}{\text{number of possible outcomes}} = \frac{4}{11}$$

Find the odds that you will randomly *not* select the letter S in the word MISSISSIPPI.

number of successes:number of failures = 7:4

REVIEW EXERCISES

Find the probability of each outcome if a computer randomly chooses a letter in the word REPRESENTING.

53. the letter S   **54.** the letter E

55. P(not N)   **56.** the letters R or P

Find the odds of each outcome if you randomly select a coin from a jar containing 90 pennies, 75 nickels, 50 dimes, and 30 quarters.

57. selecting a dime   **58.** selecting a penny

59. *not* selecting a nickel

60. selecting a nickel or a dime

● solve problems involving direct and inverse variation (Lesson 4–8)

If x varies directly as y and $x = 15$ when $y = 1.5$, find x when $y = 9$.	If y varies inversely as x and $y = 24$ when $x = 30$, find x when $y = 10$.
$\dfrac{1.5}{9} = \dfrac{15}{x}$	$\dfrac{30}{x} = \dfrac{10}{24}$
$1.5x = 9(15)$	$(30)(24) = 10x$
$1.5x = 135$	$720 = 10x$
$x = 90$	$72 = x$

Solve. Assume that y varies directly as x.

61. If $y = 15$ when $x = 5$, find y when $x = 7$.

62. If $y = 35$ when $x = 175$, find y when $x = 75$.

63. If $y = 10$ when $x = 0.75$, find x when $y = 80$.

64. If $y = 3$ when $x = 99.9$, find y when $x = 522.81$.

Solve. Assume that y varies inversely as x.

65. If $y = 28$ when $x = 42$, find y when $x = 56$.

66. If $y = 15$ when $x = 5$, find y when $x = 3$.

67. If $y = 18$ when $x = 8$, find x when $y = 3$.

68. If $y = 35$ when $x = 175$, find y when $x = 75$.

APPLICATIONS AND PROBLEM SOLVING

69. Travel Two airplanes leave Dallas at the same time and fly in opposite directions. One airplane travels 80 miles per hour faster than the other. After three hours, they are 2940 miles apart. What is the rate of each airplane? (Lesson 4–7)

A practice test for Chapter 4 is provided on page 790.

70. Coffee Anne Leibowitz owns "The Coffee Pot," a specialty coffee store. She wants to create a special mix using two coffees, one priced at $8.40 per pound and the other at $7.28 per pound. How many pounds of the $7.28 coffee should she mix with 9 pounds of the $8.40 coffee to sell the mixture for $7.95 per pound? (Lesson 4–7)

ALTERNATIVE ASSESSMENT

COOPERATIVE LEARNING PROJECT

Letter Occurrence   In this project, you will determine percentages, percent of change, and probability and odds.

There is an uneven distribution of letters which occur in words. Christofer Sholes, the inventor of the typewriter, may not have been aware of this unevenness. He gave the left hand 56 percent of all the strokes, and the two most agile fingers on the right hand hit two of the least often used letters of the alphabet, j and k.

The table below shows the percentages of the occurrence of letters in large samples.

E 12.3%	R 6.0%	F 2.3%	K 1.5%
T 9.6%	H 5.1%	M 2.2%	Q 0.2%
A 8.1%	L 4.0%	W 2.0%	X 0.2%
O 7.9%	D 3.7%	Y 1.9%	J 0.1%
N 7.2%	C 3.2%	B 1.6%	Z 0.1%
I 7.2%	U 3.1%	G 1.6%	
S 6.6%	P 2.3%	V 0.9%	

Use the second paragraph on this page to count the number of times that e, s, c, w, and k occur and compute their percentages. Compare your answers with the table. What is the probability that a letter picked at random from the paragraph is a vowel? Are the odds better that a vowel or a consonant was used in the above paragraph? What are these odds?

Write a paragraph describing similar triangles. Then determine the percentages of the occurrence of the above mentioned letters in your paragraph. Do they coincide with the percentages in the table? Determine the percent of change for the above letters in the paragraph above and the paragraph you wrote.

Follow these guidelines.

* Determine what type of data you will need in order to calculate your answers.

* Devise a chart that will be helpful in organizing the data you will need.

* Write a paragraph describing what your answers mean.

* Using the percentages from the above table, draw your idea of where the letters should be on a typewriter and explain why.

THINKING CRITICALLY

* Use an example to describe the difference between odds and probability. Explain why some events are described in terms of odds and others in terms of probability.

* Use the definitions of the trigonometric ratios to explain why, in any right $\triangle ABC$ with right angle C, $\sin A = \cos B$, and $\cos A = \sin B$.

PORTFOLIO

Even after working through a chapter of material, you can get to the end and still feel uncertain about a concept or a certain type of problem. Find a problem that you still cannot solve or that you are worried you might not be able to solve on a test. Write out the question and as much of the solution as you can until you get to the hard part. Then explain what it is that keeps you from solving the problem. Be clear and precise. Place this in your portfolio.

SELF EVALUATION

Theme learning occurs when the subject matter being studied is all related to the same theme, such as water, animals, and so on. When the course material is presented, it is all applied to that theme.

Assess yourself. Are you a person who would enjoy theme learning? Would you like to relate all your course work to one area or do you prefer having a variety of application themes? Describe a theme that would be of interest to you and tell how you would use that theme to introduce and apply the contents of a specific math subject.

CUMULATIVE REVIEW

CHAPTERS 1–4

There are nine multiple-choice questions in this section. After working each problem, write the letter of the correct answer on your paper.

1. Solve the equation $7(3b + x) = 5a$ for x.

A. $5a - 21b = x$

B. $\dfrac{5a}{7} - 21b = x$

C. $\dfrac{5a - 3b}{7} = x$

D. $\dfrac{5a - 21b}{7} = x$

2. Sales A Tampa Bay Buccaneers sweatshirt decreased in price from \$40 to \$35 during a clearance sale. Find the percent of decrease.

A. 5%

B. 12.5%

C. 14.3%

D. 114.3%

3. A caterpillar was trying to crawl out of a bucket into which it had fallen. It crawled up $1\frac{1}{4}$ feet before sliding down $\frac{3}{8}$ feet. It then crawled up $\frac{5}{6}$ feet before resting. How much farther must the caterpillar crawl to reach the top of the 2-foot bucket?

A. $\dfrac{7}{24}$ feet

B. $\dfrac{7}{8}$ feet

C. $1\frac{1}{8}$ feet

D. $1\frac{17}{24}$ feet

4. Geometry Choose the correct proportion to find the distance across the lake (AB) given $\triangle ABC$ is similar to $\triangle EDC$.

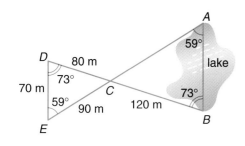

A. $\dfrac{70}{AB} = \dfrac{90}{120}$

B. $\dfrac{80}{70} = \dfrac{AB}{120}$

C. $\dfrac{70}{AB} = \dfrac{80}{120}$

D. $\dfrac{80}{90} = \dfrac{120}{AB}$

5. Kassim read 36 pages of a novel in 2 hours. At the same rate, find the number of hours it will take him to read the remaining 135 pages.

A. $3\frac{3}{4}$

B. $7\frac{1}{2}$

C. $9\frac{1}{2}$

D. $10\frac{1}{2}$

6. Choose an equivalent equation to represent "h is the product of g and the difference of the square of b and t".

A. $h - t = gb^2$

B. $h = g(b^2 - t)$

C. $h = g(b - t)^2$

D. $h + t = gb^2$

7. Probability The freshman class of Perry High School is planning an end-of-the-year dance. The dance committee decides to award door prizes. They want the odds of winning a prize to be 1:5. If they plan for 180 tickets to be sold for the dance, how many prizes will be needed?

A. 6 prizes

B. 36 prizes

C. 30 prizes

D. 150 prizes

8. Sales Raven sold her concert ticket to Sandra for 75% of the original ticket price. If Sandra paid \$26.25 for the ticket, choose the proportion that could be used to find the ticket's original price.

A. $\dfrac{75}{100} = \dfrac{26.25}{x}$

B. $\dfrac{x}{26.25} = \dfrac{75}{100}$

C. $\dfrac{x}{100} = \dfrac{75}{26.25}$

D. $\dfrac{x}{75} = \dfrac{26.25}{100}$

9. Statistics Carita's bowling scores for the first four games of a five-game series are $b + 2$, $b + 3$, $b - 2$, and $b - 1$. What must her score be for the last game to have an average of $b + 2$?

A. $b - 2$

B. $b + 2$

C. $b - 8$

D. $b + 8$

SECTION TWO: FREE RESPONSE

This section contains nine questions for which you will provide short answers. Write your answer on your paper.

10. Describe the steps required to solve the equation $4t - 5 = 7t - 23$.

11. Geometry Find the dimensions of a rectangle whose width is 5 inches less than its length and has a perimeter of 70 inches.

12. Finance In the beginning of the month, the balance in Marisa Fuente's checking account was $428.79. After writing checks totaling $1097.31, depositing 2 checks of $691.53 each, and withdrawing $100 from a bank machine, what was the new balance in Marisa's account?

13. Finance Maria Cruz wants to invest part of her $8000 savings in a bond paying 12% annual interest, and the rest of it in a savings account paying 8%. If she wants to earn a total of $760, how much money should Maria invest at each rate?

14. Weather The time t in hours, that a storm will last is given by the formula $t = \sqrt{\dfrac{d^3}{216}}$, where d is the diameter of the storm in miles. Suppose the umpires in a baseball game declared rain delay at 4:00 P.M. The storm causing the rain delay has a diameter of 12 miles. After the rain has stopped, can the game be restarted before 6:00 P.M.?

15. Statistics Find the mean, median, and mode of the data represented in the line plot below to the nearest tenth.

16. Running Raul ran a distance of 385 yards in $1\frac{3}{4}$ minutes. His younger sister, Luisa, estimated it would take her $3\frac{1}{2}$ minutes to run the same distance. If Luisa ran at a rate of 94 yards per minute, was her estimation correct? Verify your answer.

17. Government The term of office for a United States Senator is 150% as long as the term of office for the President. How long is the term of office for a United States Senator?

18. Physics Shannon weighs 126 pounds and Minal weighs 154 pounds. They are seated at opposite ends of a seesaw. Shannon and Minal are 16 feet apart, and the seesaw is balanced. How far is Shannon from the fulcrum?

SECTION THREE: OPEN-ENDED

This section contains two open-ended problems. Demonstrate your knowledge by giving a clear, concise solution to each problem. Your score on these problems will depend on how well you do the following.

- Explain your reasoning.

- Show your understanding of the mathematics in an organized manner.

- Use charts, graphs, and diagrams in your explanation.

- Show the solution in more than one way or relate it to other situations.

- Investigate beyond the requirements of the problem.

19. Geometry One angle of a triangle measures 15° more than the second. The measure of the third angle is the sum of the first two angles tripled. Find the measure of each angle.

20. Biology An adult African elephant can weigh up to 12,000 pounds. An adult blue whale can weigh as much as 37 adult African elephants. A newborn blue whale weighs one-third of an adult African elephant's weight. The newborn whale will gain 200 pounds per day until it reaches its adult weight. How many days will it take for the newborn to reach adult weight?

Graphing Relations and Functions

Objectives

In this chapter, you will:

- graph ordered pairs, relations, and equations,
- solve problems by making a table,
- identify the domain, range, and inverse of a relation,
- determine if a relation is a function,
- write an equation to represent a relation, and
- calculate and interpret the range, quartiles, and interquartile range of a set of data.

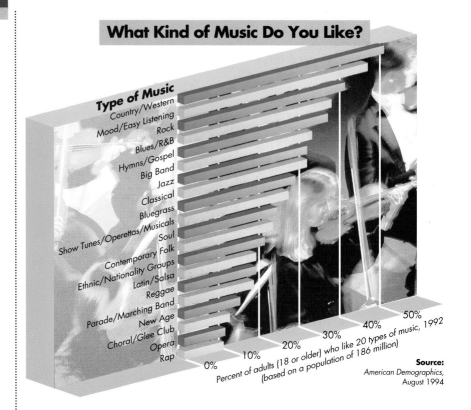

What Kind of Music Do You Like?

Type of Music

Country/Western
Mood/Easy Listening
Rock
Blues/R&B
Hymns/Gospel
Big Band
Jazz
Classical
Bluegrass
Show Tunes/Operettas/Musicals
Soul
Contemporary Folk
Ethnic/Nationality Groups
Latin/Salsa
Reggae
Parade/Marching Band
New Age
Choral/Glee Club
Opera
Rap

0%   10%   20%   30%   40%   50%

Percent of adults (18 or older) who like 20 types of music, 1992 (based on a population of 186 million)

Source:
American Demographics,
August 1994

What kinds of music do you listen to most often? Most teenagers have predictable tastes in music—listening to and buying popular rock and country/western favorites. As you become an adult, you may learn to enjoy listening to many other types of music. Who knows, you may become a fan of 12th century Gregorian chants, or perhaps even Zydeco!

TIME *Line*

2400 B.C. Place value is used in the Sumerian number system in Mesopotamia.

1591 François Viète's *Isagoge in Artem Analyticam* (Introduction to the Analytic Art) is the first work in mathematics to use letters for variables and constants.

2500 B.C.   2400   2300   A.D. 1425   1450   1475   1500   1525   1550   1575   1600   1625   1650

A.D. 1434 Jan van Eyck, a painter in the Netherlands, creates the brilliant double portrait *"The Anrolfini Marriage."*

Chapter Project

Work in groups of three to conduct your own music survey.

- Each person in the group should survey 20 people of varying ages to determine their musical tastes. Use the graph on the previous page for ideas.

- Combine your results and compare them to the graph. Does your survey match the national results?

- Create a bar graph to show your results for each type of music. Also create bar graphs for the five most popular types of music you surveyed showing the age distribution for each type of music. Then create a line graph comparing male and female musical tastes in relation to age groups.

- Compare your graphs, make some conclusions, and report your findings.

- Do you think certain types of music are associated with different cultures? You may want to include your thoughts on this in your report.

By the young age of 10, **Jamie Knight** was already an accomplished jazz singer. She has even performed at New York City's Carnegie Hall, following in the footsteps of Ella Fitzgerald and Mel Torme. Jamie is from Philadelphia, Pennsylvania, and has been performing since the age of 6. She has appeared at jazz festivals and on *Good Morning America*. She also has her own CD.

1846 Adolphe Sax, attempting to improve on the bass clarinet, invents a new instrument, the saxophone.

1938 Ynes Mexia, who collects more than 137,600 botanical specimens for classification, dies.

| 1700 | 1725 | 1750 | 1775 | 1800 | 1825 | 1850 | 1875 | 1900 | 1925 | 1950 | 1975 | 2000 |

1750 The great Chinese work of fiction *The Dream of the Red Chamber* is written by Tsao Hsueh Chin.

1882 African American Lewis H. Latimer patents the first cost-efficient method for producing electric lights.

1994 For her work as conservator of American Indian language and culture, Vi Hilbert is honored by the National Endowment for the Arts.

The Coordinate Plane

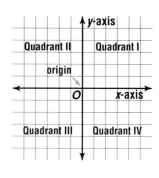

What YOU'LL LEARN

- To graph ordered pairs on a coordinate plane, and
- to solve problems by making a table.

Why IT'S IMPORTANT

You can use graphing to solve problems involving geography and cartography.

APPLICATION
Landmarks

In 1988, as part of Cincinnati's 200th birthday celebration, the city's Bicentennial Commission created parks along the Ohio River. They sold over 33,800 personalized bricks that were placed in the pathways in the parks. However, people had difficulty finding their bricks. So, in 1989, Mr. Frank Albi of Business Information Storage (BIS) developed a master directory, in which each patron's name was listed along with the section of the park and the column in which his or her brick was laid. BIS then added locator bricks to identify the columns in each section.

Perpendicular lines are lines that meet to form 90° angles.

The system used in Cincinnati helped to locate the specific site of each brick. In mathematics, points are located in reference to two perpendicular number lines called **axes.** The axes intersect at their zero points, a point called the **origin.** The horizontal number line, called the **x-axis,** and the vertical number line, called the **y-axis,** divide the plane into four **quadrants.** The quadrants are numbered as shown at the right. Notice that neither the axes nor any point on the axes are located in any quadrant. The plane containing the x- and y-axes is called the **coordinate plane.**

Unless otherwise designated, you can assume that each division on the axes represents 1 unit.

Points in the coordinate plane are named by *ordered pairs* of the form (x, y). The first number, or **x-coordinate,** corresponds to the numbers on the x-axis. The second number, or **y-coordinate,** corresponds to the numbers on the y-axis. The ordered pair for the origin is $(0, 0)$.

A point can be referred to by just a capital letter. Thus, Z can be used to mean point Z.

To find the ordered pair for point Z, shown at the right, first follow along a vertical line through Z to find the x-coordinate on the x-axis and along a horizontal line through Z to find the y-coordinate on the y-axis. The number on the x-axis that corresponds to Z is 4. The number on the y-axis that corresponds to Z is -6. Thus, the ordered pair for point Z is $(4, -6)$. This ordered pair can also be written as $Z(4, -6)$. The x-coordinate of Z is 4, and the y-coordinate of Z is -6. Point Z is located in Quadrant IV.

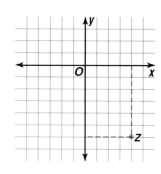

A table is often used when working with ordered pairs. It is a helpful organizational tool in problem solving.

Example ① Write ordered pairs for points *E, F, G,* and *H* shown at the right. Name the quadrant in which each point is located.

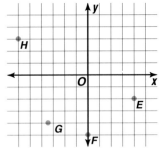

PROBLEM SOLVING

Use a Table

Use a table to help find the coordinates of each point.

LOOK BACK

You can review more about using tables in Lesson 4-7.

Point	x-coordinate	y-coordinate	Ordered Pair	Quadrant
E	4	–2	(4, –2)	IV
F	0	–5	(0, –5)	none
G	–3.5	–4	(–3.5, –4)	III
H	–6	3	(–6, 3)	II

To **graph** an ordered pair means to draw a dot at the point on the coordinate plane that corresponds to the ordered pair. This is sometimes called *plotting a point.* When graphing an ordered pair, start at the origin. The *x*-coordinate indicates how many units to move right (positive) or left (negative). The *y*-coordinate indicates how many units to move up (positive) or down (negative).

Example ② Plot the following points on a coordinate plane.

a. *N*(−3, −5)

   Start at the origin, *O.* Move left 3 units since the *x*-coordinate is −3. Then move down 5 units since the *y*-coordinate is −5. Draw a dot and label it *N.*

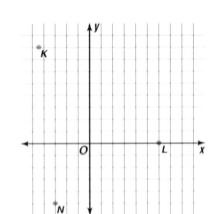

b. *K*(−4.5, 8)

   Start at the origin, *O.* Move left 4.5 units and up 8 units. Draw a dot and label it *K.*

c. *L*(6, 0)

   Start at the origin, *O.* Move right 6 units. Since the *y*-coordinate is 0, the point will be located on the *x*-axis. Draw the dot and label it *L.*

Each point in the coordinate plane can be named by exactly one ordered pair, and each ordered pair names exactly one point in the plane. Thus, there is a one-to-one correspondence between points and ordered pairs.

Completeness Property for Points in the Plane	1. **Exactly one point in the plane is named by a given ordered pair of numbers.** 2. **Exactly one ordered pair of numbers names a given point in the plane.**

On city and state maps, letters and numbers are usually used to locate landmarks on the map. The number representing the horizontal coordinate is usually listed first followed by the letter representing the vertical coordinate. However, unlike the coordinate plane where a specific point is denoted by the ordered pair, the coordinates on a map define a region or sector in which the landmark is located.

Example ③

Cartography

Use ordered pairs to name all of the sectors that Interstate Highway 35 passes through on the map of San Antonio, Texas, shown below.

F Y I

César-François Cassini, who was born in Thury, France, in 1714, was the first cartographer to create a map according to modern principles. In 1744, he directed the first National Geographic survey, the triangulation of France.

I-35 begins in the southwest corner and extends through San Antonio to the northeast corner. Part of it shares the same path as I-410.

I-35 goes through $(1, D)$, continuing through $(2, D)$, across the southeast corner of $(2, C)$, turning upward through $(3, C)$, across the southern portion of $(3, B)$, sharing the path with I-410 through $(4, B)$, and exiting the map at the eastern edge of $(4, A)$.

Communicating Mathematics

Study the lesson. Then complete the following.

1. **Draw** a coordinate plane. Label the origin, *x*-axis, *y*-axis, and the quadrants.

2. **Explain** how you can tell which quadrant a point is in by just looking at the signs of its coordinates.

3. **Copy or sketch** a map of the major thoroughfares in your community. Using a different color, draw a coordinate system on the map, making your home the origin. Find the coordinates of your school, a library, a shopping center or department store, and your favorite restaurant.

Guided Practice

Write the ordered pair for each point shown at the right. Name the quadrant in which the point is located.

4. *A*

5. *C*

6. *E*

7. *G*

Graph each point.

8. *W*(−5, 0)

9. *X*(−3, −4)

10. *Y*(3, −5)

11. *Z*(0, 4)

12. Write the ordered pair that describes a point 15 units up from and 13 units to the left of the origin.

Practice

Refer to the coordinate plane for Exercises 4–7. Write the ordered pair for each point. Name the quadrant in which the point is located.

13. *D*	14. *H*	15. *L*
16. *P*	17. *O*	18. *K*
19. *F*	20. *B*	21. *I*
22. *M*	23. *J*	24. *N*

Graph each point.

25. *A*(2, −1) 26. *B*(−3, −3) 27. *C*(0, 3.5) 28. *D*(5, −2)

29. *E*(−3, 0) 30. *F*(3, 5) 31. *G*(−3, −4) 32. *H*(−5, −2)

33. *I*(0, −2) 34. *J*($-\frac{1}{2}$, 1) 35. *K*(4, 0) 36. *L*(4, 2)

Graph each point. Then connect the points in alphabetical order and identify the figure.

37. *A*(2, 0), *B*(2, 3), *C*(1, 3), *D*(−0.5, −1), *E*(−2, 3), *F*(−3, 3), *G*(−3, −3), *H*(−2, −3), *I*(−2, 0), *J*(−0.5, −3), *K*(1, 0), *L*(1, −3), *M*(2, −3), *N*(2, 0)

38. *A*(9, 0.5), *B*(5, 1), *C*(2, 1), *D*(−1, 5), *E*(−2, 5), *F*(−1, 1), *G*(−4, 1), *H*(−5, 3), *I*(−6, 3), *J*(−5.5, 1), *K*(−5.5, 0), *L*(−6, −2), *M*(−5, −2), *N*(−4, 0), *P*(−1, 0), *Q*(−2, −4), *R*(−1, −4), *S*(2, 0), *T*(5, 0), *U*(9, 0.5)

Programming

39. The graphing calculator program at the right will plot a group of points on the same coordinate plane. You will need to set an appropriate window so that you will be able to see all of the points. After each point is plotted, press ENTER to plot another point. To end the program, press 2nd QUIT while you are at the home screen. Use this program to check the points you plotted in Exercises 37 and 38.

```
PROGRAM:PLOTPTS
:ClrDraw
:Lbl 1
:Prompt X,Y
:Pt-On(X,Y)
:Pause
:Goto 1
```

Critical Thinking

40. Describe the possible locations, in terms of quadrants or axes, for the graph of (x, y) if x and y satisfy the given conditions.
  a. $xy > 0$ **b.** $xy < 0$ **c.** $xy = 0$

Applications and Problem Solving

41. Geography On a map of the United States, degrees of longitude and latitude can be used to form ordered pairs that name points on the map. Longitude lines run north and south and should be the first coordinates of the ordered pairs. Latitude lines run east and west and should be the second coordinates of the ordered pairs. (On this map, the numbers showing degrees of longitude are listed along the horizontal axis, and those showing degrees of latitude are listed along the vertical axis.)

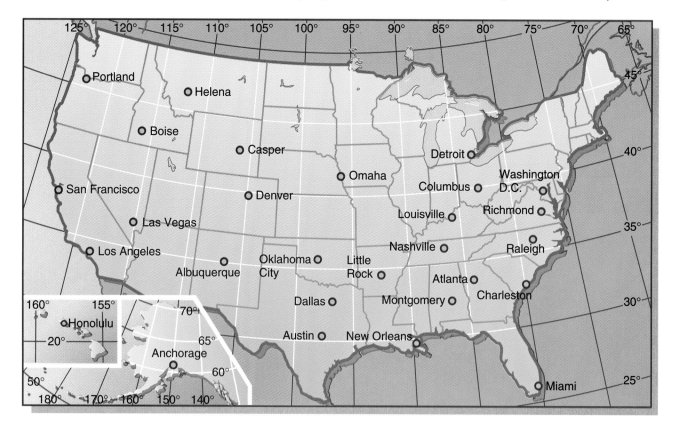

  a. Name the city at $(90°, 30°)$.
  b. Name the state in which $(120°, 45°)$ is located.
  c. Estimate the latitude and longitude of our nation's capital to the nearest 5°.
  d. What state capital has its location at $(157°, 21°)$?
  e. What is the approximate latitude and longitude of your community?
  f. Look at the longitude and latitude lines on a globe and compare them to the ones shown on this map. What do you find?

42. Tourism Diana works for a travel agency and has been asked to highlight key landmarks on a map of the Miami area for a customer. Use the map of Miami at the left to answer the following questions.

   a. What causeway is in sector (3, M)?

   b. In what sector is the city of Coral Gables?

   c. Which interstate goes from sector (2, L) to (2, O)?

   d. Name all of the sectors through which Highway 826 passes.

43. Draw a Diagram Make a seating chart of the desks in your mathematics classroom. Create a coordinate plane on your diagram so that each desk corresponds to an ordered pair in the plane.

   a. What are the coordinates of your desk in the room?

   b. Name another student in your class. What is the coordinate of this person's desk?

Mixed Review

44. If y varies directly as x, and $y = 3$ when $x = 15$, find y when $x = -25$. (Lesson 4–8)

45. Two trains leave York at the same time, one traveling north, the other south. The first train travels at 40 miles per hour and the second at 30 miles per hour. In how many hours will the trains be 245 miles apart? (Lesson 4–7)

46. Use a calculator to find the measure of $\angle A$ to the nearest degree if $\sin A = 0.2756$. (Lesson 4–3)

47. Find the mean, median, and mode for {5, 9, 1, 2, 3}. (Lesson 3–7)

Simplify each expression.

48. $\sqrt{\dfrac{16}{25}}$ (Lesson 2–8)

49. $\dfrac{7a + 35}{-7}$ (Lesson 2–7)

50. Agriculture Agribusiness technicians assist in the organization and management of farms. They may prepare graphs of weather data for a given region. For example, the graph at the right displays normal maximum, minimum, and mean temperatures for the region around Des Moines, Iowa. (Lesson 1–9)

   a. What is the normal maximum temperature for October?

   b. What is the normal range of temperatures for May?

51. Simplify $7(5a + 3b) - 4a$. (Lesson 1–8)

5-2A Graphing Technology
Relations

A Preview of Lesson 5-2

A **relation** is a set of ordered pairs. You can graph a relation using a graphing calculator. Points can be plotted on the coordinate plane using the DRAW menu.

Example 1 **Plot the following points on a coordinate plane:** $M(4, 7)$, $A(-10, 25)$, $T(-17, -17)$, and $H(23, -11)$.

Begin with a viewing window of $[-47, 47]$ by $[-31, 31]$ using a scale of 10 for each axis. This can be done quickly by entering ZOOM 6 ZOOM 8 ENTER . *This sets the scale to integer values.*

Clear the coordinate plane by turning off all stat plots and clearing the Y= list.

If you press 2nd DRAW ▶ *1 before pressing* GRAPH , *the prompt Pt-On(appears. To plot a point from this prompt, enter the coordinates of the point, a closing parenthesis, and* ENTER . *Then proceed with the cursor for the other points.*

- To plot the first point, enter GRAPH 2nd DRAW ▶ 1. *The cursor appears at the origin and X = 0 and Y = 0 appears at the bottom of the screen.*

- Now use the arrow keys to move the cursor to the coordinates of the point you wish to graph. *The coordinates of each position of the cursor appear at the bottom of the viewing screen.*

- Press ENTER to put a dot at that point.

- Continue to move the cursor and press ENTER to graph the other three points.

- When finished, you can clear the screen by pressing 2nd DRAW 1.

If you do not need a range as great as the one you get by pressing ZOOM 8, you can set the WINDOW values in the usual manner or press ZOOM 6. Once you set your new window and try the method shown in Example 1, you will find that you cannot plot points at exact integer values by using the cursor. However, there is another way to plot exact points.

Example 2 **Plot** $G(-4, 5)$, $R(1, 6)$, $A(-3, -6)$, $P(5, -8)$, **and** $H(0, 7)$ **in the standard viewing window.**

- Set the standard viewing window. **Enter:** ZOOM 6

- Return to the home screen by pressing 2nd QUIT .

- You can graph all of the points by linking commands into a series using a colon.

Enter: 2nd DRAW ▶ 1 *Pt-On(appears on the screen.*

This enters the first point. Now repeat the steps without pressing ENTER to enter all the other points. Finally, press ENTER to graph the points.

You can also use the Line(feature on the DRAW menu to connect the points. This can help you analyze the points in a relation. You must first clear the screen by pressing $\boxed{\text{2nd}}$ $\boxed{\text{DRAW}}$ 1 $\boxed{\text{ENTER}}$ before beginning a new graph.

Example 3 **Determine if the graphs of (3, 5), (−3, −5), and (0, 7) are on the same line.**

If you connect the points and only a segment appears, the three points are probably on the same line.

Enter: $\boxed{\text{2nd}}$ $\boxed{\text{DRAW}}$ 2 3 $\boxed{,}$ 5

$\boxed{,}$ $\boxed{(-)}$ 3 $\boxed{,}$ $\boxed{(-)}$ 5 $\boxed{)}$

$\boxed{\text{2nd}}$ $\boxed{:}$ $\boxed{\text{2nd}}$ $\boxed{\text{DRAW}}$ 2 $\boxed{(-)}$

3 $\boxed{,}$ $\boxed{(-)}$ 5 $\boxed{,}$ 0 $\boxed{,}$ 7 $\boxed{)}$

$\boxed{\text{2nd}}$ $\boxed{:}$ $\boxed{\text{2nd}}$ $\boxed{\text{DRAW}}$ 2

3 $\boxed{,}$ 5 $\boxed{,}$ 0 $\boxed{,}$ 7 $\boxed{)}$ $\boxed{\text{ENTER}}$

Because the resulting figure is a triangle, the three points are not on the same line.

EXERCISES

Use a graphing calculator to plot each point in the viewing window [−47, 47] by [−31, 31].

1. $X(10, 10)$ 2. $Y(0, -6)$ 3. $Z(-26, 11)$
4. $Q(-17, -19)$ 5. $T(-21, 4)$ 6. $P(-20, 0)$

7. Explain why, when you graph the points in Exercises 2 and 6, you do not see them on the screen.

8. **a.** Plot $H(9, 9)$ and $J(29, 9)$.
    b. Find two additional points E and F so that these four points form the vertices of a square. Are there other possible coordinates of E and F?
    c. Plot E and F to complete the square.

TECHNOLOGY
Tip

To make your graph look like a square, press $\boxed{\text{ZOOM}}$ 5 after setting the window and before using DRAW.

Relations

F Y I

The U.S. Congress passed the Rare and Endangered Species Act in 1966, setting up rules by which the Fish and Wildlife Service can list and protect species in danger of extinction.

What is the trend over time?

CONNECTION
Biology

There are 1.4 million classified species of microorganisms, invertebrates, plants, fish, birds, reptiles, amphibians, and mammals. Currently, 930 species worldwide have been listed as *endangered*. A species becomes endangered when its numbers are so low that the species is in danger of extinction.

In the United States, the manatee, an aquatic mammal, is considered to be endangered. Although manatees were once common throughout the coastal areas of North America, they now live mainly in Florida. In the table below, you will find the number of manatees that have been found dead since 1976.

Manatee Mortality In Florida									
Year	1976	1977	1978	1979	1980	1981	1982	1983	1984
Number of Manatees	62	114	84	77	63	116	114	81	128
Year	1985	1986	1987	1988	1989	1990	1991	1992	1993
Number of Manatees	119	122	114	133	168	206	174	163	145

Source: Florida Marine Research Institute

The manatee data could also be represented by a set of ordered pairs, as shown in the list below. Each first coordinate is the year, and the second coordinate is the number of manatees. Each ordered pair can then be graphed.

{(1976, 62), (1977, 114), (1978, 84), (1979, 77), (1980, 63), (1981, 116), (1982, 114), (1983, 81), (1984, 128), (1985, 119), (1986, 122), (1987, 114), (1988, 133), (1989, 168), (1990, 206), (1991, 174), (1992, 163), (1993, 145)}

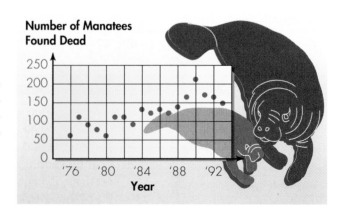

Number of Manatees Found Dead

Recall that a *relation* is a set of ordered pairs, like the one shown on the previous page. The set of first coordinates of the ordered pairs is called the *domain* of the relation. The domain usually contains the *x*-coordinates. The set of second coordinates is called the *range* of the relation. It usually contains the *y*-coordinates.

Definition of the Domain and Range of a Relation	The domain of a relation is the set of all first coordinates from the ordered pairs in the relation. The range of the relation is the set of all second coordinates from the ordered pairs.

For the relation representing the manatee data, the domain and range are as follows.

Domain = {1976, 1977, 1978, 1979, 1980, 1981, 1982, 1983, 1984, 1985, 1986, 1987, 1988, 1989, 1990, 1991, 1992, 1993}

Range = {62, 114, 84, 77, 63, 116, 81, 128, 119, 122, 133, 168, 206, 174, 163, 145}

An <u>element</u> is a member of a set.

In addition to ordered pairs, tables, and graphs, a relation can be represented by a **mapping**. A mapping illustrates how each element of the domain is paired with an element in the range. For example, the relation {(3, 3), (−1, 4), (0, −4)} can be modeled in each of the following ways.

Ordered Pairs   **Table**   **Graph**   **Mapping**

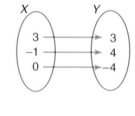

(3, 3)

(−1, 4)

(0, −4)

x	y
3	3
−1	4
0	−4

Example **①** Represent the relation shown in the graph at the right as

a. a set of ordered pairs,

b. a table, and

c. a mapping.

d. Then determine the domain and range.

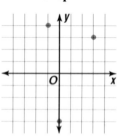

a. The set of ordered pairs for the relation is
{(−3, 3), (−1, 2), (1, 1), (1, 3), (3, −2), (4, −2)}.

b.

x	y
−3	3
−1	2
1	1
1	3
3	−2
4	−2

c.

When giving the domain and range, if a value is repeated, you list it only once.

d. The domain for this relation is {−3, −1, 1, 3, 4}, and the range is {−2, 1, 2, 3}.

In real-life situations, you may need to select a range of values for the *x*- or *y*-axis that does not begin with 0 and does not have units of 1.

Example **2** **In order to protect endangered species, there is a limit to the number of animals that can be exported to other countries. The table below shows yearly quotas for the number of wild crocodiles that can be exported from Indonesia.**

CONNECTION

Biology

a. **Determine the domain and range of the relation.**

b. **Graph the data.**

c. **What conclusions might you make from the graph of the data?**

Maximum Exportation of Crocodiles from Indonesia									
Year	1986	1987	1988	1989	1990	1991	1992	1993	1994
Number of Crocodiles	2000	2000	4000	4000	3000	3000	2700	1500	1500

Source: *Traffic USA*, World Wildlife Fund, August 1992

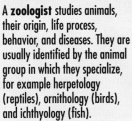

a. The domain is {1986, 1987, 1988, 1989, 1990, 1991, 1992, 1993, 1994}. The range is {1500, 2000, 2700, 3000, 4000}.

Values are usually listed in numerical order.

b. The values for the *x*-axis need to go from 1986 to 1994. It is not efficient to begin the scale at 0. The values on the *y*-axis need to include values from 1500 to 4000. You can include 0 and use units of 500.

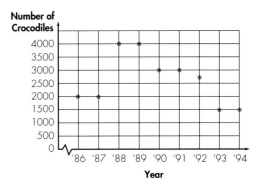

c. While the number of crocodiles that could be exported rose in 1988, it has decreased since 1991. This may indicate that crocodiles in Indonesia are even more endangered.

A jagged line is used to represent the values prior to 1986 that have been omitted.

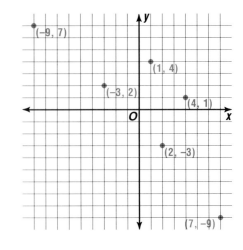

The **inverse** of any relation is obtained by switching the coordinates in each ordered pair.

Relation	Inverse
(1, 4)	(4, 1)
(−3, 2)	(2, −3)
(7, −9)	(−9, 7)

Notice that the domain of the relation becomes the range of the inverse. Similarly, the range of the relation becomes the domain of the inverse.

| **Definition of the Inverse of a Relation** | Relation **Q** is the inverse of relation **S** if and only if for every ordered pair (**a, b**) in **S**, there is an ordered pair (**b, a**) in **Q**. |

Example **3** Express the relation shown in the mapping as a set of ordered pairs. Write the inverse of the relation and draw a mapping to model the inverse.

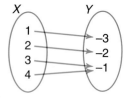

The mapping shows the relation $\{(1, -3), (2, -2), (3, -1), (4, -1)\}$.

The inverse of the relation is $\{(-3, 1), (-2, 2), (-1, 3), (-1, 4)\}$.

The mapping at the right shows the inverse.

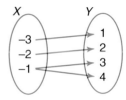

Relations and their inverses display some interesting characteristics, as you will discover below.

MODELING MATHEMATICS

Relations and Inverses

Materials: grid paper   straightedge   colored pencils

There is a special relationship between a relation and its inverse, which you can discover by paper folding.

Step 1 Graph the relation $\{(4, 5), (-3, 6), (-5, -3), (4, -7), (5, 0), (0, -3)\}$ on a coordinate plane using one color of pencil.

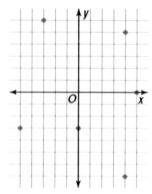

Connect the points in order using the same color pencil.

Step 2 Use a different color pencil to graph the inverse of the relation, connecting its points in order.

Step 3 Fold the graph paper through the origin so that the positive y-axis lies on top of the positive x-axis. Hold the paper up to a light so that you can see all of the points you graphed.

Your Turn

a. What do you notice about the location of the points you graphed when you looked at the folded paper?

b. Unfold the paper. Describe the pattern formed by the lines connecting the points in the relation and its inverse.

c. What do you think are the ordered pairs that represent the points on the fold line? Describe these in terms of x and y.

d. How could you graph the inverse of a function without writing ordered pairs first?

Communicating Mathematics

Study the lesson. Then complete the following.

1. **Explain** why it is important to identify the domain and range values of a relation when graphing the relation.

2. Refer to the application at the beginning of the lesson. Explain what the manatee graph suggests about patterns in the rate of manatee deaths.

3. **State** the relationship between the domain and range of a relation and the domain and range of its inverse.

 MODELING MATHEMATICS

4. Graph the relation {(0, 5), (2, 3), (1, −4), (−3, 3), (−1, −2)}. Draw a line that goes through the points (−3, −3) and (3, 3). Use this line to graph the inverse of the relation without writing the ordered pairs of the inverse.

Guided Practice

State the domain and range of each relation.

5. {(0, 2), (1, −2), (2, 4)}

6. {(−4, 2), (−2, 0), (0, 2), (2, 4)}

Express the relation shown in each table, mapping, or graph as a set of ordered pairs. Then state the domain, range, and inverse of the relation.

7.

x	y
1	3
2	4
3	5
5	7

8.

x	y
1	4
3	−2
4	4
6	−2

9.

10.

11.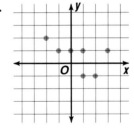

12–14. Graph the relations shown in Exercises 7–9.

15. **Economics** The graph at the right shows the unemployment rate in the United States from January 1992 to December 1994.

a. Name three ordered pairs shown by the graph.

b. Estimate the least value in the range.

c. What conclusions might you make from the graph?

U.S. Unemployment Rate 1992–1994

Source: Wall Street Journal, Dec. 5, 1994

Practice

State the domain and range of each relation.

16. {(1, 3), (2, 5), (1, −7), (2, 9), (3, 3)}

17. {(1, 7), (−2, 7), (3, 7), (−5, 7)}

18. {3.1, −1), (−4.7, 3.9), (2.4, −3.6), (−9, 2)}

19. $\left\{\left(\frac{1}{2}, \frac{1}{4}\right), \left(1\frac{1}{2}, -\frac{2}{3}\right), \left(-3, \frac{2}{5}\right), \left(-5\frac{1}{4}, -6\frac{2}{7}\right)\right\}$

Express the relation shown in each table, mapping, or graph as a set of ordered pairs. Then state the domain, range, and inverse of the relation.

20.

x	y
0	4
1	5
2	6
3	6

21.

x	y
6	4
4	−2
3	4
1	−2

22.

x	y
−4	2
−2	0
0	2
2	4

23.

24.

25.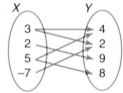

26. temperature of boiled water as it cools

time (minutes)	0	5	10	15	20	25	30
temperature (°C)	100	90	81	73	66	60	55

27. cost of car repairs

time (hours)	0	1	2	3
cost (dollars)	25	50	75	100

28. distance traveled at a rate of 55 mph

time (hours)	1.25	3.75	4.5	5.5	6
distance (miles)	68.75	206.25	247.5	302.5	330

29.

30.

31.

32.

33.

34.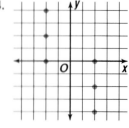

Draw a mapping and a graph for each relation.

35. $\{(8, 2), (4, 2), (8, -9), (7, 5), (-3, 2)\}$

36. $\{(3, 3), (2, 7), (-3, 3), (1, 3), (4, 1)\}$

37. $\{(6, 0), (6, -4), (4, -3), (5, -3)\}$

Graphing Calculator

Graph each relation on a graphing calculator.

a. State the WINDOW settings that you used.

b. Write the coordinates of the inverse. Then graph the inverse.

c. Name the quadrants in which each point of the relation and its inverse lies.

38.

x	y
1992	77
1993	200
1994	550
1995	880

39.

x	y
0	10
2	-8
6	6
9	-4

40.

x	y
-1	18
-2	23
-3	28
-4	33

41. Review your answers to part c of Exercises 38–40. What conclusions can you draw about the quadrant in which the inverse of a point will lie?

Critical Thinking

42. Write a relation with five ordered pairs such that the relation is its own inverse. Then describe the graph of the relation and its inverse.

Applications and Problem Solving

43. Retail Sales The graph at the right shows U.S. retail sales in billions of dollars. The amounts for each month have been adjusted to account for seasonal increases in sales.

a. Approximate the least and greatest values in the range.

b. In your own words, describe what the pattern formed by the points represents.

c. Compare this graph with the one in Exercise 15. How do you think unemployment rates might be compared with retail spending trends? Explain.

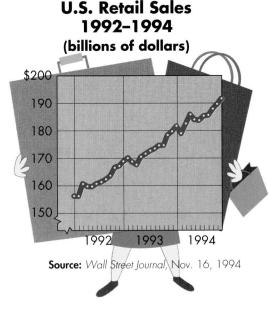

U.S. Retail Sales 1992–1994
(billions of dollars)

Source: *Wall Street Journal,* Nov. 16, 1994

44. Make a Table Mandy Chin is saving money to pay for the first year of her car insurance, which will cost about $1200. She already has $500 in her savings account. She gets a job working at a fast-food restaurant and plans to save $45 per week.

a. Make a table to show the number of weeks and how much money Mandy will have in her account after each week. For example, on Week 1, she will have $545. Graph the results.

b. When will she have enough money to pay her insurance premium?

45. Probability There are 36 different possible outcomes when you roll two dice. The outcomes can be expressed as ordered pairs, such as (4, 3).

   a. Graph all 36 possible outcomes. What is the domain and range?

   b. What is the domain and range of the inverse? What do you notice about the relation and its inverse?

   c. How many different sums are possible when two dice are rolled? Draw a line plot showing how many ways each sum can be rolled. Describe your line plot.

   d. Graph the outcomes that have a sum of 7.

   e. What is the probability of rolling a sum of 7? Explain.

Mixed Review

46. Graph $A(6, 2)$, $B(-3, 6)$, and $C(-5, -4)$ on the same coordinate plane. (Lesson 5–1)

47. If y varies inversely as x, and $y = 8$ when $x = 24$, find y when $x = 6$. (Lesson 4–8)

48. Entertainment At Backintime Cinema, tickets for adults cost $5.75 and tickets for children cost $3.75. How many of each kind of ticket were purchased if 16 tickets were bought for $68? (Lesson 4–7)

49. Use $I = prt$ to find I if $p = \$8000$, $r = 6\%$, and $t = 1$ year. (Lesson 4–4)

50. Solve $E = mc^2$ for m. (Lesson 3–6)

51. Space Science Halley's Comet flashes through the sky every 76.3 years. It last appeared in 1986. In what year of the 23rd century is Halley's Comet expected to reappear? (Lesson 2–6)

52. Agriculture Refer to the graph in Exercise 50 on page 259. (Lesson 1–9)

   a. What is the normal minimum temperature for March?

   b. What happens to the temperature after July?

53. Simplify $4[1 + 4(5x + 2y)]$. (Lesson 1–8)

WORKING ON THE

In·ves·ti·ga·tion

Refer to the Investigation on pages 190–191.

In tracking populations of fish, the Department of Natural Resources and Wildlife often chart their findings in a graph. They use these graphs to monitor the populations of the different types of fish. In this way they track which species are on the increase and which are in decline. This can be especially important in monitoring species that are rare or in danger of extinction.

1 Use the information in your chart to make a table of ordered pairs (x, y) such that x represents the number of the casting and y represents your estimate for the number of fish in the lake.

2 Draw a mapping for this relation.

3 Graph the ordered pairs on a coordinate grid. Describe the graph.

Add the results of your work to your Investigation Folder.

5–3A Graphing Technology Equations

A Preview of Lesson 5–3

Ordered pairs can be used to represent solutions to equations in two variables. Thus, the solution set of an equation for a defined domain is a relation. We can use a graphing calculator to investigate the relation and determine the range when the domain is given.

Example **Solve $y = 2.5x + 4$ if the domain is $\{-8, -6, -4, -2, 0, 2, 4, 6, 8\}$.**

Method 1

Store the elements of the domain as a list of values in L1.

Enter: [2nd] [{] [(−)] 8 [,] [(−)] 6 [,] [(−)] 4 [,] [(−)] 2 [,]

0 [,] 2 [,] 4 [,] 6 [,] 8 [2nd] [}] [STO▸] [2nd]

[L1] [ENTER]   {-8 -6 -4 -2 0 ...

To calculate the range values, compute the right-hand side of the equation, substituting L1 for x.

Enter: 2.5 [2nd] [L1] [+] 4 [ENTER]   {-16 -11 -6 -1 ...

Press the arrow keys to scroll right or left through the display to see all the values in the range.

The solution set is $\{(-8, -16), (-6, -11), (-4, -6), (-2, -1), (0, 4), (2, 9), (4, 14), (6, 19), (8, 24)\}$.

Method 2

Store the equation in the Y= menu of the calculator and view the table of values it creates.

Enter: [Y=] 2.5 [X,T,θ,n] [+] 4 [2nd] [TblSet] [(−)] 8 [ENTER] 2

[2nd] [TABLE]

ΔTbl means the increments in which the x values will appear in the table.

A table with columns X and Y_1 appears on the screen. The values in the Y_1 column represent the corresponding range values for each x value in the domain. This table helps you find the ordered pairs that are solutions for the equation. The solution set is $\{(-8, -16), (-6, -11), (-4, -6), (-2, -1), (0, 4), (2, 9), (4, 14), (6, 19), (8, 24)\}$, which agrees with the list in Method 1.

EXERCISES

Use a graphing calculator to solve each equation if the domain is $\{-3, -2, -1, 0, 1, 2, 3\}$.

1. $y = 4x - 7$ **2.** $y = x^2 + 11$ **3.** $1.2x - y = 6.8$ **4.** $6x + 2y = 12$

5-3

Equations as Relations

What YOU'LL LEARN

- To determine the range for a given domain, and
- to graph the solution set for the given domain.

Why IT'S IMPORTANT

You can use equations to explore the relations in physical science and anatomy.

CONNECTION
Physical Science

Are you as "light" as a feather or as "heavy" as a rock? The *density* of an object is the measure of its "lightness" or "heaviness." Density is defined as the mass per unit of volume. The chart at the right shows the densities of some common substances.

Substance	Density (g/cm³)
air	0.0013
aluminum	2.7
gasoline	0.7
gold	19.3
lead	11.3
mercury	13.5
silver	10.5
steel	7.8
water	1.0

F Y I

Water has a density of 1.0. Any substance with a density less than 1.0 will float in water. Any substance with a density greater than 1.0 will sink in water.

An equation relating density, mass, and volume is $m = DV$, where m is the mass of a substance, D is the density, and V is the volume. For example, the equation for gold is $m = 19.3V$ because the density of gold is 19.3 g/cm³. This equation is called an **equation in two variables,** m and V.

No one knows exactly when gold was first discovered. However, there have been gold artifacts dug up in Ur in Mesopotamia (modern day Iraq) that date as early as 3500 B.C. Suppose we wanted to know the mass of four gold artifacts with volumes of 10 cm³, 50 cm³, 100 cm³, and 150 cm³. We can use a table to find ordered pairs (V, m) that satisfy the equation.

V	D · V	m	Ordered Pair
10	19.3(10)	193	(10, 193)
50	19.3(50)	965	(50, 965)
100	19.3(100)	1930	(100, 1930)
150	19.3(150)	2895	(150, 2895)

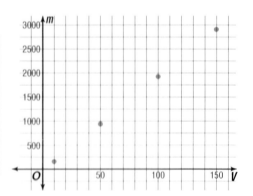

These four ordered pairs are graphed at the right. Since each ordered pair satisfies the equation, each ordered pair is a solution of the equation.

Definition of the Solution of an Equation in Two Variables	If a true statement results when the numbers in an ordered pair are substituted into an equation in two variables, then the ordered pair is a solution of the equation.

Since the solutions of an equation in two variables are ordered pairs, such an equation describes a relation. In an equation involving x and y, the set of x values is the domain of the relation. The set of corresponding y values is the range of the relation.

Example 1 Solve each equation for the given domain values. Graph the solution set.

a. $y = 4x$ if the domain is $\{-3, -2, 0, 1, 2\}$

Make a table. The values of x come from the domain. Substitute each value of x into the equation to determine the corresponding values of y.

Domain x	4x	Range y	Ordered Pair (x, y)
-3	$4(-3)$	-12	$(-3, -12)$
-2	$4(-2)$	-8	$(-2, -8)$
0	$4(0)$	0	$(0, 0)$
1	$4(1)$	4	$(1, 4)$
2	$4(2)$	8	$(2, 8)$

Then graph the solution set $\{(-3, -12), (-2, -8), (0, 0), (1, 4), (2, 8)\}$.

b. $y = x + 6$ if the domain is $\{-4, -3, -1, 2, 4\}$

Make a table. Find the values of y in the range by substituting the values of x from the domain into the equation.

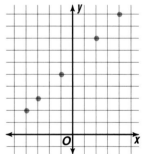

x	x + 6	y	(x, y)
-4	$-4 + 6$	2	$(-4, 2)$
-3	$-3 + 6$	3	$(-3, 3)$
-1	$-1 + 6$	5	$(-1, 5)$
2	$2 + 6$	8	$(2, 8)$
4	$4 + 6$	10	$(4, 10)$

Then graph the solution set $\{(-4, 2), (-3, 3), (-1, 5), (2, 8), (4, 10)\}$.

It is often helpful to solve an equation for y before substituting each domain value into the equation. This makes creating a table of values easier.

Example 2 Solve $8x + 4y = 24$ if the domain is $\{-2, 0, 5, 8\}$.

First solve the equation for y in terms of x.

$8x + 4y = 24$

$\quad 4y = 24 - 8x$ *Subtract 8x from each side.*

$\quad\;\; y = 6 - 2x$ *Divide each side by 4.*

Now substitute each value of x from the domain to determine the corresponding values of y in the range.

x	6 − 2x	y	(x, y)
-2	$6 - 2(-2)$	10	$(-2, 10)$
0	$6 - 2(0)$	6	$(0, 6)$
5	$6 - 2(5)$	-4	$(5, -4)$
8	$6 - 2(8)$	-10	$(8, -10)$

Then graph the solution set $\{(-2, 10), (0, 6), (5, -4), (8, -10)\}$.

You can enter selected *x* values in the TABLE feature of the graphing calculator, and it will calculate the corresponding *y* values for a given equation.

- Solve the equation for *y* and enter the equation into the Y= list.

- Press 2nd TblSet and use the arrow keys to go to Indpnt:. Highlight Ask.

- Press 2nd TABLE . Scroll up to the top of the X list. Enter the first value of the domain and press ENTER . The corresponding Y value appears in the second column. Enter the other values of the domain and record the range values.

Your Turn

a. Use a graphing calculator to find the corresponding *y* values for $y = -5x + 12$ if $x = \{-4, -2, 0, 1, 3, 5\}$. Write the solutions as ordered pairs.

b. Why do you think the Ask feature on the TABLE menu is helpful in finding the ordered pairs that are solutions for equations?

LOOK BACK

You can look back to Lesson 1-9 to review independent and dependent variables.

Variables other than *x* and *y* are often used in equations representing real situations. The domain contains values represented by the *independent variable*. Graph the domain values on the horizontal axis. The range contains the corresponding values represented by the *dependent variable*, determined by the given equation. Graph the range values on the vertical axis.

When you solve an equation for a given variable, that variable becomes the dependent variable. That is, its value depends upon the domain values chosen for the other variable.

Example ③

INTEGRATION

Geometry

The equation for the perimeter of a rectangle is $2w + 2\ell = P$. Suppose the perimeter of a rectangle is 24 centimeters.

a. Solve the equation for ℓ.

b. State the independent and dependent variables and determine the domain and range values for which this equation makes sense.

c. Choose five values for *w* and find the corresponding values of ℓ.

a. First substitute 24 for *P* in the equation. Then solve for ℓ.

$$2w + 2\ell = P$$
$$2w + 2\ell = 24$$
$$2\ell = 24 - 2w \qquad \textit{Subtract 2w from each side.}$$
$$\frac{2\ell}{2} = \frac{24 - 2w}{2} \qquad \textit{Divide each side by 2.}$$
$$\ell = 12 - w$$

b. Since the value of ℓ depends on the value of *w*, ℓ is the dependent variable and *w* is the independent variable. Since distance can only be positive, the values in the domain and range must both be greater than zero.

(continued on the next page)

c. When choosing values for w, you can choose any number greater than zero. Suppose we choose the domain $\{1, 2, 3, 4, 5\}$. Make a table of values.

w	12 − w	ℓ	(ℓ, w)
1	12 − 1	11	(1, 11)
2	12 − 2	10	(2, 10)
3	12 − 3	9	(3, 9)
4	12 − 4	8	(4, 8)
5	12 − 5	7	(5, 7)

The solution set for the domain we chose is $\{(1, 11), (2, 10), (3, 9), (4, 8), (5, 7)\}$. *Check to see if each solution satisfies the equation $2w + 2\ell = 24$.*

CHECK FOR UNDERSTANDING

Communicating Mathematics

Study the lesson. Then complete the following.

1. Refer to the connection at the beginning of the lesson.
   a. Which of the substances listed will float on water? on mercury?
   b. State the domain and range of the graphed relation.

2. **Show** why $(1, -1)$ is or is not a solution of $x + 2y = 3$.

3. **Determine** which of the points shown in the graph at the right represent ordered pairs that are solutions for $y = 2x + 1$.

4. **Explain** why, in Example 3, you only consider values greater than zero.

5. Why should you solve for y before finding ordered pairs that are solutions for an equation?

6. In your own words, explain the difference between an independent variable and a dependent variable and how you know on which axis to graph each of them.

Guided Practice

Solve each equation for the domain given in the table.

7. $y = 2x + 3$

x	y	(x, y)
−2		
−1		
0		
1		
2		
3		

8. $a = \dfrac{3b - 5}{2}$

b	a	(b, a)
−5		
−2		
0		
2		
5		

Which ordered pairs are solutions of each equation?

9. $1 + 5y = 2x$ **a.** $(-7, -3)$ **b.** $(7, 3)$ **c.** $(2, 1)$ **d.** $(-2, -1)$

10. $11 - 2y = 3x$ **a.** $(1, 3)$ **b.** $(3, 1)$ **c.** $(5, -2)$ **d.** $(-1, 4)$

11. Find the solution set for $x + 2y = 14$ if the domain is $\{-2, -1, 0, 1, 2\}$. Then graph the solution set.

12. **Geometry** The formula for the area of a rectangle is $A = \ell w$. Suppose the area of a rectangle is 36 square meters.

    a. Solve the equation for ℓ.

    b. Choose five values for w and find the corresponding values of ℓ.

EXERCISES

Practice

Which ordered pairs are solutions of each equation?

13. $3a + b = 8$ **a.** $(4, -4)$ **b.** $(8, 0)$ **c.** $(2, 2)$ **d.** $(3, 1)$

14. $y = 3x$ **a.** $(6, 2)$ **b.** $(-2, -6)$ **c.** $(0, 0)$ **d.** $(-15, -5)$

15. $3a - 8b = -4$ **a.** $(0, 0.5)$ **b.** $(4, 2)$ **c.** $(2, 0.75)$ **d.** $(2, 4)$

16. $x = 3y - 7$ **a.** $(2, -1)$ **b.** $(2, 4)$ **c.** $(-1, 2)$ **d.** $(2, 3)$

17. $2y + 4x = 8$ **a.** $(0, 2)$ **b.** $(-3, 0.5)$ **c.** $(2, 0)$ **d.** $(-2, 1)$

18. $3x + 3y = 0$ **a.** $(1, -1)$ **b.** $(2, -2)$ **c.** $(-1, 1)$ **d.** $(-2, 2)$

Solve each equation if the domain is $\{-3, -2, 0, 3, 6\}$.

19. $y = 4x$ 20. $y = 3x - 1$ 21. $2x + 2y = 14$

22. $x = 4 + y$ 23. $3x = 13 - 2y$ 24. $y = 4 - 5x$

25. $5x + 3 = y$ 26. $5x - 10y = 40$ 27. $2x + 5y = 3$

Make a table and graph the solution set for each equation and domain.

28. $y = 3x$ for $x = \{-3, -2, -1, 0, 1, 2, 3\}$

29. $y = 2x + 1$ for $x = \{-5, -3, 0, 1, 3\ 6\}$

30. $3x - 2y = 5$ for $x = \{-3, -1, 2, 4, 7\}$

31. $5x = 8 - 4y$ for $x = \{-2, -1, 0, 1, 3, 4, 5\}$

32. **Geometry** A regular pentagon has five angles that have equal measures. Suppose the measure of one angle is $(a + b)°$. The sum of the measures of the angles of any pentagon is $540°$.

    a. Write an equation for the sum of the angle measures of a regular pentagon.

    b. Solve the equation for a.

    c. Choose five values for b and find the corresponding values of a.

33. **Geometry** Two angles that are supplementary have measures whose sum is $180°$. Suppose the measures of two supplementary angles are $(x + y)°$ and $(2x + 3y)°$.

    a. Write an equation for the sum of the measures of these two angles.

    b. Solve the equation for y.

    c. Choose five values for y and find the corresponding values of x.

Find the domain for each equation if the range is $\{-2, -1, 0, 2, 3\}$.

34. $y = x + 7$ 35. $y = 3x$

36. $6x - y = -3$ 37. $5y = 8 - 4x$

Make a table and graph the solution set for each equation.

38. $y = x^2 - 3x - 10$ if the domain is $\{-3, -1, 0, 1, 3, 5\}$

39. $y = x^3$ if the domain is $\{-2, -1, 0, 1, 2\}$

40. $y = 3^x$ if the domain is $\{1, 2, 3, 4\}$

Graphing Calculator

Use a graphing calculator to find the solution set for each domain.

41. $y = 1.4x - 0.76$ for $x = \{-2.5, -1.75, 0, 1.25, 3.33\}$

42. $y = 3.5x + 12$ for $x = \{-125, -37, -6, 12, 57, 150\}$

43. $3.6y + 12x = 60$ for $x = \{-100, -30, 0, 120, 360, 720\}$

44. $75y + 25x = 100$ for $x = \{-10, -5, 0, 5, 10, 15\}$

Critical Thinking

45. Find the domain values of each relation if the range is $\{0, 16, 36\}$.

    a. $y = x^2$ **b.** $y = |4x| - 16$ **c.** $y = |4x - 16|$

46. Compare the graphs you drew in Exercises 38–40 with the other graphs in this lesson.

    a. What is different about the pattern of the points for these relations?

    b. Study the equations associated with the graphs. How do you think you could predict the pattern of the points by looking at the equation?

Applications and Problem Solving

47. Physical Science Refer to the connection at the beginning of the lesson. Suppose you have an unknown substance with a mass of 378 grams and a volume of 36 cm^3 and a second unknown substance with a mass of 87.5 grams and a volume of 125 cm^3.

    a. Solve the formula $m = DV$ for D.

    b. Identify the unknown substances.

48. Demographics The time T (in years) needed for a population to double is calculated using the formula $T = \frac{70}{R}$, where R represents the percent growth rate of the population.

    a. Determine the domain and range values for which this equation makes sense.

    b. Use the formula to approximate the doubling time to the nearest year for the continents shown in the graph at the left.

    c. The population of the United States in 1990 was 248,709,873. If the U.S. maintains the same growth rate as North America, what will be the population of the U.S. in the year 2000? in 2010? in 2020?

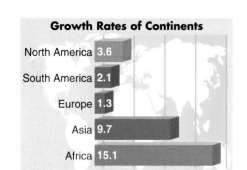

Growth Rates of Continents

North America 3.6
South America 2.1
Europe 1.3
Asia 9.7
Africa 15.1

0   4   8   12   16
Percent

49. Anatomy The formula for relating the shoe size S and the length of a woman's foot in inches L is $S = 3L - 22$. The formula for relating a man's shoe size with his foot length is $S = 3L - 26$. Copy and complete each table below to determine shoe sizes given lengths of feet.

Women		
L	S	(L, S)
$9\frac{1}{3}$		
$9\frac{5}{6}$		
$10\frac{1}{6}$		
$10\frac{2}{3}$		

Men		
L	S	(L, S)
$11\frac{1}{3}$		
$11\frac{5}{6}$		
$12\frac{1}{3}$		
$12\frac{5}{6}$		

50. **Make a Table** Winona Brownsman is doing small jobs for her rich aunt. Her aunt has asked Winona how she wishes to be paid for her services. Winona has two choices.

- *Plan A:* Winona can be paid one dollar the first day of the month and an additional dollar for each day for a month. For example, she will be paid $1 on Day 1, $2 on Day 2, $3 on Day 3, and so on.

- *Plan B:* Winona will be paid one cent on the first day and for each succeeding day, double the amount of the previous day. For example she will be paid 1¢ on Day 1, 2¢ on Day 2, 4¢ on Day 3, 8¢ on Day 4, and so on.

a. Use two tables to determine which plan would be more profitable.

b. On which day will the better plan become greater than the other?

Mixed Review

51. **Basketball** The table at the right shows the number of years that specific NBA basketball players have been playing and the average number of points each player made during the 1993–1994 season. (Lesson 5–2)

Player	Years	Points
Charles Barkley	10	21.6
Glen Rice	5	21.1
Chris Mullin	9	16.8
Jamal Mashburn	1	19.2
Scottie Pippen	7	22.0
Dominique Wilkins	12	26.0

Source: *Hawes Fantasy Basketball Guide, 1994–1995*

a. Plot the points and describe the graph.

b. What do you think the graph suggests about the relationship of the number of years played and the point average?

52. Graph each point below. Then connect the points in alphabetical order and identify the figure. (Lesson 5–1)

$$A(0, 5), B(4, -3), C(-5, 2), D(5, 2), E(-4, -3), F(0, 5)$$

53. **Employment** Hugo's wages vary directly with the time he works. If his wages for 4 days are $110, how much will they be for 17 days? (Lesson 4–8)

54. What number decreased by 80% is 14? (Lesson 4–5)

55. **Finance** If Li Fong had earned one fourth of a percent more in annual interest on an investment, the interest for one year would have been $45 greater. How much did he invest at the beginning of the year? (Lesson 4–4)

56. Solve $\frac{1}{3}a - 2b = -9c$ for a. (Lesson 3–6)

57. **Geometry** The perimeter of a rectangle is 148 inches.

a. Write an equation to represent this situation.

b. Find its dimensions if the length is 17 inches greater than three times the width. (Lesson 3–3)

58. Find a number between $-\frac{8}{17}$ and $\frac{1}{9}$. (Lesson 2–4)

59. Evaluate $3x^3 - 2y$ if $x = 0.2$ and $y = 4$. (Lesson 1–3)

5–4A Graphing Technology
Linear Relations and Functions

A Preview of Lesson 5–4

The graphing calculator is a powerful tool that can be used to study a wide variety of graphs. In many cases, equations are graphed in the standard viewing window, [−10, 10] by [−10, 10]. To set the standard viewing window, press [ZOOM] 6. You do not need to press [GRAPH] after using [ZOOM] 6.

The examples below illustrate how you can use the calculator to graph linear relations and functions.

Example **Graph 2*y* − 6*x* = 8 in the standard viewing window.**

First solve the equation for *y*.

$$2y - 6x = 8$$
$$2y = 6x + 8$$
$$y = 3x + 4$$

Note that the graph of this equation is a line.

Then enter the equation into the calculator, set the standard viewing window, and graph.

Enter: [Y=] 3 [X,T,θ,*n*] [+] 4   *Enters the equation.*

       [ZOOM] 6   *Sets the standard window.*

Example **Graph *y* = −*x* + 20 in the standard viewing window.**

Enter: [Y=] [(−)] [X,T,θ,*n*] [+] 20

       [ZOOM] 6

An equation whose graph is a non-vertical line is called a linear function.

What happened? None of the graph is shown in the standard viewing window. Pressing the WINDOW key and changing the viewing window settings to [−5, 25] by [−5, 25] with scale settings of 5 for each axis will allow you to see where the graph crosses both the *x*- and *y*-axes.

When the equation in Example 2 is graphed in the [−5, 25] by [−5, 25] viewing window, it is considered a **complete graph.** A complete graph shows all of the important characteristics of the graph on the screen. These include the origin and the points at which the graph crosses the *x*- and *y*-axes.

Example ③ **Use a graphing calculator to draw a graph that represents the solutions to this problem. Then list three of the solutions.**

A second number is two more than the opposite of the first.

Let x represent the first number, and let y represent the second number. The equation $y = -x + 2$ describes the relationship between the numbers for any first number x. Select the standard viewing window, enter the equation, and press $\boxed{\text{GRAPH}}$ to view the graph of the equation. A complete graph is displayed.

You can find sample solutions for the equation in two ways.

Method 1

Press the $\boxed{\text{TRACE}}$ key. The cursor appears as a flashing square and the approximate coordinates of the location of the cursor appears at the bottom of the screen. Use the right and left arrows to move the cursor along the line. New coordinates appear for each location of the cursor. These ordered pairs are approximate solutions for the equation. Sample solutions are $(0, 2)$, $(0.42553191, 1.5744681)$, and $(-0.6382979, 2.6382979)$.

Method 2

You can obtain exact solutions by pressing $\boxed{\text{2nd}}$ $\boxed{\text{TABLE}}$. A table of x and y values for the equation appears.

Use the up and down arrow keys to scroll through the list of values. Sample solutions are $(-3, 5)$, $(0, 2)$, and $(3, -1)$.

EXERCISES

Use a graphing calculator to graph each equation in the standard viewing window. Sketch each graph on a sheet of paper.

1. $y = 3x - 6$
2. $y = 0.5x - 4$
3. $2x - 3y = 5$
4. $5x + y = 8$
5. $2y - 2x = 3$
6. $-10x - 2y = -14$

Graph each linear function.
a. **Determine a viewing window that shows a complete graph.**
b. **Sketch the graph on a sheet of paper, noting the scale used for each axis.**
c. **Name three solutions for the equation.**

7. $0.1x + y = 10$
8. $y = 3x + 17$
9. $y = 0.01x + 0.02$
10. $200x + y = 150$

Graphing Linear Equations

CONNECTION
Health

The manner in which you burn Calories depends on your weight and the activity you are doing. The chart at the right shows the number of Calories burned per minute per kilogram (C/min/kg).

Manuel weighs 70 kilograms and wants to know how many Calories he burns playing football. The formula $C = wtr$, where w is his weight in kilograms, t is the time in minutes, and r is C/min/kg, can be used to determine how many Calories he burns.

$$C = wtr$$
$$C = 70 \cdot t \cdot 0.132$$
$$C = 9.24t$$

The number of Calories burned is dependent upon how long he exercises. So, t is the independent variable and C is the dependent variable.

Activity	C/min/kg
Basketball	0.138
Cycling (leisure)	0.064
Cycling (racing)	0.169
Dancing (aerobic)	0.135
Dancing (normal)	0.075
Drawing	0.036
Eating	0.023
Football	0.132
Free weights	0.086
Golf	0.085
Gymnastics	0.066
Jumping rope	0.162
Playing drums	0.066
Playing flute	0.035
Playing horn	0.029
Playing piano	0.040
Playing trumpet	0.031
Nautilus® training	0.092
Running (7.2 min/km)	0.135
Running (5.0 min/km)	0.208
Running (3.7 min/km)	0.252
Sitting Quietly	0.021
Swimming	0.156
Walking	0.080
Writing	0.029
Word processing	0.027

t	$9.24t$	C	(t, C)
0	9.24(0)	0	(0, 0)
10	9.24(10)	92.4	(10, 92.4)
20	9.24(20)	184.8	(20, 184.8)
30	9.24(30)	277.2	(30, 277.2)
40	9.24(40)	369.6	(40, 369.6)

When you graph the ordered pairs, a pattern begins to form. The points seem to lie in a line.

What is the range for this relation?

Suppose the domain of $C = 9.24t$ is the set of positive real numbers. There would be an infinite number of ordered pairs that are solutions for the equation. If you graphed all the solutions, they would form a line. The line shown in the graph at the right represents all the solutions for $C = 9.24t$.

Since the graph of $C = 9.24t$ is a line, $C = 9.24t$ is called a **linear equation.**

Linear equations may contain one or two variables with no variable having an exponent other than 1.

Definition of a Linear Equation in Standard Form	A linear equation is an equation that can be written in the form $Ax + By = C$, where A, B, and C are any real numbers, and A and B are not both zero.

Example **1** Determine whether each equation is a linear equation. If so, identify A, B, and C.

a. **$4x = 7 + 2y$**

First, rewrite the equation so that both variables are on the same side of the equation.

$$4x = 7 + 2y$$

$4x - 2y = 7$ *Subtract 2y from each side.*

The equation is now in the form $Ax + By = C$, where $A = 4$, $B = -2$, and $C = 7$. This is a linear equation.

b. **$2x^2 - y = 7$**

The exponents of the variables in a linear equation must be 1. Since the exponent of x is 2, this is not a linear equation.

c. **$x = 12$**

This equation can be rewritten as $x + 0y = 12$. Therefore, it is a linear equation in the form $Ax + By = C$, where $A = 1$, $B = 0$, and $C = 12$.

To graph a linear equation, it is often helpful to make a table of ordered pairs that satisfy the equation. Then graph the ordered pairs and connect the points with a line.

Example **2** Graph each equation.

a. **$y = 8x - 4$**

Select five values for the domain and make a table. Then graph the ordered pairs and connect them to draw the line.

x	$8x - 4$	y	(x, y)
-2	$8(-2) - 4$	-20	$(-2, -20)$
-1	$8(-1) - 4$	-12	$(-1, -12)$
0	$8(0) - 4$	-4	$(0, -4)$
1	$8(1) - 4$	4	$(1, 4)$
2	$8(2) - 4$	12	$(2, 12)$

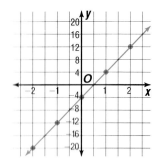

From the graph, you can find other ordered pairs that are solutions, for example, (2.5, 16).

The range of this relation is the set of real numbers.

(continued on the next page)

b. $2x + 5y = 10$

In order to find values for y more easily, solve the equation for y.

$2x + 5y = 10$

$\qquad 5y = 10 - 2x$ *Subtract 2x from each side.*

$\qquad y = \dfrac{10 - 2x}{5}$ *Divide each side by 5.*

Now make a table and draw the graph.

x	$\dfrac{10 - 2x}{5}$	y	(x, y)
-10	$\dfrac{10 - 2(-10)}{5}$	6	$(-10, 6)$
-5	$\dfrac{10 - 2(-5)}{5}$	4	$(-5, 4)$
0	$\dfrac{10 - 2(0)}{5}$	2	$(0, 2)$
5	$\dfrac{10 - 2(5)}{5}$	0	$(5, 0)$
10	$\dfrac{10 - 2(10)}{5}$	-2	$(10, -2)$

Using graphs is a good way to make comparisons in real-life situations.

Example ③

CONNECTION
Health

Refer to the connection at the beginning of the lesson. Carmen Delgado is running in a cross-country meet at the rate of 5 min/km. Her weight is 50 kg. How does the number of Calories she is burning compare to those that she would be burning if she were at home word processing her research paper or if she were playing golf with her dad?

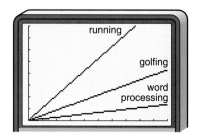

Explore   Look up the Calories burned for each of the three types of activities in the chart on page 280.

TECHNOLOGY *Tip*

Graphing calculators are helpful when comparing equations that are not easily graphed with pencil and paper.

Plan   Write the equation for each activity using the formula $C = wtr$ and 50 for w.

Running	$C = 50(0.208)t$ or $C = 10.4t$
Word processing	$C = 50(0.027)t$ or $C = 1.35t$
Golfing	$C = 50(0.085)t$ or $C = 4.25t$

Solve   Use a graphing calculator to graph all three equations. Enter the equations as Y_1, Y_2, and Y_3. Let x represent the t in each equation. Make a sketch of each graph and label each with its activity.

For each value of t, Carmen burns many more Calories running than she does golfing or word processing. Carmen would burn more Calories golfing than she would word processing.

Examine   Think about the physical movement required for each activity and how much energy that would take. Does the result make sense?

CHECK FOR UNDERSTANDING

Communicating Mathematics

Study the lesson. Then complete the following.

1. **Explain** why the ordered pair $(-1, 3)$ is a solution of $y = -2x + 1$.

2. **Describe** the graph of a linear equation in the form $Ax + By = C$ for which
   a. $A = 0$ **b.** $B = 0$ **c.** $C = 0$

3. **Show** how the graph of $y = 2x + 1$ for the domain $\{-1, 0, 2, 3\}$ differs from the graph of $y = 2x + 1$ for the domain of all real numbers.

4. **Explain** why the x and y values used to set the viewing window in Example 3 do not include negative numbers.

 MATH JOURNAL

5. **Assess Yourself**   During the next 48 hours, record three activities that you perform and the length of each activity. Determine how many Calories you burned in each activity. In which activity did you burn the most Calories? Explain why you think this activity burned the most Calories.

Guided Practice

Determine whether each equation is a linear equation. If an equation is linear, rewrite it in the form $Ax + By = C$.

6. $3x - 5y = 0$ 7. $2x = 6 - y$

8. $3x^2 + 3y = 4$ 9. $3x = 7 - 2y$

Graph each equation.

10. $3x + y = 4$ 11. $4x + 3y = 12$ 12. $\frac{1}{2}x = 8 - y$

13. $x = 6$ 14. $y = -5$ 15. $x - y = 0$

EXERCISES

Practice

Determine whether each equation is a linear equation. If an equation is linear, rewrite it in the form $Ax + By = C$.

16. $\frac{3}{x} + \frac{4}{y} = 2$ 17. $\frac{3}{5}x - \frac{2}{3}y = 5$

18. $x + y^2 = 25$ 19. $x + \frac{1}{y} = 7$

20. $3y + 2 = 0$ 21. $2y = 3x - 8$

22. $5x - 7 = 0$ 23. $2x + 5x = 7y$

24. $4x^2 - 3x = y$ 25. $3m = 2n$

26. $\frac{x}{2} = 10 + \frac{2y}{3}$ 27. $8a - 7b = 2a - 5$

Graph each equation.

28. $x + 6 = -5$ **29.** $y = 3x + 1$ **30.** $6x + 7 = -14y$

31. $2x + 7y = 9$ **32.** $y + 3 = 4$ **33.** $x - 6 = -\frac{1}{3}y$

34. $8x - y = 16$ **35.** $3x + 3y = 12$ **36.** $6x = 24 - 6y$

37. $x - \frac{7}{2} = 0$ **38.** $3x - 4y = 60$ **39.** $4x - \frac{3}{8}y = 1$

40. $2.5x + 5y = 7.5$ **41.** $x + 5y = 16$ **42.** $y + 0.25 = 2$

43. $\frac{4x}{3} = \frac{3y}{4} + 1$ **44.** $y + \frac{1}{3} = \frac{1}{4}x - 3$ **45.** $\frac{3x}{4} + \frac{y}{2} = 6$

Each table below represents points on a linear graph. Copy and complete each table.

46.

x	y
0	?
1	5
2	6
3	7
4	8
5	?

47.

x	y
10	?
5	-2.5
0	0
-5	2.5
-10	5
-15	?

48.

x	y
0	0
3	6
6	12
9	?
12	?
15	?

49.

x	y
-6	5
-4	?
-2	7
0	8
2	?
4	?

Graphing Calculator

Graph each equation using a graphing calculator. Determine a WINDOW setting to use so that a complete graph is shown. Make a sketch of each graph, noting the scale used on each axis.

50. $y = 2x + 4$ **51.** $4x - 9y = 45$ **52.** $27x + 75y = 100$

53. $17y = 22$ **54.** $0.2x - 9.7y = 8.9$ **55.** $\frac{1}{2}x - \frac{2}{3}y = 10$

56. Clear the screen. Press $\boxed{\text{2nd}}$ $\boxed{\text{DRAW}}$ 4 3 $\boxed{\text{ENTER}}$. Describe the graph. Explain how you could use this approach to graph $x = -25$.

Critical Thinking

57. The graphs of each group of equations form a family of graphs. Graph each family on the same coordinate plane. Then write an explanation of the similarities and differences that exist in the graphs of each family.

a. $y = 2x$ $y = 2x + 5$ $y = 2x - 9$ $y = 2x + 14$

b. $y = -3x$ $y = -3x + 4$ $y = -3x - 10$ $y = -3x + 7$

Applications and Problem Solving

58. Science As a thunderstorm approaches, you see lightning as it occurs, but you hear the accompanying sound of the thunder a short time afterward. The distance y, in miles, that sound travels in t seconds is given by the equation $y = 0.21t$.

  a. Determine the domain and range values for which this equation makes sense.

  b. Graph the equation.

  c. Use the graph to estimate how long it will take you to hear the thunder from a storm that is 3 miles away.

59. Employment Elva Duran works as a sales representative for Quasar Electronics. She receives a salary of $1800 per month plus a 6% commission on monthly sales over her target. Her sales in July will be $800, $1300, or $2000 over target, depending on when her orders are filled by the company's distribution center.

   a. Graph ordered pairs that represent each of her possible incomes y for the three sales figures x.

   b. Will her total income for July be more than $1850 if the equation that represents her income is $y = 1800 + 0.06x$? Explain.

60. Weight Training Hector Farentez, who weighs 68.2 kg, wanted to increase his body size by beginning a weight-training program. A local fitness club suggests that he begin his program using Nautilus® equipment before he attempts the free weights. The Nautilus® equipment includes various machines, each of which exercises a certain group of muscles. After several weeks of training, Hector has expanded his workout time from 30 minutes to 1 hour.

   a. Use the information in the table at the beginning of the lesson to determine a formula for determining how many Calories Hector is burning. Graph the equation showing each 5-minute interval.

   b. Later in his training at a weight of 72.3kg, Hector plans to spend one-half hour on Nautilus® and one-half hour on free weights. Will he use more or fewer Calories in this type of workout than an hour of Nautilus®? Explain.

Mixed Review

61. Solve $8x + 2y = 6$ for y. (Lesson 5–3)

62. Draw a mapping for the relation $\{(2, 7), (-3, 7), (2, 4)\}$. (Lesson 5–2)

63. Geometry (Lesson 5–1)

   a. Graph (3, 2), (6, 2), and (6, 5).

   b. If these three points represent the vertices of a square, what would be the coordinates of the fourth point?

   c. Assume that each unit on each axis represents one inch. What is the perimeter of the square?

64. Probability A gumball machine has 24 cherry gumballs, 5 apple gumballs, 18 grape gumballs, 14 orange gumballs, and 9 licorice gumballs. What is the probability of getting an orange gumball? (Lesson 4–6)

65. Jimmy got a discount of $4.50 on a new radio. The discounted price was $24.65. What was the percent of discount to the nearest percent? (Lesson 4–5)

66. **Running** In cross-country meets, the team score is determined by totaling the finish numbers of the five runners on the team. One member of a cross-country team placed fourth, which means a score of 4 points for the team. The next four finishers on the team placed in consecutive order, but farther behind. The team score was 70. In what places did the other members finish? (Lesson 3–5)

67. **Miniature Golf** In miniature golf, balls bounce off the walls of the course at the same angle at which they hit. If a ball strikes the wall at 30°, what is the measure of the angle between the two paths of the ball? (Lesson 3–4)

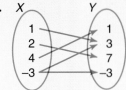

30°   30°

68. Two meters of copper tubing weigh 0.25 kilograms. How much do 50 meters of the same tubing weigh? (Lesson 3–3)

69. Simplify $\dfrac{\frac{-3}{4}}{-36}$. (Lesson 2–7)

70. Find $-21 + 52$. (Lesson 2–3)

SELF TEST

1. Graph $A(-6, 6)$, $B(3, -5)$, $C(0, 2)$, and $D(-3, -1)$. (Lesson 5–1)

Express each relation as a set of ordered pairs. Then state the domain, the range, and the inverse of the relation. (Lesson 5–2)

2.

3.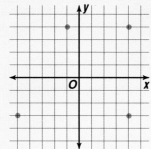

Solve each equation if the domain is $\{-2, -1, 0, 1, 3\}$. Then graph the solution set.
(Lesson 5–3)

4. $3y + 6x = 12$ for y 5. $2a + 3b = 9$ for b 6. $5r = 8 - 4s$ for s

Graph each equation. (Lesson 5–4)

7. $y = x - 1$ 8. $y = 2x - 1$ 9. $3x + 2y = 4$

10. **Geometry** The formula for the area of a trapezoid is $A = \frac{1}{2}h(b_1 + b_2)$, where b_1 and b_2 are the lengths of the bases and h is the height. Find the height of the trapezoid shown at the right if its area is 40 m². (Lesson 5–3)

5 m

h

10 m

Functions

What YOU'LL LEARN

- To determine whether a given relation is a function, and
- to find the value of a function for a given element of the domain.

Why IT'S IMPORTANT

You can use functions to solve problems involving accounting, Earth science, and health.

APPLICATION
Air Travel

During certain times of the year, airlines offer lower rates to fly to selected cities. The advertisement below shows discount ticket fares. The mileage between the cities is also listed for those people who are enrolled in frequent flier programs.

Plan Your Vacation Now!

From	To	Regular Fare	Discount Fare	Mileage
Atlanta	New York	$ 89	$ 79	892
Baltimore	Houston	149	129	1454
Boston	Greensboro	109	79	786
Cleveland	Philadelphia	94	79	441
Dayton	Houston	109	99	1178
Greensboro	Miami	84	79	814
Houston	Jacksonville	104	99	875
Houston	Miami	109	99	1231
Los Angeles	San Antonio	139	129	1277

Sample sale prices. Coach class each way. Round-trip purchase required. Seats are limited and may not be available on all flights/days.

Suppose we let r represent the regular fare, d the discount fair, and m the mileage. The relation whose ordered pairs are of the form (r, d) is graphed below on the left. The relation whose ordered pairs are of the form (m, d) is graphed below on the right.

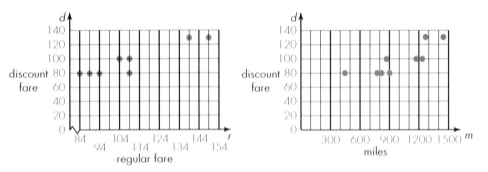

Notice that in the graph on the left, when $r = 109$, there is more than one value of d, 79 and 99. However, in the other graph, there is exactly one value of d for each value of m. Relations with this characteristic are called **functions**.

Definition of Function	**A function is a relation in which each element of the domain is paired with *exactly* one element of the range.**

Example ❶ **Determine whether each relation is a function. Explain your answer.**

a. {(2, 3), (3, 0), (5, 2), (−1, −2), (4, 1)}

Since each element of the domain is paired with exactly one element of the range, this relation is a function.

b.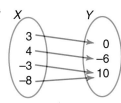

This mapping represents a function since, for each element of the domain, there is *only one* corresponding element in the range.

It does not matter if two elements of the domain are paired with the same element in the range.

c.

x	y
4	−1
5	2
5	3
6	6
−1	1

This table represents a relation that is not a function. The element 5, in the domain, is paired with both 2 and 3 in the range.

There are several ways to determine whether an equation represents a function.

Example ❷ **Determine whether $x − 4y = 12$ is a function.**

Method 1: Make a table of solutions.

First solve for y.

$x − 4y = 12$

$\quad −4y = −x + 12$ *Subtract x from each side.*

$\quad\quad y = \frac{1}{4}x − 3$ *Divide each side by −4.*

Next make a table of values like the one at the right.

It appears that for any given value of x, there is only one value for y that will satisfy the equation. Therefore, the equation $x − 4y = 12$ is a function.

x	y
−8	−5
−4	−4
−2	−3.5
0	−3
2	−2.5
4	−2
8	−1

Method 2: Graph the equation.

Since the equation is in the form $Ax + By = C$, the graph of the equation will be a line. Graph the ordered pairs from Method 1 and connect them with a line.

Now place your pencil at the left of the graph to represent a vertical line. Slowly move the pencil to the right across the graph.

For each value of x, this vertical line passes through no more than one point on the graph. Thus, the line represents a function.

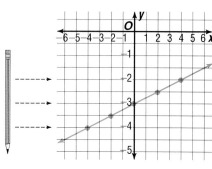

Using a pencil to see if a graph represents a function is one way to perform the **vertical line test.**

Vertical Line Test for a Function	If any vertical line passes through no more than one point of the graph of a relation, then the relation is a function.

Example ③ **Use the vertical line test to determine if each relation is a function.**

a.   b.   c.

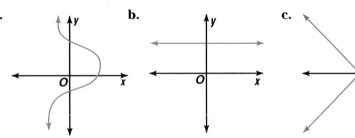

Graph b is the only relation to pass the vertical line test. Thus, it is the only function. With graphs a and c, a vertical line intersects the graph in more than one point. Thus, they are *not* functions.

Letters other than f are also used for names of functions. For example, g(x) and h(x) are sometimes used.

The ordered pair (4, f(4)) is a solution of the function f.

Equations that are functions can be written in a form called **functional notation.** For example, consider the equation $y = 3x - 7$.

equation	**functional notation**
$y = 3x - 7$	$f(x) = 3x - 7$
	The symbol f(x) is read "f of x."

In a function, x represents the elements of the domain and $f(x)$ represents the elements of the range. Suppose you want to find the value in the range that corresponds to the element 4 in the domain. This is written $f(4)$ and is read "f of 4." The value of $f(4)$ is found by substituting 4 for x in the equation. So, $f(4) = 3(4) - 7$ or 5.

Example ④ **If $f(x) = 2x - 9$, find each value.**

a. $f(6)$

$$f(6) = 2(6) - 9$$
$$= 12 - 9$$
$$= 3$$

b. $f(-2)$

$$f(-2) = 2(-2) - 9$$
$$= -4 - 9$$
$$= -13$$

c. $f(k + 1)$

$$f(k + 1) = 2(k + 1) - 9$$
$$= 2k + 2 - 9$$
$$= 2k - 7$$

The functions we have studied thus far have been linear functions. Many functions are not linear. However, you can find values of functions in the same way.

Example ⑤ **If $h(z) = z^2 - 4z + 9$, find each value.**

a. $h(-3)$

$$h(-3) = (-3)^2 - 4(-3) + 9$$
$$= 9 + 12 + 9 \text{ or } 30$$

(continued on the next page)

b. $h(5c)$

$h(5c) = (5c)^2 - 4(5c) + 9$ *Substitute 5c for z.*

$= 5 \cdot 5 \cdot c \cdot c - 4 \cdot 5 \cdot c + 9$

$= 25c^2 - 20c + 9$

c. $5[h(c)]$

$5[h(c)] = 5[(c)^2 - 4(c) + 9]$ *5[h(c)] means 5 times the value of h(c).*

$= 5 \cdot c^2 - 5 \cdot 4c + 5 \cdot 9$ *Distributive property*

$= 5c^2 - 20c + 45$

Notice that when comparing parts b and c, $5[h(c)] \neq h(5c)$.

Functions are often used in solving real-life problems.

Example 6

CONNECTION

Health

The normal systolic blood pressure **S** is a function of the age **a** of the individual. That is, a person's normal blood pressure depends on how old the person is. To determine the normal systolic blood pressure of an individual, you can use the equation $S = 0.5a + 110$, where **a** represents age in years.

a. Write the equation in functional notation.

b. Find $S(10)$, $S(30)$, $S(50)$, and $S(70)$.

c. Graph the function. Name the independent and dependent quantities.

d. Use the graph of the function to estimate whether blood pressure increases or decreases with age. Then estimate the blood pressure of an 80-year-old person.

**Best Selling
Prescription Drugs
in the World**

Brand Name	Used for
1. Zantac	ulcers
2. Vasotec	hypertension*
3. Capoten	hypertension*
4. Voltaren	arthritis
5. Tenormin	hypertension*

*high blood pressure

a. Let $S(a)$ represent the function. The equation becomes $S(a) = 0.5a + 110$.

b. Make a table with 10, 30, 50, and 70 as values of a.

a	$S(a) = 0.5a + 110$	$S(a)$	$(a, S(a))$
10	$S(10) = 0.5(10) + 110$	115	(10, 115)
30	$S(30) = 0.5(30) + 110$	125	(30, 125)
50	$S(50) = 0.5(50) + 110$	135	(50, 135)
70	$S(70) = 0.5(70) + 110$	145	(70, 145)

c. Use the ordered pairs from the table to graph the function. Age is the independent quantity, and systolic blood pressure is the dependent quantity.

d. The graph indicates that as you age, your blood pressure is expected to rise. An 80-year-old person would expect to have a systolic pressure of about 150.

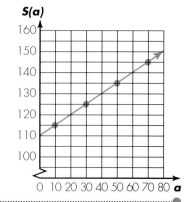

Communicating Mathematics

Study the lesson. Then complete the following.

1. **Explain** any differences between a relation and a function.

2. **Describe** how the phrase "is a function of" relates to the mathematical definition of function. For example, the diameter of a tree is a function of its age.

3. **Write** how you would find $g(1)$ if $g(x) = 3x + 12$.

4. **You Decide** Is the statement "All linear equations are functions" true? Support your answer with examples or counterexamples.

5. In your own words, explain how and why the vertical line test works.

Guided Practice

Determine whether each relation is a function.

6.

7.

x	y
−2	3
5	3
5	4
4	0

8.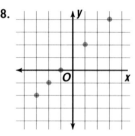

9. $\{(-3, 1), (-1, 3), (1, -2), (3, 2)\}$

10. $\{(3, 1), (-2, 2), (1, -1), (1, 2), (-3, 1), (-2, 5)\}$

11. $y + 5 = 7x$

12. $2x^2 + 3y^2 = 36$

13. The graph of $y^2 = x + 4$ is shown at the right. Use the vertical line test to determine if this relation is a function.

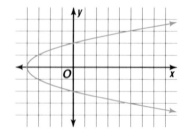

14. Which of the following graphs represent a function?

a. $y = |x - 1|$

b. $x = |y| + 1$

LOOK BACK

You can refer to Lesson 2-3 for more information on absolute value, $|x|$.

c. $y = |x| + 1$

d. $x = 1 - |y|$

If $h(x) = 3x + 2$, find each value.

15. $h(-4)$ 16. $h(2)$ 17. $h(w)$ 18. $h(r - 6)$

Practice **Determine whether each relation is a function.**

19.

a	b
−3	3
−2	3
0	4
2	4
4	4

20.

r	s
−4	3
−4	2
−4	1
4	0
4	−1
4	−2

21.

22.

23.

24.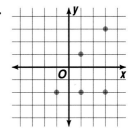

25. {(6, 3), (5, −2), (2, 3), (12, −12)}   **26.** {(4, 5), (3, −2), (−2, 5), (4, 7)}

27. {(5, −1), (6, −1), (−8, −1), (0, −1)}   **28.** {(4, −2), (−4, −2), (9, −2), (0, −2)}

29. $y = -15$ **30.** $x = 13$ **31.** $y = 3x - y$

32. $y = |x|$ **33.** $x = |y|$ **34.** $yx = 36$

If $f(x) = 4x + 2$ and $g(x) = x^2 - 2x$, find each value.

35. $f(-4)$ **36.** $g(4)$ **37.** $g\left(\frac{1}{5}\right)$ **38.** $f\left(\frac{3}{4}\right)$

39. $g(3.5)$ **40.** $f(6.2)$ **41.** $3[f(-2)]$ **42.** $-6[g(0.4)]$

43. $g(3b)$ **44.** $f(2y)$ **45.** $-3[g(1)]$ **46.** $3[g(2w)]$

47. $5[g(a^2)]$ **48.** $f(c + 3)$ **49.** $6[f(p - 2)]$ **50.** $-3[f(5w + 2)]$

51. a. Create and graph a relation that is a function and has an inverse that is also a function.

  b. Create and graph a relation that is a function and has an inverse that is not a function.

Choose the graph that best represents the information given. Explain your choice. Determine whether the graph represents a function.

52. Communication   A call to London, England costs $1.78 for the first minute and $1.00 for each additional minute. Fractions of a minute are charged as an entire minute.

a.

b.

c.

53. Manufacturing A company charges $9.95 for each custom-printed T-shirt. If you order 8 or more, but less than 15, the charge is $8.25 each. If you order 15 or more, the charge is $6.50 each.

a.

b.

c.

Critical Thinking

54. Let the domain of $f(x) = x^2 + 3x + 2$ be $\{-5, -4, -3, -2, -1, 0, 1, 2\}$.

a. Graph the function, the ordered pairs that make up the inverse of the function, and the line $y = x$ on the same coordinate plane.

b. Is the inverse of this function a function?

c. How are the graphs of the function and its inverse related to the line $y = x$?

Applications and Problem Solving

55. Metallurgy Since pure gold is very soft, other metals are often added to gold to make an alloy that is stronger and more durable. The relative amount of gold in a piece of jewelry is measured in karats. The formula for the relationship is $g = \dfrac{25k}{6}$, where k represents the number of karats and g is the percent of gold in the piece.

a. Determine the domain and range values for which this function makes sense.

b. Graph the function and describe the graph.

c. How many karats are in a ring made of pure gold?

56. Earth Science Earth's interior is composed of four levels: the crust, the mantle, the outer core, and the inner core. The inner core is believed to be solid while the outer core is molten. The temperature in the first 62 miles below Earth's surface can be approximated with the formula $T = 35d + S$, where T is the temperature in degrees Fahrenheit, d is the distance in miles, and S is the temperature (°F) on the surface.

a. Suppose the temperature at the surface is 75°F. Find the temperature at the lower edge of the crust if the crust is 30 miles deep at that point.

b. Russian scientists are credited with drilling the deepest hole ever on the Kola Peninsula, which is located near the Arctic Circle in northwest Russia about 250 miles from Finland. Drilling began in 1970, but is only 7 miles deep thus far. What is the temperature at the bottom of this hole if the outside temperature is 0°F?

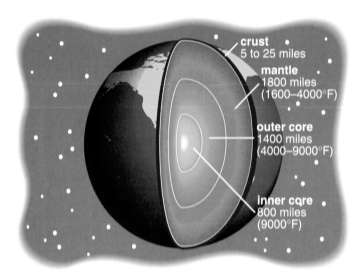

crust
5 to 25 miles

mantle
1800 miles
(1600–4000°F)

outer core
1400 miles
(4000–9000°F)

inner core
800 miles
(9000°F)

c. Why do you think that there is a range of values given for the depth of the crust?

57. Accounting The following is adapted from an article by Gail A. Eisner, CPA (certified public accountant), in *The Mathematics Teacher*.

> Early in my accounting career, the partner who was directly supervising me, a very bright, young CPA, called me into his office and asked how a person could possibly calculate a bonus if the company's formula required that the bonus be 15 percent of the net profits *after* the bonus has been deducted? I showed him the linear equation $B = 0.15(P - B)$, where B is the bonus, and P (a known quantity) represents the profits before subtracting the bonus.

  a. Suppose the profits are $2000. Determine the amount of the bonus.

  b. Solve the equation for B and graph the resulting equation.

Mixed Review

58. Graph $y - x = -5$. (Lesson 5–4)

59. Solve $3a - b = 7$ if the domain is $\{-3, -2, 4, 6\}$. (Lesson 5–3)

60. State the inverse of the relation $\{(-1, 1), (-5, 9), (4, 6)\}$. (Lesson 5–2)

61. Probability A card is selected at random from a deck of 52 cards. What is the probability of selecting a jack, queen, or king? (Lesson 4–6)

62. 44 is what percent of 89? (Lesson 4–4)

63. For what angle are the sine and cosine equal? (Lesson 4–3)

64. Trigonometry Find the measure of the third angle of a triangle if the other angles have measures $167°$ and $4°$. (Lesson 3–4)

65. Solve $\frac{z}{-4} - 9 = 3$. (Lesson 3–3)

66. Budgeting The total of Jon Young's gas and electric bills was $210.87. His electric bill was $95.25. How much was his gas bill? (Lesson 3–2)

67. Sports The players with the most runs batted in (RBI) for the National League are listed below. (Lesson 2–2)

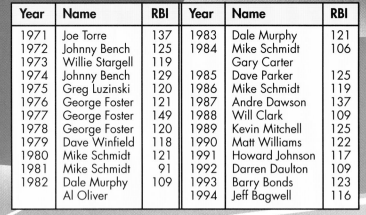

Year	Name	RBI	Year	Name	RBI
1971	Joe Torre	137	1983	Dale Murphy	121
1972	Johnny Bench	125	1984	Mike Schmidt	106
1973	Willie Stargell	119		Gary Carter	
1974	Johnny Bench	129	1985	Dave Parker	125
1975	Greg Luzinski	120	1986	Mike Schmidt	119
1976	George Foster	121	1987	Andre Dawson	137
1977	George Foster	149	1988	Will Clark	109
1978	George Foster	120	1989	Kevin Mitchell	125
1979	Dave Winfield	118	1990	Matt Williams	122
1980	Mike Schmidt	121	1991	Howard Johnson	117
1981	Mike Schmidt	91	1992	Darren Daulton	109
1982	Dale Murphy	109	1993	Barry Bonds	123
	Al Oliver		1994	Jeff Bagwell	116

Source: *The World Almanac*, 1995

  a. Make a line plot of the data.

  b. What was the greatest number of RBIs during one season?

  c. What was the least number of RBIs during one season?

  d. What was the most frequently-occurring number of RBIs during one season?

  e. How many of the players had from 119 to 129 RBIs?

Writing Equations from Patterns

5-6

What YOU'LL LEARN

- To write equations to represent relations, given some of the solutions for the equations.

Why IT'S IMPORTANT

You can write equations to represent relations in scuba diving and aquatics.

APPLICATION
Scuba Diving

As scuba divers descend, the pressure of the water increases. Scuba divers can determine their depth by the pressure. Pressure can be expressed in atmospheres. An atmosphere is equivalent to 14.7 psi (pounds per square inch) of pressure. The table below shows the relationship between atmospheres of pressure and ocean depth.

Pressure (atmospheres)	1	2	3	4	5
Depth of Ocean (feet)	0	33	66	99	132

Suppose we let p represent the atmospheres of pressure, and let d represent the ocean depth. The relation shown in the table can also be represented by a graph. Graph the points and observe the pattern.

Notice that when the ordered pairs are graphed, they form a linear pattern. You can use a straightedge to draw the line through these points. Points that lie in a linear pattern can be described by an equation. Look at the relationship between the domain and range to find a pattern that can be described by an equation.

F Y I

In 1943, Jacques-Yves Cousteau and E. Gagnon developed the Aqua-Lung™, to free divers from hoses connected to air sources on the surface. It is called *scuba* gear, which is an acronym for *self-contained underwater breathing apparatus*.

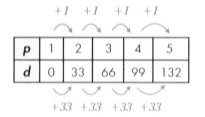

Notice that the differences in the d values are 33 times the differences of the corresponding p values. It seems that this relationship can be described by the equation $d = 33p$. Check to see if this equation is correct by substituting values of p into the equation.

p	1	2	3	4	5
$33p$	33	66	99	132	165

These values do not match those in the first chart. However, note that each corresponding value for d using the equation is 33 more than the d in the relation. Thus, we need to adjust our equation to compensate for this difference. The equation that describes this relation is $d = 33p - 33$. Since this relation is also a function, we can write this equation as $d(p) = 33p - 33$.

Example 1

Plot the points in the relation shown in the table. Then write an equation in functional notation for the relation.

x	−4	−8	−12	−16	−20
y	1	2	3	4	5

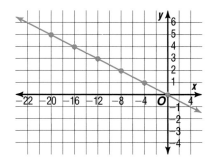

Since the points form a linear pattern, we know that there is a linear equation that describes this relation.

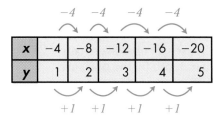

x	−4	−8	−12	−16	−20
y	1	2	3	4	5

$$\frac{range\ differences}{domain\ differences} = \frac{1}{-4}$$

The differences in the y values are one-fourth the differences in the x values. This pattern implies that the equation $y = -\frac{1}{4}x$ may describe the relation. Check the equation for the domain values in the relation.

Check: If $x = -4$, then $y = -\frac{1}{4}(-4)$ or 1.   ✔

If $x = -8$, then $y = -\frac{1}{4}(-8)$ or 2.   ✔

The equation that represents the relation is $y = -\frac{1}{4}x$. Since this relation is also a function, we can write this equation in functional notation as $f(x) = -\frac{1}{4}x$.

Sometimes you may need to find the equation for a graph.

Example 2

Write an equation in functional notation for the relation graphed at the right.

First make a table of ordered pairs for several points on the graph.

k	1	2	4	5	6	9
m	8	6	2	0	−2	−8

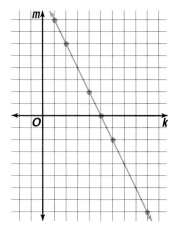

Now find the common differences of the domain and range.

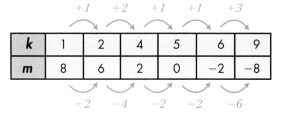

k	1	2	4	5	6	9
m	8	6	2	0	−2	−8

Notice that there is not a consistent change in the domain and range values. However, you can still find the ratio of the range difference to the domain difference for each ordered pair. There are three possible ratios.

$$\frac{\text{range differences}}{\text{domain differences}} \qquad \frac{-2}{1} = -2 \qquad \frac{-4}{2} = -2 \qquad \frac{-6}{3} = -2$$

Since all of the ratios are the same, it appears that the differences in the m values are -2 times that of the k values. This pattern suggests that $m = -2k$. Check the equation to see if it is correct.

Check: If $k = 2$, then $m = -2(2)$ or -4. But the range value for $k = 2$ is 6, a difference of 10. Try some other values in the domain to see if the same difference occurs.

k	1	4	5	6	9
$-2k$	-2	-8	-10	-12	-18
m	8	2	0	-2	-8

m is 10 more than $-2k$.

This pattern suggests that 10 should be added to one side of the equation in order to correctly describe the relation. Thus, the equation for this relation is $m = -2k + 10$. Check this equation.

Check: If $k = 2$, then $m = -2(2) + 10$ or 6. ✔
If $k = 9$, then $m = -2(9) + 10$ or -8. ✔

Therefore, $m = -2k + 10$ describes this relation. Since this relation is also a function, we can write this equation in functional notation as $f(k) = -2k + 10$.

Many times you can generalize the relationships among data you have collected with an equation.

EXPLORATION

S P R E A D S H E E T S

In the relation $\{(-2, 2), (-1, 5), (0, 8), (1, 11), (2, 14)\}$, the range difference is 3 when the domain difference is 1. You might suggest that the equation of the relation is $y = 3x$. Let's use spreadsheet software to check this equation.

- Each cell of a spreadsheet is named by a letter and number. The letter refers to the column and the number of the row. Enter each x value into cells A1 through A5. Enter the y values into cells B1 through B5.

- In cell C1, enter the formula A1∗3. This formula means take the value in cell A1 and multiply it by 3. Copy this formula to cells C2 through C5. The spreadsheet will automatically change the formula so that the appropriate cell in column A will be used.

- In cell D1, enter the formula B1 − C1. This formula will subtract the range value of the equation from the range value of the relation. Copy this formula to cells D2 through D5.

Column D tells what number to add to the equation so that it is correct. In this case, the entries in column D are 8s. The correct equation should be $y = 3x + 8$.

Your Turn

a. What formula would you use in cell C1 to test the first formula found in Example 2? Use a spreadsheet to check all the values in the relation.

b. What number would you expect to appear in column D if the equation you found is correct for a given relation?

Example **3**

APPLICATION

Hardware

F Y I

This pattern for nails does not hold for all penny sizes. For example, a 60-penny nail is only 6 inches long.

Pennies are units for measuring the lengths of nails. Nails measured in pennies can range from 2-penny nails to 60-penny nails. The graph at the right illustrates the lengths of several penny nails.

a. The length of a nail as a function of its penny status can be modeled by a linear function for 2- through 10-penny nails. Write an equation in functional notation for this relationship.

b. Find the lengths of a 3-penny nail, a 5-penny nail, and a 10-penny nail.

a. We can see from the graph that the points form a linear pattern. Thus, there is a linear equation that describes the relation for 2- through 10-penny nails.

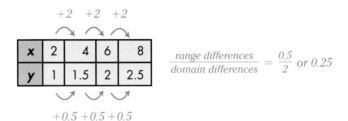

$$\frac{\text{range differences}}{\text{domain differences}} = \frac{0.5}{2} \text{ or } 0.25$$

This pattern suggests that the equation $f(x) = 0.25x$ might model this situation. However, if you check the domain values with this equation, you will find that the values of $f(x)$ are off by 0.5. The correct equation is $f(x) = 0.25x + 0.5$.

b. *3-penny nail*   *5-penny nail*   *10-penny nail*

$f(3) = 0.25(3) + 0.5$ $f(5) = 0.25(5) + 0.5$ $f(10) = 0.25(10) + 0.5$

$\qquad = 1.25$ $\qquad = 1.75$ $\qquad = 3$

A 3-penny nail is 1.25 inches long, a 5-penny nail is 1.75 inches long, and a 10-penny nail is 3 inches long.

CHECK FOR UNDERSTANDING

Communicating Mathematics

Study the lesson. Then complete the following.

1. **Analyze** the graphs in Examples 1 and 2 along with the ratios of the range differences to the domain differences. Then complete these sentences.

   a. If the ratio is positive, then the line slants___?___.

   b. If the ratio is negative, then the line slants___?___.

2. **Explain** how you can determine if an equation correctly represents a relation given as a table.

3. **You Decide** When the teacher was explaining Example 2, Narissa said she had another way to find the equation without calculating three ratios. She said she found other points that lie on the line so the differences were common. Is Narissa's method valid? Explain.

4. **Write** a linear equation in functional notation that has both (1, 1) and (0, 3) as solutions. Is this the only linear equation that has these two solutions? Explain.

Write an equation in functional notation for each relation.

5.

x	3	4	5	6	7
f(x)	12	14	16	18	20

6.

x	2	4	6	8	10
f(x)	−4	−3	−2	−1	0

Write an equation for each graphed relation.

7.

8.

9. The table at the right represents values for the function N(m). Copy and complete the table at the right. Explain how you determined the missing value.

m	8	4	2	0
N(m)	−10	−8	−7	?

10. Chemistry Most substances contract when they freeze. However, water expands in volume when it freezes. Eleven cubic feet of water becomes 12 cubic feet of ice, 33 cubic feet of water becomes 36 cubic feet of ice, and 66 cubic feet of water becomes 72 cubic feet of ice.

a. Make a graph of these data.

b. Write a functional equation for the relationship between the volume of water and the corresponding volume of ice.

Calvin and Hobbes
by Bill Watterson

EXERCISES

Copy and complete the table for each function.

11.

x	1	2	3	4	5
f(x)	12	24	36		

12.

x	−4	−2	0	2	4
g(x)	−2	−1			2

13.

x	−3	−1	1	2	4
h(x)	18			−7	−17

14.

x	−2	0	2	4	6
p(x)			2	3	4

Write an equation for each relation.

15.

x	1	2	3	4	5
f(x)	5	10	15	20	25

16.

n	1	2	3	4	5
f(n)	1	4	7	10	13

17.

x	-2	-1	1	2	4
g(x)	13	12	10	9	7

18.

n	-4	-2	0	2	4
m(n)	-11	-3	5	13	21

19.

x	0	6	12	18	24
h(x)	-2	0	2	4	6

20.

n	-4	0	4	6	8
m(n)	26	18	10	6	2

21.   **22.**   **23.**

24.   **25.**   **26.**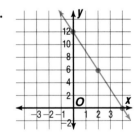

27. $\{(6, -4), (2, -12), (1, -24), (-3, 8), (-6, 4)\}$

28. $\{(-3, 10), (-2, 5), (-1, 2), (0, 1), (1, 2), (2, 5), (3, 10)\}$

29. $\{(-3, -27), (-1, -1), (2, 8), (3, 27), (10, 1000)\}$

30. $\{2, 12), (1, 48), (-1, 48), (-2, 12), (-4, 3)\}$

Critical Thinking

31. The y- and x-intercepts are those points at which a graph intersects the y- and x-axes, respectively. Use functional notation to describe these points.

32. Suppose the ratio of range differences to domain differences in a linear function $g(x)$ is $\frac{2}{3}$ and the graph of $g(x)$ goes through the point $(-3, 2)$. Draw the graph of $g(x)$ and explain why you drew it the way you did.

Applications and Problem Solving

33. Aquatics The table below illustrates the relationship between atmospheres of pressure and the depth of fresh water.

Pressure (atmospheres)	1	2	3	4	5
Depth of Fresh Water (ft)	0	34	68	102	136

a. Write an equation in functional notation for the relation.

b. Refer to the application at the beginning of the lesson. How does the pressure in fresh water compare to the pressure in ocean water?

34. Archaeology Archaeologists sometimes use lengths of heavy-duty wire to hold together the bones that compose the spinal columns of various vertebrates. Suppose a 70-foot length of wire weighs 23 pounds.

a. If this sample of wire is representative of all lengths of this wire, draw a graph showing the weight of the wire as a function of its length.

b. Use your graph to estimate the weight of 50 feet of wire to the nearest pound.

c. Write an equation to describe the relationship between the length of a wire and its weight.

d. Evaluate the equation in part c for $\ell = 50$. How does this result compare with your estimate in part b?

35. Health Alex Knockelmann is concerned about the number of Calories that she needs to intake during the basketball season so she will not lose weight. She wants to eat as many Calories before practice as she will use during practice. The table below shows some fast foods from restaurants near her school and their Calories.

Fast Food	Calories
Arby's® Ham and Cheese	380
Arby's® Roast Beef	350
Burger King's® Whopper™ with Cheese	760
Burger King's® Regular Fries	240
Burger King's® Chocolate Milk Shake	380
Wendy's® Cheeseburger	577
Wendy's® Double Cheeseburger	797
Wendy's® Fries	327
Wendy's® Frosty™	390
Pepsi® (20 oz)	241
7-Up® (20 oz)	219

a. Before practice, Alex went to Wendy's. She ate a double cheeseburger, fries, and a Frosty™. What was her Calorie intake for this meal?

b. In Lesson 5–4, you learned that playing basketball burns 0.138 C/min/kg. At Alex's weight of 80 kg, determine how many Calories she would burn each minute.

c. Make a table to show the number of Calories that Alex burns each minute for the first 10 minutes of practice. Graph the points.

d. Let C represent the number of Calories burned and t represent the number of minutes. Write a functional equation to represent the relation. Is it a linear function?

e. Use the graph and your equation to predict the number of Calories Alex would use during two hours of practice. Did she eat enough food so that she had Calories remaining at the end of practice?

36. Biology The function $f(t) = \frac{t}{0.2} - 32$ shows that the number of times a cricket chirps in an hour is a function of the temperature in degrees Celsius. *(Lesson 5–5)*

a. If it is 12°C outside, how many times will a cricket chirp in an hour?

b. If a cricket chirps 54 times in 3 hours, what is the outside temperature?

37. State the domain and range of {(1, 6), (3, 4), (5, 2)}. *(Lesson 5–2)*

38. What number is 16% of 50? *(Lesson 4–4)*

39. Trigonometry Given the similar triangles at the right, find *s*. *(Lesson 4–2)*

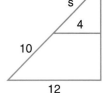

40. Solve $x + (-7) = 36$. *(Lesson 3–1)*

41. Graph {0, 3, 7} on a number line. *(Lesson 2–1)*

42. Personal Finance Many credit cards offer a protection plan so you can pay off your balance if you lose your job or have an accident and cannot work. The table below shows the monthly premium (in cents per $100 outstanding balance) for the plan on a Discover® card in the United States and its territories. Round each amount to the nearest cent and make a stem-and-leaf plot of these fees. *(Lesson 1–4)*

Area	Fee	Area	Fee	Area	Fee	Area	Fee	Area	Fee	Area	Fee
AK	68.2	FL	75	LA	74	NC	64.6	OK	75	UT	68.9
AL	66	GA	75	MA	63.5	ND	64.8	OR	69.8	VA	59.3
AR	66	HI	57.5	MD	75	NE	68	PA	22.5	VI	66
AZ	73.7	IA	58.8	ME	62	NH	46.9	PR	74	VT	56.7
CA	75	ID	60	MI	66	NJ	64.2	RI	67	WA	60
CO	68.1	IL	75	MN	32.3	NM	74	SC	71.9	WI	59
CT	53.2	IN	62.1	MO	75	NV	74	SD	66	WV	74
DC	66	KS	68.3	MS	74	NY	45.3	TN	75	WY	73.1
DE	66	KY	75	MT	74	OH	75	TX	42.7		

WORKING ON THE

In·ves·ti·ga·tion

Refer to the Investigation on pages 190–191.

Many of the prcedures from one area of science can be applied to other areas of science. Estimating populations is not reserved just for fish. Conservationists also keep track of species of birds, insects, plants, and other wildlife in a similar way.

1 Design a different experiment that illustrates the capture-recapture method. Complete your experiment a few times to see if it is as accurate as the one you have completed.

2 What was the range of your estimates using the method described in the Investigation? What was the range of your estimates using your method? Which do you think is the most reliable method?

Add the results of your work to your Investigation Folder.

5–7A Graphing Technology
Measures of Variation

A Preview of Lesson 5–7

The graphing calculator is a powerful tool for observing patterns and analyzing data that come from research or experimentation. Many of the measures of variation you will learn about in Lesson 5–7 can be calculated by using a graphing calculator.

Before analyzing any data, you must know how to enter it into a list.

Example **①** **The Glencoe Publishing Golf Tournament is held every September to raise money for a scholarship fund. The list below includes the scores for the employees of the manufacturing department. Enter the data and find the range of these scores.**

139, 99, 105, 115, 88, 91, 105, 80, 102, 101, 103, 95, 99, 77, 112

First enter the data.

Method 1 Assign the data to List 1 (L1) from the home screen.

Enter: 2nd { 139 , 99 , 105 , 115 , 88 , 91

, 105 , 80 , 102 , 101 , 103 , 95 , 99

, 77 , 112 2nd } STO▶ 2nd L1 ENTER

Method 2 Enter the data directly into the list display.

Before entering your data, you must clear the L1 list of previously entered data.

Enter: STAT 4 2nd L1 ENTER

Enter each score into the L1 list.

Enter: STAT 1

The cursor appears in the L1 list. Enter the data one by one pressing ENTER after each entry, including the last one.

Then find the range.

The range of the scores is the difference of the maximum and minimum values in the list.

Enter: 2nd QUIT 2nd LIST ▶ 2 2nd L1) —

2nd LIST ▶ 1 2nd L1) ENTER *62*

The range of the data is 62.

LOOK BACK

You can review another method for using the graphing calculator to find the median of a set of data in Lesson 3–7.

You can use the calculator to sort the data from least to greatest by pressing

STAT 2 2nd L1

ENTER *. You can view the*
list by pressing STAT

ENTER *.*

If you do not use 2nd
QUIT *to leave the LIST*
menu, the range will be
added to L1, which alters
your data.

The data can be arranged into four groups that have approximately the same number of data in each group. The points separating these groups are called **quartiles** and are symbolized by Q_1, Q_2, and Q_3. The median of the set is Q_2. The **interquartile range** is $Q_3 - Q_1$. Half of the data lie in this range. You can use a graphing calculator to help you find each of these measures.

Example ❷ Find the quartiles and the interquartile range of the golf data in Example 1.

Before finding the measures you want, you must first make sure the calculator is set to refer to the correct set of data. Our data is in L_1.

Enter: STAT ▶ 3 and highlight L_1 under 1-Var Stats. Make sure that the frequency is 1.

Enter: STAT ▶ 1 ENTER

Use the down arrow key to scroll to the end of the statistics. Your screen should look like the one at the right. This tells you there are 15 items of data, the least value (minX) is 77, the greatest value (maxX) is 139, the lower quartile (Q_1) is 91, the median is 101, and the upper quartile (Q_3) is 105.

```
1-Var Stats
↑n=15
 minX=77
 Q1=91
 Med=101
 Q3=105
 maxX=139
```

Use Q_3 and Q_1 to find the interquartile range: $Q_3 - Q_1 = 105 - 91$ or 14.

EXERCISES

Remember to change the Setup when calculating the measures for each set of data.

Enter each set of data into the list given. Then find the lower quartile, median, upper quartile, range, and interquartile range of the data.

1. 12, 17, 16, 23, 18 in L_1

2. 56, 45, 37, 43, 10, 34 in L_2

3. 77, 78, 68, 96, 99, 84, 65 in L_3

4. 30, 90, 40, 70, 50, 100, 80, 60 in L_4

5. 3, 3.2, 6, 45, 7, 26, 1, 3.4, 4, 5.3, 5, 78, 8, 21, 5 in L_5

6. 85, 77, 58, 69, 62, 73, 55, 82, 67, 77, 59, 92, 75, 69, 76 in L_6

7. Look at the range and the interquartile range of each set of data. What do you think it means if the range is great but the interquartile range is small?

8. Suppose you entered a set of data and the interquartile range was 0. What does that mean?

Integration: Statistics
Measures of Variation

APPLICATION
Climate

Justin Williams has been promoted and has the option of relocating either to Columbus, Ohio, or San Francisco, California. The Williams family lives in Tampa, Florida, where the average temperature is about 72°F and it is usually warm all year long. His family's wardrobe includes basically all lightweight apparel. He wondered if moving would mean they would have to get lots of new clothes. He looked up the average high temperature for the two cities and found the following information in a table of average high temperatures of U.S. cities.

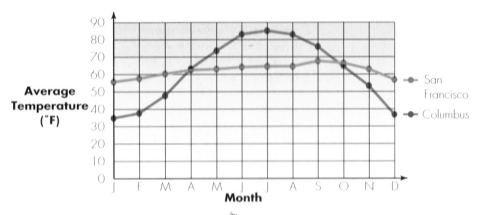

	Columbus	San Francisco
Jan.	35°	
Feb.	38°	56°
Mar.	49°	59°
Apr.	62°	60°
May	73°	61°
June	81°	63°
July	84°	64°
Aug.	83°	64°
Sept.	77°	65°
Oct.	65°	69°
Nov.	51°	68°
Dec.	39°	63°
		57°

Source: U.S. National Oceanic and Atmospheric Administration

Mr. Williams calculated the mean and median of the temperatures of each city. The mean high temperature for Columbus was 63.5°, and for San Francisco, it was 62.4°. The median high temperature for Columbus was 63.5°, and for San Francisco, it was 63°. He figured that meant the climate in the two cities was just about the same. But then he decided to take another look at the temperatures and find another way to analyze them.

The months and temperatures for each city form a relation. Mr. Williams decided to graph the relation for each city. He used different colors for each city.

He noticed that the pattern in the temperatures was very different. This case shows that measures of central tendency may not give an accurate enough description of the data. Often **measures of variation** are also used to describe the distribution of the data.

One of the most common measures of variation is the **range.** Unlike the other definition of range in this chapter, the range of a set of data is a measure of the spread of the data.

Definition of Range	The range of a set of data is the difference between the greatest and the least values of the set.

Let's find the range for each set of temperatures. Make a table to help organize the data.

City	Greatest Temperature	Least Temperature	Range
Columbus	84°	35°	84° − 35° or 49°
San Francisco	69°	56°	69° − 56° or 13°

The range of temperatures for Columbus is greater than that for San Francisco. This means that temperatures in Columbus vary more than temperatures in San Francisco. If his family moves to Columbus, he will definitely need some new clothes for their colder winters.

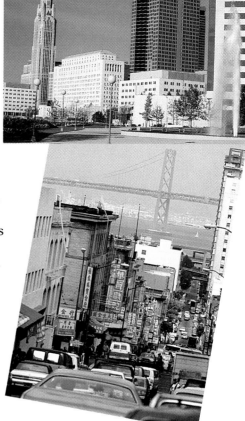

Another commonly used measure of variation is called the **interquartile range.** In a set of data, the **quartiles** are values that divide the data into four equal parts. Statisticians often use Q_1, Q_2, and Q_3 to represent the three quartiles. Remember that the median separates the data into two equal parts. Q_2 is the median. Q_1 is the **lower quartile.** It divides the lower half of the data into two equal parts. Likewise, Q_3 is the **upper quartile.** It divides the upper half of the data into two equal parts. The difference between the upper and lower quartiles is the interquartile range (IQR).

The abbreviations LQ and UQ are often used to represent the lower quartile and upper quartile.

Definition of Interquartile Range	The difference between the upper quartile and the lower quartile of a set of data is called the interquartile range. It represents the middle half, or 50%, of the data in the set.

Example

APPLICATION

Football

1 The table below shows the heights and weights of 13 veteran players on the 1994 Dallas Cowboys football team. Find the median, upper and lower quartiles, and the interquartile range for the weights of the players.

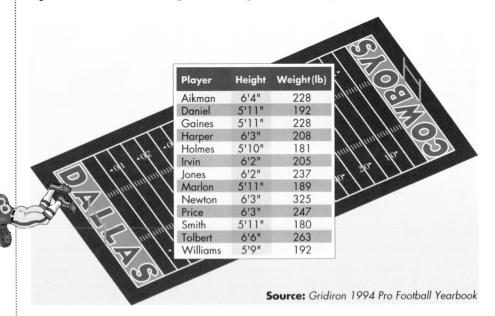

Player	Height	Weight (lb)
Aikman	6'4"	228
Daniel	5'11"	192
Gaines	5'11"	228
Harper	6'3"	208
Holmes	5'10"	181
Irvin	6'2"	205
Jones	6'2"	237
Marlon	5'11"	189
Newton	6'3"	325
Price	6'3"	247
Smith	5'11"	180
Tolbert	6'6"	263
Williams	5'9"	192

Source: *Gridiron 1994 Pro Football Yearbook*

Order the 13 weights from least to greatest. Then find the median.

180   181   189   192   192   205   208   228   228   237   247   263   325
↑
median

Remember that the median is the middle number when the data are arranged in numerical order.

The lower quartile is the median of the lower half of the data, and the upper quartile is the median of the upper half of the data. If the median is an item in the set of data, it is not included in either half.

180   181   189 ⌐ 192   192   205   208   228   228   237 ⌐ 247   263   325
↑
$Q_1 = 190.5$ $Q_2 =$ median $Q_3 = 242$

The interquartile range is $242 - 190.5$ or 51.5. Therefore, the middle half, or 50% of the weights of the football players, varies within 51.5 pounds.

In Example 1, one weight, 325 pounds, is much greater than the others. In a set of data, a value that is much greater or much less than the rest of the data can be called an **outlier.** An outlier is defined as any element of a set of data that is at least 1.5 interquartile ranges greater than the upper quartile or less than the lower quartile.

An outlier will not affect the quartiles, but it will affect the mean.

The interquartile range of the football data is 51.5. So 1.5(51.5) is 77.25.

113.25   319.25

⟵ −77.25   +77.25 ⟶

180   181   189 | 192   192   205   208   228   228   237 | 247   263 ┆ 325
↓   ↑   ↓
$Q_1 = 190.5$ $Q_2 =$ median $Q_3 = 242$

From the diagram above, you can see that the only item of data occurring beyond the dashed lines is 325. So, this is the only outlier.

Example **2**

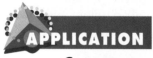

APPLICATION
Commerce

LOOK BACK

You can review stem-and-leaf plots in Lesson 1-4.

The double stem-and-leaf plot below represents the tonnage of domestic and foreign imports being delivered at the top 19 busiest ports in the United States during 1992.

Domestic	Stem	Foreign
[5	10	
	9	[5
	8	
5	7	3
5	6	
	5	
8⬚0	4	1 4
8 1	3	6⬚6
8 5]△3 [1	2	0⬚0 3 4 5 5]△5 [5 9 9
9 9 7⬚5 2 0	1	8 8
9 8]	0	8]

$5|2 = 25,000,000$ tons $7|3 = 73,000,000$ tons

Source: Army Corps of Engineers, U.S. Dept. of Defense

a. The brackets group the values in the lower half and the values in the upper half. What do the triangles contain? What do the boxes contain?

b. Find the interquartile ranges of the domestic and the foreign tonnage.

c. Which type of imports had a more consistent tonnage—the domestic or the foreign?

d. Find any outliers.

a. The triangles contain the medians.
The boxes contain the lower and upper quartiles.

b. **Domestic**

Q_1 is 15 and Q_3 is 40.

$IQR = Q_3 - Q_1$

$= 40 - 15$ or 25

Foreign

Q_1 is 20 and Q_3 is 36.

$IQR = Q_3 - Q_1$

$= 36 - 20$ or 16

c. The foreign imports had more consistent tonnage because it had the smaller interquartile range.

d. Outliers are those values that lie 1.5 IQRs above Q_3 or below Q_1.

Domestic

$1.5(IQR) + Q_3 = 1.5(25) + 40$ or 77.5

$Q_1 - 1.5\,(IQR) = 15 - 1.5(25)$ or -22.5

Since $105 > 77.5$, 105 is an outlier. There are no outliers below -22.5.

Foreign

$1.5(IQR) + Q_3 = (1.5)(16) + 36$ or 60

$Q_1 - 1.5\,(IQR) = 20 - (1.5)(16)$ or -4

Since $95 > 60$ and $73 > 60$, 95 and 73 are outliers.

There are no outliers below -4.

Communicating Mathematics

Study the lesson. Then complete the following.

1. Refer to the application at the beginning of the lesson. Compare the interquartile ranges of the temperatures in San Francisco and Columbus.

2. **Describe** how the mean is affected by an outlier.

3. **Predict** the following.
   a. If you were to measure the height of all of the students in your school, in which group of students might you find outliers?
   b. If you were to measure the weight of all of the students in your school, in which group of students might you find outliers?

4. **Explain** the difference between range as defined in Lesson 5–2 and range as defined in this lesson.

MODELING MATHEMATICS

5. Along with your classmates, line up according to your heights.
   a. "Fold" the line in half by having the shortest person meet the tallest person. What is the median height?
   b. Do the same with each half of the line. What is lower quartile and upper quartile of the heights of the students in your classroom?

Guided Practice

Find the range, median, upper quartile, lower quartile, and interquartile range of each set of data. Identify any outliers.

6. 16, 24, 11, 17, 19

7. 43, 45, 56, 37, 11, 34

8.

Top 10 Jobs of the Future	
Job	Average Salary
Civil engineer	$55,800
Computer system analyst	42,700
Electrical engineer	59,100
Geologist	50,800
High school teacher	32,500
Pharmacist	47,500
Physical therapist	37,200
Physician	148,000
Psychologist	53,000
School principal	57,300

Source: *American Careers*, Fall 1994

9.

Stem	Leaf
0	0 2 3
1	1 7 9
2	2 3 5 6
3	3 4 4 5 9
4	0 7 8 8

2 | 2 = 22

10. **Football** Refer to the data in Example 1. Find the median, upper and lower quartiles, and the interquartile range for the heights of the players.

Practice

Find the range, median, upper quartile, lower quartile, and interquartile range for each set of data.

11. 77, 78, 68, 96, 99, 84, 65

12. 17°, 46°, 18°, 22°, 18°, 21°, 19°

13. 2, 4, 6, 8, 2, 4, 6, 8, 10, 0

14. 89, 56, 75, 82, 64, 73, 87, 92

15. 30.8, 29.9, 30.0, 31.0, 30.1, 30.5, 30.7, 31.0

16. 78, 2, 3.4, 4, 45, 7, 5.3, 5, 3, 1, 3.2, 6, 26, 8, 5

17. 1050, 1175, 835, 1075, 1025, 1145, 1100, 1125, 975, 1005, 1125, 1095, 1075, 1055

18.

Stem	Leaf
5	3 6 8
6	5 8
7	0 3 7 7 9
8	1 4 8 8 9
9	9

$9\,|\,9 = 9900$

19.

Stem	Leaf
19	3 5 5
20	2 2 5 8
21	5 8 8 9 9 9
22	0 1 7 8 9
23	2

$19\,|\,3 = \$193$

20.

Stem	Leaf
5	0 3 7 9
6	1 3 4 5 5 6
7	1 5 6 6 9
8	1 2 3 5 8
9	2 5 6 9

$5\,|\,0 = 5.0\ cm$

Graphing Calculator

21. **a.** Use a graphing calculator to find the median, upper and lower quartiles, range, and interquartile range of the number of books given in the table below.

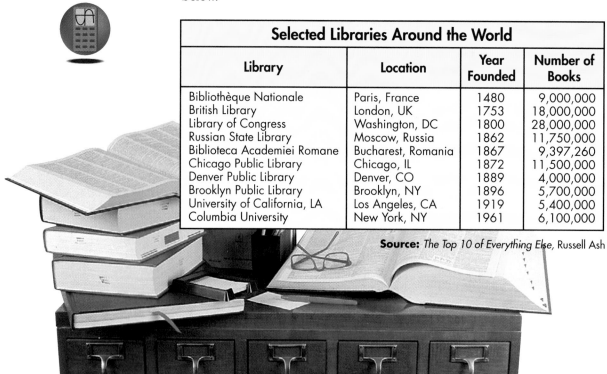

Selected Libraries Around the World			
Library	Location	Year Founded	Number of Books
Bibliothèque Nationale	Paris, France	1480	9,000,000
British Library	London, UK	1753	18,000,000
Library of Congress	Washington, DC	1800	28,000,000
Russian State Library	Moscow, Russia	1862	11,750,000
Biblioteca Academiei Romane	Bucharest, Romania	1867	9,397,260
Chicago Public Library	Chicago, IL	1872	11,500,000
Denver Public Library	Denver, CO	1889	4,000,000
Brooklyn Public Library	Brooklyn, NY	1896	5,700,000
University of California, LA	Los Angeles, CA	1919	5,400,000
Columbia University	New York, NY	1961	6,100,000

Source: *The Top 10 of Everything Else,* Russell Ash

b. Would it make sense to find these values for the years in which these libraries were founded? Why or why not?

Critical Thinking

22. **a.** Write an example of a set of 19 numbers that has a range of 60, a median of 40, an interquartile range of 16, and one outlier.

b. Write an example of a set of 19 numbers that has a range of 20, a median of 40, an interquartile range of 11, and no outliers.

Find the range, median, upper quartile, lower quartile, and interquartile range for each set of data.

23. Entertainment

Long-Running Broadway Plays (as of 7/17/94)	
Play	Performances
42nd Street	3,486
Annie	2,377
Cats	4,917
Chorus Line	6,137
Fiddler on the Roof	3,242
Grease	3,388
Hello Dolly	2,844
Les Miserables	3,005
Life with Father	3,224
My Fair Lady	2,717
Oh, Calcutta	5,959
Phantom of the Opera	2,717
Tobacco Road	3,182

Source: *Variety*, 1994

24. Transportation

Popular Passenger Cars in the U.S. (1993)	
Name of Car	Number Sold
Chevrolet Cavalier	273,617
Chevrolet Lumina	219,683
Ford Escort	269,034
Ford Taurus	360,448
Ford Tempo	217,644
Honda Accord	330,030
Honda Civic	255,579
Pontiac Grand Am	214,761
Saturn	229,356
Toyota Camry	299,737

Source: American Automobile Manufacturers Association

25. Education

Enrollment at Selected Colleges in the U.S. (1993–1994)	
College	Enrollment
Baylor University	12,194
Brown University	7,655
California Institute of Technology	1,977
Columbia University	3,441
Dartmouth College	5,475
Georgia State University	23,651
Howard University	10,736
Massachusetts Institute of Technology	9,790
Princeton University	6,444
Stanford University	14,002
Yale University	10,844

Source: *Peterson's Guides,* © 1994

26. Nutrition

Calories in Foods, per serving	
Food	Calories
Apple	100
Banana	130
Bread	60
Cupcake	200
Donut	120
Ice cream	185
Orange juice	85
Peach	50
Peanut butter	190
Potato	100
Soft taco	225
Whole milk	133

Source: U.S. Dept. of Agriculture

Phantom of the Opera

Hello Dolly

27. Sports The double stem-and-leaf plot below represents the ages of the top 20 male and female golfers for the 1994 season, according to *Golfer's Almanac,* 1995.

Male	Stem	Female
8	1	
9 9 8 8 5	2	3 4 5 5 9 9
8 8 7 6 6 5	3	1 2 2 2 4 4 5 8 9 9 9
6 2 0 0	4	0 4 6
4 3 1 0	5	

4 | 5 = 54 years old   4 | 6 = 46 years old

a. Find the ranges, quartiles, and interquartile ranges of the average golfer's age for male and female golfers.

b. Identify any outliers.

c. Compare the ranges and interquartile ranges for male and female golfers. What can you conclude from these statistics?

28. Entertainment The total sales of movie rentals as of April 1994 for selected Academy Award winners are listed below.

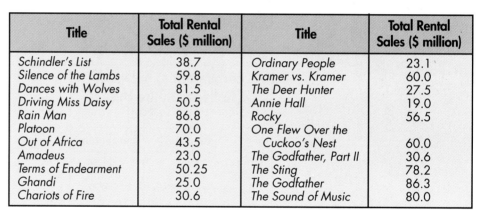

Title	Total Rental Sales ($ million)	Title	Total Rental Sales ($ million)
Schindler's List	38.7	Ordinary People	23.1
Silence of the Lambs	59.8	Kramer vs. Kramer	60.0
Dances with Wolves	81.5	The Deer Hunter	27.5
Driving Miss Daisy	50.5	Annie Hall	19.0
Rain Man	86.8	Rocky	56.5
Platoon	70.0	One Flew Over the	
Out of Africa	43.5	Cuckoo's Nest	60.0
Amadeus	23.0	The Godfather, Part II	30.6
Terms of Endearment	50.25	The Sting	78.2
Ghandi	25.0	The Godfather	86.3
Chariots of Fire	30.6	The Sound of Music	80.0

Source: *Variety,* May 1994

a. Find the range, median, upper and lower quartiles, and interquartile range for the total rental sales.

b. Identify any outliers.

c. The least value on the list of all-time top 20 movie rentals as of April, 1994, is $96.3 million. What can you conclude about Academy Award winners in relation to this data?

29. a. List the grades you have made during this grading period in mathematics. Determine the range, the quartiles, and the interquartile range of the scores.

b. List the grades you have made during this grading period in English. Determine the range, the quartiles, and the interquartile range of the scores.

c. In which of the two courses are your grades more consistent? Why do you think this is the case?

Mixed Review

30. Geology The underground temperature of rocks varies with their depth below the surface. The temperature at the surface is about 20°C. At a depth of 2 km, the temperature is about 90°C, and at a depth of 10 km, the temperature is about 370°C. (Lesson 5–6)

a. Write an equation to describe this relationship.

b. Use the equation to predict the temperature at a depth of 13 km.

31. Which ordered pairs are solutions to $5 - 1.5x = 2y$? (Lesson 5–3)

   a. $(0, 1)$ **b.** $(8, 2)$ **c.** $\left(4, -\dfrac{1}{2}\right)$ **d.** $(2, 1)$

32. Ecology For every 20 grams of smog produced by a typical automobile in an hour, gasoline-powered lawn mowers produce 50 grams. If it takes Jodi 1.5 hours to mow her lawn, how long would she be able to drive her car and produce the same amount of smog? (Lesson 4–1)

33. Sports The table at the right shows how many millions of people participated in in-line skating and mountain biking in 1991 and 1992. Write an equation to represent each situation. Then solve. (Lesson 3–1)

   a. How many more people were participating in in-line skating in 1992 than in 1991?

   b. How many more people were mountain biking in 1992 than in 1991?

Year	1991	1992
In-Line Skating	6.2	9.4
Mountain Biking	6.0	6.9

Source: *Men's Fitness*

34. Simplify $3(-4) + 2(-7)$. (Lesson 2–6)

Record Cold Temperatures

The excerpt below appeared in an article in *Mother Earth News* in January, 1995. All of the temperatures are given in °F.

HOW COLD DOES IT GET IN ALASKA? The all-time low temperature is −80 degrees at Prospect Creek Camp on January 23, 1971. Think of it: that temperature is to 0 degrees what 0 degrees is to a typical summer afternoon. Still, you might say, Alaska is up there near the North Pole. Surely, down here in the "lower 48," we never approach such levels of incredible frigidity. Yes we do. On January 20, 1954, the temperature in Rogers Pass, Montana, nudged −70 degrees.... Is there anywhere that temperatures can get colder than in Alaska? Siberia can get a little colder. But there is one place on Earth that is much more frigid: Antarctica. Earth's all-time record low was set at Vostok Station in Antarctica in 1983: −128 degrees. ■

Antarctica

1. How much colder is Earth's record low temperature than a temperature of 80°F in summer?

2. The wind-chill factor is a measure of how cold we feel when we experience wind along with a low temperature. Do you think that the wind-chill factor also applies to nonliving objects such as bicycles and cars? Why or why not?

3. The hottest temperature recorded on Earth was 136°F at Azizia, Tripolitaina, in northern Africa, on September 13, 1922. If every country's record high and low temperatures for the 20th century were entered in a computer, what would be the range of the data?

4. If the interquartile range of temperatures in a country is very small, what does that tell you about the climate of that country?

In·ves·ti·ga·tion

Go Fish!

Refer to the Investigation on pages 190–191.

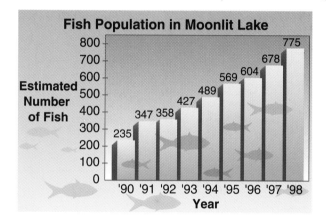

Sampling is a research method frequently used in many fields. Manufacturers use sampling to determine the quality of their products. Surveys are a way of sampling the opinions of the public about items they encounter in everyday life such as what television programs they watch or what food is their favorite. Capture-recapture is only one of many methods used in sampling a population.

Analyze

You have conducted experiments and organized your data in various ways. It is now time to analyze your findings and state your conclusions.

PORTFOLIO ASSESSMENT

You may want to keep your work on this Investigation in your portfolio.

1 What are the strengths and weaknesses of the capture-recapture method of sampling?

2 How many samples do you think give the best results? Why?

3 Write a one-page report on how the capture-recapture method is used in real life. Be sure to include references for your report.

Write

Imagine that the fish population in Moonlit Lake had been estimated each year for nine years and that these estimates are reflected in the following graph.

Fish Population in Moonlit Lake

Estimated Number of Fish vs. Year

235, 347, 358, 427, 489, 569, 604, 678, 775 ('90–'98)

Notice that the fish population is increasing over time. Suppose city council member Aubrey Howard wanted to show other members of the city council that the growth in the fish population has been rapid. Mr. Howard might choose to find a best-fit line for the graph.

4 Plot the information from the graph as ordered pairs on a coordinate plane.

5 Use two vertical lines to separate the data into three sets of equal size.

6 Locate a point in each set whose x- and y-coordinates represent the medians of the x- and y-coordinates of all the points in that set. In other words, locate the middle point of the three sets.

7 Now take a straightedge and align it with the outer two median points. Then slide it one third of the distance to the median point of the center set of points. Then draw the line. This is a *best-fit line*.

8 What might Mr. Howard's argument be? Write a one-page speech for the council member, using the graph as evidence for your opinion.

VOCABULARY

After completing this chapter, you should be able to define each term, property, or phrase and give an example or two of each.

Algebra

axes   (p. 254)

complete graph   (p. 278)

completeness property for points in the plane (p. 256)

coordinate plane   (p. 254)

domain   (p. 263)

equation in two variables (p. 271)

function   (p. 287)

functional notation   (p. 289)

graph   (p. 255)

inverse of a relation (p. 264)

linear equation   (p. 280)

linear equation in standard form   (p. 281)

linear function   (p. 278)

mapping   (p. 263)

origin   (p. 254)

quadrants   (p. 254)

range   (p. 263)

relation   (pp. 260, 263)

solution of an equation in two variables   (p. 271)

vertical line test   (p. 289)

x-axis   (p. 254)

x-coordinate   (p. 254)

y-axis   (p. 254)

y-coordinate   (p. 254)

Statistics

interquartile range (pp. 304, 306)

lower quartile   (p. 306)

measures of variation (p. 306)

outlier   (p. 307)

quartiles   (pp. 304, 306)

range   (p. 306)

upper quartile   (p. 306)

Problem Solving

use a table   (p. 255)

UNDERSTANDING AND USING THE VOCABULARY

Choose the letter of the term that best matches each statement or phrase.

1. In the coordinate plane, the axes intersect at the __?__.

2. A __?__ is a set of ordered pairs.

3. __?__ are graphed on a coordinate plane.

4. In a coordinate system, the __?__ is a horizontal line.

5. In the ordered pair, A(2, 7), 7 is the __?__.

6. The coordinate axes separate a plane into four __?__.

7. An equation whose graph is a non-vertical straight line is called a __?__.

8. In the relation {(4, −2), (0, 5), (6, 2), (−1, 8)}, the __?__ is the set {−1, 0, 4, 6}.

9. The domain contains values represented by the __?__.

a. domain

b. independent variable

c. linear function

d. ordered pairs

e. origin

f. quadrants

g. relation

h. *x*-axis

i. *y*-axis

j. *y*-coordinate

SKILLS AND CONCEPTS

OBJECTIVES AND EXAMPLES	REVIEW EXERCISES

Upon completing this chapter, you should be able to:

Use these exercises to review and prepare for the chapter test.

- graph ordered pairs on a coordinate plane (Lesson 5–1)

Graph $T(3, -2)$ and state the quadrant in which the point is located.

$T(3, -2)$ is located in Quadrant IV.

Graph each point.

10. $A(4, 2)$

11. $B(-1, 3)$

12. $C(0, -5)$

13. $D(-3, -2)$

Write the ordered pair for each point shown at the right. State the quadrant in which the point is located.

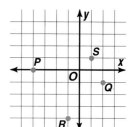

14. P

15. Q

16. R

17. S

- identify the domain, range, and inverse of a relation (Lesson 5–2)

Determine the domain, range, and inverse of $\{(6, 6), (4, -3), (6, 0)\}$.

The domain is $\{4, 6\}$.

The range is $\{-3, 0, 6\}$.

The inverse is $\{(6, 6), (-3, 4), (0, 6)\}$.

State the domain and range of each relation.

18. $\{(4, 1), (4, 6), (4, -1)\}$

19. $\{(-3, 5), (-3, 6), (4, 5), (4, 6)\}$

20. $\{(-2, 1), (-5, 1), (-7, 1)\}$

21. $\{(-3, 1), (-2, 0), (-1, 1), (0, 2)\}$

Draw a mapping and a graph for each relation. State the inverse of each relation.

22. $\{(4, 4), (-3, 5), (4, -1), (0, 3)\}$

23. $\{(0, 2), (3, -1), (2, 2), (-2, -1)\}$

- determine the range for a given domain (Lesson 5–3)

Solve $2x + y = 8$ if the domain is $\{3, 2, 1\}$.

Solve for y.

$2x + y = 8$
$\quad y = 8 - 2x$

x	$8 - 2x$	y	(x, y)
3	$8 - 2(3)$	2	(3, 2)
2	$8 - 2(2)$	4	(2, 4)
1	$8 - 2(1)$	6	(1, 6)

Solve each equation if the domain is $\{-4, -2, 0, 2, 4\}$.

24. $y = 4x + 5$

25. $x - y = 9$

26. $3x + 2y = 9$

27. $4x - 3y = 0$

Make a table and graph the solution set for each equation and domain.

28. $y = 7 - 3x$ for $x = \{-3, -2, -1, 0, 1, 2, 3\}$

29. $5x - y = -3$ for $x = \{-2, 0, 2, 4, 6\}$

OBJECTIVES AND EXAMPLES

• graph linear equations   (Lesson 5–4)

Graph $y = 3x - 4$.

x	3x − 4	y	(x, y)
0	3(0) − 4	−4	(0, −4)
1	3(1) − 4	−1	(1, −1)
2	3(2) − 4	2	(2, 2)

• determine whether a given relation is a function (Lesson 5–5)

Is {(3, 2),(5, 3), (4, 3), (5, 2)} a function?

Because there are two values of y for one value of x, 5, the relation is *not* a function.

• find the value of a function for a given element of the domain   (Lesson 5–5)

Given $g(x) = 2x - 1$, find $g(-6)$.

$$g(-6) = 2(-6) - 1$$
$$= -12 - 1$$
$$= -13$$

• write an equation to represent a relation, given some of the solutions for the equation   (Lesson 5–6)

Write an equation for the relation given in the chart below.

a	−2	−1	0	1	2
b	20	17	14	11	8

The equation is $b = 14 - 3a$.

REVIEW EXERCISES

Graph each equation.

30. $y = -x + 2$

31. $x + 5y = 4$

32. $2x - 3y = 6$

33. $5x + 2y = 10$

34. $\frac{1}{2}x + \frac{1}{3}y = 3$

35. $y - \frac{1}{3} = \frac{1}{3}x + \frac{2}{3}$

Determine whether each relation is a function.

36.

a	b
−2	6
3	−2
3	0
4	6

37.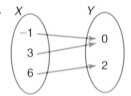

38. {(3, 8), (9, 3), (−3, 8), (5, 3)}

39. $x - y^2 = 4$

40. $xy = 6$

41. $3x - 4y = 7$

If $g(x) = x^2 - x + 1$, find each value.

42. $g(2)$

43. $g(-1)$

44. $g\left(\frac{1}{2}\right)$

45. $g(a + 1)$

46. $g(-2a)$

47. $2g(a - 3)$

Write an equation for each relation.

48.

x	0	1	2	3	4
y	5	8	11	14	17

49.

x	2	4	5	7	10
y	−2	0	1	3	6

50.

x	3	6	9	12	15
y	−1	−3	−5	−7	−9

OBJECTIVES AND EXAMPLES

- calculate and interpret the range, quartiles, and interquartile range of sets of data (Lesson 5–7)

Find the range, median, upper quartile, lower quartile, and interquartile range for the set of data below.

    25, 20, 30, 24, 22, 26, 28, 29, 19

Order the set of data from least to greatest.
    19 20 22 24 25 26 28 29 30

The range is $30 - 19 = 11$.

The median is the middle number, 25.

The lower quartile is $\frac{20 + 22}{2}$ or 21.

The upper quartile is $\frac{28 + 29}{2}$ or 28.5.

The interquartile range is $28.5 - 21$ or 7.5.

REVIEW EXERCISES

Find the range, median, upper quartile, lower quartile, and interquartile range for each set of data.

51. 30, 90, 40, 70, 50, 100, 80, 60

52. 3, 3.2, 45, 7, 2, 1, 3.4, 4, 5.3, 5, 78, 8, 21, 5

53. 85, 77, 58, 69, 62, 73, 55, 82, 67, 77, 59, 92, 75, 69, 76

54. The average annual snowfall, in inches, for each of 12 northeastern cities is listed below.

| 111.5 | 70.7 | 59.8 | 68.6 | 63.8 | 254.8 |
| 64.3 | 82.3 | 91.7 | 88.9 | 110.5 | 77.1 |

APPLICATIONS AND PROBLEM SOLVING

55. Finance To save for a new bicycle, Ralph begins a savings plan. Ralph's savings are described by the equation $s = 5w + 56$, where s represents his total savings in dollars and w represents the number of weeks since the start of the savings plan. (Lesson 5–4)

  a. Draw a graph of this equation.

  b. Use the graph to determine how much money Ralph had already saved when the plan started.

56. Car Rental The cost of a one-day car rental from A-1 Car Rental is $41 if you drive 100 miles, $51.80 if you drive 160 miles, and $63.50 if you drive 225 miles. (Lesson 5–6)

  a. Write an equation to describe this relationship.

  b. Use the equation to determine the per-mile charge.

57. Entertainment The ratings of the top 20 favorite syndicated television programs for 1993–94 are listed below. (Lesson 5–7)

Program	Rating
Action Pack Network	6.1
Baywatch	6.5
Cops	5.7
Current Affair	6.6
Donahue	5.1
Entertainment Tonight	8.4
Family Matters	5.9
Hard Copy	6.7
Inside Edition	7.3
Jeopardy!	12.6
Married With Children	6.9
National Geographic on Assignment	7.5
Oprah Winfrey	9.7
Roseanne	7.9
Sally Jessy Raphael	5.2
Star Trek	10.8
Star Trek: Deep Space Nine	8.2
Wheel of Fortune	14.7
Wheel of Fortune-Weekend	7.2
World Wrestling Federation	5.7

  a. Find the range, quartiles, and interquartile range for the ratings.

  b. Identify any outliers.

A practice test for Chapter 5 is provided on page 791.

ALTERNATIVE ASSESSMENT

COOPERATIVE LEARNING PROJECT

Graphs and Business   In this chapter, you learned how to graph relations and determine whether the relation is a linear equation. Once a relation has been graphed, the process of analyzing it to determine more applicable information is necessary in order to completely understand the problem or situation.

In this project, imagine that your family owns a catering business and you want to look at the amount that you charge for banquets. You have been asked to present a report to the other members of your family that shows recommended charges and how they will affect your profit. After some research, you find that the initial basic charge for a banquet should be $250 plus $4 per person. Your food costs per person are $2.50, and your other costs such as labor, time, etc. total $1.00 per person.

Graph the various charges determined by the number of persons. Graph the various costs determined by the number of persons.

Follow these steps to organize your presentation.

- Determine the charges for various banquet sizes.
- Determine the costs for various banquet sizes.
- Determine how to plot and display your graphs appropriately.
- Write an algebraic model that describes the total charge for a banquet.
- Write an algebraic model that describes the total cost for a banquet.
- Write several paragraphs for your presentation and incorporate your algebraic models and graphs to help your presentation.

THINKING CRITICALLY

- Use one set of data to construct two graphs in which each represents the data, but portrays different meanings.
- Create a relation that is a function and one that is not a function. Compare and contrast the two relations.

PORTFOLIO

You have been asked to show another student how to graph an equation, but it must be completely written and nonverbal. Select one of the graphing assignments from this chapter and list the steps involved in graphing this equation. Be sure to include your steps in the appropriate order.

When you have completed this paper, give it to another student in your class and have him or her read it and follow the written steps to graph a different equation. Collect his or her graph and check for accuracy. Place both of these in your portfolio.

SELF EVALUATION

Graphs can be used to analyze functions or data. To *analyze* means to separate or distinguish all the parts of something in order to discover more information about the complex whole element.

Assess yourself. How analytical are you? Do you delve deeper into a problem and look for smaller components that will help you understand the larger problem, or do you look at the big picture and attempt to solve the problem within the larger view of things? Describe two or three ways in which you could divide a complex problem or situation in mathematics and/or in your daily life in order to analyze it.

In·ves·ti·ga·tion

Smoke Gets In Your Eyes

MATERIALS NEEDED

stopwatch

grid paper

balloons

tape measure

string

In the 1950s and '60s, scientists piled up a mountain of evidence on the life-threatening health consequences of smoking. Recently there have been studies done on the effects of secondhand smoke. Secondhand smoke is the smoke others breathe when they are close to a person who is smoking. In 1993, the Environmental Protection Agency (EPA) estimated that secondhand smoke is responsible for several thousands of cases of lung cancer among nonsmokers each year in the U.S. Passive smoke joins a select company of only about a dozen other environmental pollutants in this risk category.

As an aware citizen and health-conscious teenager, you understand this problem and want to start a campaign to limit the effects of secondhand smoke. In this Investigation, you will gather evidence about how much air is inhaled when you breathe and compare that to smoke in the room. You will also examine the financial and health aspects of this problem. Work in groups of three. Make an Investigation Folder in which you can store all of your work on this Investigation for future use.

Adult Cigarette Use in the United States

Source: U.S. Department of Health and Human Resources

GROUP MEMBER	1	2	3
Number of breaths			
Circumference of balloon			
Radius of balloon			
Volume of balloon			
Volume of each breath			

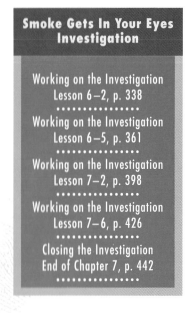

CIGARETTE SMOKE IS HARMFUL!

LUNGS AT WORK NO SMOKING

AMERICAN LUNG ASSOCIATION®

COLLECT THE DATA

1 Begin by copying the chart above onto a sheet of paper.

2 Stretch each balloon. Have each member of the group blow up at least three round balloons to full size and let the air out.

3 Give each team member time to rest.

4 Cut a piece of string that is approximately 30 inches long. Tie the string so that it can form a circle that is approximately 8 inches in diameter when laid on a flat surface.

5 One member of the group should hold the string hoop around the balloon and another member should blow the balloon up again. The remaining group member should count and record the number of puffs it takes to blow up each balloon to the size of the hoop. Each group member should blow up one balloon.

6 Measure each inflated balloon to estimate its circumference at the fullest part. Find the approximate radius of the balloon by using the formula $r = \frac{C}{2\pi}$. Record the radius.

ANALYZE THE DATA

7 Display your data on a graph.

8 Calculate the average number of breaths it takes to fill a balloon.

9 The volume of the spherical balloon can be estimated by the formula $V = \frac{4}{3}\pi r^3$. Compute the volume of each balloon. Then estimate the volume of each breath.

You will continue working on this Investigation throughout Chapters 6 and 7.

Be sure to keep your chart and materials in your Investigation Folder.

Smoke Gets In Your Eyes Investigation

Working on the Investigation
Lesson 6–2, p. 338

Working on the Investigation
Lesson 6–5, p. 361

Working on the Investigation
Lesson 7–2, p. 398

Working on the Investigation
Lesson 7–6, p. 426

Closing the Investigation
End of Chapter 7, p. 442

Analyzing Linear Equations

Objectives

In this chapter, you will:

- find the slope of a line, given the coordinates of two of its points,
- write linear equations in point-slope, standard, and slope-intercept forms,
- draw a scatter plot and find the equation of a best-fit line for the data,
- solve problems by using models,
- graph linear equations, and
- use slope to determine if two lines are parallel or perpendicular.

Immigration on the Rise

Foreign-born residents as share of U.S. population

Source: *U.S. News and World Report,* September 25, 1995

Almost $20 billion a year is spent on bilingual programs in U.S. schools. These programs were launched in 1968 to help immigrant children learn English. The percentage of foreign-born Americans is rising. More than 22 states have enacted laws proclaiming English as the official language. What are your views on making English the official language of the U.S.?

TIME *Line*

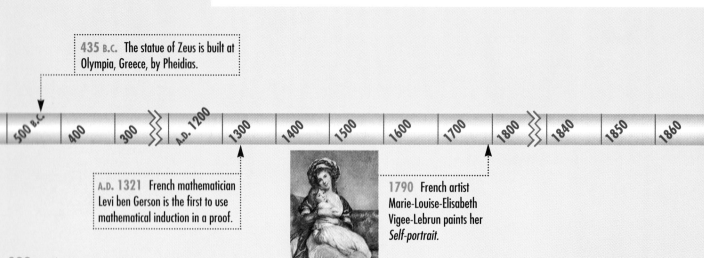

435 B.C. The statue of Zeus is built at Olympia, Greece, by Pheidias.

A.D. 1321 French mathematician Levi ben Gerson is the first to use mathematical induction in a proof.

1790 French artist Marie-Louise-Elisabeth Vigee-Lebrun paints her *Self-portrait.*

PEOPLE IN THE
NEWS

O ne student from Webster, Texas, **Jorge Arturo Pineda Aguilar,** is a member of the new immigrant minority. He and his family are from Monterrey, Mexico, and recently moved to Texas. At age 13, Jorge is one of the recent winners of the 1995 Youth Honor Awards presented by *Skipping Stones,* a multicultural children's magazine. His 1995 winning essay, written in both Spanish and English, deals with racial and ethnic prejudice.

Jorge regrets that so many people have ill-feelings about someone who doesn't speak English. He writes, "All of us are human beings and the color or nationality of a person should not be an issue. What should matter are the feelings of each person. We have to be together—work together for a common well being."

Chapter Project

T he table below shows the per capita income and the total non-English speaking population (ages 5 years or older) in 15 selected states.

State	1993 Per Capita Income ($)	Non-English Speaking Population
AL	17,234	107,866
AR	16,143	60,781
FL	20,857	2,098,315
IL	22,582	1,499,112
IN	19,203	245,826
KY	17,173	86,482
LA	16,667	391,994
NC	17,488	240,866
NM	16,297	493,999
NY	24,623	3,908,720
OK	17,020	145,798
TN	17,666	131,550
TX	19,189	3,970,304
VA	21,634	418,521
WV	16,209	44,203

The population is given according to the 1990 U.S. Census.

Round the data to the nearest thousand. Graph the data for each state by plotting the per capita income data along the horizontal axis and the non-English-speaking population data along the vertical axis.

- What pattern, if any, do you notice?
- Can a best-fit line be drawn? If so, what might be the equation of this line?
- Can a valid relationship exist between these data? Why or why not?

1968 The U.S. Postal Service issues a stamp honoring the life of Chief Joseph of the Nez Perce Nation.

1973 Marion Wright Edelman founds the Children's Defense Fund.

1880 1890 1900 1910 1920 1930 1940 1950 1960 1970 1980 1990 2000

1926 Mexican artist Diego Rivera paints *The Tortilla Maker.*

1995 Twins Chris and Courtney Salthouse from Chamblee, Georgia *both* score a perfect 1600 on their SATs.

MODELING MATHEMATICS

6-1A Slope

Materials: geoboard   rubber bands

A Preview of Lesson 6–1

The **slope** of a line segment is the ratio of the change in the vertical distance between the endpoints of the segment to the change in the horizontal distance. You can use a geoboard to model a line segment and calculate its slope.

Activity   **Model line segment *AB* whose endpoints are *A*(1, 2) and *B*(3, 5). Then find the slope of segment *AB*.**

Step 1   Suppose the pegs of the geoboard represent ordered pairs on a coordinate plane. Let's define the lower left peg as the point representing the ordered pair (1, 1). Locate the pegs that represent (1, 2) and (3, 5). Place a rubber band around these pegs to model segment *AB*.

Step 2   Use a different color rubber band to show the horizontal distance from the *x* value of *A* to the *x* value of *B*. Use another color rubber band to show the vertical distance from the *y* value of *A* to the *y* value of *B*.

Step 3   As you move from point *A* to point *B*, you can go right and then up. These are positive directions on the coordinate plane. So, the red rubber band represents 2 units. The green rubber band represents 3 units. The slope of segment *AB* is the ratio of the *y* distance to the *x* distance or $\frac{3}{2}$.

Model   1. Use a geoboard to find the slope of a segment whose endpoints are at (5, 2) and (1, 4).

Write   2. For the activity above and Exercise 1, write ratios that compare the differences in the *y* values of the ordered pairs to the differences in the *x* values of the ordered pairs. How do these compare with the slopes of the respective segments?

3. What value could you assign to the lower left peg if you were modeling segment *CD* with endpoints *C*(3, 5) and *D*(6, 7)?

4. Write a rule for finding the slope of any line segment shown on a coordinate plane.

5. Do you think the rule applies to ordered pairs that involve negative numbers? Explain your answer and give examples.

Slope

APPLICATION
Construction

Alan and Mabel Wong bought a lot in San Francisco on which to build a house. The lot is located on a street having an 11% grade. The length of the sidewalk along the lot is 43 feet. City ordinances state that there must be a 5-foot clearance on each side of the house to the property line. The house they would like to build is 32 feet wide. Will they be able to build their house on the lot they bought? *This problem will be solved in Example 4.*

What do you think of when you hear the word *slope*? You might think of how something slants either uphill or downhill. Suppose you placed a ladder against a wall. Then you moved the base of the ladder out about a foot. What happens to the slant of the ladder? Suppose you move it again. What happens? Observe the pattern in the diagram below.

Notice that the top of the ladder moves down the wall when the bottom of the ladder moves out. As the top moves down the wall, the slant, or slope, becomes less steep. If you continued to move the ladder out, eventually it would lie flat on the ground, and there would be no slant at all.

The steepness of the line representing the ladder is called the **slope** of the line. It is defined as the ratio of the **rise,** or vertical change, to the **run,** or horizontal change, as you move from one point on the line to another. The graph at the right shows a line that passes through the origin and the point at (5, 4).

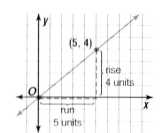

$$\text{slope} = \frac{\text{change in } y \text{ (rise)}}{\text{change in } x \text{ (run)}}$$

$$= \frac{4}{5}$$

So, the slope of this line is $\frac{4}{5}$.

Definition of Slope	The slope *m* of a line is the ratio of the change in the **y-coordinates** to the corresponding change in the **x-coordinates**.

For simplicity, we will refer to the change in the y-coordinates as the y change and the corresponding change in the x-coordinates as the x change.

Example **Determine the slope of each line.**

a.

$$\frac{y \text{ change}}{x \text{ change}} = \frac{3}{2}$$

$$m = \frac{3}{2}$$

b.

$$\frac{y \text{ change}}{x \text{ change}} = \frac{4}{-3}$$

$$m = \frac{4}{-3}$$

c.

$$\frac{y \text{ change}}{x \text{ change}} = \frac{0}{2}$$

$$m = 0$$

d.

$$\frac{y \text{ change}}{x \text{ change}} = \frac{3}{0}$$

Since division by 0 is undefined, the slope is undefined.

Let's analyze the slopes of the graphs in Example 1.
- Graph a slopes upward from left to right and has a positive slope.
- Graph b slopes downward from left to right and has a negative slope.
- Graph c is a horizontal line and has a slope of 0.
- Graph d is a vertical line and its slope is undefined.

These observations are true of other lines that have the same characteristics.

Since a line is made up of an infinite number of points, you can use any two points on that line to find the slope of the line. So, we can generalize the definition of slope for any two points on the line.

Determining Slope Given Two Points	Given the coordinates of two points, (x_1, y_1) and (x_2, y_2), on a line, the slope *m* can be found as follows: $$m = \frac{y_2 - y_1}{x_2 - x_1}, \text{ where } x_1 \neq x_2.$$

y_2 is read "y sub 2." The 2 is called a subscript.

Example **Determine the slope of the line that passes through (2, −5) and (7, −10).**

Method 1
Let $(2, -5) = (x_1, y_1)$ and $(7, -10) = (x_2, y_2)$.

$$m = \frac{y_2 - y_1}{x_2 - x_1}$$

$$= \frac{-10 - (-5)}{7 - 2}$$

$$= \frac{-5}{5} \text{ or } -1$$

Method 2
Let $(7, -10) = (x_1, y_1)$ and $(2, -5) = (x_2, y_2)$.

$$m = \frac{y_2 - y_1}{x_2 - x_1}$$

$$= \frac{-5 - (-10)}{2 - 7}$$

$$= \frac{5}{-5} \text{ or } -1$$

The slope of the line is −1.

As you can see in Example 2, it does not matter which ordered pair is selected as (x_1, y_1).

You can use a spreadsheet to calculate the slopes of several lines very quickly.

A spreadsheet is a table of cells that can contain text (labels) or numbers and formulas. Each cell is named by the column and row in which it is located. In the spreadsheet below, cells A1, B1, C1, D1, E1, and F1 contain labels.

a. Enter the formula (E2−C2)/(D2−B2) into cell F2.

b. Copy the formula to all of the cells in the F column. When you do, the spreadsheet automatically changes the formula to correspond to the values in that row. That is, for F3 the formula becomes (E3−C3)/(D3−B3), for F4 it becomes (E4−C4)/(D4−B4), and so on.

	===A===	===B===	===C===	===D===	===E===	===F===
1	LINE	X1	Y1	X2	Y2	SLOPE
2	A	2	3	4	6	1.5
3						
4						
5						

Your Turn

a. Enter the following ordered pairs from each line in the spreadsheet.

line a	$(5, 5), (11, 11)$	line e	$(-1, 5), (6, 3)$
line b	$(4, -4), (3, 5)$	line f	$(5, 3), (4, 0)$
line c	$(6, -1), (4, -1)$	line g	$(2, 3), (11, 14)$
line d	$(11, 3), (1, 1)$	line h	$(10, -5), (10, 3)$

b. Use the calculate command to have the spreadsheet compute the slope.

c. Use the ordered pairs to graph each line on grid paper. Label each line.

d. From the results of your spreadsheet, make lists of lines with positive slope, negative slope, and 0 slope.

e. Compare the lists with the graphs of the lines. Write a statement to summarize your observations.

If you know the slope of a line and the coordinates of one of the points on a line, you can find the coordinates of other points on that line.

Example ❸ **Determine the value of r so the line through $(r, 6)$ and $(10, -3)$ has a slope of $-\dfrac{3}{2}$.**

$$m = \frac{y_2 - y_1}{x_2 - x_1}$$

$$-\frac{3}{2} = \frac{-3 - 6}{10 - r} \quad \begin{array}{l} (x_1, y_1) = (r, 6) \\ (x_2, y_2) = (10, -3) \end{array}$$

$$-\frac{3}{2} = \frac{-9}{10 - r}$$

$$-3(10 - r) = -9(2) \quad \textit{Find the cross products}$$

$$-30 + 3r = -18 \quad \textit{Solve for r}$$

$$3r = 12$$

$$r = 4$$

You can use slope with other mathematical skills to solve real-world problems like the one presented at the beginning of this lesson.

Example **4**

Construction

LOOK BACK

You can refer to Lesson 4-3 for more information on trigonometric functions.

Refer to the application at the beginning of the lesson. Will the Wongs be able to build their house on their lot?

The width of the house and the two 5-foot clearances equal $32 + 2(5)$ or 42 feet. Thus, the horizontal measure between the property lines must be at least 42 feet.

We can use trigonometric ratios to determine if there is adequate room for the house. A grade of 11 percent means that the rise is 11 feet while the run is 100 feet. We can use this information to find the measure of $\angle A$.

$\tan A = \dfrac{11}{100}$ $\leftarrow$ *opposite*
  $\leftarrow$ *adjacent*

$\tan A = 0.11$

$A \approx 6.3°$ *Use a calculator.*

We can use the measure of $\angle A$ to find the horizontal measure x of the property.

$\cos A = \dfrac{x}{43}$ $\leftarrow$ *adjacent*
  $\leftarrow$ *hypotenuse*

$43(\cos 6.3°) = x$ *Multiply each side by 43.*

$42.74 = x$ *Use a calculator.*

The lot is 42.74 feet wide. The house and clearances will fit the 43-foot lot. So, the Wongs can build their house on the lot they bought.

CHECK FOR UNDERSTANDING

Communicating Mathematics

Study the lesson. Then complete the following.

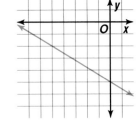

1. **Explain** how you would find the slope of the line in the graph at the right.

2. **Describe** what it means if a road has an 8% grade.

3. Would it matter if you subtracted the x-coordinates in the opposite order from the way you subtracted the y-coordinates?

4. **Draw** the graph of a line having each slope.
   a. a positive slope
   b. a negative slope
   c. a slope of 0
   d. undefined slope

5. **Explain** why the formula for determining the slope using two points does not apply to vertical lines.

MODELING MATHEMATICS

6. Use a geoboard to model line segments with each slope. Make a sketch of your model.
   a. 4
   b. $\dfrac{3}{4}$
   c. 0
   d. undefined
   e. $-\dfrac{3}{2}$

Guided Practice

Determine the slope of each line.

7.

8.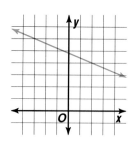

Determine the slope of the line that passes through each pair of points.

9. $(7, -4), (9, -1)$ 10. $(5, 7), (-2, -3)$ 11. $(0.75, 1), (0.75, -1)$

Determine the value of _r_ so the line that passes through each pair of points has the given slope.

12. $(6, -2), (r, -6), m = -4$ 13. $(9, r), (6, 3), m = -\dfrac{1}{3}$

14. **Architecture**   The slope of the roof line of a building is often referred to as the *pitch* of the roof.

   a. Use a millimeter ruler to find the rise and run of the roof of the tower on the horse barn in Versailles, Kentucky, shown at the right.

   b. Then write the pitch as a decimal.

EXERCISES

Practice

Determine the slope of each line.

15. a

16. b

17. c

18. d

19. e

20. f

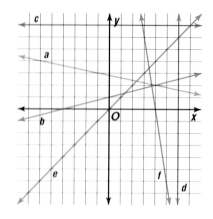

Determine the slope of the line that passes through each pair of points.

21. $(2, 3), (9, 7)$ 22. $(-3, -4), (5, -1)$ 23. $(2, -1), (5, -3)$

24. $(2, 6), (-1, 3)$ 25. $(-5, 4), (-5, -1)$ 26. $(-2, 3), (8, 3)$

27. $(4, -5), (4, 2)$ 28. $\left(2\dfrac{1}{2}, -1\dfrac{1}{2}\right), \left(-\dfrac{1}{2}, \dfrac{1}{2}\right)$ 29. $\left(\dfrac{3}{4}, 1\dfrac{1}{4}\right), \left(-\dfrac{1}{2}, -1\right)$

Determine the value of *r* so the line that passes through each pair of points has the given slope.

30. $(5, r), (2, -3), m = \frac{4}{3}$

31. $(-2, 7), (r, 3), m = \frac{4}{3}$

32. $(4, -5), (3, r), m = 8$

33. $(6, 2), (9, r), m = -1$

34. $(4, r), (r, 2), m = -\frac{5}{3}$

35. $(r, 5), (-2, r), m = -\frac{2}{9}$

Draw a line through the given point that has the given slope.

36. $(3, -1), m = \frac{1}{3}$

37. $(-2, -3), m = \frac{4}{3}$

38. $(4, -2), m = -\frac{2}{5}$

39. The points $A(12, -4)$ and $B(6, 8)$ lie on a line. Find the coordinates of a third point that lies on line AB. Describe how you determined the coordinates of this point.

Critical Thinking

40. Carpentry A carpenter was a member of a crew building a roof that is 30 feet long at the base. The roof had a pitch of 4 inches for every foot of length along the base. His task on the crew was to put in vertical supports every 16 inches along the base. He would climb up the ladder, measure 16 inches horizontally, and then measure the vertical height to the roof. He then climbed down the ladder, sawed the piece he needed, and went back up the ladder to put it in place. He wondered if there wasn't some way he could figure out how long each support would be ahead of time so he wouldn't have to climb the ladder so many times. Explain how you can use what you have learned in this chapter to help him out.

30 ft

Applications and Problem Solving

41. Road Construction The Castaic Grade in Southern California has a grade of 5%. The length of the roadway is 5.6 miles. What is the change in elevation from the top of the grade to the bottom of the grade in feet?

42. Carpentry Julio Mendez is a carpenter. He is building a staircase between the first and second floors of a house, a height of 9 feet. The *tread*, or depth of each step, must be 10 inches. The slope of the staircase cannot exceed $\frac{3}{4}$.

riser

tread

9 ft

a. How many steps should he plan?

b. What is the measure of the riser, or height of each step?

43. Driving The Eisenhower Tunnel in Colorado was completed in 1973 and is the fourth longest road tunnel in the United States. The eastern entrance of the Eisenhower Tunnel is at an elevation of 11,080 feet. The tunnel is 8941 feet long and has an upgrade of 0.895% toward the western end. What is the elevation of the western end of the tunnel?

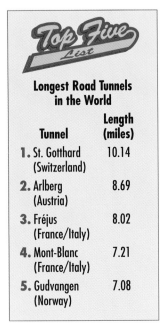

Longest Road Tunnels in the World

Tunnel	Length (miles)
1. St. Gotthard (Switzerland)	10.14
2. Arlberg (Austria)	8.69
3. Fréjus (France/Italy)	8.02
4. Mont-Blanc (France/Italy)	7.21
5. Gudvangen (Norway)	7.08

44. Architecture Use a millimeter ruler to estimate the pitch, or slope, of each object.

a.

Uro Indian huts of Peru

b.

S. Miniato, Florence, Italy

c.

Great Pyramid at Giza, Eqypt

Mixed Review

45. Statistics Find the range, median, upper and lower quartiles, and interquartile range for the set of data. (Lesson 5–7)
3, 3.2, 6, 45, 7, 26, 2, 3.4, 4, 5.3, 5, 78, 8, 1, 5

46. Patterns Copy and complete the table below. (Lesson 5–5)

n	1	2	3	4	5
$f(n)$			12	11	10

47. Use the formula $I = prt$ to find r if $I = \$2430$, $p = \$9000$, and $t = 2$ years 3 months. (Lesson 4–4)

48. Solve $\dfrac{6}{14} = \dfrac{7}{x-3}$. (Lesson 4–1)

Solve each equation. Check your solution.

49. $\dfrac{2}{3}x + 5 = \dfrac{1}{2}x + 4$ (Lesson 3–5)   **50.** $3x - 8 = 22$ (Lesson 3–3)

51. Animals The average lifespans of 20 different animals are listed below. (Lesson 2–2)

Animal	Lifespan (years)	Animal	Lifespan (years)
Baboon	20	Mouse	3
Camel	12	Opossum	1
Cow	15	Pig	10
Elephant	40	Rabbit	5
Fox	7	Sea Lion	12
Gorilla	20	Sheep	12
Hippopotamus	25	Squirrel	10
Kangaroo	7	Tiger	16
Lion	15	Wolf	5
Monkey	15	Zebra	15

a. How many animals live between 7 and 16 years, inclusive?

b. Make a line plot of the average lifespans of the animals from part a.

c. Which number occurred most frequently?

d. How many animals have an average lifespan of at least 20 years?

52. Evaluate $29 - 3(9 - 4)$. (Lesson 1–3)

Writing Linear Equations in Point-Slope and Standard Forms

What YOU'LL LEARN

- To write linear equations in point-slope form, and
- to write linear equations in standard form.

Why IT'S IMPORTANT

You can write linear equations to solve problems involving geometry.

CONNECTION
Geography

If you lived in Miami, Florida, and moved to Denver, Colorado, what adjustments do you think you might have to make? Obviously, the weather is different, but did you know you would also have to adjust to living at a higher altitude? As the altitude increases, the oxygen in the air decreases. This can affect your breathing and cause dizziness, headache, insomnia, and loss of appetite.

Over time, people who move to higher elevations experience acclimatization, or the process of getting accustomed to the new climate. Their bodies develop more red blood cells to carry oxygen to the muscles. Generally, long-term acclimatization to an altitude of 7000 feet takes about 2 weeks. After that, it takes 1 week for each additional 2000 feet of altitude. The graph at the right shows the acclimatization for altitudes greater than 7000 feet.

What are the independent and dependent variables of this relation?

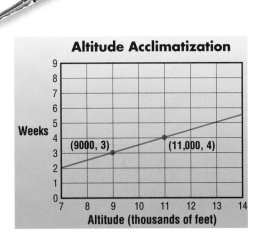

Altitude Acclimatization

Weeks (9000, 3) (11,000, 4)

Altitude (thousands of feet)

We can use any two points on the graph to find the slope of the line. For example, let (9000, 3) and (11,000, 4) represent (x_1, y_1) and (x_2, y_2), respectively.

$$m = \frac{y_2 - y_1}{x_2 - x_1}$$

$$= \frac{4 - 3}{11,000 - 9000} \text{ or } \frac{1}{2000}$$

Suppose we let (x, y) represent any other point on the line. We can use the slope and one of the given ordered pairs to write an equation for the line.

$$m = \frac{y_2 - y_1}{x_2 - x_1}$$

$$\frac{1}{2000} = \frac{y - 3}{x - 9000} \qquad m = \frac{1}{2000}, (x_1, y_1) = (9000, 3), (x_2, y_2) = (x, y)$$

$$\frac{1}{2000}(x - 9000) = y - 3 \qquad \textit{Multiply each side by } (x - 9000).$$

$$y - 3 = \frac{1}{2000}(x - 9000) \qquad \textit{Reflexive property } (=)$$

Since this form of the equation was generated using the coordinates of a known point and the slope of the line, it is called the **point-slope form.**

Point-Slope Form of a Linear Equation	For a given point (x_1, y_1) on a nonvertical line having slope m, the point-slope form of a linear equation is as follows. $$y - y_1 = m(x - x_1)$$

You can write an equation in point-slope form for the graph of any nonvertical line if you know the slope of the line and the coordinates of one point on that line.

Example Write the point-slope form of an equation for each line.

a. a line that passes through $(-3, 5)$ and has a slope of $-\frac{3}{4}$

$$y - y_1 = m(x - x_1)$$
$$y - 5 = -\frac{3}{4}[x - (-3)] \quad \text{\textit{Replace } } x_1 \text{ \textit{with} } -3,$$
$$y - 5 = -\frac{3}{4}(x + 3) \quad \quad y_1 \text{ \textit{with 5, and} } m \text{ \textit{with} } -\frac{3}{4}.$$

An equation of the line is $y - 5 = -\frac{3}{4}(x + 3)$.

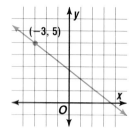

b. a horizontal line that passes through $(-6, 2)$

Horizontal lines have a slope of 0.
$$y - y_1 = m(x - x_1)$$
$$y - 2 = 0[x - (-6)]$$
$$y - 2 = 0$$
$$y = 2$$

An equation of the line is $y = 2$.

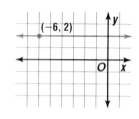

F Y I

Tibet has one of the highest altitudes in the world. Nomads in Tibet tend their yaks for 5 months at altitudes of 14,000–17,000 feet above sea level. Then they return to their permanent homes, which are located 12,000 to 14,000 feet above sea level.

You can also write the equations of horizontal lines by using the y-coordinate as in Example 1b.

A vertical line has a slope that is undefined, so you cannot use the point-slope form of an equation. However, you can write the equation of a vertical line by using the coordinates of the points through which it passes. Suppose the line passes through $(3, 5)$ and $(3, -2)$. The equation of the line is $x = 3$, since the x-coordinate of every point on the line is 3.

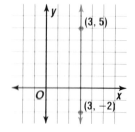

Any linear equation can also be expressed in the form $Ax + By = C$, where A, B, and C are integers and A and B are not both zero. This is called the **standard form** of a linear equation. *Usually, $A \geq 0$.*

Standard Form of a Linear Equation	The standard form of a linear equation is $Ax + By = C$, where A, B, and C are integers, $A \geq 0$, and A and B are not both zero.

Linear equations that are written in point-slope form can be rewritten in standard form.

Example **2** **Write $y + 5 = -\frac{5}{4}(x - 2)$ in standard form.**

$$y + 5 = -\frac{5}{4}(x - 2)$$

$4(y + 5) = 4 \cdot \left(-\frac{5}{4}\right)(x - 2)$ *Multiply each side by 4 to eliminate the fraction.*

$4y + 20 = -5x + 10$ *Distributive property*

$4y = -5x - 10$ *Subtract 20 from each side.*

$5x + 4y = -10$ *Add 5x to each side.*

The standard form of the equation is $5x + 4y = -10$.

You can write an equation of a line if you know the coordinates of two points on that line.

Example **3** **Write the point-slope form and the standard form of an equation of the line that passes through $(-8, 3)$ and $(4, 5)$.**

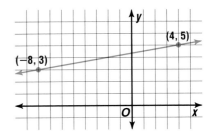

First determine the slope of the line.

$$m = \frac{5 - 3}{4 - (-8)} \text{ or } \frac{1}{6}$$

You can use either point for (x_1, y_1) in the point-slope form.

Method 1: Use $(-8, 3)$.

$$y - 3 = \frac{1}{6}(x - (-8))$$

$$y - 3 = \frac{1}{6}(x + 8)$$

Method 2: Use $(4, 5)$.

$$y - 5 = \frac{1}{6}(x - 4)$$

Both $y - 3 = \frac{1}{6}(x + 8)$ and $y - 5 = \frac{1}{6}(x - 4)$ are valid equations for the point-slope form of an equation for the line passing through $(-8, 3)$ and $(4, 5)$. Now let's find the standard form of each equation.

Method 1

$$y - 3 = \frac{1}{6}(x + 8)$$

$6(y - 3) = 6 \cdot \frac{1}{6}(x + 8)$ *Multiply each side by 6.*

$6y - 18 = x + 8$

$6y = x + 26$ *Add 18 to each side.*

$-x + 6y = 26$ *Subtract x from each side.*

$x - 6y = -26$ *Multiply by −1.*

Method 2

$$y - 5 = \frac{1}{6}(x - 4)$$

$6(y - 5) = 6 \cdot \frac{1}{6}(x - 4)$ *Multiply each side by 6.*

$6y - 30 = x - 4$

$6y = x + 26$ *Add 30 to each side.*

$-x + 6y = 26$ *Subtract x from each side.*

$x - 6y = -26$ *Multiply by −1.*

Regardless of which point was used to find the point-slope form, the standard form of the equation, when all common factors are removed, always results in the same equation.

Example **4** Write the equations of the lines containing the sides of parallelogram *ABCD* in point-slope and standard forms.

INTEGRATION

Geometry

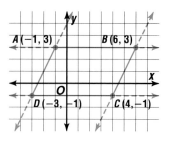

Since the lines containing segments *AB* and *CD* are horizontal, the slope of each line is 0.

The equation of $\overleftrightarrow{AB}$ is $y = 3$.

The equation of $\overleftrightarrow{CD}$ is $y = -1$.

Now use the coordinates of the points to find the slopes and equations of the lines containing the other two sides.

Use $A(-1, 3)$ and $D(-3, -1)$ to find the slope of $\overline{AD}$.

$$m = \frac{3 - (-1)}{-1 - (-3)}$$

$$= \frac{4}{2} \text{ or } 2$$

Let's use $A(-1, 3)$ for (x_1, y_1) in the point-slope form.

$$y - y_1 = m(x - x_1)$$
$$y - 3 = 2[x - (-1)]$$
$$y - 3 = 2(x + 1)$$

Now find the standard form of this equation.

$$y - 3 = 2(x + 1)$$
$$y - 3 = 2x + 2$$
$$y = 2x + 5$$
$$2x - y = -5$$

Use $B(6, 3)$ and $C(4, -1)$ to find the slope of $\overline{BC}$.

$$m = \frac{3 - (-1)}{6 - 4}$$

$$= \frac{4}{2} \text{ or } 2$$

Let's use $B(6, 3)$ for (x_1, y_1) in the point-slope form.

$$y - y_1 = m(x - x_1)$$
$$y - 3 = 2(x - 6)$$

Now find the standard form of this equation.

$$y - 3 = 2(x - 6)$$
$$y - 3 = 2x - 12$$
$$y = 2x - 9$$
$$2x - y = 9$$

The equations of the lines containing the sides of the parallelogram *ABCD* are $y = 3$, $y = -1$, $2x - y = -5$, and $2x - y = 9$.

CHECK FOR UNDERSTANDING

Communicating Mathematics

Study the lesson. Then complete the following.

1. **Explain** what x_1 and y_1 in the slope-intercept form of an equation represent.

2. **Compare and contrast** the graphs of $x = 6$ and $y = 6$.

3. **Explain** why *A* and *B* in the standard form of a linear equation cannot both be zero.

4. **You Decide** A line contains the points $P(-3, 2)$ and $Q(2, 5)$. Elena thinks that $R(12, 12)$ is also on the line. Use the point-slope form to determine if she is correct.

Guided Practice

State the slope and the coordinates of a point through which the line represented by each equation passes.

5. $y + 3 = 4(x - 2)$ 6. $y - (-6) = -\frac{2}{3}(x + 5)$ 7. $3(x + 7) = y - 1$

Write the point-slope form of an equation of the line that passes through the given point and has the given slope.

8. $(9, 1), m = \frac{2}{3}$ **9.** $(-2, 4), m = -3$ **10.** $(-3, 6), m = 0$

Write each equation in standard form.

11. $y + 3 = -\frac{3}{4}(x - 1)$ **12.** $y - \frac{1}{2} = \frac{5}{6}(x + 2)$ **13.** $y - 3 = 2(x + 1.5)$

Write the point-slope form and the standard form of an equation of the line that passes through each pair of points.

14. $(-6, 1), (-8, 2)$ **15.** $(-1, -2), (-8, 2)$ **16.** $(4, 8), (-2.5, 8)$

17. Geometry Write the standard form of the equation of the line containing the hypotenuse of the right triangle shown at the right.

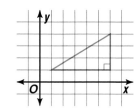

EXERCISES

Practice   **Write the point-slope form of an equation of the line that passes through the given point and has the given slope.**

18. $(3, 8), m = 2$ **19.** $(4, 5), m = 3$ **20.** $(-4, -3), m = 1$

21. $(-6, 1), m = -4$ **22.** $(0, 5), m = 0$ **23.** $(1, 3), m = -2$

24. $(3, 5), m = \frac{2}{3}$ **25.** $(8, -3), m = \frac{3}{4}$ **26.** $(-6, 3), m = -\frac{2}{3}$

Write the standard form of an equation of the line that passes through the given point and has the given slope.

27. $(2, 13), m = 4$ **28.** $(-5, -3), m = 4$

29. $(-4, 6), m = \frac{3}{2}$ **30.** $(-2, -7), m = 0$

31. $(8, 2), m = -\frac{2}{5}$ **32.** $(-5, 5), m = $ undefined

Write the point-slope form of an equation of the line that passes through each pair of points.

33. $(-5, 2), (4, -1)$ **34.** $(6, 1), (7, -4)$

35. $(-8, -1), (6, 5)$ **36.** $(2, 3), (5, 1)$

37. $(4, -2), (8, -2)$ **38.** $(2.5, 3), (-0.5, -4.5)$

Write the standard form of an equation of the line that passes through each pair of points.

39. $(6, 5), (12, -3)$

40. $(-2, -7), (1, 2)$

41. $(-5, 9), (3, -2)$

42. $(0.7, -1.3), (-0.4, 3.1)$

43. $\left(-\frac{2}{3}, -\frac{5}{6}\right), \left(\frac{3}{4}, -\frac{1}{3}\right)$

44. $\left(-2, 7\frac{2}{3}\right), \left(-2, \frac{16}{5}\right)$

Critical Thinking

45. A line contains the points at $(9, 1)$ and $(5, 5)$. Write a convincing argument that the same line intersects the x-axis at $(10, 0)$.

Applications and Problem Solving

46. Geometry Three lines intersect to form triangle ABC, as shown at the right.

a. Write the equation for each line in point-slope form.

b. Write the equation for each line in standard form.

47. Draw a Diagram The Americans with Disabilities Act (ADA) of 1990 states that ramps should have at least a 12-inch run for each rise of 1 inch, with a maximum rise of 30 inches. The post office in Meyersville is changing one of its entrances to accommodate a wheelchair ramp. The distance from the street to the building is 18 feet. The current sidewalk has steps up to the entrance. The entranceway is 30 inches above ground level.

a. Determine if a wheelchair ramp that meets ADA standards can be placed over the existing sidewalk. Explain why or why not.

b. Draw a diagram of an alternative plan for the ramp.

Mixed Review

48. Aviation An airplane flying over Albuquerque at an elevation of 33,000 feet begins its descent to land at Santa Fe, 50 miles away. If the elevation of Santa Fe is 7000 feet, what should be the approximate slope of descent, expressed as a percent? (Lesson 6–1)

49. Statistics The stem-and-leaf plot at the right represents the cost per cup of various brands of coffee. (Lesson 5–7)

a. Find the range and interquartile range for the costs.

b. Identify any outliers.

Stem	Leaf
0	6 6 6 6 8 9 9 9 9
1	0 2 3 4 5 7 7 8 8
2	4 8 9
3	0 2 $1|2 = \$0.12$

50. Draw a mapping for $\{(-6, 0), (-1, 2), (-3, 4)\}$. (Lesson 5–2)

51. If $y = 12$ when $x = 3$, find y when $x = 7$. Assume that y varies directly as x. (Lesson 4–8)

52. **Motion** At 1:30 P.M., an airplane leaves Tucson for Baltimore, a distance of 2240 miles. The plane flies at 280 miles per hour. A second airplane leaves Tucson at 2:15 P.M., and is scheduled to land in Baltimore 15 minutes before the first airplane. At what rate must the second airplane travel to arrive on schedule? (Lesson 4–7)

53. **Work Backward** In Lupita's garden, all of the flowers are either pink, yellow, or white. Given any three of the flowers, at least one of them is pink. Given any three of the flowers, at least one of them is white. Can you say that given any three of the flowers, at least one of them must be yellow? Why? (Lesson 3–3)

54. **Printing** A customer orders a job requiring 2500 sheets of paper. He has a choice of ordering a full carton of 3000 sheets at $27.50 per thousand or "breaking" a carton (ordering exactly the number of sheets required) and paying $38.40 per thousand. Which would be least expensive in this case? (Lesson 2–6)

55. Name the property illustrated by $(a + 3b) + 2c = a + (3b + 2c)$. (Lesson 1–8)

WORKING ON THE
In·ves·ti·ga·tion

Refer to the Investigation on pages 320–321.

Smoke Gets In Your Eyes

The amount of smoke exhaled into a room on each puff of a cigarette is comparable to the amount of air you exhale in a deep breath. The average smoker takes 10 puffs per cigarette.

1 Suppose that each breath into a balloon approximates the smoke from each exhale during smoking. How many cigarettes are represented by the air in the balloon?

2 Make a graph showing the number of breaths and the cumulative smoke in the air from smoking a cigarette. Describe the graph, and write a function that fits the data graphed.

3 Graph the relationship between the radius of the balloon and the circumference of the balloon. Write an equation to represent this relationship.

4 Find the volume of each balloon in milliliters if 1 in$^3 \approx 16.4$ milliliters. The average at-rest exhale is approximately 500 milliliters. How does the volume of the balloon compare with the volume of air you would expect from the number of breaths it took to inflate the balloon? What accounts for the difference?

Add the results of your work to your Investigation Folder.

6-3

Integration: Statistics

Scatter Plots and

Education

Best-Fit Lines

What YOU'LL LEARN

- To graph and interpret points on a scatter plot,
- to draw and write equations for best-fit lines and make predictions by using those equations, and
- to solve problems by using models.

Why IT'S IMPORTANT

You can use scatter plots to display data, examine trends, and make predictions.

If you get a good score on your SAT does that mean you have a good chance of graduating from college? The table below shows the average SAT (Scholastic Assessment Test) scores for freshmen at selected universities and the graduation rate of those students.

School (State)	Average SAT	Graduation Rate
Baylor University (TX)	1045	69%
Brandeis University (MA)	1215	81%
Case Western Reserve University (OH)	1235	65%
College of William and Mary (VA)	1240	90%
Colorado School of Mines (CO)	1200	78%
Georgia Institute of Technology (GA)	1240	68%
Lehigh University (PA)	1140	88%
New York University (NY)	1145	69%
Penn State University, Main Campus (PA)	1096	61%
Pepperdine University (CA)	1070	64%
Rensselaer Polytechnic Institute (NY)	1190	68%
Rutgers at New Brunswick (NJ)	1110	74%
Tulane University (LA)	1168	71%
University of Florida (FL)	1135	64%
University of North Carolina at Chapel Hill (NC)	1045	81%
University of Texas at Austin (TX)	1135	62%
Wake Forest University (NC)	1250	86%

Source: *America's Best Colleges, June, 1995*

To determine if there is a relationship between SAT scores and graduation rates, we can display the data points in a graph called a **scatter plot.** In a scatter plot, the two sets of data are plotted as ordered pairs in the coordinate plane. In this example, the independent variable is the average SAT score, and the dependent variable is the graduation rate. The scatter plot is shown at the left. The graph indicates that higher SAT scores do not necessarily result in higher graduation rates.

One way to solve real-world problems is to **use a model.** Scatter plots are an excellent way to model

Lesson 6–3 **INTEGRATION** *Statistics Scatter Plots and Best-Fit Lines* **339**

Look for a relationship between *x* and *y* in the graphs below.

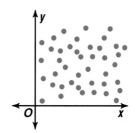

In this graph, *x* and *y* have a **positive correlation**. That is, the values are related in the same way. As *x* increases, *y* increases.

In this graph, *x* and *y* have a **negative correlation**. That is, the values are related in opposite ways. As *x* increases, *y* decreases.

In this graph, *x* and *y* have *no correlation*. In this case, *x* and *y* are not related and are said to be *independent*.

Example ❶ The table below shows 13 of the fastest-growing cities in the United States and their latitude and longitude.

Fastest-Growing Cities	Ranking	North Latitude	West Longitude
Austin, TX	9	30°	98°
Bakersfield, CA	1	35°	119°
Colorado Springs, CO	13	39°	105°
Durham, NC	8	36°	79°
Fresno, CA	2	37°	120°
Laredo, TX	10	28°	100°
Las Vegas, NV	3	36°	115°
Raleigh, NC	6	36°	79°
Reno, NV	12	40°	120°
Sacramento, CA	11	39°	121°
San Bernardino, CA	7	34°	117°
Stockton, CA	5	38°	121°
Tallahassee, FL	4	30°	84°

Source: Bureau of the Census, National Oceanic and Atmospheric Administration

Problem Solving

Use a Model

a. Draw one scatter plot to represent the correlation between the rankings and each city's latitude, and another to represent the correlation between the rankings and each city's longitude.
b. Is there a correlation between the cities' locations and their popularity?

a. The scatter plots are shown below.

b. Both of the correlations are fairly weak. However, the correlation between the rankings and latitude is stronger than the correlation between the rankings and longitude because the data approximate a line a little more. It is impossible to tell whether either correlation is positive or negative. Thus, we could say that there is a weak correlation between the cities' location and popularity.

The best-fit line may not pass through any of the data points.

You can use a scatter plot to make predictions. To help you do this, you can draw a line, called a **best-fit line**, that passes close to most of the data points. Use a ruler to draw a line that is close to most or all of the points. Then use the ordered pairs representing points on the line to make predictions.

A best-fit line shows if the correlation between two variables is *strong* or *weak*. The correlation is strong if the data points come close to, or lie on, the best-fit line. The correlation is weak if the data points do not come close to the line.

Example ②

APPLICATION
Animals

FYI

Almost 40% of U.S. households have at least one dog kept as a pet, accounting for over 50 million dogs. According to the American Kennel Club, the most popular dogs in America are Labrador retrievers.

The best-fit line drawn in this example is arbitrary. You may draw another best-fit line that is equally as valid.

Dogs age differently than humans do. You may have heard someone say that a dog ages 1 year for every 7 human years. However, that is not the case. The table at the right shows the relationship between dog years and human years.

Dog Years	Human Years
1	15
2	24
3	28
4	32
5	37
6	42
7	47

Source: *National Geographic World,* January, 1995

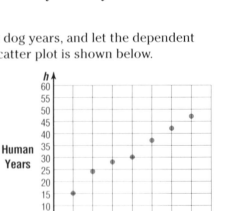

a. Draw a scatter plot to model the data and determine what relationship, if any, exists in the data.

b. Draw a best-fit line for the scatter plot.

c. Find an equation for the best-fit line.

d. Use the equation to determine how many human years are comparable to 13 dog years.

a. Let the independent variable d be dog years, and let the dependent variable h be human years. The scatter plot is shown below.

The plot seems to indicate that there is nearly a linear relationship between dog years and human years.

There also appears to be a positive correlation between the two variables. As x increases, y increases.

b. No line will pass through all of the data points. However, the best-fit line describes the trend of the data. Draw a line that seems to represent the data.

The slope of the best-fit line is positive. Thus, the correlation between the two variables is also positive.

c. As you can see, the best-fit line we drew passes through three of the data points: (4, 32), (5, 37), and (6, 42). Use two of these points to write the equation of the line.

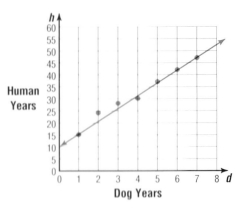

(continued on the next page)

First, find the slope.

$$\frac{y_2 - y_1}{x_2 - x_1} = \frac{37 - 32}{5 - 4} \text{ or } 5 \quad (x_1, y_1) = (4, 32) \text{ and } (x_2, y_2) = (5, 37)$$

Then, use the point-slope form.

$h - y_1 = m(d - x_1)$

$h - 32 = 5(d - 4)$ *Let (4, 32) represent (x_1, y_1).*

$h - 32 = 5d - 20$ *Distributive property*

$5d - h = -12$

d. To determine how many human years are comparable to 13 dog years, let $d = 13$ and solve for h.

$5d - h = -12$

$5(13) - h = -12$

$h = 77$

A dog that is 13 years old is comparable to a human that is 77 years old.

A **regression line** is the most accurate best-fit line for a set of data, and can be determined with a graphing calculator or computer. A graphing calculator assigns each regression line an r value. This value ($-1 \le r \le 1$) measures how closely the data are related. -1 indicates a strong negative correlation and 1 indicates a strong positive correlation.

LOOK BACK

Refer to Lesson 5-7A for more information on entering data into lists on a graphing calculator.

EXPLORATION
GRAPHING CALCULATORS

The table at the right shows the years of rate increases and the fares charged to ride the subway in New York City.

a. Use the Edit option on the STAT menu to enter the year data in L1 and the fare data into L2.

b. Use the window [1950, 1995] with a scale factor of 5 and [0, 1.5] with a scale factor of 0.25. Make sure any equations are deleted from the Y= list. Then press 2nd STAT PLOT 1. Make sure the following items are highlighted: On, the first type of graph (scatter plot), L1 as the Xlist, and L2 as the Ylist. Press GRAPH, and describe the graph.

c. Then find an equation for the regression line.
Enter: STAT ▶ 5 2nd L1 , 2nd L2 ENTER.
The screen displays the equation $y = ax + b$ and gives values for a, b, and r. Record these values on your paper.

d. Now graph the best-fit line.
Enter: Y= VARS 5 ▶ ▶ 7 GRAPH

Year	Fare
1953	$0.15
1966	0.20
1970	0.30
1972	0.35
1975	0.50
1981	0.75
1984	0.90
1986	1.00
1990	1.15
1992	1.25

Source: *New York Times, July 25, 1993*

Your Turn

a. How well do you feel the graph of the equation fits the data? Justify your answer.

b. Do any of the data points lie on the best-fit line? If so, name them.

c. Remove the ordered pair (1953, $0.15) from the set of data and repeat the process. What impact has removing this ordered pair had on the fit of the line to the data? (*Hint:* Compare the r values.)

d. Use the equation and graph to predict the fares for the year 2000.

Communicating Mathematics

Study the lesson. Then complete the following.

1. **Explain** how to determine whether a scatter plot has a positive or negative correlation.

2. **Draw** sketches of scatter plots that have each type of correlation.
   a. very strong positive   b. very strong negative   c. no correlation

3. **Describe** a situation that illustrates each type of plot in Exercise 2.

4. What does the *r* value on a graphing calculator tell you? Give examples.

5. What situations in your own life would present a negative correlation?

Guided Practice

Explain whether a scatter plot for each pair of variables would probably show a *positive*, *negative*, or *no* correlation between the variables.

6. money earned by a restaurant worker and time spent on the job

7. heights of fathers and sons

8. grades in school and how many times you miss class

9. **Archery**   The graph at the right shows the results when three student archers shot at a target nine times each.

   a. As the archers take more and more shots, what happens to the accuracy of the shots?

   b. Is there a correlation between the variables? Is it positive or negative?

   c. Which data points seem to be extremes from the rest of the data?

10. **Biology**   The table below shows the average body temperature of 14 insects at a given air temperature.

   a. Draw a scatter plot and a best-fit line for these data.

   b. Write an equation for the best-fit line.

   c. What conclusion might you draw from these data?

Practice

Explain whether a scatter plot for each pair of variables would probably show a *positive*, *negative*, or *no* correlation between the variables.

11. your age and weight from ages 1 to 20

12. temperature of a cup of coffee and the time it sits on a table

13. the amount a mail carrier earns each day and the weight of the mail delivered each day

Explain whether a scatter plot for each pair of variables would probably show a *positive*, *negative*, or *no* correlation between the variables.

14. the distance traveled and the time driving

15. a person's height and their birth month

16. the amount of snow on the ground and the daily temperature

17. the amount of time you exercise and the amount of calories you burn

18. the time water boils and the amount of water in the pot

19. the number of files stored on a disk and the amount of memory left on the disk

Determine whether a best-fit line should be drawn for each set of data graphed below. Explain your reasoning.

20.   21.   22.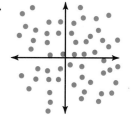

23. A test contains 20 true-false questions. Which scatter plot below best illustrates all the possible combinations of the numbers of correct and incorrect answers if the horizontal axis represents the number of responses that are correct and the vertical axis represents the number that are incorrect? Explain your choice.

a.   b.   c.

Graphing Calculator

24. The average January and July temperatures (°F) for 14 cities in various parts of the United States are listed below.

Jan.	21	35	42	32	51	7	30	21	20	40	28	45	56	39
July	71	79	79	77	82	70	73	73	70	86	81	90	69	82

a. Use a graphing calculator to enter the January temperatures as L1 and the July temperatures as L2 and create a scatter plot.

b. Describe the pattern of the points in the scatter plot.

c. Record the equation of the regression line and the *r* value.

d. Write a sentence that generally describes the relationship between the January temperature in a city and its July temperature.

Critical Thinking

25. Different correlation values are acceptable for different situations. Give an example for each situation. Explain your reasoning.

a. When is a strongly positive correlation a bad thing?

b. When is a mildly positive correlation an acceptable outcome?

c. When would a strongly negative correlation be the desired outcome?

d. When would no correlation be acceptable?

26. Many verbal expressions can be translated into algebraic expressions or equations. How do you think you could illustrate the expression "The bigger they are, the harder they fall" as a scatter plot?

27. Education   Refer to the application at the beginning of the lesson.
  a. Describe the correlation of the SAT scores to the graduation rate.
  b. Copy the graph and draw a best-fit line. What is the equation of the best-fit line you drew?

28. Biology   Scientists use samples to generalize relationships they observe in nature. The table at the right shows the length and weight of several humpback whales. A long ton is about 2240 pounds.

Length (feet)	Weight (long tons)
40	25
42	29
45	34
46	35
50	43
52	45
55	51

  a. Make a scatter plot of these data. Does a relationship exist? If so, what is the relationship?
  b. Use your scatter plot as a model to predict what you think a humpback whale 51 feet long might weigh.
  c. How long might a whale that weighs between 30 long tons and 35 long tons be?
  d. A 46-foot long humpback whale weighing 36 long tons was observed. Does this alter your conclusion about the correlation of the data? Explain.

29. Test Scores   The scores received on the first two tests of the grading period for 30 students in algebra class are listed below as ordered pairs.

30, 43	58, 57	55, 61	4, 71	32, 27	68, 59
54, 47	38, 27	56, 47	72, 63	50, 23	70, 67
60, 53	73, 79	55, 68	19, 58	38, 41	71, 59
68, 75	74, 83	58, 67	66, 73	72, 67	42, 57
71, 69	70, 89	94, 59	8, 84	84, 71	82, 73

  a. Make a scatter plot of the data.
  b. What correlation, if any, do you observe in the data?
  c. Draw a best-fit line and write the equation of the line.

30. Use a Model   Collect 10 round objects around your home. Treat each of them as a circle and measure the diameter and circumference in millimeters.
  a. Use a scatter plot as a model to relate the diameter and circumference of each object.
  b. Write an equation for the best-fit line.
  c. The actual formula that relates the circumference and diameter of a circle is $C = \pi d$. How does this formula compare with your equation?

31. Air Travel   A Boeing 747 takes off from the runway, and 20 minutes after liftoff, the plane is 30,000 feet from the ground. At 22 minutes after liftoff, the plane is 45,000 feet from the ground.   (Lesson 6–2)
  a. Write an equation in standard form to represent the flight of the plane during this time, with x = time in minutes, and y = altitude in feet.
  b. Use the equation to find the plane's altitude 18 minutes after takeoff.
  c. Can this equation be used to determine the height of the plane at any time during the flight? If so, explain how.

32. Solve $6b - a = 32$ if the domain is $\{-2, -1, 0, 2, 5\}$.   (Lesson 5–3)

33. Trigonometry   Use a calculator to solve $\cos W = 0.2598$.   (Lesson 4–3)

34. Simplify $1.2(4x - 5y) - 0.2(-1.5x + 8y)$.   (Lesson 2–6)

35. Solve $8.2 - 6.75 = m$.   (Lesson 1–5)

Writing Linear Equations in Slope-Intercept Form

- To determine the *x*- and *y*-intercepts of linear graphs from their equations,
- to write equations in slope-intercept form, and
- to write and solve direct variation equations.

Why IT'S IMPORTANT

You can write linear equations to solve problems involving swimming and banking.

APPLICATION

Banking

Have you seen advertisements like the one at the right in the newspaper? The bank is offering free checking. However, in the fine print of this ad, you will find that the free checking only occurs if you maintain a minimum daily balance of at least $2000.

Suppose this bank charges a monthly service fee of $3 plus 10¢ for each check or withdrawal transaction if you fail to maintain a minimum balance of $2000. The graph at the right shows the line that represents this situation. *You will write an equation for this line in Exercise 57.*

The coordinates at which a graph intersects the axes are known as the **y-intercept** and the **x-intercept**. Since this graph intersects the *y*-axis at $(0, 3)$, the *y*-intercept is 3. If you were to extend the graph to the left, you would find that it crosses the *x*-axis at $(-30, 0)$, so the *x*-intercept is -30.

Notice that the x-coordinate of the ordered pair for the y-intercept and the y-coordinate of the ordered pair for the x-intercept are both 0.

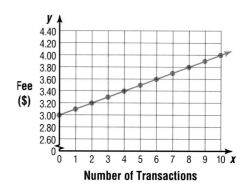

CHECK PLUS

Free Checking
You can have unlimited check writing at no cost and a **no annual fee** CheckCard when you keep the required minimum balance ** on deposit in your **Check Plus** Checking Account and another deposit account *** AND your first 50 printed checks are **FREE!**

*Only one checking account per customer qualifies for **Check Plus**. ** You will be assessed a fee every month your combined minimum balance falls below $2000. *** savings, CD, IRA, or money market accounts*

y

4.40
4.20
4.00
3.80
3.60
Fee   3.40
($)   3.20
3.00
2.80
2.60
0

0 1 2 3 4 5 6 7 8 9 10 **X**
Number of Transactions

Example **1** **Find the *x*- and *y*-intercepts of the graph of $3x + 4y = 6$.**

Remember that the *x*-intercept is the point at which $y = 0$.

$$3x + 4y = 6$$
$$3x + 4(0) = 6 \qquad \textit{Let y = 0.}$$
$$3x = 6$$
$$x = \frac{6}{3} \text{ or } 2 \qquad \textit{Divide each side by 3.}$$

Now let $x = 0$ to find the *y*-intercept.

$$3x + 4y = 6$$
$$3(0) + 4y = 6 \qquad \textit{Let x = 0.}$$
$$4y = 6$$
$$y = \frac{6}{4} \text{ or } \frac{3}{2} \qquad \textit{Divide each side by 4.}$$

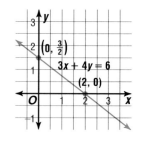

The *x*-intercept is 2, and the *y*-intercept is $\frac{3}{2}$. This means that the graph crosses the *x*-axis at $(2, 0)$ and the *y*-axis at $\left(0, \frac{3}{2}\right)$. You can use these points to graph the equation $3x + 4y = 6$.

Consider the graph at the right. The line crosses the *y*-axis at $(0, b)$, so its *y*-intercept is *b*. Write the point-slope form of an equation for this line using $(0, b)$ as (x_1, y_1).

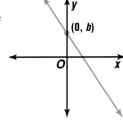

$y - y_1 = m(x - x_1)$ *Point-slope form*

$y - b = m(x - 0)$ *Replace (x_1, y_1) with $(0, b)$.*

$y - b = mx$

$y = mx + b$ *Add b to each side.*

This form of an equation is called the **slope-intercept form** of a linear equation, because in this form the slope and *y*-intercept are easily identified.

$$y = mx + b$$
$$\uparrow \quad \nwarrow$$
slope   y-intercept

Slope-Intercept Form of a Linear Equation	Given the slope *m* and the *y*-intercept *b* of a line, the slope-intercept form of an equation of the line is $y = mx + b$.

If you know the slope and *y*-intercept of a line, you can write an equation of the line in slope-intercept form. This equation can also be written in standard form.

Example **Write an equation of a line in slope-intercept form if the line has a slope of $\frac{2}{3}$ and a *y*-intercept of 6. Then write the equation in standard form.**

$y = mx + b$

$y = \frac{2}{3}x + 6$ *$b = 6, m = \frac{2}{3}$*

Now rewrite the equation in standard form.

$y = \frac{2}{3}x + 6$

$3y = 2x + 18$ *Multiply by 3 to eliminate the fraction.*

$-2x + 3y = 18$ *Subtract 2x from each side.*

or

$2x - 3y = -18$ *Multiply by −1 so that A is positive.*

Let's compare the slope-intercept form with the standard form of a linear equation. First solve $Ax + By = C$ for *y*.

$Ax + By = C$

$Ax - Ax + By = C - Ax$ *Subtract Ax from each side.*

$By = -Ax + C$ *Commutative property of addition*

$y = -\frac{A}{B}x + \frac{C}{B}$ *Divide each side by B.*
$\qquad\quad \uparrow \qquad \uparrow$
slope   y-intercept

Thus, you can identify the slope and *y*-intercept from an equation written in standard form by using $m = -\frac{A}{B}$ and $b = \frac{C}{B}$.

Example 3 Find the slope and y-intercept of the graph of $5x - 3y = 6$.

Method 1

In $5x - 3y = 6$, which is in standard form, $A = 5$, $B = -3$, and $C = 6$.
Find the slope.

$$m = -\frac{A}{B}$$

$$= -\frac{5}{-3} \text{ or } \frac{5}{3}$$

Find the y-intercept.

$$b = \frac{C}{B}$$

$$= \frac{6}{-3} \text{ or } -2$$

Method 2

Solve for y to find the slope-intercept form.

$$5x - 3y = 6$$

$$5x - 5x - 3y = 6 - 5x$$

$$-3y = -5x + 6$$

$$\frac{-3y}{-3} = \frac{-5x}{-3} + \frac{6}{-3}$$

$$y = \frac{5}{3}x + (-2)$$

$$m = \frac{5}{3}, b = -2$$

In Lesson 6–2, you learned how to use the point-slope form to find an equation of a line passing through two given points. Now you have another tool that you can use for the same situation.

Example 4 Write the slope-intercept and standard forms of the equation for a line that passes through $(-3, -1)$ and $(6, -4)$.

First find the slope.

$$m = \frac{y_2 - y_1}{x_2 - x_1}$$

$$= \frac{-4 - (-1)}{6 - (-3)} \quad (x_1, y_1) = (-3, -1) \text{ and } (x_2, y_2) = (6, -4)$$

$$= \frac{-3}{9} \text{ or } -\frac{1}{3}$$

Method 1

In $Ax + By = C$, the slope is $-\frac{A}{B}$.
If $-\frac{A}{B} = -\frac{1}{3}$, then $A = 1$ and $B = 3$.
So $Ax + By = C$ becomes $1x + 3y = C$.

To find the value of C, substitute either ordered pair into the equation. For example, choose $(6, -4)$.

$$x + 3y = C$$
$$6 + 3(-4) = C \quad (x, y) = (6, -4)$$
$$-6 = C$$

An equation in standard form for the line is $x + 3y = -6$.

Solve the equation for y to find the slope-intercept form.

$$x + 3y = -6$$
$$3y = -x - 6$$
$$y = -\frac{1}{3}x - \frac{6}{3}$$
$$y = -\frac{1}{3}x - 2$$

Method 2

Use one of the points in the slope-intercept form to find b.

$$y = mx + b$$
$$-4 = -\frac{1}{3}(6) + b \quad (x, y) = (6, -4)$$
$$-4 = -2 + b$$
$$-2 = b$$

An equation in slope-intercept form for the line is $y = -\frac{1}{3}x - 2$.

Now write the equation in standard form.

$$y = -\frac{1}{3}x - 2$$
$$3y = 3\left(-\frac{1}{3}x - 2\right)$$
$$3y = -x - 6$$
$$x + 3y = -6$$

Check: Make sure that both points satisfy the equation.

$$x + 3y = -6 \qquad\qquad\qquad x + 3y = -6$$
$$-3 + 3(-1) \overset{?}{=} -6 \quad (x, y) = (-3, -1) \qquad 6 + 3(-4) \overset{?}{=} -6 \quad (x, y) = (6, -4)$$
$$-6 = -6 \quad ✔ \qquad\qquad\qquad\qquad -6 = -6 \quad ✔$$

A special case of the slope-intercept form occurs when the y-intercept is 0. When $m = k$ and $b = 0$, $y = mx + b$ becomes $y = kx$. You may recognize this as the equation for *direct variation*. Another way to write this equation is $\frac{y}{x} = k$.

Example ⑤

Penny Dean holds the record for the fastest time swimming the English Channel. The 23-year-old Californian swam the 21-mile distance in $7\frac{2}{3}$ hours on July 29, 1978. At the same rate, how long would it take her to swim the Catalina Channel, a 26-mile distance near Los Angeles?

Explore Make a table to relate the information given in the problem.

Channel	Distance	Time
English	21 miles	$7\frac{2}{3}$ hours
Catalina	26 miles	? hours

Plan The formula that relates time t and distance d is $d = rt$, where r is the rate at which she travels. This formula is an example of direct variation. Use the information from the English Channel swim to find her rate. Then apply that rate to the Catalina Channel swim.

Solve For her English Channel swim, $d = 21$ miles, and $t = 7\frac{2}{3}$ or $\frac{23}{3}$ hours. Find the rate r.

$$d = rt$$
$$21 = r\left(\frac{23}{3}\right) \quad d = 21, r = \frac{23}{3}$$
$$\frac{3}{23}(21) = \left(\frac{23}{3}\right)\left(\frac{3}{23}\right)r$$
$$\frac{63}{23} = r$$

Now use the same formula and the rate r to find the time for the Catalina Channel swim.

$$d = rt$$
$$26 = \frac{63}{23}t$$
$$\frac{23}{63}(26) = \frac{23}{63}\left(\frac{63}{23}\right)t$$
$$9.5 \approx t \quad \textit{Use a calculator.}$$

At the same rate that she swam the English Channel, Ms. Dean could swim the Catalina Channel in 9.5 hours.

Examine You can also use a proportion to solve this problem since the rate r will be the same in each situation. If $d = rt$, then $r = \frac{d}{t}$, a form of the direct variation equation.

$$\frac{d_1}{d_2} = \frac{t_1}{t_2}$$
$$\frac{21}{26} = \frac{\frac{23}{3}}{t}$$
$$21t = \frac{23}{3}(26) \quad \textit{Find the cross products.}$$
$$t = \frac{23 \cdot 26}{3 \cdot 21} \quad \textit{Divide each side by 21.}$$
$$t \approx 9.5 \quad \textit{Use a calculator.}$$

The answer checks.

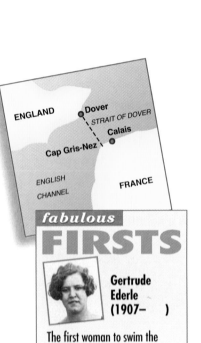
CHECK FOR UNDERSTANDING

Communicating Mathematics

Study the lesson. Then complete the following.

1. **Explain** how to find the intercepts of a line given its equation. Then find the intercepts of the standard form $Ax + By = C$.

2. **Write** a sentence to explain how direct variation and the slope-intercept form of a linear equation are related.

3. **Describe** two ways in which a direct variation problem may be solved.

4. Why can't you write the equation of a vertical line in slope-intercept form?

5. You Decide Taka wrote $y = 3.5x - 8$ as an equation of a line having a slope of 3.5 and a *y*-intercept of −8. Chuma wrote $7x - 2y = 16$ as an equation of the same line. Who is correct, and why?

6. List all the types of information you can use to write an equation of a line. Be sure to include examples of each.

Guided Practice

7. For the line shown in the graph at the right:
   a. state the slope,
   b. state the *x*- and *y*-intercepts, and
   c. write an equation in slope-intercept form.

Find the x- and y-intercepts of the graph of each equation.

8. $3x + 4y = 24$

9. $\frac{3}{4}x - 2y + 7 = 0$

Write an equation in slope-intercept form of a line with the given slope and y-intercept. Then write the equation in standard form.

10. $m = -4, b = 5$

11. $m = \frac{2}{3}, b = -10$

Find the slope and y-intercept of the graph of each equation.

12. $y = -3x + 7$

13. $2x + y = -4$

14. $x = 3y - 2$

Write an equation in standard form for a line that passes through each pair of points.

15. $(8, 1), (-2, 3)$

16. $(5, 7), (1, 9)$

17. Write a direct variation equation that has $(3, 11)$ as a solution. Then find *y* when $x = 12$.

18. Construction The roof line of a house starts 12 feet above the ground and has a slope of $\frac{1}{4}$.
   a. If the outer wall of the house represents the *y*-axis and the floor represents the *x*-axis, find the equation, in slope-intercept form, of the line that represents the roof.
   b. If you are standing inside the house 6 feet from the wall, how high is the roof at that point?

12 ft

EXERCISES

Practice

For each line graphed at the right:

a. state the slope,

b. state the *x*- and *y*-intercepts, and

c. write an equation in slope-intercept form.

19. *p*

20. *q*

21. *r*

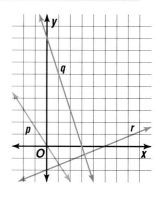

Find the *x*- and *y*-intercepts of the graph of each equation.

22. $5x - 3y = -12$ **23.** $4x + 7y = 8$ **24.** $5y - 2 = 2x$

25. $y - 6x = 5$ **26.** $4y - x = 3$ **27.** $3y = 18$

Write an equation in slope-intercept form of a line with the given slope and *y*-intercept. Then write the equation in standard form.

28. $m = 3, b = 5$ **29.** $m = 7, b = -2$ **30.** $m = -6, b = 0$

31. $m = -1.5, b = 3.75$ **32.** $m = \frac{1}{4}, b = -10$ **33.** $m = 0, b = -7$

Find the slope and *y*-intercept of the graph of each equation.

34. $3y = 2x - 9$ **35.** $5x + 4y = 10$

36. $4x - \frac{1}{3}y = -2$ **37.** $\frac{2}{3}x + \frac{1}{6}y = 2$

38. $5(x - 3y) = 2(x + 3)$ **39.** $4(3x + 9) - 3(5y + 7) = 11$

Write an equation in standard form for a line that passes through each pair of points.

40. $(-3, -5), (4, 5)$ **41.** $(7, -2), (-4, -2)$ **42.** $(-6, 1), (4, -2)$

43. $(3, 5), (3, -6)$ **44.** $(7, 4), (-5, 9)$ **45.** $(2, 9), (-5, 9)$

Write a direct variation equation that has each ordered pair as a solution. Then find the missing value.

46. $(-5, 8), (?, 11)$ **47.** $(24, 11), (36, ?)$ **48.** $(-17, 3), (?, 15)$

49. Write an equation in slope-intercept form of a line with slope $\frac{2}{3}$ and an *x*-intercept of 4.

50. Find the coordinates of a point on the graph of $5x - 3y = 10$ in which the *x*-coordinate is 5 greater than the *y*-coordinate.

51. Find the coordinates of a point on the graph of $3x + 7y = -16$ in which the *x*-coordinate is 3 times the *y*-coordinate.

52. Write the equation in standard form of the line with an *x*-intercept of 7 and a *y*-intercept of -2.

Programming

TECHNOLOGY Tip

The DrawF, Text(, and Vertical commands are located in the DRAW menu.

53. The program below finds the slope of a line, given the coordinates of two of the points through which it passes. Then it displays the graph of the line and gives the equation of the line in slope-intercept form.

```
PROGRAM:SLOPE
: ClrDraw                    : DrawF AX+B
: Input "X1=",Q              : Text(2,5,"EQUATION IS"
: Input "Y1=",R              : Text(8,5,"Y=",A,"X+",B)
: Input "X2=",S              : Else
: Input "Y2=",T              : ClrHome
: If Q-S≠0                   : Text(2,5,"SLOPE UNDEFINED")
: Then                       : Text(8,5,"EQUATION OF LINE")
: (R-T)/(Q-S)→A              : Text(15,5,"X=",Q)
: R-AQ→B                     : Vertical Q
```

   a. Use the program to write an equation of a line that passes through $(8.57, -3.82)$ and $(11.09, 1.31)$. Round numbers to the nearest hundredth.

   b. Use the program to check your equations in Exercises 40–45.

54. The *x*-intercept of a line is *p*, and the *y*-intercept is *q*. Write an equation of the line.

55. Chemistry   Charles' Law states that when the pressure is constant, the volume of a gas is directly proportional to the temperature on the Kelvin scale. Write an equation for each situation and solve.

  a. If the volume of a gas is 35 ft^3 at 290 K, what is the volume at 350 K?

  b. If the volume of a gas is 200 ft^3 at 300 K, at what temperature is the volume 180 ft^3?

56. Life Expectancy   The life expectancies at birth of the average male and female born between 1970 and 2000 (estimated) are shown in the table at the right.

  a. Write an equation for the line that passes through (1970, 67.1) and (2000, 73.2).

  b. Write an equation for the line that passes through (1970, 74.7) and (2000, 80.2).

Life Expectancy		
Year Born	Males (years)	Females (years)
1970	67.1	74.7
1975	68.8	76.6
1980	70.4	77.8
1985	71.1	78.2
1990	71.8	78.8
1995	72.8	79.7
2000 (est.)	73.2	80.2

  c. Graph the ordered pairs for male life expectancy and the equation from part a.

  d. On the same graph, use a different color pencil to graph the ordered pairs for female life expectancy and the equation from part b.

  e. Write a paragraph to describe the relationship between the points and lines you graphed.

  f. Use the data to predict the life expectancy for males and females in the year 2100.

57. Banking   Refer to the application at the beginning of the lesson.

  a. Write an equation to represent the line that shows the service fees charged by the bank for an account that maintains a daily balance that is less than $2000.

  b. Suppose you are not able to maintain $2000 in an account and you write about 25 checks each month. Use the equation from part a to determine whether you should open an account at this bank or go to the bank across the street that charges a flat $5 each month.

58. Statistics   The table below shows the keyboarding speeds of 12 students in words per minute (wpm) and their weeks of experience.   (Lesson 6–3)

Experience (weeks)	4	7	8	1	6	3	5	2	9	6	7	10
Keyboarding Speed (wpm)	33	45	46	20	40	30	38	22	52	44	42	55

  a. Make a scatter plot of these data.

  b. Draw a best-fit line for the data. Find the equation of the line.

  c. Use the equation to predict the keyboarding speed of a student after a 12-week course.

  d. Why can't this equation be used to predict the speed for any number of weeks of experience?

59. Determine the slope of the line that passes through $(14, 3)$ and $(-11, 3)$. (Lesson 6–1)

60. Write an equation for the function shown in the chart below. (Lesson 5–6)

m	1	2	4	5	6	9
n	9	6	0	−3	−6	−15

61. Graph $A(5, -2)$ on a coordinate plane. (Lesson 5–1)

62. Probability A card is selected at random from a deck of 52 cards. What is the probability of selecting a black card? (Lesson 4–6)

63. Construction When building a roof, a 5-foot support is to be placed at point *B* as shown on the diagram at the right. Find the length of the support that is to be placed at point *A*. (Lesson 4–2)

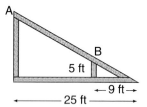

64. Solve $8x + 2y = 6$ for y. (Lesson 3–6)

65. Solve $-36 = 4z$. (Lesson 3–2)

66. Evaluate $|a + k|$ if $a = -5$ and $k = 3$. (Lesson 2–3)

67. Simplify $4(3x + 2) + 2(x + 3)$. (Lesson 1–7)

SELF TEST

Determine the slope of the line that passes through each pair of points. (Lesson 6–1)

1. $(-7, 10)$ and $(-2, 5)$ **2.** $(-6, 3)$ and $(-12, 3)$ **3.** $(-5, 7)$ and $(-5, -15)$

4. Determine the value of r so the line through $(r, 3)$ and $(6, -2)$ has a slope of $-\dfrac{5}{2}$. (Lesson 6–1)

Write the point-slope form of an equation of the line that passes through the given point and has the given slope. (Lesson 6–2)

5. $(-6, 4)$, $m = \dfrac{1}{2}$ **6.** $(-12, 12)$, $m = 0$

7. Write the standard form of an equation of the line that passes through $(-3, 4)$ and $(2, 3)$. (Lesson 6–2)

8. Computers A newspaper article stated that with advancements in technology, computers would become cheaper and cheaper. Is this true? The table at the right shows the average cost of a computer system for each year from 1991 to 1994. (Lesson 6–3)

Year	Average Cost ($)
1991	1100
1992	1143
1993	1183
1994	1219

a. Make a scatter plot of these data.

b. Draw a best-fit line and write an equation that describes it.

c. Does the scatter plot show a positive, negative, or no correlation between the variables?

d. Write what this correlation means in words.

e. With each year, computer systems become faster at handling more and more data. How can you use this fact to substantiate the newspaper article claim?

9. Find the *x*- and *y*-intercepts of the graph of $7x + 3y = -42$. (Lesson 6–4)

Source: *Vitality*, 1995

10. Write the slope-intercept form of an equation for the line that passes through $(0, -3)$ and $(6, 0)$. (Lesson 6–4)

6-5A Graphing Technology
Parent and Family Graphs

A Preview of Lesson 6-5

What is a family? Generally, a family is a group of people that are related either by birth, marriage, or adoption. Graphs can also form families. A **family of graphs** includes graphs and equations of graphs that have at least one characteristic in common. That characteristic differentiates the group of graphs from other groups.

Families of linear graphs often fall into two categories—those with the same slope or those with the same intercept. The **parent graph** is the simplest of the graphs in a family. For many linear functions, the parent graph is of the form $y = mx$, where m is any number.

A graphing calculator is a useful tool in studying a group of graphs to determine if they form a family.

Example **Graph $y = x$, $y = 2x$, and $y = 4x$ in the standard viewing window. Describe any similarities and differences among the graphs. Write a description of the family.**

Clear the Y= list of all other equations.

Enter: [Y=] [X,T,θ,n] [ENTER] 2 [X,T,θ,n] [ENTER] 4 [X,T,θ,n]
[ENTER] [ZOOM] 6

These three graphs form a family in which the slope of each graph is positive and each graph goes through the origin. However, each graph has a different slope. The parent graph is the graph of $y = x$. The graph of $y = 4x$ is the steepest, and the graph of $y = x$ is the least steep.

You might describe this family of graphs as lines that pass through the origin. Another description might be lines that have 0 as their y-intercept.

You can use braces to enter equations that have a common characteristic. Since all of the equations in Example 1 are of the form $y = mx$, you can use braces to enter the different values of m using one step as follows.

Enter: [Y=] [2nd] [{] 1 [,] 2 [,] 4 [2nd] [}] [X,T,θ,n] [GRAPH]

This tells the calculator to graphs equations for which the coefficients of x are 1, 2, and 4 and the y-intercept is 0.

Example **Graph $y = x$, $y = x + 3$, and $y = x - 3$ in the standard viewing window. Describe any similarities and differences among the graphs. Write a description of the family.**

You can enter each of the graphs into the Y= list as you did in Example 1 or use braces to enter the equations in one step. Notice that all three equations can be written in the form $y = x + b$. The braces can be used to enter the different values of b.

Enter: [Y=] [X,T,θ,n] [+] [2nd] [{] 0 [,]
3 [,] [(−)] 3 [2nd] [}] 6 [ZOOM]

The parent graph in this example is the graph of y = x.

The slope of each graph is positive. However, each graph has a different y-intercept.

The graph of $y = x$ has a y-intercept of 0.
The graph of $y = x + 3$ has a y-intercept of 3.
The graph of $y = x - 3$ has a y-intercept of -3.

Since the slope of each line is 1, we can describe this family as linear graphs whose slope is 1.

A function that is closely related to linear functions is the **absolute value function**.

Example ③ Graph $y = |x|$, $y = |x + 2|$, and $y = |x| + 4$ on the same screen. Describe any similarities and differences among the graphs.

Enter: [Y=] [2nd] [ABS] [X,T,θ,n]
[ENTER] [2nd] [ABS] [(]
[X,T,θ,n] [+] [2] [)] [ENTER]
[2nd] [ABS] [X,T,θ,n] [+] [4]
[ZOOM] [6]

The parent graph in this example is the graph of y = |x|.

Each graph is shaped like the letter v. The graph of $y = |x + 2|$ is shaped like the graph of $y = |x|$, but is shifted 2 units to the left. The graph of $y = |x| + 4$ is shaped like the graph of $y = |x|$, but is shifted 4 units up.

EXERCISES

Graph each set of equations on the same screen. Describe any similarities or differences among the graphs. State what the graphs have in common.

1. $y = -x$
$y = -2x$
$y = -4x$

2. $y = |x|$
$y = 2|x|$
$y = 0.5|x|$

3. $y = -x$
$y = -x + 2$
$y = -x - 3$

4. Write a sentence comparing the graphs of equations with a positive coefficient of x and graphs with a negative coefficient of x.

Sketch a graph on paper to represent how you think the graph of each equation will appear. Describe any similarities to the graphs in Examples 1 and 2 and Exercises 1–3. Then use a graphing calculator to confirm your prediction and describe how accurate your sketch was.

5. $y = x - 5$ 　　6. $y = -|x| + 6$ 　　7. $y = 0.1x$ 　　8. $y = \frac{1}{3}x + 4$

9. Write an equation of a line whose graph lies between the graphs of $y = -x + 2$ and $y = -x + 3$.

10. Write a paragraph explaining how the values of m and b in the slope-intercept form affect the graph of the equation. Include several graphs with your paragraph.

Graphing Linear Equations

6-5

What YOU'LL LEARN

- To graph a line given any linear equation.

Why IT'S IMPORTANT

You can graph linear equations to show trends in fields like health and physical science.

New York City — Barcelona, Spain

CONNECTION
Earth Science

Solome was flying from New York to Spain to be an exchange student for the spring quarter. After climbing to cruising level, the pilot announced that they were at an altitude of 13,700 meters and the outside temperature was about $-76°C$. You may think that since you get closer to the sun as you get farther from Earth, you might get warmer as altitude increases. However, as altitude increases, the air gets thinner and colder.

The air temperature above Earth on a day when the temperature at ground level is 15°C can be calculated using the formula $a + 150t = 2250$, where a is the altitude in meters and t is the temperature in degrees Celsius.

Suppose you wanted to graph $a + 150t = 2250$. In Chapter 5, you learned that you could graph a linear equation by finding and then graphing several ordered pairs that satisfy the equation.

Using the x- and y-intercepts is a convenient way to find ordered pairs to graph a linear equation. In $a + 150t = 2250$, these can be called the a-intercept and the t-intercept.

Find the a-intercept.

$$a + 150t = 2250$$
$$a + 150(0) = 2250 \quad \textit{Let t = 0.}$$
$$a = 2250$$

Find the t-intercept.

$$a + 150t = 2250$$
$$0 + 150t = 2250 \quad \textit{Let a = 0.}$$
$$t = 15$$

Now graph (2250, 0) and (0, 15), the ordered pairs for the intercepts. Draw the line representing the equation by connecting the points. Thus, you can graph a line if you know two points on that line.

Use the equation to confirm the pilot's announcement that the temperature was $-76°C$ at 13,700 meters altitude.

$$a + 150t = 2250$$
$$13{,}700 + 150t = 2250$$
$$150t = -11{,}450$$
$$t \approx -76.33$$

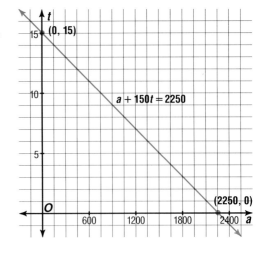

So, the pilot was correct in saying that the outside temperature was about $-76°C$.

You can also graph a line if you know its slope and a point on the line. The forms of linear equations you have learned in this chapter can help you find the slope and a point on the line.

Example **1** Graph $y - 1 = 3(x + 2)$.

This equation is in point-slope form. The slope is 3, and a point on the line is $(-2, 1)$.

Graph the point $(-2, 1)$. Remember that the slope represents the change in y and x.

$$\frac{3}{1} = \frac{\text{change in } y}{\text{change in } x}$$

Thus, from the point $(-2, 1)$, you can move up 3 and right 1. Draw a dot. You can repeat this process to find another point on the graph. Draw a line connecting the points. This line is the graph of $y - 1 = 3(x + 2)$.

Check: Check $(-1, 4)$ to make sure that it satisfies the equation.

$$y - 1 = 3(x + 2)$$
$$4 - 1 \stackrel{?}{=} 3(-1 + 2)$$
$$3 = 3 \quad \checkmark$$

You can also use the slope-intercept form to help you graph an equation.

Example **2** Graph $\frac{3}{4} x + \frac{1}{2} y = 4$.

Solve the equation for y to find the slope-intercept form.
$$\frac{3}{4}x + \frac{1}{2}y = 4$$

$$\frac{1}{2}y = -\frac{3}{4}x + 4 \quad \textit{Subtract } \tfrac{3}{4}x \textit{ from each side.}$$

$$y = -\frac{3}{2}x + 8 \quad \textit{Multiply each side by 2.}$$

The slope-intercept form of this equation tells us that the slope is $-\frac{3}{2}$ and the y-intercept is 8. Graph the y-intercept, and then use the slope to find another point on the line. Draw the line.

Be sure to check your second point to make sure it satisfies the *original* equation.

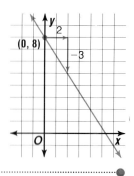

You frequently need to rewrite equations before you can determine the best way to graph them.

Example **3** Graph $\frac{4}{5}(2x - y) = 6x + \frac{2}{5}y - 10$.

First, simplify the equation.
$$\frac{4}{5}(2x - y) = 6x + \frac{2}{5}y - 10$$

$$4(2x - y) = 30x + 2y - 50 \quad \textit{Multiply each side by 5 to eliminate the fractions.}$$

$$8x - 4y = 30x + 2y - 50$$

$$0 = 22x + 6y - 50 \quad \textit{Add } -8x \textit{ and } 4y \textit{ to each side.}$$

$$50 = 22x + 6y \quad \textit{Add 50 to each side.}$$

$$25 = 11x + 3y \quad \textit{Divide each side by 2.}$$

$$11x + 3y = 25 \quad \textit{Standard form} \qquad \textit{(continued on the next page)}$$

The equation is now in standard form. You can choose one of several methods to graph this equation.

ordered pairs

If $x = 2$, $y = 1$.

If $x = -1$, $y = 12$.

Graph $(2, 1)$ and $(-1, 12)$.

intercepts

If $x = 0$, $y = \frac{25}{3}$.

If $y = 0$, $x = \frac{25}{11}$.

Graph $\left(0, \frac{25}{3}\right)$ and $\left(\frac{25}{11}, 0\right)$.

slope-intercept form

$11x + 3y = 25$

$$3y = -11x + 25$$
$$y = -\frac{11}{3}x + \frac{25}{3}$$

The y-intercept is $\frac{25}{3}$, and the slope is $-\frac{11}{3}$.

Notice that the result is the same regardless of the method you choose. Choose the method that is easiest for you.

It is often helpful to make a sketch of the graph of an equation when you are trying to observe patterns and make predictions.

Example **4** The mule deer likes to feed upon the bitterbrush plant, which is native to the dry regions of western North America. The diameter of each bitterbrush twig is related to the length of the twig by the function $\ell = 1.25 + 8.983d$, where ℓ is the length (inches) of the twig, d is its diameter in inches, and $d \geq 0.1$.

CONNECTION

Biology

a. Graph the function. Name the independent and dependent variables.

b. Describe the relationship between a twig's diameter and length.

c. What would be the length of a twig if its diameter is 0.5 inch?

a. In this equation, d is the independent variable, and ℓ is the dependent variable. Since d must be a nonnegative value, it may be easier to choose two values of d and find the corresponding value of ℓ to graph. A calculator may be helpful.

Let $d = 0.1$

$\ell = 1.25 + 8.983(0.1)$
$\quad = 2.1483$

Let $d = 1$.

$\ell = 1.25 + 8.983(1)$
$\quad = 10.233$

Graph $(0.1, 2.148)$ and $(1, 10.233)$. Connect them with a line.

b. As the diameter of the twig increases, the length of the twig increases.

c. Let $d = 0.5$ and evaluate the equation.

$\ell = 1.25 + 8.983(0.5)$

$\quad = 5.7415$

A 0.5-inch diameter twig is about 5.74 inches long.

Check to see if this point is on the graph.

CHECK FOR UNDERSTANDING

Communicating Mathematics

Study the lesson. Then complete the following.

1. **Write** a counterexample to the statement "You must always know two points that satisfy an equation in order to graph the equation."

2. **Explain** why the value of d cannot be zero in Example 4.

3. **Illustrate** how you graph a line if you know a point on the line and its slope.

4. **Describe** two ways in which you can interpret the slope $-\frac{3}{4}$ when using it to find another point on the graph of a line.

MATH JOURNAL

5. **Assess Yourself**   List all the possible ways you know to graph the equation of a line. Which of these methods do you prefer and why?

Guided Practice

6. Graph $3x - 8y = 12$ by using the x- and y-intercepts.

7. Graph $y + 5 = -2(x + 1)$ by using the slope and a point on the line.

8. Graph $\frac{2}{3}x + \frac{1}{2}y = 3$ by using the slope and y-intercept.

Graph each equation.

9. $y = 2x - 3$ 10. $y = \frac{2}{5}x - 4$ 11. $6y + 12 = 18$ 12. $5x = 9$

13. **Health**   Did you know that your fingernails grow twice as fast as your toenails? Your fingernails grow about $\frac{1}{8}$ inch each month. Suppose the nail on your index finger is $\frac{1}{4}$ inch long.

   a. Graph the line representing your fingernail's growth for a year.

   b. How long would your nail be at the end of one year?

   c. How much do your toenails grow each month?

   d. Graph the line representing your toenail's growth for one year.

   e. What does the slope of each line represent?

EXERCISES

Practice

Graph each equation by using the x- and y-intercepts.

14. $y = 5x - 10$ 15. $6x - y = 9$ 16. $\frac{1}{2}x - \frac{2}{3}y = -6$

Graph each equation by using the slope and a point on the line.

17. $y - 2 = 3(x - 5)$
18. $y + 6 = -\frac{3}{2}(x + 5)$
19. $2(x - 3) = y + \frac{3}{2}$

Graph each equation by using the slope and y-intercept.

20. $y = -\frac{3}{4}x + 4$
21. $2x - 3y = -7$
22. $5y + 3 = -5$

Graph each equation.

23. $y = 3x - 5$
24. $6y + 5 = 5y + 3$
25. $3(y + 4) = -2x$

26. $y = \frac{2}{3}x + 1$
27. $5x + 2y = 20$
28. $\frac{2}{3}x - y = 4$

29. $6x - y = 8$
30. $15x - 29y = 429$
31. $y = 0.17x + 1.75$

32. $y - 3 = \frac{2}{3}(x - 6)$
33. $y + 1 = -2(x + 3)$

Graphing Calculator

Graph each group of equations on the same screen. Describe the graphs in terms of a family of graphs.

34. $y = 2x + 4$
$2x - 3y = -12$
$y + 5 = 3(x + 3)$

35. $12x - 3y = 15$
$y = 4x - 11$
$y + x = 5x$

Critical Thinking

36. An equation belongs to the same family of graphs as $y = 3x + 5$, but it has a y-intercept of 7. How could you graph the equation without finding the equation first?

Applications and Problem Solving

37. **Health** In cardiopulmonary resuscitation (CPR) on an adult, you initially give two breaths before you begin a cycle of one breath for every five pumps on the heart per minute. Each pump of the heart takes about a second. Thus, in a minute (60 seconds), you give about $60 \div 5$ or 12 breaths. The number of breaths can be described by the equation $B = 2 + 12t$, where t represents the amount of time in minutes and B is the total number of breaths after pumping starts.

a. Graph the equation.

b. It is recommended that CPR be sustained until an emergency squad can get to the patient. If the squad takes 10 minutes to get to the patients, how many breaths should the patient have received?

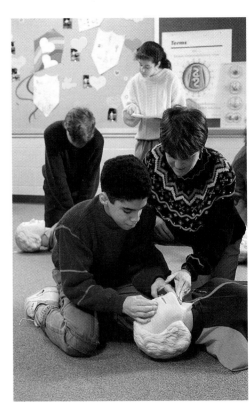

38. Physical Science   The length of an object's shadow depends on its height. At a certain time of day on a particular day of the year, the lengths of shadows can be calculated by the formula $S = 1.5h$, where h is the height and S is the length of the shadow.

   a. Graph the equation.

   b. At the time of day for which the formula is valid, the oak tree in your back yard is casting a 30-foot shadow. What is the height of the tree?

Mixed Review

39. Write an equation in slope-intercept form of the line with y-intercept 12 and slope the same as the line whose equation is $2x - 5y - 10 = 0$.   (Lesson 6–4)

40. Draw a mapping for the relation $\{(1, 3), (2, 3), (2, 1), (3, 2)\}$.   (Lesson 5–2)

41. Travel   Art leaves home at 10:00 A.M., driving at 50 miles per hour. At 11:30 A.M., Jennifer leaves home, going in the same direction, at 45 miles per hour. When will they be 100 miles apart?   (Lesson 4–7)

42. Geometry   Find the supplement of 87°.   (Lesson 3–4)

43. Find the square root of 3.24.   (Lesson 2–8)

44. Find $-13 + (-9)$.   (Lesson 2–1)

45. Write a verbal expression for $m - 1$.   (Lesson 1–1)

WORKING ON THE
In·ves·ti·ga·tion

Refer to the Investigation on pages 320–321.

Smoke Gets In Your Eyes

Which of your group members was closest to the average in number of breaths it took to blow up a balloon? Use that group member to conduct another experiment. For this activity, assume that each breath into a balloon approximates the exhaled smoke from one puff of a cigarette.

1 Have that member simulate the amount of smoke from each exhale during smoking by blowing into a balloon. Use a tape measure to determine the circumference of the balloon after each breath. Record each measurement.

2 Calculate the volume after each breath. Graph the relationship of the volume to each breath.

Describe the graph. Does the graph represent a function? If so, describe the domain and range.

3 There are 20 cigarettes in each pack. If a smoker smokes an entire pack of cigarettes, how many balloons would be filled? What would be the volume of smoke-filled air?

4 Graph the data relating the cumulative smoke in the air for each cigarette smoked in a pack of cigarettes.

5 If a smoker smokes an entire carton of cigarettes (10 packs/carton), how many balloons would be filled? What would be the volume of smoke-filled air?

Add the results of your work to your Investigation Folder.

Integration: Geometry
Parallel and Perpendicular Lines

What YOU'LL LEARN

- To determine if two lines are parallel or perpendicular by their slopes, and

- to write equations of lines that pass through a given point, parallel or perpendicular to the graph of a given equation.

Why IT'S IMPORTANT

You can use parallel and perpendicular lines to solve problems involving health and construction.

APPLICATION
World Cultures

Kite-making has been a national sport in some Far East countries since ancient times. In September, Chinese families celebrate the Ninth Day of the Ninth Moon as Kites' Day. Entire families go outside and fly kites. These kites may be simple diamond shapes or more complex shapes like fish, birds, dragons, and colorfully-dressed people.

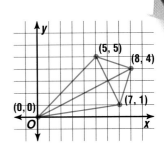

The outline of a simple kite is shown on the coordinate plane at the right. The ordered pairs for the tips of the kite are labeled. We know that the two slats that support that kite meet at right angles. Lines that intersect at right angles are called **perpendicular lines.** *You will learn more about these lines later in the lesson.*

Lines in the same plane that never intersect are called **parallel lines.** Parallel lines can form families of graphs. In Lesson 6–5A, you learned that families of graphs can either have the same slope or the same intercept. The graph at the right shows a family of graphs represented by the following equations.

$y = 2x$ *The parent graph is the graph of y = 2x.*

$y = 2x + 3$

$y = 2x - 1$

The slope of each line is 2. Notice that the lines seem to be parallel. In fact, they are parallel.

Definition of Parallel Lines in a Coordinate Plane	If two nonvertical lines have the same slope, then they are parallel. All vertical lines are parallel.

A **parallelogram** is a quadrilateral in which opposite sides are parallel. We can use slope to determine if quadrilaterals graphed on a coordinate plane are parallelograms. *The slope of a segment is the same as the slope of the line containing that segment.*

Example ❶ **Determine whether quadrilateral _ABCD_ is a parallelogram if its vertices are _A_(−5, −3), _B_(5, 3), _C_(7, 9), and _D_(−3, 3).**

Explore   Graph the four vertices. Connect the vertices to form the quadrilateral.

Plan   Use each pair of vertices to find the slope of each segment.

Use $m = \dfrac{y_2 - y_1}{x_2 - x_1}$.

AB means a line segment whose endpoints are A and B.

Solve

$\overline{AB}$: $m = \dfrac{3-(-3)}{5-(-5)} = \dfrac{6}{10}$ or $\dfrac{3}{5}$

$\overline{BC}$: $m = \dfrac{9-3}{7-5} = \dfrac{6}{2}$ or 3

$\overline{CD}$: $m = \dfrac{3-9}{-3-7} = \dfrac{-6}{-10}$ or $\dfrac{3}{5}$

$\overline{AD}$: $m = \dfrac{3-(-3)}{-3-(-5)} = \dfrac{6}{2}$ or 3

$\overline{AB}$ and $\overline{CD}$ have the same slope, $\dfrac{3}{5}$. $\overline{BC}$ and $\overline{AD}$ have the same slope, 3. Both pairs of opposite sides are parallel. Thus, _ABCD_ is a parallelogram.

Examine   Another definition of a parallelogram is that its opposite sides are congruent, or equal in length. Use a millimeter ruler to measure each side. $\overline{AB}$ and $\overline{CD}$ are approximately 37 millimeters long. $\overline{BC}$ and $\overline{AD}$ are approximately 20 millimeters long. The figure is probably a parallelogram.

You can write the equation of a line parallel to another line if you know a point on the line and the equation of the other line.

Example ❷ **Write an equation in slope-intercept form of the line that passes through (4, 0) and is parallel to the graph of _4x − 3y = 2_.**

First find the slope by rewriting the equation in slope-intercept form.

$4x - 3y = 2$

$-3y = -4x + 2$

$y = \dfrac{4}{3}x - \dfrac{2}{3}$

The slope is $\dfrac{4}{3}$.

Method 1

Use point-slope form.

$y - y_1 = m(x - x_1)$

$y - 0 = \dfrac{4}{3}(x - 4)$ $m = \dfrac{4}{3}, x_1 = 4,$ and $y_1 = 0$

$y = \dfrac{4}{3}x - \dfrac{16}{3}$

This equation is in slope-intercept form.

An equation for the line is $y = \dfrac{4}{3}x - \dfrac{16}{3}$.

Method 2

Use the slope-intercept form.

$y = mx + b$

$0 = \dfrac{4}{3}(4) + b$

$0 = \dfrac{16}{3} + b$

$-\dfrac{16}{3} = b$

Now substitute the values for _m_ and _b_ into the slope-intercept form.

$y = \dfrac{4}{3}x - \dfrac{16}{3}$

Sometimes graphs can be drawn in a way that leads you to the wrong conclusion. Mathematics can help you determine if a graph is misleading.

Example ③

Health

The American Heart Association recommends that men keep their weight within a certain "healthy weight band" as shown in the graph at the right. Are the lines parallel as they appear in the graph?

From the graph, we can determine that the endpoints of the minimum edge are (94, 58) and (148, 74). The endpoints of the maximum edge are (110, 58) and (175, 74). Find the slopes of each segment.

$$m_1 = \frac{74 - 58}{148 - 94} = \frac{16}{54} \text{ or about } 0.30$$

$$m_2 = \frac{74 - 58}{175 - 110} = \frac{16}{65} \text{ or about } 0.25$$

The segments do not have the same slope. Therefore, they are not parallel as they appear in this graph.

We have seen that the slopes of parallel lines are equal. What is the relationship of the slopes of perpendicular lines?

Example ④

Kite-Making

Refer to the application at the beginning of the lesson.

a. **Write equations of the lines containing the slats of the kite in slope-intercept form.**

b. **Determine the relationship between the slopes of perpendicular lines.**

a. The endpoints of the longer slat are (0, 0) and (8, 4). The endpoints of the shorter slat are (5, 5) and (7, 1). Find the slope of each slat and then use the point-slope form to find an equation for each slat.

long slat

$$m = \frac{4 - 0}{8 - 0} = \frac{4}{8} \text{ or } \frac{1}{2}$$

Let $(x_1, y_1) = (0, 0)$.

$$y - y_1 = m(x - x_1)$$
$$y - 0 = \frac{1}{2}(x - 0)$$
$$y = \frac{1}{2}x$$

short slat

$$m = \frac{1 - 5}{7 - 5} = \frac{-4}{2} \text{ or } -2$$

Let $(x_1, y_1) = (5, 5)$.

$$y - y_1 = m(x - x_1)$$
$$y - 5 = -2(x - 5)$$
$$y - 5 = -2x + 10$$
$$y = -2x + 15$$

Equations of the slats are $y = \frac{1}{2}x$ and $y = -2x + 15$.

b. The slats of the kite are perpendicular. The slopes of the slats are $\frac{1}{2}$ and -2. The two slopes are negative reciprocals of each other. Note that their product is $\frac{1}{2}(-2)$ or -1.

These results suggest the following definition.

Definition of Perpendicular Lines in a Coordinate Plane	If the product of the slopes of two lines is -1, then the lines are perpendicular. In a plane, vertical lines and horizontal lines are perpendicular.

The slopes of perpendicular lines are negative reciprocals of each other.

You can use a right triangle and a coordinate grid to model the slopes of perpendicular lines.

MODELING MATHEMATICS

Perpendicular Lines on a Coordinate Plane

Materials: grid paper   scissors

a. A scalene triangle is one in which no two sides are equal. Cut out a scalene right triangle ABC so that $\angle C$ is the right angle. Label the vertices and the sides as shown at the right.

b. Draw a coordinate plane on the grid paper. Place $\triangle ABC$ on the coordinate plane so that A is at the origin and side b lies along the positive x-axis.

c. Name the coordinates of B.

d. What is the slope of side c?

Your Turn

a. Rotate the triangle $90°$ counterclockwise so that A is still at the origin and side b is along the positive y-axis.

b. Name the coordinates of B.

c. What is the slope of c?

d. What is the relationship between the first position of c and the second?

e. What is the relationship between the slopes of c in each position?

You can use your knowledge of perpendicular lines to write equations of lines perpendicular to a given line.

Example   **Write the slope-intercept form of an equation that passes through $(8, -2)$ and is perpendicular to the graph of $5x - 3y = 7$.**

First find the slope of the given line.

$$5x - 3y = 7$$
$$-3y = -5x + 7$$
$$y = \frac{5}{3}x - \frac{7}{3}$$

The slope of the line is $\frac{5}{3}$. The slope of the line perpendicular to this line is the negative reciprocal of $\frac{5}{3}$, or $-\frac{3}{5}$.

Use the point-slope form to find the equation.

$$y - y_1 = m(x - x_1)$$
$$y - (-2) = -\frac{3}{5}(x - 8) \quad m = -\frac{3}{5}, (x_1, y_1) = (8, -2)$$
$$y + 2 = -\frac{3}{5}x + \frac{24}{5} \quad \text{Distributive property}$$
$$y = -\frac{3}{5}x + \frac{14}{5} \quad \text{Subtract 2 or } \frac{10}{5} \text{ from each side.}$$

An equation for the line is $y = -\frac{3}{5}x + \frac{14}{5}$.

Communicating Mathematics

Study the lesson. Then complete the following.

1. **Describe** the relationship between the slopes of each of the following.

   a. two parallel lines

   b. two perpendicular lines

2. **Explain** what negative reciprocals are. Give an example.

3. **Check** the results of Examples 2 and 5 by graphing the original equation and the answer equation for each example on the same coordinate plane. Describe what happens.

MODELING MATHEMATICS

4. Refer to the Modeling Mathematics activity. Why do you think the right triangle should be scalene?

5. Repeat the activity with a different size right triangle.

   a. Are the results different?

   b. What do you think would happen if you used a triangle that was not a right triangle?

Guided Practice

State the slopes of the lines parallel to and perpendicular to the graph of each equation.

6. $6x - 5y = 11$ 7. $y = \frac{2}{3}x - \frac{4}{5}$ 8. $3x = 10y - 3$

Determine whether the graphs of each pair of equations are parallel, perpendicular, or neither.

9. $3x - 7y = 1, 7x + 3y = 4$ 10. $5x - 2y = 6, 4y - 10x = -48$

Write an equation in slope-intercept form of the line that passes through the given point and is parallel to the graph of each equation.

11. $(9, -3), 5x - 6y = 2$ 12. $(0, 4), 2y = 5x - 7$ 13. $(7, -2), x - y = 0$

Write an equation in slope-intercept form of the line that passes through the given point and is perpendicular to the graph of each equation.

14. $(8, 5), 7x + 4y = 23$ 15. $(0, 0), 9y = 3 - 5x$ 16. $(-2, 7), 2x - 5y = 3$

17. **Geometry**   The diagonals of a square are segments that connect the opposite vertices. Use what you've learned in this lesson to determine the relationship between the diagonals $\overline{AC}$ and $\overline{BD}$ of the square graphed at the right.

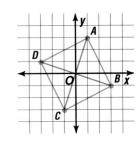

Practice

Determine whether the graphs of each pair of equations are parallel, perpendicular, or neither.

18. $y = -2x + 11$
$y + 2x = 23$

19. $3y = 2x + 14$
$2x - 3y = 2$

20. $y = -5x$
$y = 5x - 18$

21. $y = 0.6x + 7$
$3y = -5x + 30$

22. $y = 3x + 5$
$y = 5x + 3$

23. $y + 6 = -5$
$y + x = y + 7$

Write an equation in slope-intercept form of the line that passes through the given point and is parallel to the graph of each equation.

24. $(2, -7), y = x - 2$

25. $(2, 3), y = x + 5$

26. $(-5, -4), 2x + 3y = -1$

27. $(5, 6), 8x - 7y = 23$

28. $(1, -2), y = -2x + 7$

29. $(0, -5), 5x - 2y = -7$

30. $(5, -4), x - 3y = 8$

31. $(-3, 2), 2x - 3y = 6$

32. $(2, -1), y = -0.5x + 2$

Write an equation in slope-intercept form of the line that passes through the given point and is perpendicular to the graph of each equation.

33. $(6, -13), 2x - 9y = 5$

34. $(-3, 1), y = \frac{1}{3}x + 2$

35. $(6, -1), 3y + x = 3$

36. $(6, -2), y = \frac{3}{5}x - 4$

37. $(0, -1), 5x - y = 3$

38. $(8, -2), 5x - 7 = 3y$

39. $(4, -3), 2x - 7y = 12$

40. $(3, 7), y = \frac{3}{4}x - 1$

41. $(3, -3), 3x + 7 = 2x$

Write an equation of the line having the following properties.

42. passes through $(-5, 3)$ and is perpendicular to the x-axis

43. is parallel to the graph of $y = \frac{5}{4}x - 3$ and passes through the origin.

44. has an x-intercept of 3 and is perpendicular to the graph of $5x - 3y = 2$

45. has a y-intercept of -6 and is parallel to the graph of $x - 3y = 8$

Critical Thinking

46. Lines a, b, and c lie on the same coordinate plane. Line a is perpendicular to line b and intersects it at $(3, 6)$. Line b is perpendicular to line c, which is the graph of $y = 2x + 3$.
  a. Write the equation of lines a and b.
  b. Graph the three lines.
  c. What is the relationship of lines a and c?

Applications and Problem Solving

47. Geometry A rhombus is a parallelogram that has perpendicular diagonals. Determine if quadrilateral $ABCD$ with $A(-2, 1)$, $B(3, 3)$, $C(5, 7)$, and $D(0, 5)$ is a rhombus. Explain.

48. Construction A large roller is being used to flatten new asphalt on a hill that has a slope of $-\frac{1}{10}$. The steering mechanism for the front roller is perpendicular to the hill.

a. What is the slope of the steering mechanism?

b. If the bottom of the hill has coordinates $(9, 3)$ and the point directly below the steering mechanism is at $(-1, 4)$, write the equations of the lines representing the hill and the steering mechanism.

steering mechanism

49. Geometry An angle inscribed in a circle is one in which the vertex of the angle lies on the circle and the two sides intersect the circle, as shown at the right.

a. Find the equations of the lines that contain the sides of $\angle Z$.

b. What type of angle is $\angle Z$?

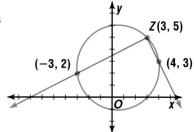

$Z(3, 5)$

$(-3, 2)$

$(4, 3)$

Mixed Review

50. Graph $7x - 2y = -7$ by using the x- and y-intercepts. (Lesson 6–5)

51. Statistics Make a scatter plot of the data below. (Lesson 6–3)

Miles Driven	200	322	250	290	310	135	60	150	180	70	315	175
Fuel Used (gallons)	7.5	14	11	10	10	5	2.3	5	6.2	3	11	6.5

52. Patterns Copy and complete the table below. (Lesson 5–4)

x	-1	2	5	8	11	14
$f(x)$		-1	-7		-19	

53. Find 4% of $6070. (Lesson 4–4)

54. Solve $16 = \frac{s - 8}{-7}$. Check your solution. (Lesson 3–3)

55. Metallurgy The gold content of jewelry is given in karats. For example, 24-karat gold is pure gold, and 18-karat gold is $\frac{18}{24}$ or 0.75 gold. (Lesson 2–7)

a. What fraction of 10-karat gold is pure gold? What fraction is *not* gold?

b. If a piece of jewelry is $\frac{2}{3}$ gold, how would you describe it using karats?

56. Find $-0.0005 + (-0.3)$. (Lesson 2–5)

57. Name the property illustrated by $1(87) = 87$. (Lesson 1–6)

58. Solve $x = 6 + 0.28$. (Lesson 1–5)

Integration: Geometry
Midpoint of a Line Segment

Maria Hernandez made a frame for one of her watercolor paintings in a class offered at the community college. Because her canvas was heavy and the wooden frame was light, her instructor suggested that she stabilize her frame by adding crossbars connecting the consecutive sides. One of the books she used as a reference showed the crossbars inserted so that the bars are placed on the sides of the frame at the **midpoint** of each side. The midpoint of a line segment is the point that is halfway between the endpoints of the segment.

Maria decided to use some grid paper to model her frame and the crossbars she is going to insert. Each square of the paper represented 2 inches of her frame. To find the midpoint of the interior vertical edge, she counted how many units it was from one corner to the next and divided by 2. The inside of the frame is 16 inches high, so the midpoint is 8 inches from the corner. She counted 4 units on the grid and placed a bullet on her drawing. She used a similar method to find the midpoint of the interior horizontal edge. The inside is 20 inches wide, so the midpoint is 10 inches from the end.

Suppose we place Maria's frame drawing on a coordinate plane. Let's list the coordinates of the corners of the frame and the midpoints of the sides. What pattern do you notice?

$A(2, 2)$
$B(22, 2)$
$C(22, 18)$
$D(2, 18)$

Midpoint of $\overline{AB}$: $(12, 2)$
Midpoint of $\overline{BC}$: $(22, 10)$
Midpoint of $\overline{CD}$: $(12, 18)$
Midpoint of $\overline{AD}$: $(2, 10)$

Notice that the coordinates of the midpoint of $\overline{AB}$ are the mean of the corresponding coordinates of A and B.

Midpoint of $\overline{AB}$: $(12, 2)$

mean of x-coordinates   *mean of y-coordinates*
of A(2, 2) and B(22, 2)   *of A(2, 2) and B(22, 2)*

$$\frac{2 + 22}{2} = 12 \qquad \frac{2 + 2}{2} = 2$$

The woodworking example uses the midpoints of horizontal and vertical line segments. The Modeling Mathematics activity on the next page applies to any line segment.

MODELING MATHEMATICS

Midpoint of a Line Segment

Materials: grid paper ✐ ruler ✍ colored pencils

- Create a coordinate plane on your grid paper.

- Use a ruler to draw any line segment on the coordinate plane. The segment should not be vertical or horizontal. Label the endpoints of the segment A and B with their coordinates.

- Hold the paper up to the light and fold it so that A and B coincide. Crease the paper at the fold.

- Unfold the paper. Label the point M at which the crease meets the segment and write the coordinates of M.

- Use a different color pencil to draw a vertical line through the lower endpoint of your segment. Then draw a horizontal line through the upper endpoint

of your segment. Label the point of intersection P.

- Write the coordinates of the midpoints of $\overline{AP}$ and $\overline{BP}$.

- How do the x-coordinate of the vertical segment and the y-coordinate of the horizontal segment compare with the coordinates of M?

Your Turn

a. Create another coordinate plane and draw a segment with a different slope from the first one you drew.

b. Repeat the activity with this segment. What are your results?

c. Write a general rule for finding the midpoint of any segment.

This activity suggests the following rule for finding the midpoint of any segment, given the coordinates of its endpoints.

Midpoint of a Line Segment on a Coordinate Plane	The coordinates of the midpoint of a line segment whose endpoints are at (x_1, y_1) and (x_2, y_2) are given by $\left(\dfrac{x_1 + x_2}{2}, \dfrac{y_1 + y_2}{2}\right)$.

Example **1** If the vertices of parallelogram $WXYZ$ are $W(3, 0)$, $X(9, 3)$, $Y(7, 10)$, and $Z(1, 7)$, prove that the diagonals bisect each other. That is, prove that they intersect at their midpoints.

Explore Graph the vertices and draw the parallelogram and its diagonals.

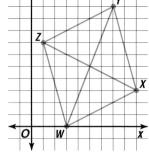

Plan Find the coordinates of the midpoints of the diagonals to see if they are the same point.

Solve Find the midpoints of $\overline{WY}$ and $\overline{XZ}$.

$W(x_1, y_1) = W(3, 0)$ $\qquad\qquad$ $X(x_1, y_1) = X(9, 3)$
$Y(x_2, y_2) = Y(7, 10)$ $\qquad\qquad$ $Z(x_2, y_2) = Z(1, 7)$

midpoint of $\overline{WY} = \left(\dfrac{3 + 7}{2}, \dfrac{0 + 10}{2}\right)$ $\quad$ midpoint of $\overline{XZ} = \left(\dfrac{9 + 1}{2}, \dfrac{3 + 7}{2}\right)$

$\qquad\qquad\qquad = \left(\dfrac{10}{2}, \dfrac{10}{2}\right)$ $\qquad\qquad\qquad\qquad = \left(\dfrac{10}{2}, \dfrac{10}{2}\right)$

$\qquad\qquad\qquad = (5, 5)$ $\qquad\qquad\qquad\qquad\qquad = (5, 5)$

Examine Since the midpoint of $\overline{XZ}$ has the same coordinates as the midpoint of $\overline{WY}$, the diagonals bisect each other.

If you know the midpoint and one endpoint of a segment, you can find the other endpoint.

Example **2** The center of a circle is $M(0, 2)$, and the endpoint of one of its radii is $A(-6, -4)$. If $\overline{AB}$ is a diameter of the circle, what are the coordinates of B?

Use the midpoint formula and substitute the values you know.

$$(x, y) = \left(\frac{x_1 + x_2}{2}, \frac{y_1 + y_2}{2} \right)$$

$$(0, 2) = \left(\frac{-6 + x_2}{2}, \frac{-4 + y_2}{2} \right) \quad \begin{array}{l} (x, y) = (0, 2), \\ (x_1, y_1) = (-6, -4) \end{array}$$

The x- and y-coordinates of the ordered pairs are equal since the ordered pairs are equal. So, two equations can be formed.

Find the x-coordinate.

$$0 = \frac{-6 + x_2}{2}$$

$$0 = -6 + x_2$$

$$6 = x_2$$

Find the y-coordinate.

$$2 = \frac{-4 + y_2}{2}$$

$$4 = -4 + y_2$$

$$8 = y_2$$

The coordinates of endpoint $B(x_2, y_2)$ are $(6, 8)$.

Check: Copy the graph of the circle and points A and M. Extend the radius through M to intersect the circle. This segment is the diameter of the circle. The intersection point is at $(6, 8)$, which are the coordinates of point B.

CHECK FOR UNDERSTANDING

Communicating Mathematics

Study the lesson. Then complete the following.

1. **Explain** how you can use the midpoint formula to find the midpoint of each of the following.

   a. a vertical line segment   **b.** a horizontal line segment

2. **Demonstrate** how to find the coordinates of vertex C of rectangle $ABCD$ if its two diagonals meet at $M(-3, 4)$ and the coordinates of A are $(0, -1)$.

3. Draw a rectangle on a coordinate plane.

   a. Find the midpoint of each side of the rectangle.

   b. Find the midpoint of the diagonals.

   c. How do the coordinates of the midpoint of the diagonal relate to the coordinates of the midpoints of the sides?

Find the coordinates of the midpoint of a segment with each pair of endpoints.

4. $A(5, -2), B(5, 8)$

5. $C(-5, 6), D(8, 6)$

6. $E(6, 5), F(14, 7)$

7. $G(-9, -6), H(-3, 8)$

Find the coordinates of the other endpoint of a segment given one endpoint and the midpoint M.

8. $A(3, 6), M(-1.5, 5)$

9. $B(-3, 7), M(-7, 7)$

10. $C(8, 11), M\left(5, \frac{17}{2}\right)$

11. $D(8, 4.5), M(8, 7.15)$

12. Archery The target for archery competition is composed of 10 rings and 5 colored circles. Hitting the bull's-eye is worth ten points, and hitting the outer ring is worth one point. A target is shown on a coordinate plane at the right. The two labeled points are the endpoints of a diameter of the target. Find the coordinates of the center of the bull's-eye.

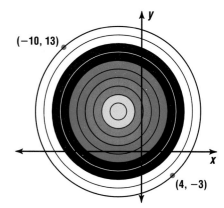

$(-10, 13)$

$(4, -3)$

EXERCISES

Find the coordinates of the midpoint of a segment with each pair of endpoints.

13. $J(6, 6), K(19, 6)$

14. $L(5, 9), M(-7, 3)$

15. $P(-5, 9), Q(7, 1)$

16. $R(7, 4), S(11, -10)$

17. $T(8, -7), U(-2, 11)$

18. $V(-8, 1.2), W(-8, 7.4)$

19. $X(-3, 9), Y(4, -7)$

20. $B(9, -4), C(2, 7)$

21. $D(4.7, -2.9), E(-3.1, 8.3)$

22. $F(a, b), G(c, d)$

23. $Y(6x, 14y), Z(2x, 4y)$

24. $M(-2w, -7v), N(6w, 2v)$

Find the coordinates of the other endpoint of a segment given one endpoint and the midpoint M.

25. $E(-7, 8), M(-7, 4)$

26. $F(4, 2), M(2, 1)$

27. $G(3, -6), M(12, -6)$

28. $H(5, 3), M(6, 4)$

29. $L(-8, 4), M\left(\frac{1}{2}, 7\right)$

30. $N(5, -9), M(8, -7.5)$

31. $R\left(\frac{1}{6}, \frac{1}{3}\right), M\left(\frac{1}{2}, \frac{1}{3}\right)$

32. $S(a, b), M\left(\frac{x+a}{2}, \frac{y+b}{2}\right)$

If P is the midpoint of segment AB, find the coordinates of the missing point.

33. $A(6.5, -8.2), P(4.4, -0.7)$

34. $P(1.2, 4.5), B(5.3, 1.9)$

35. $A(9.7, -5.4), B(3.6, 1.7)$

36. $A(-5.9, 7.2), P(-1.05, 5.85)$

37. The coordinates of the endpoints of a diameter of a circle are $(5, 8)$ and $(-7, 2)$. Find the coordinates of the center.

38. Find the coordinates of the point on $\overline{AB}$ that is one-fourth the distance from A to B for $A(8, 3)$ and $B(10, -5)$.

39. The endpoints of $\overline{AB}$ are $A(-2, 7)$ and $B(6, -5)$. Find the coordinates of P if P lies on $\overline{AB}$ and is $\frac{3}{8}$ the distance from A to B.

Critical Thinking

40. Points W, X, Y, and Z are the midpoints of the sides of quadrilateral $QUAD$. Write a convincing argument that quadrilateral $WXYZ$ is a parallelogram.

Applications and Problem Solving

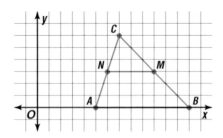

41. Geometry N is the midpoint of $\overline{AC}$ in the triangle shown at the left. M is the midpoint of $\overline{BC}$.
  a. Write the coordinates of M and N.
  b. Compare $\overline{MN}$ and $\overline{AB}$. Verify your findings.

42. Technology The screen of a TI-83 graphing calculator is composed of tiny dots called *pixels* that are turned on or off during a display. Each pixel is named by the row and column in which it lies.

  a. If a segment drawn on the screen has one endpoint at pixel $(20, 43)$ and its midpoint is at pixel $(30, 60)$, what is the pixel name for the other endpoint of the segment?
  b. What is the pixel name of the origin?
  c. How does this coordinate system differ from the coordinate system we usually use for graphing?

43. Geometry S and T are the midpoints of their respective sides in the triangle shown at the right.
  a. Find the coordinates of P and Q.

  b. If the area of triangle RST is 15.5 square units, what is the area of triangle RPQ? Explain how you arrived at your answer.

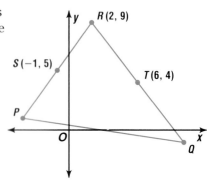

Mixed Review

44. Write an equation in slope-intercept form of the line that passes through $(8, -2)$ and is perpendicular to the graph of $5x - 3y = 7$. (Lesson 6–6)

45. Write an equation in slope-intercept form of the line that passes through $(-6, 2)$ and $(3, -5)$. (Lesson 6–4)

46. Given $g(x) = x^2 - x$, find $g(4b)$. (Lesson 5–5)

47. Determine whether $9x + (-7) = 6y$ is a linear equation. If it is, rewrite it in standard form. (Lesson 5–4)

48. Trigonometry If $\sin N = 0.6124$, use a calculator to find the measure of angle N to the nearest degree. (Lesson 4–3)

49. Solve $-27 - b = -7$. (Lesson 3–1)

50. In Mr. Tucker's algebra class, students can get extra points for finding the correct solution to the "Riddle of the Week." One week, Mr. Tucker posed this riddle. *Josh is 10 years older than his brother. Next year, he will be three times as old as his brother.* Answer the following questions. (Lesson 2–9)
  a. How old was Josh when his brother was born?
  b. If his brother's age is represented by a, what is Josh's age?
  c. India says that Josh is 12 years old and his brother is 4. Does India get the extra points for her answer? Why or why not?

51. Simplify $0.2(3x + 0.2) + 0.5(5x + 3)$. (Lesson 1–8)

Mathematics and SOCIETY

People Around the World

This excerpt is from an article in *The Amicus Journal*, Winter 1994.

THE LAST FORTY YEARS SAW THE FASTEST rise in human numbers in all previous history, from only 2.5 billion people in 1950 to 5.6 billion today....The second half of the 1990s will add an additional 94 million people per year. That is equivalent to a new United States every thirty-three months, another Britain every seven months, a Washington (DC) every six days. A whole Earth of 1800 was added in just one decade, according to United Nations Population Division statistics. After 2000, annual additions will slow, but by 2050 the United Nations expects the human race to total just over 10 billion—an extra Earth of 1980 on top of today's. ■

1. What effect does rapid population growth have on Earth's natural resources? Be specific.

2. What changes would you expect to see if the community in which you live doubles in population during your lifetime?

3. A village is deserted, but its buildings are intact and there are ample resources. Suppose 1000 refugees per day flock to this village.
  a. At this rate, how long would it take to reach the midpoint to a population of five million?
  b. Is the proposition of a village growing at this rate realistic? Why or why not?

VOCABULARY

After completing this chapter, you should be able to define each term, property, or phrase and give an example or two of each.

Algebra

family of graphs   (p. 354)
parent graph   (p. 354)
point-slope form   (p. 333)
rise   (p. 325)
run   (p. 325)
slope   (pp. 324, 325)
slope-intercept form   (p. 347)
standard form   (p. 333)
x-intercept   (p. 346)
y-intercept   (p. 346)

Problem Solving

use a model   (p. 339)

Geometry

midpoint   (p. 369)
parallel lines   (p. 362)
parallelogram   (p. 362)
perpendicular lines   (p. 362)

Statistics

best-fit line   (p. 341)
negative correlation   (p. 340)
positive correlation   (p. 340)
regression line   (p. 342)
scatter plot   (p. 339)

UNDERSTANDING AND USING THE VOCABULARY

Choose the correct term to complete each sentence.

1. The lines with equations $y = -2x + 7$ and $y = -2x - 6$ are (*parallel, perpendicular*) lines.

2. The equation $y - 2 = -3(x - 1)$ is written in (*point-slope, slope-intercept*) form.

3. If the endpoints of $\overline{AB}$ are $A(6, -3)$ and $B(2, 4)$, then the (*midpoint, slope*) of $\overline{AB}$ is at $\left(4, \frac{1}{2}\right)$.

4. The (*x-intercept, y-intercept*) of the equation $2x + 3y = -1$ is $-\frac{1}{2}$.

5. The lines with equations $y = \frac{1}{3}x + 1$ and $y = -3x - 5$ are (*parallel, perpendicular*) lines.

6. The slope of a line is defined as the ratio of the (*rise, run*), or vertical change, to the (*rise, run*), or horizontal change, as you move from one point on the line to another.

7. The equation $y = -\frac{1}{2}x + 3$ is written in (*slope-intercept, standard*) form.

8. The (*x-intercept, y-intercept*) of the equation $-x - 4y = 2$ is $-\frac{1}{2}$.

9. The (*midpoint, slope*) of the line with equation $-3y = 2x + 3$ is $-\frac{2}{3}$.

10. The equation $3x - 4y = -5$ is written in (*point-slope, standard*) form.

SKILLS AND CONCEPTS

OBJECTIVES AND EXAMPLES	REVIEW EXERCISES

Upon completing this chapter, you should be able to:

Use these exercises to review and prepare for the chapter test.

• find the slope of a line, given the coordinates of two points on the line   (Lesson 6–1)

Determine the slope of the line that passes through $(-6, 5)$ and $(3, -2)$.

$$m = \frac{y_2 - y_1}{x_2 - x_1}$$

$$= \frac{-2 - 5}{3 - (-6)}$$

$$= -\frac{7}{9}$$

The slope is $-\frac{7}{9}$.

Determine the slope of the line that passes through each pair of points.

11. $(8, 3), (2, 5)$ 12. $(-2, 5), (-2, 9)$

13. $(-3, 6), (-8, 4)$ 14. $(4, 3), (-5, 3)$

Determine the value of r so the line that passes through each pair of points has the given slope.

15. $(r, 4), (7, 3), m = \frac{3}{4}$

16. $(4, -7), (-2, r), m = \frac{8}{3}$

• write linear equations in point-slope form (Lesson 6–2)

Write the point-slope form of an equation of the line that passes through the points at $(-4, 7)$ and $(-2, 3)$.

$$m = \frac{3 - 7}{-2 - (-4)} \text{ or } -2$$

$$y - y_1 = m(x - x_1)$$

$$y - 7 = -2[x - (-4)]$$

$$y - 7 = -2(x + 4)$$

An equation of the line is $y - 7 = -2(x + 4)$.

Write the point-slope form of an equation of the line that passes through the given point and has the given slope.

17. $(4, -3), m = -2$ 18. $(8, 5), m = 5$

19. $(-5, 7), m = 0$ 20. $(6, 2), m = \frac{1}{2}$

Write the point-slope form of an equation of the line that passes through each pair of points.

21. $(0, 3), (-5, 0)$ 22. $(5, 4), (6, 3)$

23. $(4, 1), (-3, 7)$ 24. $(2, -5), (0, 4)$

• write linear equations in standard form   (Lesson 6–2)

Write the standard form of an equation of the line that passes through the points at $(6, -4)$ and $(-1, 5)$.

$$m = \frac{5 - (-4)}{-1 - 6} \text{ or } -\frac{9}{7}$$

$$y - y_1 = m(x - x_1) \quad \bigg| \quad 7(y + 4) = 7 \cdot -\frac{9}{7}(x - 6)$$

$$y - (-4) = -\frac{9}{7}(x - 6) \quad \bigg| \quad 7y + 28 = -9(x - 6)$$

$$y + 4 = -\frac{9}{7}(x - 6) \quad \bigg| \quad 7y + 28 = -9x + 54$$

$$y + 4 = -\frac{9}{7}(x - 6) \quad \bigg| \quad 9x + 7y = 26$$

An equation of the line is $9x + 7y = 26$.

Write the standard form of an equation of the line that passes through the given point and has the given slope.

25. $(4, -6), m = 3$ 26. $(1, 5), m = 0$

27. $(6, -1), m = \frac{3}{4}$ 28. $(8, 3), m = $ undefined

Write the standard form of an equation of the line that passes through each pair of points.

29. $(-2, 5), (9, 5)$ 30. $(0, 5), (-2, 0)$

31. $\left(-2, \frac{2}{3}\right), \left(-2, \frac{2}{7}\right)$ 32. $\left(-5, 7\right), \left(0, \frac{1}{2}\right)$

OBJECTIVES AND EXAMPLES

- graph and interpret points on a scatter plot; draw and write equations for best-fit lines and make predictions by using those equations (Lesson 6–3)

The scatter plot below shows a negative relationship since the line suggested by the points would have a negative slope.

Mistakes made

Hours studied

REVIEW EXERCISES

33. Draw a scatter plot of the data below. Let the independent variable be height.

Buildings in Oklahoma City	Height (feet)	Number of Stories
Liberty Tower	500	36
First National Center	493	28
City Place	440	33
First Oklahoma Tower	425	31
Kerr-McGee Center	393	30
Mid-America Tower	362	29

a. Does a relationship exist? If so, what is the relationship?

b. Use your scatter plot as a model to predict how many stories you might think a building 475 feet high would be.

c. Draw a best-fit line and write the equation of the line.

- determine the x- and y-intercepts of linear graphs from their equations and write equations in slope-intercept form   (Lesson 6–4)

Determine the slope and x- and y-intercepts of the graph of $3x - 2y = 7$.

Solve for y to find the slope and y-intercept.	Let $y = 0$ to find the x-intercept.
$3x - 2y = 7$	$3x - 2(0) = 7$
$-2y = -3x + 7$	$3x = 7$
$y = \dfrac{3}{2}x - \dfrac{7}{2}$	$x = \dfrac{7}{3}$

The slope is $\dfrac{3}{2}$, the y-intercept is $-\dfrac{7}{2}$, and the x-intercept is $\dfrac{7}{3}$.

Write an equation in slope-intercept form of a line with the given slope and y-intercept.

34. $m = 2, b = 4$

35. $m = -3, b = 0$

36. $m = -\dfrac{1}{2}, b = -9$

37. $m = 0, b = 5.5$

Find the slope and y-intercept of the graph of each equation.

38. $x = 2y - 7$ 39. $8x + y = 4$

40. $y = \dfrac{1}{4}x + 3$ 41. $\dfrac{1}{2}x + \dfrac{1}{4}y = 3$

- graph a line given any linear equation   (Lesson 6–5)

Graph $-2x - y = 5$.

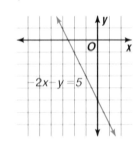

$-2x - y = 5$

Graph each equation.

42. $3x - y = 9$ 43. $5x + 2y = 12$

44. $y = \dfrac{2}{3}x + 4$ 45. $y = -\dfrac{3}{2}x - 6$

46. $3x + 4y = 6$ 47. $5x - \dfrac{1}{2}y = 2$

48. $y - 4 = -2(x + 1)$ 49. $y + 5 = -\dfrac{3}{4}(x - 6)$

OBJECTIVES AND EXAMPLES

• write equations of lines that pass through a given point, parallel or perpendicular to the graph of a given equation   (Lesson 6–6)

Write an equation of the line that is perpendicular to the graph of $2x + y = 6$ and passes through $(2, 3)$.

$$2x + y = 6$$
$$y = -2x + 6$$

The slope of this line is -2. Therefore, the slope of the line perpendicular to it is $\frac{1}{2}$.

$$y - 3 = \frac{1}{2}(x - 2) \quad m = \frac{1}{2}, (x_1, y_1) = (2, 3)$$
$$y = \frac{1}{2}x + 2$$

REVIEW EXERCISES

Write an equation of the line having the following properties

50. parallel to $4x - y = 7$ and passes through $(2, -1)$

51. perpendicular to $2x - 7y = 1$ and passes through $(-4, 0)$

52. parallel to $3x + 9y = 1$ and passes through $(3, 0)$

53. perpendicular to $8x - 3y = 7$ and passes through $(4, 5)$

54. parallel to $-y = -2x + 4$ and passes through $(5, -6)$

55. perpendicular to $5y = -x + 1$ and passes through $(2, -5)$

• find the coordinates of the midpoint of a line segment in the coordinate plane   (Lesson 6–7)

The endpoints of a segment are at $(11, 4)$ and $(9, 2)$. Find its midpoint.

$$(x, y) = \left(\frac{x_1 + x_2}{2}, \frac{y_1 + y_2}{2} \right)$$
$$= \left(\frac{11+9}{2}, \frac{4+2}{2} \right)$$
$$= (10, 3)$$

Find the coordinates of the midpoint of a segment with each pair of endpoints.

56. $A(3, 5), B(9, -3)$ 57. $A(-6, 6), B(8, -11)$

58. $A(14, 4), B(2, 0)$ 59. $A(2, 7), B(8, 4)$

60. $A(2, 5), B(4, -1)$ 61. $A(10, 4), B(-3, -7)$

Find the coordinates of the other endpoint of a segment given one endpoint and the midpoint M.

62. $A(3, 5), M(11, 7)$ 63. $A(5, 3), M(9, 7)$

64. $A(4, -11), M(5, -9)$ 65. $A(11, -4), M(3, 8)$

APPLICATIONS AND PROBLEM SOLVING

66. **Skiing**   A course for cross-country skiing is regulated so that the slope of any hill cannot be greater than 0.33. Suppose a hill rises 60 meters over a horizontal distance of 250 meters.

    a. What is the slope of the hill?

    b. Does the hill meet the requirements?

    (Lesson 6–1)

67. **Travel**   Jon Erlanger is taking a long trip. In the first two hours, he drives 80 miles. After that, he averages 45 miles per hour. Write an equation in slope-intercept form relating distance traveled and time.   (Lesson 6–4)

A practice test for Chapter 6 is provided on page 792.

68. **Entertainment**   Carolyn Parks owns stock in Star Gazer Motion Picture Company. Every other week, she graphs the closing value of a share of the stock. What is the midpoint between the highest and lowest values of the stock?   (Lesson 6–7)

ALTERNATIVE ASSESSMENT

COOPERATIVE LEARNING PROJECT

Video Rental In this chapter, you investigated graphing linear equations. You used slope and data points to graph equations. These graphs represented information that could be analyzed and extended in order to interpret further information than what was given.

In this project, you will analyze two video-store rental policies. There are two stores in Joe's neighborhood where he can rent videos: Video World and Mega Video. At one of the two stores, you have to buy a membership card first. The other one does not have these cards; you simply pay per video. Both have a daily charge (rate per day). Mega Video has a membership card that costs $15 per year and then each video that is rented costs $2 per day. Video World simply rents each video for $3.25 per day.

Which is the most economical video store for Joe to use? If he rarely rents videos, which store should he use? If he often rents videos, which store should he use?

Follow these steps to determine the store he should use.

- Determine a way to graph this information.

- Develop a linear equation for each store that describes the amount of money it costs to rent a video.

- Determine whether the graphs should be drawn separately or on the same grid.

- Investigate the graphs and determine what each graph represents.

- What factors other than price might Joe want to consider before he chooses a store?

- Write a paragraph describing various situations for Joe and the appropriate solutions.

THINKING CRITICALLY

- Give an example of an equation of a line that has the same x- and y-intercepts. What is the slope of the line?

- Give an example of an equation of a line whose x- and y-intercepts are opposites. What is the slope of the line?

- Make a conjecture about the slopes of lines whose x- and y-intercepts are equal or opposites.

- The x-intercept of a line is s, and the y-intercept is t. Write the equation of the line.

PORTFOLIO

In mathematics, there is often more than one way to solve a problem. In order to graph a line for example, various information can be used. Sometimes you use two points, sometimes you use a point and a slope, and sometimes you use x- and y-intercepts. Find an equation from your work in this chapter and describe at least three ways in which to graph it. Place this in your portfolio.

SELF EVALUATION

Checking your answer is the final step in solving a problem. This step is the most critical, but it is also the step that is most often forgotten. When an answer is finally reached, it should be checked to see whether it makes sense and also whether it can be verified.

Assess yourself. Do you use all four steps when you solve a problem? Do you reach an answer, assume that it is the correct answer, and then move on without checking it? Do you ask yourself questions about your answer to verify that it is correct and that it makes sense? Can you think of an example of a problem in your daily life in which you didn't ask yourself questions about your solution to the problem and later you discovered that the solution didn't work?

CUMULATIVE REVIEW

CHAPTERS 1–6

There are eight multiple-choice questions in this section. After working each problem, write the letter of the correct answer on your paper.

1. Name the quadrant in which point $P(x, y)$ is located if it satisfies the condition $x < 0$ and $y = -3$.

   A. I

   B. II

   C. III

   D. IV

2. Raini made a map of his neighborhood using a coordinate grid. His school has coordinates $(3, 8)$, where 3 is the number of blocks east he must walk and 8 is the number of blocks north. From school, he walks 4 blocks east and 2 blocks south to get to his friend's house. Find the coordinates of the midpoint of the line segment that could be drawn to connect his school and his friend's house.

   A. $(5, 7)$

   B. $(8, -2)$

   C. $(7, 6)$

   D. $(5, -6)$

3. Determine the slope of the line that passes through the points at $(-3, 6)$ and $(-5, 9)$.

   A. $\frac{3}{2}$

   B. $\frac{2}{3}$

   C. $-\frac{3}{2}$

   D. none of these

4. **Probability** A shipment of 100 CDs was just received at Tunes R Us. There is a 4% probability that one of the CDs was damaged during shipment, even though the package may not be cracked. If Craig buys a CD from this shipment, what are the odds that he is buying a damaged CD?

   A. 1:25

   B. $\frac{4}{100}$

   C. 96:4

   D. 1:24

5. Choose the graph that represents a function.

6. Choose the property illustrated by the statement *The quantity g plus h times b equals g times b plus h times b.*

   A. distributive property

   B. commutative property of multiplication

   C. multiplicative identity property

   D. associative property of addition

7. Choose the line that represents the graph of the equation $y = -\frac{2}{3}x + 2$.

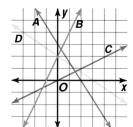

   A. line A

   B. line B

   C. line C

   D. line D

8. **Geometry** The formula Area $= \frac{1}{2}bc \sin A$ can be used to find the area of any triangle ABC. Find the area of the triangle shown below to the nearest tenth of a square centimeter if $b = 16$, $c = 9$, and $\angle A$ is a $36°$ angle.

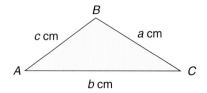

   A. 72 cm^2

   B. 42.3 cm^2

   C. 84.6 cm^2

   D. 144 cm^2

SECTION TWO: SHORT ANSWER

This section contains seven questions for which you will provide short answers. Write your answer on your paper.

9. Express the relation shown in the graph as a set of ordered pairs. Then draw a mapping of the relation.

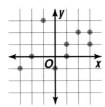

10. Write an equation of the line that passes through the point at $(2, -1)$ and is perpendicular to $4x - y = 7$.

11. The average person can live for eleven days without water, assuming a mean temperature of $60°$ F. Determine if survival is possible given daily high temperatures of $59°$, $70°$, $49°$, $62°$, $46°$, $63°$, $71°$, $64°$, $55°$, $68°$, and $54°$ F. Verify your answer.

12. Write the solution set for $5x + y = 4$, if the domain is $\{-2, -1, 0, 2, 5\}$.

13. Peanuts sell for $3.00 per pound. Cashews sell for $6.00 per pound. How many pounds of cashews should be mixed with 12 pounds of peanuts to obtain a mixture that sells for $4.20 per pound?

14. At Videoville, Eshe earns a weekly salary of $150 plus $0.30 for each video over 100 that she sells each week. If Eshe sells v videos in a week, then her total weekly salary is $C(v) = 150 + 0.30(v - 100)$ for $v > 100$. If she wants to earn $225 each week to save up for a stereo system, how many videos must she sell each week?

15. Graph the solution set {integers greater than or equal to -4} on a number line.

SECTION THREE: OPEN ENDED

This section contains two open-ended problems. Demonstrate your knowledge by giving a clear, concise solution to each problem. Your score on these problems will depend on how well you do the following.

- Explain your reasoning.
- Show your understanding of the mathematics in an organized manner.
- Use charts, graphs, and diagrams in your explanation.
- Show the solution in more than one way or relate it to other situations.
- Investigate beyond the requirements of the problem.

16. For long-distance phone calls, a telephone company charges $1.72 for a 4-minute call, $2.40 for a 6-minute call, and $5.46 for a 15-minute call. Determine the charge for a 1-minute call.

17. The table below shows the annual income and the number of years of college education for eleven people.

Income (thousands)	College Education (years)
$23	3.0
20	2.0
25	4.0
47	6.0
19	2.5
48	7.5
35	6.5
10	1.0
39	5.5
26	4.5
36	4.0

a. Make a scatter plot of the data.

b. Based on this plot, how does the number of years of college affect income?

Solving Linear Inequalities

Objectives

In this chapter, you will:

- solve inequalities,
- graph solutions of inequalities,
- graph solutions of open sentences that involve absolute value,
- solve problems by drawing a diagram, and
- use box-and-whisker plots to display and analyze data.

Increase Your Earnings Potential

	Lifetime Earnings (thousands of dollars)
Professional degree	3013
Doctorate	2142
Master's degree	1619
Bachelor's degree	1421
2-year degree	1082
Some college	993
High school graduate	821
Not a high school graduate	609

Source: *Chicago Tribune*, 1995

There is a direct relationship between lifetime earnings and educational attainment. College graduation and advanced degrees really do make a difference. Set your sights on a career and aim to be the best you can be.

TIME *Line*

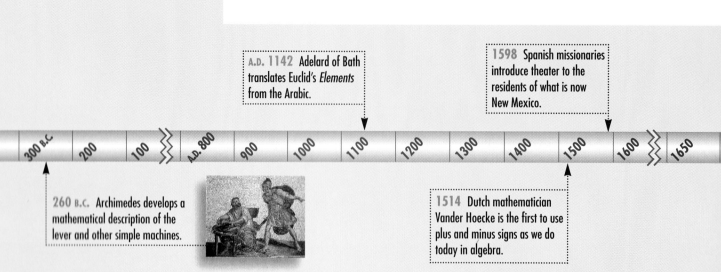

A.D. **1142** Adelard of Bath translates Euclid's *Elements* from the Arabic.

1598 Spanish missionaries introduce theater to the residents of what is now New Mexico.

300 B.C. 200 100 A.D. 800 900 1000 1100 1200 1300 1400 1500 1600 1650

260 B.C. Archimedes develops a mathematical description of the lever and other simple machines.

1514 Dutch mathematician Vander Hoecke is the first to use plus and minus signs as we do today in algebra.

Kimana plans to study either botany, genetics, or microbiology at a nearby university. She hasn't ruled out teaching as a possible career choice, but is interested in finding other businesses in which she can apply the science she will be studying.

• Visit the public library or local college to investigate careers that relate to Kimana's interests.

• Which careers might offer Kimana an opportunity to earn an amount greater than that presented in the graph for a bachelor's degree?

• Which careers require a degree for which the years studied are greater than those for a bachelor's degree? Choose a career. Write an inequality that might express the time needed to study and train for that career.

Shakema Hodge is aiming for an advanced degree in microbiology. The young scientist from St. Thomas, Virgin Islands, earned her bachelor's degree in biology at the University of the Virgin Islands. Now she is enrolled in a 5-year Ph.D. program in molecular microbiology at the University of Rochester in New York. She would love to teach high school students, believing that "to capture a young person's mind, you have to expose them to the sciences at an early age."

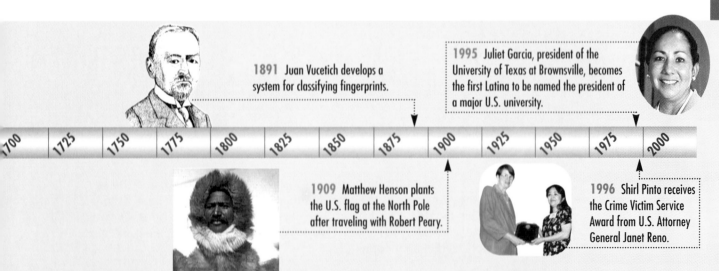

1891 Juan Vucetich develops a system for classifying fingerprints.

1995 Juliet Garcia, president of the University of Texas at Brownsville, becomes the first Latina to be named the president of a major U.S. university.

1909 Matthew Henson plants the U.S. flag at the North Pole after traveling with Robert Peary.

1996 Shirl Pinto receives the Crime Victim Service Award from U.S. Attorney General Janet Reno.

1700 1725 1750 1775 1800 1825 1850 1875 1900 1925 1950 1975 2000

Solving Inequalities by Using Addition and Subtraction

What YOU'LL LEARN

- To solve inequalities by using addition and subtraction.

Why IT'S IMPORTANT

You can use inequalities to solve problems involving nutrition and personal finance.

APPLICATION
Nutrition

In 1990, the U.S. Department of Agriculture issued new dietary guidelines. These guidelines recommend that people greatly reduce their fat intake. Your recommended calorie intake depends on your height, desired weight, and physical activity. The average 14-year-old is 5 feet 2 inches tall and weighs 107 pounds. Boys should consume about 2434 calories per day and girls 2208 calories per day to maintain this weight.

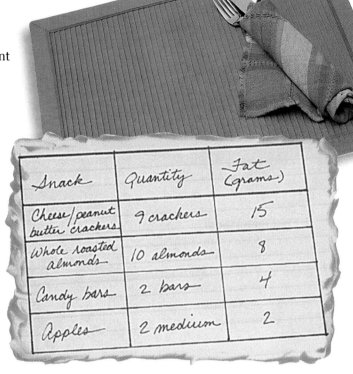

Snack	Quantity	Fat (grams)
Cheese/peanut butter crackers	9 crackers	15
Whole roasted almonds	10 almonds	8
Candy bars	2 bars	4
Apples	2 medium	2

Oliana learned in health class that no more than 30% of her calorie intake should come from fat. For her 2030-calorie-a-day diet, that means no more than 68 grams of fat. She keeps track of her snacks for one day and records their fat content, as shown in the table above. How many grams of fat can Oliana have in the other foods she eats that day and stay within the guidelines?

Let's write an inequality to represent the problem. Let g represent the remaining grams of fat that Oliana can eat that day.

grams of fat in snacks	plus	grams of fat remaining	is less than or equal to	total grams
$15 + 8 + 4 + 2$	$+$	g	$\leq$	68

That is, $29 + g \leq 68$.

The symbol $\leq$ indicates *less than or equal to*. It is used in this situation because the total number of grams of fat in Oliana's daily diet should be no greater than 68. If this were an equation, we would subtract 29 from (or add -29 to) each side. Can the same procedure be used in an inequality? *This problem will be solved in Example 1.*

Let's explore what happens if inequalities are solved in the same manner as equations. We know that $7 > 2$. What happens when you add or subtract the same quantity to each side of the inequality? We can use number lines to model the situation.

Add 3 to each side.

$$7 > 2$$
$$7 + 3 \stackrel{?}{>} 2 + 3$$
$$10 > 5$$

Subtract 4 from each side.

$$7 > 2$$
$$7 - 4 \stackrel{?}{>} 2 - 4$$
$$3 > -2$$

In each case, the inequality holds true. These examples illustrate two properties of inequalities.

Addition and Subtraction Properties for Inequalities	**For all numbers a, b, and c, the following are true.** **1. If $a > b$, then $a + c > b + c$ and $a - c > b - c$.** **2. If $a < b$, then $a + c < b + c$ and $a - c < b - c$.**

These properties are also true when $>$ and $<$ are replaced by $\geq$ and $\leq$. So, we can use these properties to obtain a solution to the application at the beginning of the lesson.

Example

1

APPLICATION

Nutrition

Preferred snack foods and
% of calories from fat

1. potato chips   58%
2. tortilla chips   47%
3. popcorn   45%
4. pretzels   8%
5. mixed nuts   80%

Refer to the application at the beginning of the lesson. Solve $29 + g \leq 68$.

$$29 + g \leq 68$$
$$29 - 29 + g \leq 68 - 29 \quad \textit{Subtract 29 from each side.}$$
$$g \leq 39 \qquad\qquad \textit{This means all numbers less than or equal to 39.}$$

The solution set can be written as {all numbers less than or equal to 39}.

Check: To check this solution, substitute 39, a number less than 39, and a number greater than 39 into the inequality.

Let $g = 39$.   Let $g = 20$.   Let $g = 40$.

$$29 + g \leq 68 \qquad\quad 29 + g \leq 68 \qquad\quad 29 + g \leq 68$$
$$29 + 39 \stackrel{?}{\leq} 68 \qquad\quad 29 + 20 \stackrel{?}{\leq} 68 \qquad\quad 29 + 40 \stackrel{?}{\leq} 68$$
$$68 \leq 68 \quad \text{true} \qquad\quad 49 \leq 68 \quad \text{true} \qquad\quad 69 \leq 68 \quad \text{false}$$

So, Oliana can have 39 or fewer grams of fat in other foods that day and stay within the dietary guidelines.

The solution to the inequality in Example 1 was expressed as a set. A more concise way of writing a solution set is to use **set-builder notation**. The solution in set-builder notation is $\{g \mid g \leq 39\}$. This is read *the set of all numbers g such that g is less than or equal to 39.*

LOOK BACK

You can refer to Lesson 1-5 for information on solution sets.

In Lesson 2–4, you learned that you can show the solution to an inequality on a graph. The solution to Example 1 is shown on the number line below.

28 29 30 31 32 33 34 35 36 37 38 39 40 41 42

The closed circle at 39 tells us that 39 is included in the inequality. The heavy arrow pointing to the left shows that it also includes all numbers less than 39.
If the inequality was <, the circle would be open.

Example **2** **Solve $13 + 2z < 3z - 39$. Then graph the solution.**

$$13 + 2z < 3z - 39$$
$$13 + 2z - 2z < 3z - 2z - 39 \quad \text{\textit{Subtract 2z from each side.}}$$
$$13 < z - 39$$
$$13 + 39 < z - 39 + 39 \quad \text{\textit{Add 39 to each side.}}$$
$$52 < z$$

Since $52 < z$ is the same as $z > 52$, the solution set is $\{z \mid z > 52\}$.

The graph of the solution contains an open circle at 52 since 52 is not included in the solution, and the arrow points to the right.

49 50 51 52 53 54 55 56

Verbal problems containing phrases like *greater than* or *less than* can often be solved by using inequalities. The following chart shows some other phrases that indicate inequalities.

Inequalities			
<	>	≤	≥
• less than • fewer than	• greater than • more than	• at most • no more than • less than or equal to	• at least • no less than • greater than or equal to

Example **3**

APPLICATION

Budgeting

Alvaro, Chip, and Solomon have earned $500 to buy equipment for their band. They have already spent $275 on a used guitar and a drum set. They are now considering buying a $125 amplifier. What is the most they can spend on promotional materials and T-shirts for the band if they buy the amplifier?

Explore At most means that they cannot go over what is left of their budget of $500. They have spent $275, so they have $225 left. Let m = the amount of money for promotional materials and T-shirts.

Plan	Total to spend	is at most	$225.
	125 + m	≤	225

Solve

$$125 + m \le 225$$

$$125 - 125 + m \le 225 - 125 \quad \textit{Subtract 125 from each side.}$$

$$m \le 100$$

The members of the band can spend $100 or less on promotional materials and T-shirts.

Examine   Since $275 + $125 + $100 = $500, Alvaro, Chip, and Solomon can spend $100 or less on promotional materials and T-shirts.

When solving problems involving equations, it is often necessary to write an equation that represents the words in the problem. This is also true of inequalities.

Example **Write an inequality for the sentence below. Then solve the inequality and check the solution.**

Three times a number is more than the difference of twice that number and three.

Three times a number	is more than	twice the number	minus	three
3x	>	2x	−	3

$$3x > 2x - 3$$

$$3x - 2x > 2x - 2x - 3 \quad \textit{Subtract 2x from each side.}$$

$$x > -3$$

The solution set is $\{x \mid x > -3\}$.

CHECK FOR UNDERSTANDING

Communicating Mathematics

Study the lesson. Then complete the following.

1. **Write** three inequalities that are equivalent to $x < -10$.

2. **Explain** what $\{w \mid w > -3\}$ means.

3. **Explain** the difference between the solution sets for $x + 24 < 17$ and $x + 24 \le 17$.

4. **Describe** how you would graph the solution to an inequality. Include examples and graphs in your explanation.

5. **Is** it possible for the solution set of an inequality to be the empty set? If so, give an example.

6. Sometimes statements we make can be translated into inequalities. For example, *In some states, you have to be at least 16 years old to have a driver's license* can be expressed as $a \geq 16$, and *Tomás cannot lift more than 72 pounds* can be translated into $w \leq 72$. Following these examples, write three statements that deal with your everyday life. Then translate each into a corresponding inequality.

Guided Practice

Match each inequality with the graph of its solution.

7. $b - 18 > -3$

a. (number line: 10 11 12 13 14 15 16 17, closed dot at 13 shaded left)

8. $10 \geq -3 + x$

b. (number line: −6 −5 −4 −3 −2 −1 0 1, closed dot at −3 shaded right)

9. $x + 11 < 6$

c. (number line: 10 11 12 13 14 15 16 17, open dot at 14 shaded right)

10. $4c - 3 \leq 5c$

d. (number line: −7 −6 −5 −4 −3 −2 −1 0, open dot at −5 shaded left)

Solve each inequality. Then check your solution.

11. $x + 7 > 2$

12. $10 \geq x + 8$

13. $y - 7 < -12$

14. $-81 + q > 16 + 2q$

Define a variable, write an inequality, and solve each problem. Then check your solution.

15. A number decreased by 17 is less than -13.

16. A number increased by 4 is at least 3.

EXERCISES

Practice

Solve each inequality. Then check your solution, and graph it on a number line.

17. $a - 12 < 6$

18. $m - 3 < -17$

19. $2x \leq x + 1$

20. $-9 + d > 9$

21. $x + \frac{1}{3} > 4$

22. $-0.11 \leq n - (-0.04)$

23. $2x + 3 > x + 5$

24. $7h - 1 \leq 6h$

Solve each inequality. Then check your solution.

25. $x + \frac{1}{8} < \frac{1}{2}$

26. $3x + \frac{4}{5} \leq 4x + \frac{3}{5}$

27. $3x - 9 \leq 2x + 6$

28. $6w + 4 \geq 5w + 4$

29. $-0.17x - 0.23 < 0.75 - 1.17x$

30. $0.8x + 5 \geq 6 - 0.2x$

31. $3(r - 2) < 2r + 4$

32. $-x - 11 \geq 23$

Define a variable, write an inequality, and solve each problem. Then check your solution.

33. A number decreased by -4 is at least 9.

34. The sum of a number and 5 is at least 17.

35. Three times a number is less than twice the number added to 8.

36. Twenty-one is no less than the sum of a number and -2.

37. The sum of two numbers is less than 53. One number is 20. What is the other number?

38. The sum of four times a number and 7 is less than 3 times that number.

39. Twice a number is more than the difference of that number and 6.

40. The sum of two numbers is 100. One number is at least 16 more than the other number. What are the two numbers?

If $3x \geq 2x + 5$, then complete each inequality.

41. $3x + 7 \geq 2x + \underline{\quad ? \quad}$

42. $3x - 10 \geq 2x - \underline{\quad ? \quad}$

43. $3x + \underline{\quad ? \quad} \geq 2x + 3$

44. $\underline{\quad ? \quad} \leq x$

Programming

45. **Geometry** For three line segments to form a triangle, the sum of the lengths of any two sides must exceed the length of the third side. Let the lengths of the possible sides be a, b, and c. Then these three inequalities must be true: $a + b > c$, $a + c > b$, and $b + c > a$. The graphing calculator program at the right uses these inequalities to determine if the three lengths can be measures of the sides of a triangle.

Use the program to determine whether segments with the given lengths can form a triangle.

Tip: To run the program again after trying one set of numbers, press ENTER .

a. 10 in., 12 in., 27 in.

b. 3 ft, 4 ft, 5 ft

c. 125 cm, 140 cm, 150 cm

d. 1.5 m, 2.0 m, 2.5 m

```
PROGRAM: TRIANGLE
: Disp "ENTER THREE LENGTHS"
: Prompt A, B, C
: If C≥A+B
: Then
: Goto 1
: End
: If B≥A+C
: Then
: Goto 1
: End
: If A≥B+C
: Then
: Goto 1
: End
: Disp "THIS IS A TRIANGLE."
: Stop
: Lbl 1
: Disp "NOT A TRIANGLE"
```

Critical Thinking

46. Using an example, show that even though $x > y$ and $t > w$, $x - t > y - w$ may be false.

47. What does the sentence $-2.4 < x < 3.6$ mean?

Define a variable, write an inequality, and solve each problem.

48. Academics   Josie must have at least 320 points in her math class to get a B. She needs a B or better to maintain her grade-point average so she can play on the basketball team. The grade is based on four 50-point tests, three 20-point quizzes, two 20-point projects, and a final exam worth 100 points. Josie's record of her grades are shown in the table below.

	Points	Total Points
Tests	40, 42, 41, 45	168
Quizzes	15, 12, 19	46
Projects	15, 18	33

a. Write an inequality to represent the range of scores on the final exam needed in order for Josie to make a B.

b. Solve the inequality and graph its solution.

49. Personal Finances   Tanaka had $75 to buy presents for his family. He bought a $21.95 shirt for his dad, a $23.42 necklace for his mother, and a $16.75 CD for his sister. He still has to buy a present for his brother.

a. How much can he spend on his brother's present?

b. What factors not stated in the problem may affect how much money he can spend?

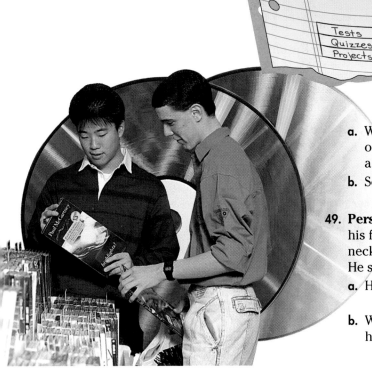

Mixed Review

50. Find the midpoint of the line segment whose endpoints are at $(-1, 9)$ and $(-5, 5)$.   (Lesson 6–7)

Write an equation in slope-intercept form for a line that satisfies each condition.   (Lesson 6–6)

51. perpendicular to $x + 7 = 3y$ that passes through $(1, 0)$

52. parallel to $\frac{1}{5} y - 3x = 2$ that passes through $(0, -3)$

53. Statistics   Find the range, quartiles, and interquartile range for the set of data at the right.   (Lesson 5–7)

Stem	Leaf
12	4 6 7 7
13	1 1 6 9
14	0 5 7
15	0 3 9 9 9
16	5 6 6

$14 \mid 0 = \$140$

54. Solve $4x + 3y = 16$ if the domain is $\{-2, -1, 0, 2, 5\}$.   (Lesson 5–3)

55. If y varies inversely as x and $y = 32$ when $x = 3$, find y when $x = 8$.   (Lesson 4–8)

56. Solve $y - \frac{7}{16} = -\frac{5}{8}$ (Lesson 3–1)

57. Replace the ___?___ with $<$, $>$, or $=$ to make $\frac{6}{13}$ ___?___ $\frac{1}{2}$ true.   (Lesson 2–4)

58. Look for a Pattern   How many triangles are shown at the right? Count only the triangles pointing upward.   (Lesson 1–2)

7-2A Solving Inequalities

MODELING MATHEMATICS

Materials: equation mat cups and counters self-adhesive note

A Preview of Lesson 7–2   You can use an equation model to solve inequalities.

Activity Model the solution for $-2x < 4$.

Step 1 Use the note to cover the equals sign on the equation mat. Then write a $<$ symbol on the note. Label 2 cups with a negative sign, and place them on the left side. Place 4 positive counters on the right.

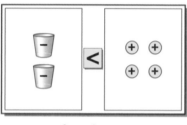

$$-2x < 4$$

Step 2 Since we cannot solve for a negative cup, we must eliminate the negative cups by adding 2 positive cups to each side. Remove the zero pairs.

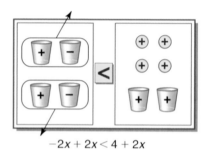

$$-2x + 2x < 4 + 2x$$

Step 3 Add 4 negative counters to each side to isolate the cups. Remove the zero pairs.

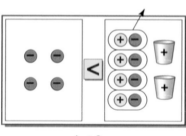

$$-4 < 2x$$

Step 4 Separate the counters into 2 groups.

$$-2 < x \text{ or } x > -2$$

Model

1. Compare the symbol and location of the variable in the original problem with those in the solution. What do you find?

2. Model the solution for $3x > 12$. What do you find? How is this different from solving $-2x < 4$?

Write

3. Write a rule for solving inequalities involving multiplication.

4. Do you think the rule applies to inequalities involving division?
   Remember that dividing by a number is the same as multiplying by its reciprocal.

Solving Inequalities by Using Multiplication and Division

What YOU'LL LEARN

- To solve inequalities by using multiplication and division.

Why IT'S IMPORTANT

You can use inequalities to solve problems involving the physical and political sciences.

CONNECTION

Physical Science

A lever can be used to multiply the effort force you exert when trying to move something. The fixed point or *fulcrum* of a lever separates the length of the lever into two sections—the *effort arm* on which the effort force is applied and the *resistance arm* that exerts the resistance force. The *mechanical advantage* of a lever is the number of times a lever multiplies that effort force.

The formula for determining the mechanical advantage *MA* of a lever can be expressed as $MA = \dfrac{L_e}{L_r}$, where L_e represents the length of the effort arm and L_r represents the length of the resistance arm.

GLOBAL CONNECTIONS

About 15,000 years ago, an important application of the lever appeared as hunters used an *atlatl*, the Aztec word for spear-thrower. This simple device was a handle that provided extra mechanical advantage by adding length to the hunter's arm. This enabled the hunter to use less motion to get greater results when hunting.

Suppose a group of volunteers is clearing hiking trails at Yosemite National Park. They need to position a lever so that a mechanical advantage of at least 7 is achieved in order to remove a boulder blocking the trail. The volunteers place the lever on a rock so they can use the rock as a fulcrum. They will need the resistance arm to be 1.5 ft long so that it is long enough to get under the boulder. What should be the length of the lever in order to move the boulder?

We need to find the length of the effort arm to find the total length. Let L_e represent the length of the effort arm. We know that 1.5 feet is the length of the resistance arm L_r. Since the mechanical advantage must be at least 7, we can write an inequality using the formula.

$MA \geq 7$

$\dfrac{L_e}{L_r} \geq 7$ *Replace MA with $\dfrac{L_e}{L_r}$.*

$\dfrac{L_e}{1.5} \geq 7$ *Replace L_r with 1.5.*   *You will solve this problem in Example 1.*

If you were solving the equation $\frac{L_e}{1.5} = 7$, you would multiply each side by 1.5. Will this method work when solving inequalities? Before answering this question, let's explore how multiplying (or dividing) an inequality by a positive or negative number affects the inequality. Consider the inequality $10 < 15$, which we know is true.

Multiply by 2.

$10 < 15$

$10(2) < 15(2)$

$20 < 30$ true

Multiply by −2.

$10 < 15$

$10(-2) < 15(-2)$ false

$-20 < -30$ false

$-20 > -30$ true

Divide by 5.

$10 < 15$

$\frac{10}{5} < \frac{15}{5}$

$2 < 3$ true

Divide by −5.

$10 < 15$

$\frac{10}{-5} < \frac{15}{-5}$ false

$-2 < -3$ false

$-2 > -3$ true

These results suggest the following.

- If each side of a true inequality is multiplied or divided by the same positive number, the resulting inequality is also true.

- If each side of a true inequality is multiplied or divided by the same negative number, the direction of the inequality symbol must be *reversed* so that the resulting inequality is also true.

Multiplication and Division Properties for Inequalities

> **For all numbers, a, b, and c, the following are true.**
>
> 1. **If c is positive and $a < b$, then $ac < bc$ and $\frac{a}{c} < \frac{b}{c}$, and**
>
>   **if c is positive and $a > b$, then $ac > bc$ and $\frac{a}{c} > \frac{b}{c}$.**
>
> 2. **If c is negative and $a < b$, then $ac > bc$ and $\frac{a}{c} > \frac{b}{c}$, and**
>
>   **if c is negative and $a > b$, then $ac < bc$ and $\frac{a}{c} < \frac{b}{c}$.**

These properties also hold for inequalities involving $\leq$ and $\geq$.

Example ❶

CONNECTION

Physical Science

Refer to the connection at the beginning of the lesson. What should the minimum length of the lever be?

$\frac{L_e}{1.5} \geq 7$

$1.5 \cdot \frac{L_e}{1.5} \geq 1.5(7)$ *Multiply each side by 1.5.*

$L_e \geq 10.5$

The effort arm must be at least 10.5 feet long.

In order to find the length of the lever, add the lengths of the effort arm and the resistance arm. The lever should be at least $10.5 + 1.5$ or 12 feet long.

Example **Solve $\frac{x}{12} \le \frac{3}{2}$.**

$$\frac{x}{12} \le \frac{3}{2}$$

$$12 \cdot \frac{x}{12} \le 12 \cdot \frac{3}{2} \qquad \textit{Multiply each side by 12.}$$

$$x \le 18 \qquad \textit{Since we multiplied by a positive number,}$$
$$\textit{the inequality symbol stays the same.}$$

The solution set is $\{x \mid x \le 18\}$.

Since dividing is the same as multiplying by the reciprocal, there can be two methods to solve an inequality that involves multiplication.

Example **3** **Solve $-3w > 27$.**

Method 1

$$-3w > 27$$

$$\frac{-3w}{-3} < \frac{27}{-3} \qquad \textit{Divide each side}$$
$$\textit{by } -3 \textit{ and change}$$
$$w < -9 \qquad > \textit{to} <.$$

Method 2

$$-3w > 27$$

$$\left(-\frac{1}{3}\right)(-3w) < \left(-\frac{1}{3}\right)(27) \qquad \textit{Multiply each side}$$
$$\textit{by } -\frac{1}{3} \textit{ and change}$$
$$w < -9 \qquad > \textit{to} <.$$

Check: Let w be any number less than -9.

$$-3w > 27$$

$$-3(-10) \overset{?}{>} 27 \qquad \textit{Suppose we select } -10.$$

$$30 > 27 \quad \text{true}$$

Numbers less than -9 compose the solution set. The solution set is $\{w \mid w < -9\}$.

Example **4** **Angelica Moreno is a sales representative for an appliance distributor. She needs at least $5000 in weekly sales of a particular TV model to qualify for a sales competition to win a trip to the Bahamas. If the TVs sell for $250 each, how many TVs will Ms. Moreno have to sell to qualify?**

Explore   Let t represent the number of TVs to be sold. At least $5000 means greater than or equal to $5000.

Plan   The price of one TV times the number of sets sold must be greater than or equal to the total amount of sales needed.

The price of one TV	times	the number of TVs sold	is at least	$5000.
$250	×	t	≥	$5000

Solve
$$250t \ge 5000$$

$$\frac{250t}{250} \ge \frac{5000}{250} \qquad \textit{Divide each side by 250.}$$

$$t \ge 20$$

Examine   Ms. Moreno must sell a minimum of 20 TVs to qualify for the contest. The amount of money from the sale of 20 TVs is $250(20) or $5000.

5 Triangle *XYZ* is not an acute triangle. The greatest angle in the triangle has a measure of $(6d)°$. What are the possible values of *d*?

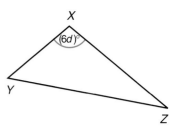

INTEGRATION
Geometry

Since $\triangle XYZ$ is not acute, the measure of the greatest angle must be 90° or larger, but less than 180°. Thus, $6d \geq 90$ and $6d < 180$.

$6d \geq 90$ $6d < 180$

$\dfrac{6d}{6} \geq \dfrac{90}{6}$ *Divide each side by 6.* $\dfrac{6d}{6} < \dfrac{180}{6}$ *Divide each side by 6.*

$d \geq 15$ $d < 30$

The value of *d* must be greater than or equal to 15 but less than 30.

CHECK FOR UNDERSTANDING

Communicating Mathematics

Study the lesson. Then complete the following.

1. **Classify** each statement as *true* or *false*. If false, explain how to change the inequality to make it true.

   a. If $x > 9$, then $-3x > -27$.

   b. If $x < 4$, then $3x < 12$.

2. **Complete** each statement.

   a. If each side of an inequality is multiplied by the same __?__ number, the direction of the inequality symbol must be reversed so that the resulting inequality is true.

   b. Multiplying by __?__ is the same as dividing by -6.

   c. An acute triangle has angles whose measures are all less than __?__.

3. **You Decide** Utina and Paige are discussing the rules for changes in the direction of the inequality symbol when solving inequalities. Paige says the rule is "Whenever you have a negative sign in the problem, the direction of the inequality symbol will change." Utina says that is not always true. Decide who is correct and give examples to support your answer.

MODELING
MATHEMATICS

Use models to solve each inequality. Write the answer in set-builder notation.

4. $3x < 15$ 5. $-6x < 18$

6. $2x + 6 > x - 7$ 7. $-4x + 8 \geq 14$

8. Use models to determine the appropriate symbol ($>$, $<$, or $=$) to complete each comparison.

   a. $x + 5$ __?__ $x - 7$

   b. $x + 3 + (x - 6)$ __?__ $(2x + 7) - 4$

   c. $2x + 8$ __?__ $2x + 6 - x$

State the number by which you multiply or divide to solve each inequality. Indicate whether the direction of the inequality symbol reverses. Then solve.

9. $-6y \geq -24$

10. $10x > 20$

11. $\frac{x}{4} < -5$

12. $-\frac{2}{7}z \geq -12$

Solve each inequality. Then check your solution.

13. $\frac{4}{5}x < 24$

14. $-\frac{v}{3} \geq 4$

15. $-0.1t \geq 3$

16. $5y > -25$

Define a variable, write an inequality, and solve each problem. Then check your solution.

17. One fifth of a number is at most 4.025.

18. The opposite of six times a number is less than 216.

19. Geometry Determine the value of s so that the area of the square is at least 144 square feet.

s ft

s ft

EXERCISES

Solve each inequality. Then check your solution.

20. $7a \leq 49$

21. $12b > -144$

22. $-5w > -125$

23. $-x \leq 44$

24. $4 < -x$

25. $-102 > 17r$

26. $\frac{b}{-12} \leq 3$

27. $\frac{t}{13} < 13$

28. $\frac{2}{3}w > -22$

29. $6 \leq 0.8g$

30. $-15b < -28$

31. $-0.049 \leq 0.07x$

32. $\frac{3}{7}h < \frac{3}{49}$

33. $\frac{12r}{-4} > \frac{3}{20}$

34. $\frac{3b}{4} \leq \frac{2}{3}$

35. $-\frac{1}{3}x > 9$

36. $\frac{y}{6} \geq \frac{1}{2}$

37. $\frac{-3m}{4} \leq 18$

Define a variable, write an inequality, and solve each problem. Then check your solution.

38. Four times a number is at most 36.

39. Thirty-six is at least one half of a number.

40. The opposite of three times a number is more than 48.

41. Three fourths of a number is at most -24.

42. Eighty percent of a number is less than 24.

43. The product of two numbers is no greater than 144. One of the numbers is -8. What is the other number?

44. Geometry Determine the value of x so that the area of the rectangle at the right is at least 918 square feet.

36 feet

$3x$ feet

45. Geometry Determine the value of y so that the perimeter of the triangle at the right is less than 100 meters.

$3y$ m $5y$ m

$6y$ m

Complete.

46. If $24m \geq 16$, then __?__ ≥ 12.

47. If $-9 \leq 15b$, then $25b$ __?__ -15.

48. If $5y < -12$, then $20y <$ __?__.

49. If $-10a > 21$, then $30a$ __?__ -63.

Critical Thinking

50. Use an example to show that if $x > y$, then $x^2 > y^2$ is not necessarily true.

Applications and Problem Solving

Define a variable, write an inequality, and solve each problem.

51. Travel The charge per mile for a compact rental car at 4-D Rentals is $0.12. Mrs. Rodriguez is on a business trip and must rent a car to attend various meetings. She has a budget of $50 per rental for mileage charges. What is the greatest number of miles Mrs. Rodriguez can travel without going over her budget?

52. Physics Refer to the application at the beginning of the lesson. A city worker needs to raise a utility access cover. She has an iron bar to use as a lever. When using a lever, the resistance force is equal to the effort force multiplied by the mechanical advantage, or $F_r = MA \cdot F_e$. The worker weighs 120 pounds, so she can supply that much effort force. All utility covers in the city weigh at least 360 pounds. What mechanical advantage does she need to lift the cover?

53. Political Science A candidate needs 5000 signatures on a petition before she can run for a township office. Experience shows that 15% of the signatures on petitions are not valid. What is the smallest number of signatures the candidate should get to end up with 5000 valid signatures?

Mixed Review

54. Define a variable, write an inequality, and solve the following problem. The difference of five times a number less four times that number plus seven is at most 34. *(Lesson 7–1)*

55. Geometry Three vertices of a square are at $(-5, -3)$, $(-5, 5)$, and $(3, -3)$. Suppose you were to inscribe a circle in this square. What would the coordinates of the center of the circle be? (Lesson 6–7)

56. Geometry Determine the slopes of the lines parallel and perpendicular to the graph of $3x - 6 = -y$. (Lesson 6–6)

57. Determine the slope of the line that passes through $(4, -9)$ and $(-2, 3)$. (Lesson 6–1)

58. Write an equation in functional notation for the relation at the right. (Lesson 5–6)

a	-2	0	2	4	6	8	10
b	-5	-3	-1	1	3	5	7

59. Budgeting JoAnne Paulsen's take-home pay is $1782 per month. She spends $325 on rent, $120 on groceries, and $40 on gas. She allows herself 12% of the remaining amount for entertainment. How much can she spend on entertainment each month? (Lesson 4–4)

60. What number is 47% of 27? (Lesson 4–4)

61. Soccer A soccer field is 75 yards shorter than 3 times its width. Its perimeter is 370 yards. Find its dimensions. (Lesson 3–5)

62. Evaluate $5y + 3$ if $y = 1.3$. (Lesson 1–3)

WORKING ON THE

In·ves·ti·ga·tion

Refer to the Investigation on pages 320–321.

Smoke Gets In Your Eyes

Currently, about 50 million Americans smoke, and each smoker averages about 30 cigarettes a day. Consider the amount of smoke produced by a single cigarette. Make some projections on the volume of cigarette smoke generated by all 50 million smoking Americans.

1 Find the volume of smoke produced by the average smoker.

2 Project the amount of smoke generated by 50 million smokers in a year.

3 Estimate the volume of air in your room at home. Write an inequality that would estimate the number of puffs p it would take to fill your room. Solve the inequality.

4 Suppose a smoker was locked in your room with several cartons of cigarettes. Would it be possible for the smoker to fill the room completely with smoke in the same concentration as the smoke exhaled? Explain your answer.

Add the results of your work to your Investigation Folder.

Solving Multi-Step Inequalities

APPLICATION
Engineering

Rosa Whitehair is a partner in an engineering consulting firm. Her fee for consulting on large construction projects is $1000 plus 10% of the design fee that the company charges its clients. Ms. Whitehair is considering two construction projects: a 50-story office building and the design of an airport terminal. She is interested in both projects, but decides to choose the one for which her fee is higher. The company that is designing the office building has agreed to pay Ms. Whitehair a flat fee of $5000. How much does the design company's fee need to be for her to choose the terminal?

What YOU'LL LEARN

- To solve linear inequalities involving more than one operation, and
- to find the solution set for a linear inequality when replacement values are given for the variables.

Why IT'S IMPORTANT

You can use inequalities to solve problems involving engineering and real estate.

Let x represent the design fee for the airport terminal.

The flat fee	plus	10% of the design fee	is more than	the flat fee for the office building.
1000	+	0.10x	>	5000

LOOK BACK

You can review solving multi-step equations in Lesson 3-3.

This inequality involves more than one operation. It can be solved by undoing the operations in reverse of the order of operations in the same way you would solve an equation with more than one operation.

$$1000 + 0.10x > 5000$$
$$1000 - 1000 + 0.10x > 5000 - 1000 \qquad \textit{Subtract 1000 from each side.}$$
$$0.10x > 4000$$
$$\frac{0.10x}{0.10} > \frac{4000}{0.10} \qquad \textit{Divide each side by 0.10.}$$
$$x > 40,000$$

If the company charges design fees higher than $40,000 for the terminal, Ms. Whitehair will choose them since her consulting fee is higher.

Example **Determine the value of *x* so that ∠*A* is acute.**

Assume that x is positive.

For ∠*A* to be acute, its measure must be less than 90°.

Thus, $3x - 15 < 90$.

$$3x - 15 < 90$$

$3x - 15 + 15 < 90 + 15$ *Add 15 to each side.*

$$3x < 105$$

$\dfrac{3x}{3} < \dfrac{105}{3}$ *Divide each side by 3.*

$$x < 35$$

For ∠*A* to be acute, *x* must be less than 35.

Sometimes inequalities, like equations, involve variables on each side of the inequality.

Example **Solve $-4w + 9 \le w - 21$.**

$$-4w + 9 \le w - 21$$

$-w - 4w + 9 \le w - 21 - w$ *Subtract w from each side.*

$$-5w + 9 \le -21$$

$-5w + 9 - 9 \le -21 - 9$ *Subtract 9 from each side.*

$$-5w \le -30$$

$\dfrac{-5w}{-5} \ge \dfrac{-30}{-5}$ *Divide each side by −5 and change ≤ to ≥.*

$$w \ge 6$$

The solution set is $\{w \,|\, w \ge 6\}$.

When we solve an inequality, the solution set usually includes all numbers for a certain criteria, such as $\{x \,|\, x > 4\}$. Sometimes a replacement set is given from which the solution set can be chosen.

Example **Determine the solution set for $3x + 6 > 12$ if the replacement set for *x* is $\{-2, -1, 0, 1, 2, 3, 4, 5\}$.**

Method 1

Substitute values into the inequality to find the values that satisfy the inequality. Try −2.

$$3x + 6 > 12$$
$$3(-2) + 6 > 12$$
$$0 > 12 \quad \text{false}$$

From this trial, we can estimate that the value of *x* must be much greater than −2 to make the inequality true. Try 2.

$$3x + 6 > 12$$
$$3(2) + 6 > 12$$
$$12 > 12 \quad \text{false}$$

From this trial, we see that values greater than 2 must be in the solution set. Try 3.

$$3x + 6 > 12$$
$$3(3) + 6 > 12$$
$$15 > 12 \quad \text{true} \qquad \text{The solution set is } \{3, 4, 5\}.$$

Method 2

Solve the inequality for all values of x. Then determine which values from the replacement set belong to the solution set.

$$3x + 6 > 12$$
$$3x + 6 - 6 > 12 - 6$$
$$3x > 6$$
$$\frac{3x}{3} > \frac{6}{3}$$
$$x > 2$$

The solution set is those numbers from the replacement set that are greater than 2. Thus, the solution set is $\{3, 4, 5\}$.

EXPLORATION

GRAPHING CALCULATORS

You can use the inequality symbols in the TEST menu on the TI-83 graphing calculator to find the solution to an inequality in one variable.

Your Turn

a. Clear the [Y=] list. Enter $3x + 6 > 4x + 9$ as Y1. (The symbol $>$ is item 3 on the TEST menu.) Press [GRAPH]. Describe what you see.

b. Use the TRACE function to scan the values along the graph. What do you notice about the values of y on the graph?

c. Solve the inequality algebraically. How does your solution compare to the pattern you noticed in part **b**?

When solving some inequalities that contain grouping symbols, remember to first use the distributive property to remove the grouping symbols.

Example **4** **Solve $5(k + 4) - 2(k + 6) \geq 5(k + 1) - 1$. Then graph the solution.**

$$5(k + 4) - 2(k + 6) \geq 5(k + 1) - 1$$
$$5k + 20 - 2k - 12 \geq 5k + 5 - 1 \qquad \textit{Distributive property.}$$
$$3k + 8 \geq 5k + 4 \qquad \textit{Combine like terms.}$$
$$3k - 5k + 8 \geq 5k - 5k + 4 \quad \textit{Subtract 5k from each side.}$$
$$-2k + 8 \geq 4$$
$$-2k + 8 - 8 \geq 4 - 8 \qquad \textit{Subtract 8 from each side.}$$
$$-2k \geq -4$$
$$\frac{-2k}{-2} \leq \frac{-4}{-2} \qquad \textit{Divide each side by} -2 \textit{ and change} \geq \textit{to} \leq.$$
$$k \leq 2$$

The solution set is $\{k \mid k \leq 2\}$.

The graph of $k \leq 2$ is shown at the right.

−4 −3 −2 −1 0 1 2 3 4 5

Communicating Mathematics

Study the lesson. Then complete the following.

1. **Explain** each step in solving $-3x + 7 < 4x - 5$.

2. **Write** an inequality that expresses the fact that when you add 3 feet to the perimeter of a square of sides with length s, the resulting perimeter does not exceed 50 feet.

3. **Write** an inequality that has no solution.

4. **Describe** how you would solve $16 - 5w > 29$ without dividing by -5 or multiplying by $-\frac{1}{5}$.

5. Refer to Example 3. Which of the two methods seems more efficient in solving the inequality for the given replacement set? Explain your selection.

MODELING MATHEMATICS

6. Use the method shown in Lesson 7–2A to model the solutions for each inequality.

   a. $3 - 4x \geq 15$ **b.** $6x - 1 < 5 + 3x$

Guided Practice

Choose the correct solution for each inequality.

7. Solve $2m + 5 \leq 4m - 1$.

   a. $m > -3$ **b.** $m < 3$ **c.** $m \geq 3$ **d.** $m \leq -3$

8. Solve $13r - 11 \geq 7r + 37$.

   a. $r < 8$ **b.** $r \geq 8$ **c.** $r \leq -8$ **d.** $r > -8$

Solve each inequality. Then check your solution.

9. $9x + 2 > 20$ 10. $-4h + 7 > 15$

11. $-2 - \frac{d}{5} < 23$ 12. $6a + 9 < -4a + 29$

Find the solution set of each inequality given the replacement set.

13. $3x - 1 > 4, \{-1, 0, 1, 2, 3\}$

14. $-7a + 6 \leq 48, \{-10, -9, -8, -7, -6, -5, -4, -3\}$

15. **Number Theory** Consider the sentence *The sum of two consecutive even integers is greater than 75.*

   a. Write an inequality for this statement.

   b. Solve the inequality.

   c. Name two consecutive even integers that meet the requirements of the statement.

Practice

Find the solution set of each inequality if the replacement set for each variable is $\{-10, -9, -8, ..., 8, 9, 10\}$.

16. $n - 3 \geq \frac{n + 1}{2}$ 17. $\frac{2(x + 2)}{3} < 4$

18. $1.3y - 12 < 0.9y + 4$ 19. $-20 \geq 8 + 7k$

Solve each inequality. Then check your solution.

20. $2m + 7 > 17$

21. $-3 > -3t + 6$

22. $-2 - 3x \geq 2$

23. $\frac{2}{3}w - 3 \leq 7$

24. $7x - 1 < 29 - 2x$

25. $8n + 2 - 10n < 20$

26. $2x + 5 < 3x - 7$

27. $5 - 4m + 8 + 2m > -17$

28. $\frac{2x - 3}{5} < 7$

29. $x < \frac{2x - 15}{3}$

30. $9r + 15 \geq 24 + 10r$

31. $6p - 2 \leq 3p + 12$

32. $4y + 2 < 8y - (6y - 10)$

33. $3(x - 2) - 8x < 44$

34. $3.1q - 1.4 > 1.3q + 6.7$

35. $-5(k + 4) \geq 3(k - 4)$

36. $5(2h - 6) - 7(h + 7) > 4h$

37. $7 + 3y > 2(y + 3) - 2(-1 - y)$

Define a variable, write an inequality, and solve each problem. Then check your solution.

38. Two thirds of a number decreased by 27 is at least 9.

39. Three times the sum of a number and 7 is greater than 5 times the number less 13.

Number Theory

40. The sum of two consecutive odd integers is at most 123. Find the pair with the greatest sum.

41. Find all sets of two consecutive positive odd integers whose sum is no greater than 18.

42. Find all sets of three consecutive positive even integers whose sum is less than 40.

Solve each inequality. Then check your solution.

43. $3x + 4 > 2(x + 3) + x$

44. $3 - 3(y - 2) < 13 - 3(y - 6)$

Graphing Calculator

45. Use the methods presented in the Exploration to solve each inequality. Use the viewing window $[-9.4, 9.4]$ by $[-5, 5]$ with scale factors of 1.

  a. $-5 - 8x \geq 59$

  b. $13x - 11 > 7x + 37$

  c. $8x - (x - 5) > x + 17$

  d. $-5(x + 4) \geq 3(x - 4)$

Critical Thinking

46. We can write the expression *the numbers between -3 and 4* as $-3 < x < 4$. What does the algebraic sentence $-3 < x + 2 < 4$ represent?

Applications and Problem Solving

Define a variable, write an inequality, and solve each problem.

47. Personal Finances A couple does not want to have a charge of more than $50 for dinner at a restaurant. A sales tax of 4% is added to the bill, and they plan to tip 15% after the tax has been added. What can the couple spend on the meal?

48. Recreation The admission fee to a video game arcade is $1.25 per person, and it costs $0.50 for each game played. Latoya and Donnetta have a total of $10.00 to spend. What is the greatest number of games they will be able to play?

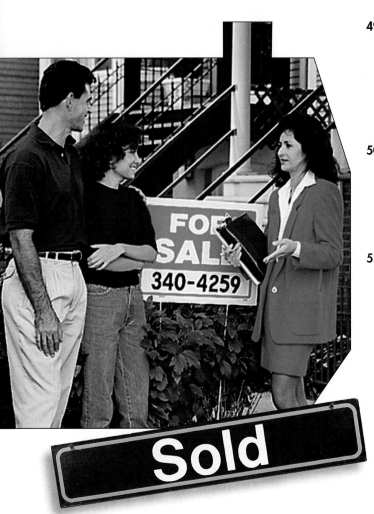

49. Fund-raising A university is running a drive to raise money. A corporation has promised to match 40% of whatever the university can raise from other sources. How much must the school raise from other sources to have a total of at least $800,000 after the corporation's donation?

50. Real Estate A homeowner is selling her house. She must pay 7% of the selling price to her real estate agent after the house is sold. To the nearest dollar, what must be the selling price of her house to have at least $90,000 after the agent is paid?

51. Labor A union worker is currently making $400 per week. His union is seeking a one-year contract. If there is a strike, the new contract will provide for a 6% raise; if there is no strike, the worker expects no increase in salary.

   a. Assuming the worker will not be paid during the strike, how many weeks could he go on strike and still make at least as much for the year as he would have made without a strike?

   b. How would your answer change if the worker was currently making $575 per week?

   c. How would your answer to part a change if the worker's union provided him with $120 per week during the strike?

Mixed Review

Solve each inequality.

52. $2r - 2.1 < -8.7 + r$ (Lesson 7–2)   **53.** $7 - 2y < -y - 3$ (Lesson 7–1)

54. Write the slope-intercept form of an equation for the line that passes through $(-12, 12)$ and $(-2, 7)$. (Lesson 6–4)

55. Write the standard form of an equation of the line that passes through $(2, 4)$ with slope $-\frac{3}{2}$. (Lesson 6–2)

56. Business The owner of No Spots City Car Wash found that if c cars were washed in a day, the average daily profit $P(c)$ was given by the formula $P(c) = -0.027c^2 + 8c - 280$. Find the values of $P(c)$ for various values of c to determine the least number of cars that must be washed each day for No Spots City to make a profit. (Lesson 5–5)

57. Find the domain for $3y - 2 = x$ if the range is $\left\{-1, 6, 0, -\frac{1}{3}, 2\right\}$. (Lesson 5–3)

58. Advertising For many years, the slogan of Crest® toothpaste has been "Four out of five dentists recommend that their patients use Crest." What are the odds that your dentist will *not* recommend you use Crest? (Lesson 4–6)

59. Jason scored the following points for his basketball team during the past ten games: 18, 32, 20, 21, 34, 9, 33, 37, 22, 25. Find the mean, median, and mode of his total points. (Lesson 3–7)

60. Find $18 - (-34)$. (Lesson 2–3)

Solving Compound Inequalities

What YOU'LL LEARN

- To solve problems by making a diagram,
- to solve compound inequalities and graph their solution sets, and
- to solve problems that involve compound inequalities.

Why IT'S IMPORTANT

You can use inequalities to solve problems involving chemistry and physics.

CONNECTION
Chemistry

The largest fish that spends its whole life in fresh water is the rare Pla beuk, found in the Mekong River in China, Laos, Cambodia, and Thailand. The largest specimen was reportedly 9 feet $10\frac{1}{4}$ inches long and weighed 533.5 pounds.

Such rare fish are sometimes displayed in aquariums. Aquariums can house freshwater or marine life and must be closely monitored to maintain the correct temperature and pH for the animals to survive. pH is a measure of acidity. To determine pH, a scale with values from 0 to 14 is used. One such scale is shown in the diagram below.

F Y I

The largest fish ever caught is a whale shark. Caught near Karachi, Pakistan in 1949, it was 41.5 feet long and weighed 16.5 tons. The smallest fish is a dwarf goby found in the Indian Ocean. Males of this species are only 0.34 inches long.

whale shark

dwarf goby

If we let p represent the value of the pH scale, we can express the different pH levels by using inequalities. For example, an acid solution will have a pH level of $p \geq 0$ and $p < 7$. When considered together, these two inequalities form a **compound inequality.** This compound inequality can also be written without using *and* in two ways.

$$0 \leq p < 7 \quad \text{or} \quad 7 > p \geq 0$$

The statement $0 \leq p < 7$ can be read *0 is less than or equal to p, which is less than 7.* The statement $7 > p \geq 0$ can be read *7 is greater than p, which is greater than or equal to 0.*

The pH levels of bases could be written as follows.

$$p > 7 \text{ and } p \leq 14 \quad \text{or} \quad 7 < p \leq 14 \quad \text{or} \quad 14 \geq p > 7$$

You will graph this inequality in Exercise 6.

You can **draw a diagram** to help solve many problems. Sometimes a picture will help you decide how to work the problem. Other times the picture will show you the answer to the problem.

Example **1**

On May 6, 1994, President Francois Mitterrand of France and Queen Elizabeth II of England officially opened the Channel Tunnel connecting England and France. After the ceremonies, a group of 36 English and French government officials had dinner at a restaurant in Calais, France, to celebrate the occasion. Suppose the restaurant staff used small tables that seat four people each, placed end to end, to form one long table. How many tables were needed to seat everyone?

Draw a diagram to represent the tables placed end to end. Use Xs to indicate where the people are sitting. Let's start with a guess, say 10 tables.

Ten tables will seat 22 people. If we use an extra table, we can seat 2 more people. Now, let's look for a pattern.

Number of tables	10	11	12	13	14	15	16	17
Number of people seated	22	24	26	28	30	32	34	36

This pattern shows that the restaurant needed 17 tables to seat all 36 officials.

The logic symbol for and is ∧. *You can write $x > 7$ and $x < 10$ as $(x > 7) \wedge (x < 10)$.*

A compound inequality containing *and* is true only if *both* inequalities are true. Thus, the graphs of a compound inequality containing *and* is the **intersection** of the graphs of the two inequalities. The intersection can be found by graphing the two inequalities and then determining where these graphs overlap. In other words, draw a diagram to solve the inequality.

Example **2**

Graph the solution set of $x \geq -2$ and $x < 5$.

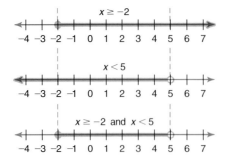

The solution set, shown in the bottom graph, is $\{x \mid -2 \leq x < 5\}$. Note that the graph of $x \geq -2$ includes the point -2. The graph of $x < 5$ does not include 5.

Example **3** **Solve $-1 < x + 3 < 5$. Then graph the solution set.**

First express $-1 < x + 3 < 5$ using *and*. Then solve each inequality.

$$-1 < x + 3 \qquad\qquad \text{and} \qquad\qquad x + 3 < 5$$
$$-1 - 3 < x + 3 - 3 \qquad\qquad\qquad\qquad x + 3 - 3 < 5 - 3$$
$$-4 < x \qquad\qquad\qquad\qquad\qquad\qquad x < 2$$

Now graph each solution and find the intersection of the solutions.

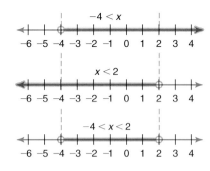

The solution set is $\{x \mid -4 < x < 2\}$.

The following example shows how you can solve a problem by using geometry, a diagram, and a compound inequality.

Example **4**

PROBLEM SOLVING
Draw a Diagram

Mai and Luis hope someday to compete in the Olympics in pairs ice skating. Each day they travel from their homes to an ice rink to practice before going to school. Luis lives 17 miles from the rink, and Mai lives 20 miles from it. If this were all the information you were given, determine how far apart Mai and Luis live.

Explore   Mai lives 20 miles from the rink and Luis lives 17 miles from the rink. We do not know the relative positions of Mai's and Luis's homes in relation to the rink.

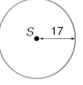

Plan   Draw a diagram of the situation. Let *S* be the location of the skating rink. Since Luis lives 17 miles from the rink, we can draw a circle with a radius of 17. Luis will live somewhere on that circle. Likewise, we can draw another circle with radius of 20 for the location of Mai's home.

(continued on the next page)

Solve Let's examine three possibilities for the locations of Mai's and Luis's homes.

(1) Mai and Luis live along the same radius from the skating rink.

(2) Mai and Luis live on opposite radii from the skating rink.

(3) Mai and Luis live somewhere other than the locations described in (1) and (2).

Let *L* and *M* represent where Luis and Mai live, respectively.

(1) the same radius

(2) opposite radii

(3) somewhere other than (1) or (2)

The distance is 20 − 17 or 3 miles.

The distance is 20 + 17 or 37 miles.

By the triangle inequality theorem, the distance must be less than 37 miles and greater than 3 miles.

The distance *d* can be described by the inequality $3 \le d \le 37$.

Examine Diagrams (1) and (2) show the least and greatest possiblities. To convince yourself that the statement with diagram (3) is always true, you may want to sketch a different triangle.

LOOK BACK

You can find more information about the triangle inequality theorem in Exercise 45 (Programming) of Lesson 7–1.

Another type of compound inequality contains the word *or* instead of *and*. A compound inequality containing *or* is true if one or more of the inequalities is true. The graph of a compound inequality containing *or* is the **union** of the graphs of the two inequalities.

Example **5** Graph the solution set of $x \ge -1$ or $x < -4$.

The last graph shows the solution set, $\{x \mid x \ge -1 \text{ or } x < -4\}$.

Example **6** **Solve $3w + 8 < 2$ or $w + 12 > 2 - w$. Graph the solution set.**

$$3w + 8 < 2 \qquad\qquad \text{or} \qquad\qquad w + 12 > 2 - w$$
$$3w + 8 - 8 < 2 - 8 \qquad\qquad\qquad w + w + 12 > 2 - w + w$$
$$3w < -6 \qquad\qquad\qquad\qquad 2w + 12 > 2$$
$$\frac{3w}{3} < \frac{-6}{3} \qquad\qquad\qquad 2w + 12 - 12 > 2 - 12$$
$$w < -2 \qquad\qquad\qquad\qquad 2w > -10$$
$$\frac{2w}{2} > \frac{-10}{2}$$
$$w > -5$$

Now graph each solution and find the union of the two.

$w < -2$

$w > -5$

                          Find the union.

The last graph shows the solution set, $\{x \mid x \text{ is a real number}\}$.

CHECK FOR UNDERSTANDING

Communicating Mathematics

Study the lesson. Then complete the following.

1. **Write** a statement that could be represented by $\$7.50 < p \leq \$18.50$.

2. **Write** a compound inequality that describes the graph below.

   ![number line from -6 to 3]
   $-6\ -5\ -4\ -3\ -2\ -1\ \ 0\ \ 1\ \ 2\ \ 3$

3. **List** two problem-solving strategies used in Example 1. Describe how each strategy helped in the solution of the problem.

4. **Describe** the difference between a compound inequality containing *and* and a compound inequality containing *or*.

5. **Name** two ways in which a picture or diagram can help you solve a problem more easily.

6. **Refer** to the connection at the beginning of the lesson. Graph the inequality that represents the pH levels of bases.

Guided Practice

Write each compound inequality without using *and*. Then graph the solution set.

7. $x < 9$ and $0 \leq x$

8. $x > -2$ and $x < 3$

Write a compound inequality for each solution set shown below.

9. ![number line from -5 to 4]
   $-5\ -4\ -3\ -2\ -1\ \ 0\ \ 1\ \ 2\ \ 3\ \ 4$

10. ![number line from -5 to 4]
   $-5\ -4\ -3\ -2\ -1\ \ 0\ \ 1\ \ 2\ \ 3\ \ 4$

11. Graph the solution set of $y > 5$ and $y < -3$. Describe what the solution set means.

Solve each compound inequality. Then graph the solution set.

12. $2 \le y + 6$ and $y + 6 < 8$

13. $4 + h \le -3$ or $4 + h \ge 5$

14. $b + 5 > 10$ or $b \ge 0$

15. $2 + w > 2w + 1 \ge -4 + w$

16. Write a compound inequality without using *and* for the following situation. Solve the inequality and then check the solution.
It costs the same to register mail containing articles with values from $0 to $100.

17. **Draw a Diagram** You can cut a pizza into seven pieces with only three straight cuts as shown at the right. Draw a diagram to show the greatest number of pieces you can make with five straight cuts.

EXERCISES

Practice

Graph the solution set of each compound inequality.

18. $m \ge -5$ and $m < 3$

19. $p < -8$ and $p > 4$

20. $s < 3$ or $s \ge 1$

21. $n \le -5$ or $n \ge -1$

22. $w > -3$ and $w < 1$

23. $x < -7$ or $x \ge 0$

Write a compound inequality for each solution set shown below.

24.
```
←|+|+◇—●+—●◇+|+|+|+→
 −6−5−4−3−2−1 0 1 2 3 4 5 6
```

25.
```
←|+|+◆—●+—+—+—+—+—●+|→
 −6−5−4−3−2−1 0 1 2 3 4 5 6
```

26.
```
←+—+—+—+—+—+◇+|◇—+—+—+→
 −6−5−4−3−2−1 0 1 2 3 4 5 6
```

27.
```
←+—+—+—+—+◆+|+|◇—+—+—+→
 −6−5−4−3−2−1 0 1 2 3 4 5 6
```

Solve each compound inequality. Then graph the solution set.

28. $4m - 5 > 7$ or $4m - 5 < -9$

29. $x - 4 < 1$ and $x + 2 > 1$

30. $y + 6 > -1$ and $y - 2 < 4$

31. $x + 4 < 2$ or $x - 2 > 1$

32. $10 - 2p > 12$ and $7p < 4p + 9$

33. $6 - c > c$ or $3c - 1 < c + 13$

34. $4 < 2x - 2 < 10$

35. $14 < 3h + 2 < 2$

36. $8 > 5 - 3q$ and $5 - 3q > -13$

37. $-1 + x \le 3$ or $-x \le -4$

38. $3n + 11 \le 13$ or $2n \ge 5n - 12$

39. $3y + 1 > 10$ and $y \ne 6$

40. $4z + 8 \ge z + 6$ or $7z - 14 \ge 2z - 4$

41. $5x + 7 > 2x + 4$ or $3x + 3 < 24 - 4x$

42. $2 - 5(2y - 3) > 2$ or $3y < 2(y - 8)$

43. $5w > 4(2w - 3)$ and $5(w - 3) + 2 < 7$

Write a compound inequality for each solution set shown below.

44.
```
←+—+—+—+—+—+—+—+—+—+—+—+→
 −6−5−4−3−2−1 0 1 2 3 4 5 6
```

45.
```
←+—+—+—+—+—+—+—+—+—+—+—+→
 −6−5−4−3−2−1 0 1 2 3 4 5 6
```

Define a variable, write a compound inequality, and solve each problem. Then check your solution.

46. When three times the distance to the finish line is increased by 5 km, the total kilometers covered in the race will be between 50 and 89 km.

47. The sum of a number and 2 is no more than 6 or no less than 10.

48. The sum of twice a number and 5 lies between 7 and 11.

49. Five less than 6 times a number is at most 37 and at least 31.

Solve each inequality. Then graph the solution set.

50. $3 + y > 2y > -3 - y$ **51.** $m > 2m - 1 > m - 5$ **52.** $\frac{5}{x} + 3 > 0$

Graphing Calculator

53. In Lesson 7–3, you learned how to use a graphing calculator to find graphically which values of x make a given inequality true. You can also use this method to test compound inequalities. The words *and* and *or* can be found in the LOGIC submenu of the TEST menu. Use this method to solve each of the following using your graphing calculator.

    a. $3 + x < -4$ or $3 + x > 4$ **b.** $-2 \le x + 3$ and $x + 3 < 4$

Critical Thinking

54. For what values of a does the compound inequality $-a < x < a$ have no solution?

55. Write a compound inequality for the solution set shown below.

    -6 -5 -4 -3 -2 -1 0 1 2 3 4 5 6

Applications and Problem Solving

56. Draw a Diagram There are eight houses on McArthur Street, all in a row. These houses are numbered from 1 to 8. Allison, whose house number is greater than 2, lives next door to her best friend, Adrienne. Belinda, whose house number is greater than 5, lives two doors away from her boyfriend, Benito. Cheri, whose house number is greater than Benito's, lives three doors away from her piano teacher, Mr. Crawford. Darryl, whose house number is less than 4, lives four doors away from his teammate, Don. Who lives in each house?

Define a variable, write a compound inequality, and solve each problem.

57. Physics According to Hooke's Law, the force F (in pounds) required to stretch a certain spring x inches beyond its natural length is given by $F = 4.5x$. If forces between 20 and 30 pounds, inclusive, are applied to the spring, what will be the range of the increased lengths of the stretched spring?

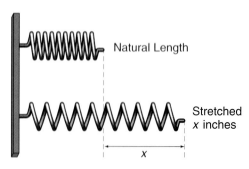

Natural Length

Stretched x inches

x

58. Statistics Clarissa must have an average of 92, 93, or 94 points to receive a grade of A– in social studies. She earned 92, 96, and 88 on the first three tests of the grading period. What range of scores on the fourth test will give her an A–?

Define a variable, write an inequality, and solve each problem.

59. Three fourths of a number decreased by 8 is at least 3.   (Lesson 7–3)

60. A number added to 23 is at most 5.   (Lesson 7–1)

61. If $7x - 2 \le 9x + 3$, then $2x \ge$ _?_ .   (Lesson 7–1)

62. Geometry   Line a is perpendicular to a line that is perpendicular to a line whose equation is $3x - 7y = -3$. If all lines are in the same plane, what is the slope of line a?   (Lesson 6–6)

63. Graph $x = 2y - 4$ using the x- and y-intercepts.   (Lesson 6–5)

64. Write an equation in functional notation for the relation graphed at the right.   (Lesson 5–6)

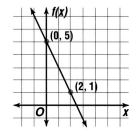

65. Sarah is visiting a friend about two miles away from her house. She sees a hot-air balloon directly above her house. Sara estimates that the angle of elevation formed by the hot-air balloon is about 15°. About how high is the hot-air balloon?   (Lesson 4–3)

66. Solve $\dfrac{4m - 3}{-2} = 12$.   (Lesson 3–3)

67. Simplify $-17px + 22bg + 35px + (-37bg)$.   (Lesson 2–5)

68. Evaluate $3ab - c^2$ if $a = 6$, $b = 4$, and $c = 3$.   (Lesson 1–3)

SELF TEST

Solve each inequality. Then check your solution.   (Lessons 7–1, 7–2, and 7–3)

1. $y + 15 \ge -2$

2. $-102 > 17r$

3. $5 - 6n > -19$

4. $\dfrac{11 - 6w}{5} > 10$

5. $7(g + 8) < 3(g + 12)$

6. $0.1y - 2 \le 0.3y - 5$

7. Choose an equivalent statement for $4 \ge x - 1 \ge -3$.   (Lesson 7–4)

   a. $5 \ge x \ge -4$ **b.** $3 \ge x \ge -4$ **c.** $5 \ge x \ge -2$ **d.** $3 \ge x \ge -2$

8. Solve $8 + 3t < 2$ or $-12 < 11t - 1$. Then graph the solution set.   (Lesson 7–4)

9. Sports   Jennifer has scored 18, 15, and 30 points in her last three starts on the junior varsity girls basketball team. How many points must she score in her next start so that her four-game average is greater than 20 points?   (Lesson 7–3)

10. Draw a Diagram   Suppose you roll a die two times.   (Lesson 7–4)

   a. Draw a diagram to show the possibilities of rolling an even number the first time and an odd number the second time.

   b. How many of the possibilities have a sum of 7?

Integration: Probability
Compound Events

What YOU'LL LEARN

• To find the probability of a compound event.

Why IT'S IMPORTANT

You can use tree diagrams to solve problems involving business, sports, and games.

A group of college students are planning a trip to visit Virgin Islands National Park, where they will study the Carib Indian relics and the remnants of the forts built by the Danes. The Carib Indians were the original occupants of the islands, but had died or left by the early 1600s. The Danes formally claimed the islands in 1666, and they remained under Danish control until 1917.

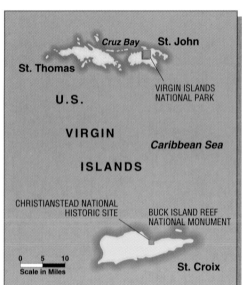

The group's advisor must plan how the group will get to the Virgin Islands. From their college, they will travel to Miami, Florida, by car, bus, train, or plane. Then to travel to St. Thomas in the Virgin Islands, they could take a plane or a ship. Suppose the advisor picks a mode of transportation at random. What is the probability that they will travel by car first and then fly?

To calculate this probability, you need to know all of the possible ways to get to St. Thomas. One method for finding this out is to draw a **tree diagram.** The tree diagram below shows how to get from the college to Miami and then to St. Thomas. The last column details all the possible combinations or **outcomes** of transportation.

Note that the number of outcomes is the product of the number of ways to Miami (4) and the number of ways to the Virgin Islands (2).

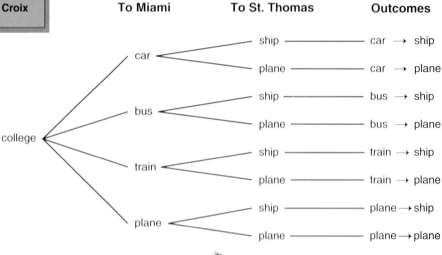

To Miami	To St. Thomas	Outcomes
car	ship	car → ship
	plane	car → plane
bus	ship	bus → ship
	plane	bus → plane
train	ship	train → ship
	plane	train → plane
plane	ship	plane → ship
	plane	plane → plane

college

Since the mode of transportation is chosen at random, you can assume that all combinations of travel are equally likely. Since there are 8 outcomes, the probability of traveling by car first and then flying is $\frac{1}{8}$ or 0.125.

This problem is an example of finding the probability of a **compound event.** A compound event consists of two or more **simple events.** Choosing a car, plane, train, or bus for the first part of the trip is a simple event. Then, selecting a plane or ship for the second stage of the trip is another simple event. The selection of a mode of transportation for each part of the trip is a compound event.

How is a compound event similar to a compound inequality?

Example Use the tree diagram for the application at the beginning of the lesson to answer each question.

    a. What is the probability that the group will take a ship to get to St. Thomas?

    b. What is the probability that the group will travel by plane for both parts of the trip?

    a. Since 4 of the 8 outcomes involve taking a ship, the probability is $\frac{4}{8}$ or 0.5.

    b. Since only 1 outcome out of 8 involves taking a plane for both parts of the trip, the probability is $\frac{1}{8}$ or 0.125.

LOOK BACK

You can find more information on probability in Lesson 4-6.

Refer to the application at the beginning of the lesson. Because one choice *does not affect* the others, we say that these are **independent events.** If the outcome of an event *does affect* the outcome of another event, we say that these are **dependent events.**

Example Booker T. Washington High School is having its annual Spring Carnival. The ninth grade class has decided to have a game booth. To win a small stuffed animal, a player will have to draw 2 marbles of the same color from a box containing 3 marbles—1 red, 1 white, and 1 yellow. First a marble is drawn, put back in the box, and then a second marble is drawn.

    a. What are the possible outcomes?

    b. What is the probability of winning the game?

    a. First, let's draw a tree diagram to see the possibilities of drawing 2 marbles of the same color.

Since the first marble is replaced, the outcome of the first selection does not affect the outcome of the second selection. Thus, they are independent events.

1st selection	2nd selection	Outcomes
red	red	red, red
	white	red, white
	yellow	red, yellow
white	red	white, red
	white	white, white
	yellow	white, yellow
yellow	red	yellow, red
	white	yellow, white
	yellow	yellow, yellow

There are 9 outcomes.

b. There are 3 ways out of 9 to draw the same color twice. The probability for each draw is equally likely, so the probability is $\frac{3}{9}$ or $0.\overline{3}$.

Probability can also play an important role in determining the possible outcomes of a sporting event.

Example **3**

APPLICATION

Basketball

The Houston Rockets and the New York Knicks are going to play a best two out of three exhibition game series.

a. What are the possible outcomes of the series?

b. Assuming that the teams are equally matched, what is the probability that the the series will end in two games?

a. The tree diagram below shows the possible outcomes of the first two games.

1st game winner	2nd game winner	Outcomes
Rockets	Rockets	Rockets win in two games.
	Knicks	3rd game is required.
Knicks	Rockets	3rd game is required.
	Knicks	Knicks win in two games.

There are 4 possible outcomes

b. The series can end in two games in two ways. The probability is $\frac{2}{4}$ or 0.5.

CHECK FOR UNDERSTANDING

Communicating Mathematics

Study the lesson. Then complete the following.

1. **Draw** a tree diagram to represent the outcomes of tossing a coin twice.

2. **Describe** a compound event that involves three stages. Draw a tree diagram for such an event.

3. **Explain** the difference between a simple event and a compound event.

4. **Compare and contrast** the tree diagram used in Example 2 and the one used in Example 3.

Mᴀᴛʜ Jᴏᴜʀɴᴀʟ

5. Write a paragraph describing how probabilities are used in real life. Give specific examples.

Guided Practice

6. **Games** In one turn in the game of Yahtzee®, there are 13 different categories you try to complete by rolling five dice. On the first try of your turn, you roll all five dice. Then you may pick up any or all of the dice and roll again on a second and third try. To get a Yahtzee, you must have the same number appearing on all five dice. What is the probability that you roll a Yahtzee on the first try of your turn?

7. **Dining** For lunch at the 66 Diner, you can select one item from each of the following categories for express service guaranteed in 15 minutes.

Entree	Side Dish	Beverage
Burger Sandwich Taco Pizza	Soup Salad French Fries	Lemonade Soft Drink

a. Draw a tree diagram showing all of the meal combinations.

b. What is the probability that a customer will have soup with the meal?

c. What is the probability of selecting a burger with French fries?

d. What is the probability of having pizza with a salad and a soft drink?

EXERCISES

Applications and Problem Solving

8. **Video Games** In a computer video game, you have a choice of five roads to collect an important clue to solve a mystery. At the end of each road, there are two doors. The clue can only be found behind one door.

a. Draw a tree diagram to show the possibilities you have to find the clue.

b. What is the probability that you will find the clue?

9. **Travel** Three different airlines fly from Bowling Green to Lexington. Those same three airlines and two other airlines fly from Lexington to Louisville. There are no direct flights from Bowling Green to Louisville.

a. How many ways can a traveler book flights from Bowling Green to Louisville?

b. What is the probability that flights booked at random from Bowling Green to Louisville use the same airline?

10. **Business** On Wednesday, Ralph and Linda were talking about what they were going to wear on Friday since their company had declared it T-Shirt Day. Ralph said he only had three T-shirts he could wear to work. One was a solid color, one was an Earth Day shirt, and the other was a shirt with the company name and logo on it. He also said he would either wear tan pants or jeans. Linda is going to wear jeans and her company T-shirt. If Ralph selects a pair of pants and a T-shirt at random, what is the probability that he, too, will be wearing jeans and the company T-shirt?

11. **Food** Kita and Jason are working on the school newspaper's final layout for October. They decide that after they finish, they will go get pizza. They have a coupon for a large pizza with three toppings for $8.99. In discussing what the three toppings should be, they find they have four favorite toppings, and choosing three that both agreed upon was going to be difficult. After some discussion, they decided that pepperoni was the one topping they both agreed upon and they would put the names of the other three toppings in a bag and choose two at random. If the other three toppings are mushrooms, olives, and sausage, what is the probability they will get a pizza with mushrooms?

12. **Games** Twister® is a game composed of a mat with four rows, each having six circles of the same color. One person uses a spinner and calls out *hand* or *foot*, *right* or *left*, and a color (*blue, red, green,* or *yellow*). Each player must then place that particular body part on a circle of that color. The caller continues to call out body parts and colors with the players moving in the appropriate manner. If a player falls down in the process, he or she is disqualified. The winner is the last player still in position.

a. Draw a tree diagram to show the possibilities for body parts and colors.

b. What is the probability that the caller will spin and say *"right hand yellow?"*

13. **Playing Cards** The deck of cards for Rook® is composed of 57 cards. There are four suits (red, yellow, black, and green), each containing the numbers 1 to 14, and a rook card. Suppose there are two piles of five cards each with the cards turned face down. This first pile contains a red 3, a black 3, a red 5, a red 14, and a yellow 10. The second pile contains a green 5, a red 10, a black 10, a green 1, and a yellow 14. You are asked to draw a card from each pile.

a. List all the possibilities of drawing two cards.

b. What is the probability that both cards are red?

c. What is the probability that both cards are 10s?

d. What is the probability that both cards are green?

e. What is the probability that the sum of the cards is at least 15?

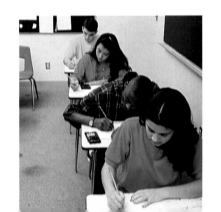

14. **Testing** Marty is taking his final exam in algebra. He takes too long answering the calculation part of the exam to properly evaluate the last five questions, which are true-false. To keep from getting points off for not answering these questions at all, he guesses at them just as the bell rings.

a. Draw a tree diagram to show the possible answer outcomes.

b. What is the probability that he got at least two of the questions correct if the answers were T, F, F, T, F?

c. Suppose Marty answered all the questions as true. Is this a good strategy to try to get the most correct without reading the question? Explain.

15. **Babies** According to the National Center for Health Statistics, each time a baby is born, the probability that it is a boy is always 51.3% and a girl is 48.7%, regardless of the sex of any previous births in the family. Suppose a couple has four children.
    a. What is the probability they are all girls?
    b. What is the probability there are one girl and three boys?
    c. How many children are there in your family? Find the probability of your family situation.

16. **Computers** The results of a survey in which students were asked how they used their computers are shown in the Venn diagram at the right. Based on these findings, what is the probability that a student, chosen at random, will use a computer only for word processing, and a second student will use it only for games?

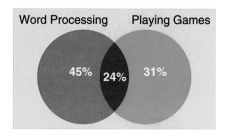

Word Processing   Playing Games

45%   24%   31%

Critical Thinking

17. Every working day Adrianna either bicycles or drives to work. If she oversleeps, she drives to work 80% of the time. If she does not oversleep, she bicycles 80% of the time. She oversleeps 20% of the time. What is the probability that she will drive to work on any particular day?

18. A bag of marbles contains 2 yellow marbles, 1 blue marble, and 3 red marbles. Suppose you select one marble at random and *do not* put it back in the bag. Then you select another marble.
    a. Draw a tree diagram to model this situation. How many possible outcomes are there?
    b. What is the probability of selecting a red marble first and then a yellow one?

Mixed Review

19. **Statistics** Amy's scores on the first three of four 100-point biology tests were 88, 90, and 91. To get an A− in the class, her average must be between 88 and 92, inclusive, on all tests. What score must she receive on the fourth test to get an A− in biology? (Lesson 7–4)

20. Define a variable, write an inequality, and solve the following. Two thirds of a number is more than 99. (Lesson 7–2)

21. Determine the slope and *y*-intercept of the graph of $3x - y = 9$. (Lesson 6–4)

22. Solve $8r - 7t + 2 = 5(r + 2t) - 9$ for *r*. (Lesson 5–3)

23. **Optics** Peripheral vision is the ability to see around you when you keep your head and eyes pointed straight ahead. The average person with good peripheral vision can see about 180° when they are still. For a person in a moving vehicle, peripheral vision decreases as speed increases, as shown in the illustration below. (Lesson 3–4)

a. If Seth Lytle is driving at 65 mph, approximately how large is his angle of vision?

b. About how large an angle is he not seeing compared to his stationary angle of vision?

c. If there were a deer standing 50° to the right of Seth's center view and he was driving at 45 mph, would he be able to see the deer?

Computers and Chess

The excerpt below appeared in an article in the January 1995 issue of *Discover* magazine.

HOW SMART HAVE COMPUTERS BECOME? Very smart, at least by one measure: this past August, a computer defeated the human world champion at chess for the first time. Russian grand master Garry Kasparov was outwitted by Genius 2, a computer program designed by an English physicist, Richard Lang, who calls himself a mediocre chess player.... Chess programs like Lang's work by focusing on what computers do best: long and intricate calculations. Before deciding on a move, the program examines each of the roughly 36 possibilities in an average game situation. For each of these 36 "branches," it looks 16 moves ahead, determining its opponent's possible countermoves, its own possible responses to each countermove, and so on. Sixteen moves ahead, the number of possible board configurations is huge. The genius of Genius 2 lies in the way it prunes the possibility tree, ruling out bad moves from the start. But the software also had help from the hardware; Lang's computer uses a speedy Intel Pentium microprocessor that can execute 166 million instructions per second. "We wouldn't have won (the computer world championship) had we not had this processor," Lang concedes. ■

1. Many chess-playing programs are written by people who are not top-level chess players. How then can these programs defeat very good players?

2. If you want to improve your chess game by practicing against a computer, would you choose a program that is below your playing level, at your level, a little above your level, or far above your level? Why?

3. Do you think it's fair to match human chess players against computers? Why or why not?

4. How might a computer use probability in determining the proper move to make?

Solving Open Sentences Involving Absolute Value

What YOU'LL LEARN

- To solve open sentences involving absolute value and graph the solutions.

Why IT'S IMPORTANT

You can use open sentences to solve problems involving space exploration, law enforcement, and entertainment.

fabulous FIRSTS

Sally Ride (1951–)

Sally Ride was the first American woman in space. She was a member of the *Challenger STS-7* crew that was launched June 18, 1983. She has a Ph.D. in physics from Stanford University.

The solution set can also be written as $\{x \mid x = -4 \text{ or } x = 4\}$.

APPLICATION

Space Exploration

On Tuesday, February 7, 1995, the space shuttle *Discovery* maneuvered within 37 feet of the Russian space station *Mir*, 245 miles above Earth. To accomplish this feat, *Discovery* had to be launched within 2.5 minutes of a designated time (12:45 A.M.). This time period is known as the *launch window*. If t represents the time elapsed since the launch countdown began and there are 300 minutes scheduled from the beginning of the countdown to blast-off, you can write the following inequality to represent the launch window.

$|300 - t| \leq 2.5$ *The difference between 300 minutes and the actual time elapsed since the countdown began must be less than or equal to 2.5 minutes.*

We use absolute value because $300 - t$ cannot be negative.

There are three types of open sentences that can involve absolute value. They are as follows, when n is nonnegative.

$$|x| = n \qquad |x| < n \qquad |x| > n$$

First let's consider the case of $|x| = n$.

If $|x| = 4$, this means that the difference from 0 to x is 4 units.

Therefore, if $|x| = 4$, then $x = -4$ or $x = 4$. The solution set is $\{-4, 4\}$. So, if $|x| = n$, then $x = -n$ or $x = n$.

Equations involving absolute value can be solved by graphing them on a number line or by writing them as a compound sentence and solving it.

Example **Solve $|x - 3| = 5$.**

Method 1: Graphing

$|x - 3| = 5$ means the distance between x and 3 is 5 units. To find x on the number line, start at 3 and move 5 units in either direction.

The solution set is $\{-2, 8\}$.

Method 2: Compound Sentence

$|x - 3| = 5$ also means $x - 3 = 5$ or $-(x - 3) = 5$.

$$x - 3 = 5 \quad\quad \text{or} \quad -(x - 3) = 5$$
$$x - 3 + 3 = 5 + 3 \quad\quad\quad x - 3 = -5 \quad\quad \textit{Multiply each side by } -1.$$
$$x = 8 \quad\quad\quad\quad x - 3 + 3 = -5 + 3$$
$$x = -2$$

This verifies the solution set.

Now let's consider the case of $|x| < n$. Inequalities involving absolute value can also be represented on a number line or as compound inequalities. Let's examine $|x| < 4$.

$|x| < 4$ means that the distance from 0 to x is less than 4 units.

When the inequality is < or ≤, the compound sentence uses <u>and</u>.

Therefore, $x > -4$ and $x < 4$. The solution set is $\{x \mid -4 < x < 4\}$. So, if $|x| < n$, then $x > -n$ and $x < n$.

Example ② Solve $|3 + 2x| < 11$ and graph the solution set.

$|3 + 2x| < 11$ means $3 + 2x < 11$ and $3 + 2x > -11$.

$$3 + 2x < 11 \quad\quad \text{and} \quad 3 + 2x > -11$$
$$3 - 3 + 2x < 11 - 3 \quad\quad 3 - 3 + 2x > -11 - 3$$
$$2x < 8 \quad\quad\quad\quad 2x > -14$$
$$\frac{2x}{2} < \frac{8}{2} \quad\quad\quad\quad \frac{2x}{2} > \frac{-14}{2}$$
$$x < 4 \quad\quad\quad\quad x > -7$$

The solution set is $\{x \mid x > -7$ and $x < 4\}$, which can be written as $\{x \mid -7 < x < 4\}$.

Now graph the solution set.

Finally let's examine $|x| > 4$. This means that the distance from 0 to x is greater than 4 units.

When the inequality is > or ≥, the compound sentence uses <u>or</u>.

Therefore, $x < -4$ or $x > 4$. The solution set is $\{x \mid x < -4$ or $x > 4\}$. So, if $|x| > n$, then $x < -n$ or $x > n$.

Example ③ **Solve** $|5 + 2y| \geq 3$ **and graph the solution set.**

$|5 + 2y| \geq 3$ means $5 + 2y \leq -3$ or $5 + 2y \geq 3$.

$5 + 2y \leq -3$ or $5 + 2y \geq 3$

$5 - 5 + 2y \leq -3 - 5$ $5 - 5 + 2y \geq 3 - 5$

$$2y \leq -8 \qquad\qquad 2y \geq -2$$

$$\frac{2y}{2} \leq \frac{-8}{2} \qquad\qquad \frac{2y}{2} \geq \frac{-2}{2}$$

$$y \leq -4 \qquad\qquad y \geq -1$$

The solution set is $\{y \mid y \leq -4 \text{ or } y \geq -1\}$.

$$\begin{array}{ccccccccccc} -7 & -6 & -5 & -4 & -3 & -2 & -1 & 0 & 1 & 2 \end{array}$$

Organizations such as OSHA set standards for buildings to meet the needs of those using the building. Building code standards are often written as maximums or minimums that must be met. These standards can be written as inequalities.

Example ④

There are a number of specifications in the building industry that address the needs of physically-challenged persons. For example, hallways in hospitals must have handrails. The handrails must be placed within a range of 2 inches from a height of 36 inches.

a. Write an open sentence that involves absolute value to represent the range of acceptable heights for hallway handrails.

b. Find and graph the corresponding compound sentence.

a. Let h be an acceptable height for the handrail. Then h can differ from 36 inches by no more than 2 inches. Write an open sentence that represents the range of acceptable heights.

    h differs from 36   *by less than or equal to*   *2.*

         $|h - 36|$ $\leq$ 2

Now, solve $|h - 36| \leq 2$ to find the compound sentence.

$h - 36 \geq -2$ and $h - 36 \leq 2$

$h - 36 + 36 \geq -2 + 36$ $h - 36 + 36 \leq 2 + 36$

$\qquad h \geq 34 \qquad\qquad\qquad\qquad h \leq 38$

The compound sentence is $34 \leq h \leq 38$.

b. The graph of this sentence is shown below.

So, the handrail must be placed between 34 and 38, inclusive, inches from the floor.

Communicating Mathematics

Study the lesson. Then complete the following.

1. **Describe** two ways you can solve an open sentence involving an absolute value.

2. **Explain** the difference in the solution for $|x + 7| > 4$ and the solution for $|x + 7| < 4$.

3. If $x < 0$ and $|x| = n$, describe n in terms of x.

4. **You Decide** Jamila's teacher said they should work out each problem and test points to determine whether their solutions are correct. Jamila thinks that she knows the solution for problems like $|w + 5| < 0$, where the absolute value quantity is always less than 0, without testing points. Is she correct? Why or why not?

Guided Practice

Choose the replacement set that makes each open sentence true.

5. $|x - 7| = 2$
   a. $\{9, -5\}$
   b. $\{5, -9\}$
   c. $\{5, 9\}$
   d. $\varnothing$

6. $|x - 2| > 4$
   a. $\{x \mid -2 < x < 6\}$
   b. $\{x \mid x = 6 \text{ or } x = 2\}$
   c. $\{x \mid x > 6 \text{ or } x > -2\}$
   d. $\{x \mid x < -2 \text{ or } x > 6\}$

State which graph below matches each open sentence in Exercises 7–10.

7. $|y| = 3$

8. $|y| > 3$

9. $|y| < 3$

10. $|y| \leq 3$

Solve each open sentence. Then graph the solution set.

11. $|m| \geq 5$

12. $|n| < 6$

13. $|r + 3| < 6$

14. $|8 - t| \geq 3$

For each graph, write an open sentence involving absolute value.

15.

16.

Practice

Solve each open sentence. Then graph the solution set.

17. $|y - 2| = 4$

18. $|3 - 3x| = 0$

19. $|7x + 2| = -2$

20. $|w + 8| \geq 1$

21. $|2 - y| \leq 1$

22. $|t + 4| \geq 3$

23. $|4y - 8| < 0$

24. $|2x + 5| < 4$

25. $|3e - 7| < 2$

26. $|3x + 4| < 8$

27. $|1 - 3y| > -2$

28. $3 + |x| > 3$

29. $|8 - (w - 1)| \leq 9$

30. $|6 - (11 - b)| = -3$

31. $|2.2y - 1.1| = 5.5$

32. $|3r - 0.5| \geq 5.5$

33. $\left|\frac{2 - 3x}{5}\right| \geq 2$

34. $\left|\frac{1}{2} - 3p\right| < \frac{7}{2}$

Express each statement in terms of an inequality involving absolute value. Do *not* try to solve.

35. The diameter of the lead in a pencil p must be within 0.01 millimeters of 1 millimeter.

36. The cruise control of a car set at 55 mph should keep the speed s within 3 mph of 55 mph.

37. A liquid at 50°C will change to a gas or liquid if the temperature t increases or decreases more than 50°C.

For each graph, write an open sentence involving absolute value.

38.

```
<-+-●-+-+-+-+-+-+-●-+->
 -3-2-1 0 1 2 3 4 5 6 7
```

39.

```
<-+-+-●-+-+-+-+-+-●-+->
-6-5-4-3-2-1 0 1 2 3 4
```

40.

```
<-+-+-+-●-+-+-+-●-+-+->
 -5-4-3-2-1 0 1 2 3 4 5
```

41.

```
<-+-+-+-+-●-+-+-●-+-+->
-4 -3 -2 -1 0 1 2 3 4 5
```

42.

```
<-+-+-◇-+-+-+-+-◇-+-+->
-6-5-4-3-2-1 0 1 2 3 4
```

43.

```
<-+-+-●-+-+-+-+-+-●-+->
 3 4 5 6 7 8 9 10 11 12 13
```

44. Find all integer solutions of $|x| < 4$.

45. Find all integer solutions of $|x| \leq 2$.

46. If $a > 0$, how many integer solutions exist for $|x| < a$?

47. If $a > 0$, how many integer solutions exist for $|x| \leq a$?

Critical Thinking

48. Solve $|y - 3| = |2 + y|$.

49. Under what conditions is $-|a|$ negative? positive?

50. Suppose $8 \leq x \leq 12$. Write an absolute value inequality that is equivalent to this compound inequality.

Applications and Problem Solving

51. **Probability** Suppose $|x| \leq 6$ and x is an integer. Find the probability of $|x|$ being a factor of 18.

52. **Chemistry** For hydrogen to be a liquid, its temperature must be within 2°C of −257°C. What is the range of temperatures for this substance to remain a liquid?

53. **Law Enforcement**   A radar gun used to determine the speed of passing cars must be within 7 mph of the actual speed of a selected car. If a highway patrol officer reads a speed of 59 mph for a car, does he have irrefutable evidence that the car was speeding in a 55 mph zone? Explain your reasoning and include a graph.

54. **Space Exploration**   Refer to the application at the beginning of the lesson. Find how much time can elapse from the beginning of the countdown to remain within the launch window.

55. **Entertainment**   Luis Gomez is a contestant on the *Price is Right*. He must guess within $1500 of the actual price of a Jeep Cherokee without going over in order to win the vehicle. The actual price of the Jeep is $18,000. What is the range of guesses in which Luis can win the vehicle?

56. **Spending**   The graph below shows the spending power of kids aged 3 to 17.

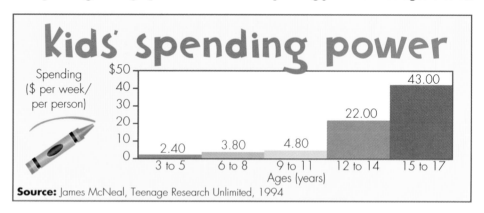

Source: James McNeal, Teenage Research Unlimited, 1994

a. Write an inequality that represents the spending power of 3- to 17-year olds. Then write an absolute value inequality that describes their spending.

b. Write an inequality that represents the spending power of 12- to 17-year olds. Then write an absolute value inequality that describes their spending.

c. Keep a record of how much money you spend in a week. How does your spending compare with the data in this graph?

Mixed Review

57. **Draw a Diagram**   The Sanchez family acts as a host family for foreign exchange students during each quarter of the year. Suppose it is equally likely that they get a boy or a girl each quarter.   (Lesson 7–5)

a. Draw a tree diagram to represent the possible orders of boys (B) and girls (G) during the four quarters of the year. List the possible outcomes.

b. From the tree diagram, what is the probability that all of the students will be girls? boys?

c. From the tree diagram, what is the probability that they will host two boys and two girls?

58. Peter wants to buy Crystal an engagement ring. He wants to spend between $1700 and $2200. If he goes shopping during a 12%-off sale to celebrate the store's 12th anniversary, what would his price range be? (Lesson 7–4)

59. Solve $10x - 2 \geq 4(x - 2)$. (Lesson 7–3)

Solve each inequality.

60. $396 > -11t$ (Lesson 7–2)

61. $-11 \leq k - (-4)$ (Lesson 7–1)

62. Find the coordinates of the midpoint of the line segment whose endpoints are at $(-4, 1)$ and $(10, 3)$. (Lesson 6–7)

63. Graph $2x - 9 = 2y$. (Lesson 5–4)

64. Architecture Julie is making a model of a house for her drafting class. One part of the house has a triangle like the one below. She wants the base of the triangle to measure 2 inches. What will be the measures of the other two sides? (Lesson 4–2)

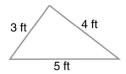

3 ft   4 ft

5 ft

65. Solve $2m + \frac{3}{4}n = \frac{1}{2}m - 9$ for m. (Lesson 3–6)

66. Simplify $\frac{-42r + 18}{3}$. (Lesson 2–7)

WORKING ON THE

In·ves·ti·ga·tion

Refer to the Investigation on pages 320–321.

Smoke Gets In Your Eyes

Smoking increases the chance of health problems. Constant exposure to secondhand smoke also puts the nonsmoker at risk. According to a statement from a tobacco company, a nonsmoker working among smoking coworkers inhales the smoke of 1.25 cigarettes per month. A restaurant worker in the smoking section breathes just 2 cigarettes per month. The Environmental Protection Agency (EPA) disagrees with these data.

1 Write an equation to express the equivalent number of cigarettes one would breathe if he/she worked in the smoking section of a restaurant for one year.

2 A lawsuit involving the tobacco industry and the EPA states that the EPA did not use the standard significance level of 5% in evaluating their data from various studies. A 5% level means that their numerical conclusions have a 5% margin of error. How does the margin of error relate to absolute value?

3 In January, 1995, the average price of a pack of cigarettes was $2.06, of which 56¢ is federal and state excise tax. The price of cigarettes is on the rise. Write an inequality to represent how much is spent by the average smoker in a year. Solve the inequality.

4 How much tax revenue is generated by the sale of cigarettes to the estimated number of smokers in the United States?

Add the results of your work to your Investigation Folder.

Integration: Statistics
Box-and-Whisker Plots

APPLICATION
Olympics

In 1992, the site of the Summer Games of the XXV Olympiad was Barcelona, Spain. More than 14,000 athletes from 172 nations competed for medals in 257 events. The table at the right shows the number of gold medals won by the top 16 medal-winning teams.

We can describe these data using the mean, median, and mode. We can also use the median, along with the quartiles and interquartile range, to obtain a graphic representation of the data. A type of diagram, or graph, that shows quartiles and extreme values of data is called a **box-and-whisker plot.**

1992 SUMMER OLYMPICS Top 16 Medal Winners		
Team	Number of Medals Won	
	Gold	Total
Unified Team (formerly USSR)	45	112
USA	37	108
Germany	33	82
China	16	54
Cuba	14	31
Hungary	11	30
South Korea	12	29
France	8	29
Australia	7	27
Spain	13	22
Japan	3	22
Britain	5	20
Italy	6	19
Poland	3	19
Canada	6	18
Romania	4	18

Source: *The World Almanac, 1995*

Box-and-whisker plots are sometimes called box plots.

LOOK BACK

You can refer to Lesson 3-7 for information on measures of central tendency and Lesson 5-7 for measures of variation.

The median is not included in either half of the data.

Suppose we wanted to make a box-and-whisker plot of the numbers of gold medals won by each of the nations in the table. First, arrange the data in numerical order. Next, compute the median and quartiles. Also, identify the extreme values.

$$3 \quad 3 \quad 4 \quad 5 \quad 6 \quad 6 \quad 7 \quad \boxed{8 \quad 11} \quad 12 \quad 13 \quad 14 \quad 16 \quad 33 \quad 37 \quad 45$$

median (Q2)

The median for this set of data is the average of the eighth and ninth values.

$$\text{median} = \frac{8 + 11}{2} \text{ or } 9.5$$

Recall that the *lower quartile* (Q_1) is the median of the lower half of the distribution of values. The *upper quartile* (Q_3) is the median of the upper half of the data.

$$\boxed{3 \quad 3 \quad 4 \quad 5 \quad 6 \quad 6 \quad 7 \quad 8} \qquad \boxed{11 \quad 12 \quad 13 \quad 14 \quad 16 \quad 33 \quad 37 \quad 45}$$

$$Q_1 \qquad\qquad\qquad\qquad\qquad Q_3$$

$$Q_1 = \frac{5 + 6}{2} \text{ or } 5.5 \qquad\qquad Q_3 = \frac{14 + 16}{2} \text{ or } 15$$

The **extreme values** are the least value (LV), 3, and the greatest value (GV), 45.

 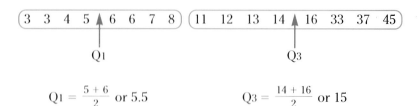

Now we have the information we need to draw a box-and-whisker plot.

Step 1 Draw a number line. Assign a scale to the number line that includes the extreme values. Plot dots to represent the extreme values (LV and GV), the upper and lower quartile points (Q3 and Q1), and the median (Q2).

Step 2 Draw a box to designate the data falling between the upper and lower quartiles. Draw a vertical line through the point representing the median. Draw a segment from the lower quartile to the least value and one from the upper quartile to the greatest value. These segments are the **whiskers** of the plot.

The median line will not always divide the box into equal parts.

Even though the whiskers are different lengths, each whisker contains at least one fourth of the data while the box contains one half of the data. Compound inequalities can be used to describe the data in each fourth. Assume that the replacement set for x is the set of data.

1st fourth	$\{x \mid x < 5.5\}$
2nd fourth	$\{x \mid 5.5 < x < 9.5\}$
3rd fourth	$\{x \mid 9.5 < x < 15\}$
4th fourth	$\{x \mid x > 15\}$

TECHNOLOGY Tip

You can use your graphing calculator to find the values you need to make a box-and-whisker plot. Refer to Lesson 5–7A for instructions on finding these values.

Step 3 Before finishing the box-and-whisker plot, check for outliers. In Lesson 5–7, you learned that an outlier is any element of the set of data that is at least 1.5 interquartile ranges above the upper quartile or below the lower quartile. Recall that the *interquartile range* (IQR) is the difference between the upper and lower quartiles, or in this case, $15 - 5.5$ or 9.5.

$x \geq Q3 + 1.5(IQR)$　　　　or　　　　$x \leq Q1 - 1.5(IQR)$

$x \geq 15 + 1.5(9.5)$　　　　　　　　　　$x \leq 5.5 - 1.5(9.5)$

$x \geq 15 + 14.25$　　　　　　　　　　　$x \leq 5.5 - 14.25$

$x \geq 29.25$　　　　　　　　　　　　　$x \leq -8.75$

Step 4 If x is an outlier in this set of data, then the outliers can be described as $\{x \mid x \leq -8.75 \text{ or } x \geq 29.5\}$. In this case, there are no data less than -8.75. However, 45, 37, and 33 are greater than 29.25, so they are outliers. We now need to revise the box-and-whisker plot. Outliers are plotted as isolated points, and the right whisker is shortened to stop at 16.

Refer to the application at the beginning of the lesson. Use the box-and-whisker plot for the gold medals to answer each question.

a. What percent of the teams won between 6 and 15 gold medals?

b. What does the box-and-whisker plot tell us about the upper half of the data compared to the lower half?

a. The box in the plot indicates 50% of the values in the distribution. Since the box goes from 5.5 to 15, we know that 50% of the teams won between 6 and 15 gold medals.

b. The upper half of the data is spread out while the lower half is fairly clustered together.

CHECK FOR UNDERSTANDING

Communicating Mathematics

Study the lesson. Then complete the following.

1. **Explain** how to determine the scale of the number line in a box-and-whisker plot.

2. **Describe** which two points the two whiskers of a box-and-whisker plot connect.

3. What does Q_2 represent?

4. Refer to the box-and-whisker plot at the right. Assume that LV, Q_1, Q_2, Q_3, and GV are whole numbers.

   a. What percent of the data is between 85 and 90?

   b. Between what two values does the middle 50% of the data lie?

   c. What outliers are represented in the box-and-whisker plot?

5. **Describe** what characteristics of a set of data you can gather from its box-and-whisker plot. What are some things you cannot gather from a box-and-whisker plot?

Guided Practice

6. The table at the right shows the birthrates for 15 selected countries.

   a. Find the median, upper quartile, lower quartile, and interquartile range for each year's data.

   b. Are there any outliers? If so, name them.

   c. Draw a box-and-whisker plot for each set of data on the same number line.

   d. Which set of data seems to be more clustered? Why?

Birth Rates for Selected Countries
(per 1000 population)

Country	1985	1992
Australia	15.7	15.1
Cuba	18.0	14.5
Denmark	10.6	13.1
France	13.9	12.9
Hong Kong	14.0	11.9
Israel	23.5	21.5
Italy	10.1	9.9
Japan	11.9	9.7
Netherlands	12.3	13.0
Panama	26.6	23.3
Poland	18.2	13.4
Portugal	12.8	11.4
Singapore	16.6	17.7
Switzerland	11.6	12.6
United States	15.7	15.7

Source: United Nations, *Monthly Bulletin of Statistics,* May 1994

7. Refer to box-and-whisker plots A and B.

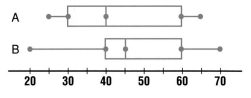

   a. Estimate the least value, greatest value, lower quartile, upper quartile, and median for each plot. Assume that these values are whole numbers.

   b. Which set of data contains the least value?

   c. Which plot has the greatest interquartile range?

   d. Which plot has the greatest range?

EXERCISES

Applications and Problem Solving

8. Manufacturing   The box-and-whisker plots at the right show the results of testing the useful life of 10 light bulbs each from two manufacturers.

   a. Which test had the most varied results?

   b. Were there any outliers? If so, for which brand?

   c. How would you compare the medians of the tests?

   d. Based on this plot, from which manufacturer would you buy your light bulbs? Why?

9. Meteorology   Meteorologists keep track of temperatures for four 90-day periods during the year to predict the trends in weather for future years. The following low temperatures were recorded during a 2-week cold snap in Indianapolis during 1993.

$$30°, 20°, 2°, 12°, 5°, 4°, 17°, 7°, 6°, 16°, 5°, 0°, 5°, 16°$$

   a. Find the median, upper quartile, lower quartile, and interquartile range.

   b. Are there any outliers? If so, name them.

   c. Draw a box-and-whisker plot of the data.

10. Football   The table below shows the American Football Conference's leading quarterbacks in touchdowns for the 1993 season. Make a box-and-whisker plot for the data.

1993 AFC Individual Leaders in Passing		
Player	**Team**	**Touchdowns**
Steve DeBerg	Miami	7
John Elway	Denver	25
Boomer Esiason	N.Y. Jets	16
John Friesz	San Diego	6
Jeff George	Indianapolis	8
Jeff Hostetler	L.A. Raiders	14
Jim Kelly	Buffalo	18
Scott Mitchell	Miami	12
Joe Montana	Kansas City	13
Warren Moon	Houston	21
Neil O'Donnell	Pittsburgh	14
Vinny Testaverde	Cleveland	14

Source: *The World Almanac,* 1995

11. **Demographics** According to the 1990 Census, the American Indian population in the United States is 1.959 million. Many American Indian people live on reservations or trust lands. The stem-and-leaf plot shows the number of reservations in the 34 states that have them.

Stem	Leaf
0	1 1 1 1 1 1 1 1 1 1 1 1 2
•	3 3 3 3 3 4 4 4 4 7 7 8 8 9
1	1 4 9
2	3 5 7
9	6 $9 \mid 6 = 96$

a. Make a box-and-whisker plot of these data.
b. Describe the distribution of the data.
c. Why do you think there are so many outliers?
d. The mean of these data is about 8.8. How does this compare with the median?

12. **Environment** The table below shows the number of hazardous waste sites in 25 states.

Hazardous Waste Sites in the United States (selected states)									
State	No.	State	No.	State	No.	State	No.	State	No.
AL	13	FL	57	LA	13	OH	38	TN	17
AR	12	GA	13	MD	13	OK	11	TX	30
CA	96	IL	37	MI	77	OR	12	VA	25
CO	18	IN	33	NY	85	PA	101	WA	56
CT	16	KY	20	NC	22	SC	24	WV	6

Source: Environmental Protection Agency, May 1994

a. Make a box-and-whisker plot of the data.
b. What is the median number of waste sites for the states listed?
c. Which states, if any, are outliers?
d. Which half of the data is more widely dispersed?

13. **Education** The graph below shows the average American College Testing (ACT) Program mathematics scores for students from 1985–1993.

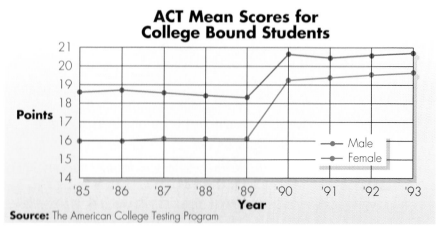

ACT Mean Scores for College Bound Students

Source: The American College Testing Program

(continued on the next page)

a. Make a box-and-whisker plot of the scores for the male students and another for the female students using the same scale. Compare the plots.

b. In a particular year, an entirely new ACT Assessment was given that emphasized rhetorical skills, advanced mathematics items, and a new reading test. From the data on the graph, in which year do you think this occurred? Why?

14. History Did you know that there have been more U.S. vice-presidents than presidents? As of 1995, there have been 41 presidents and 45 vice-presidents. Some presidents had more than one vice-president, and some had none. The line plot below shows the ages of vice-presidents on their inauguration days.

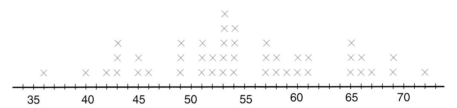

Source: *The World Almanac, 1995*

a. Make a box-and-whisker plot of these data.

b. The ages for presidents on their inauguration days have the following statistics: LV = 42, Q1 = 51, Q2 = 55, Q3 = 58, GV = 68, and 69 is an outlier. Make a box-and-whisker plot to represent these ages.

c. Which set of data is more clustered? Explain your answer.

d. Fourteen vice-presidents went on to become presidents. Which ages of vice-presidents were definitely not those who went on to become president?

Critical Thinking

15. The box-and-whisker plots shown below picture the distribution of test scores in two algebra classes taught by two different teachers. If you could select one of the classes to be in, which one would it be, and why?

Mixed Review

16. Travel Greg's car gets between 18 and 21 miles per gallon of gasoline. If his car's tank holds 15 gallons, what is the range of distances that Greg can drive his car on one tank of gas? (Lesson 7–6)

17. Solve $2m - 3 > 7$ or $2m + 7 > 9$. (Lesson 7–4)

18. Write the standard form of an equation of the line that passes through $(4, 7)$ and $(1, -2)$. (Lesson 6–2)

19. Graph $(0, 6)$, $(8, -1)$, and $(-3, 2)$. (Lesson 5–1)

20. Solve $\frac{6}{x - 3} = \frac{3}{4}$. (Lesson 4–1)

7–7B Graphing Technology
Box-and-Whisker Plots

An Extension of Lesson 7–7

You can use a graphing calculator to compare two sets of data by using a double box-and-whisker plot. The plots that the calculator draws, however, do not account for outliers. If you want to use a graphing calculator to help you sketch a plot, it will be necessary for you to check for outliers and adjust the graph as needed.

In an experiment, the ability of boys and girls to identify objects held in their left hands versus those held in their right hands was tested. The left side of the body is controlled by the right side of the brain and vice versa. The results of the experiment found that the boys did not identify objects with their right hand as well as with their left. The girls could identify objects equally as well with either hand.

Texas Instruments decided to test this premise by conducting the same tests with 12 male and 10 female employees, chosen at random. Thirty small objects were selected and separated into two groups—one for the right hand and one for the left hand. Blindfolded employees felt each of the objects with the prescribed hand and tried to identify them. The results are presented in the chart below.

> **LOOK BACK**
>
> For more information on entering data into lists on the graphing calculator, see Lesson 5-7A.

Each employee had a score for left and right hands.

Correct Responses			
Female Left Hand	Female Right Hand	Male Left Hand	Male Right Hand
8	4	7	12
9	3	8	6
12	7	7	12
11	12	5	12
10	11	7	7
8	11	8	11
12	13	11	12
7	12	4	8
9	11	10	12
11	12	14	11
		13	9
		5	9

Source: *TI-82 Graphics Calculator Guidebook*

Make a double box-and-whisker plot of the data to compare the results of Female Left Hand with Female Right Hand using a graphing calculator.

Step 1   Clear lists L1, L2, L3, and L4. Enter the data from each column of the table into lists L1, L2, L3, and L4, respectively.

Step 2   Select the box-and-whisker plot and define which list will be used.

Enter: [2nd] [STAT PLOT] 1 [ENTER]   *Turns plot on.*

[▼] [▶] [▶] [ENTER]   *Selects box-and-whisker plot.*

If L1 is not highlighted in the Xlist, use the down arrow and [ENTER] to highlight it and make sure the frequency is set for 1.

Repeat the process to assign Plot2 as a box-and-whisker plot using L2.

Step 3   Clear the Y= list. Set the WINDOW settings for Xscl = 1, Ymin = 0, and Yscl = 0. Ignore the other settings. Press [ZOOM] 9 to select ZoomStat. This sets the other settings and displays the box-and-whisker plots. It will only display those plots that you have turned on.

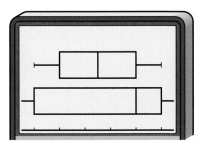

Step 4   Use [TRACE] to examine the minX (least value), Q1 (lower quartile), Med (median), Q3 (upper quartile), and maxX (greatest value).

EXERCISES

1. Which set of data does the upper plot represent?

2. Observe the two graphs. Does it appear that the females guessed correctly more often with the left hand or the right? How do you know?

3. Reset your calculator to define Plot1 as L3 and Plot 2 as L4 to examine the males' data. What do you observe in these plots?

4. Reset the calculator to compare the left-hand results of males and females. Were the males or females better at guessing with their left hands?

5. Reset the calculator to compare the right-hand results of males and females. Which group seemed more adept at identifying objects with their right hands?

6. How do the results of this experiment compare with the study of boys and girls mentioned at the beginning of this lesson? What reasons may account for any discrepancies?

7–8A Graphing Technology
Graphing Inequalities

A Preview of Lesson 7–8

Inequalities in two variables can be graphed on a graphing calculator using the "Shade(" command, which is option 7 on the DRAW menu. You must enter *two* functions to activate the shading since the calculator always shades between two specified functions. The first function entered defines the lower boundary of the region to be shaded. The second function defines the upper boundary of the region. The calculator graphs both functions and shades between the two.

Example **Graph $y \geq 2x - 3$ in the standard viewing window.**

The inequality refers to points at which y is *greater than or equal to* $2x - 3$. This means we want to shade above the graph of $y = 2x - 3$. Since the calculator screen shows only part of the coordinate plane, we can use the top of the screen, Ymax or 10, as the upper boundary and $2x - 3$ as the lower boundary.

Before using the "Shade(" option, be sure to clear any equations stored in the Y list, and press ZOOM *6 for the standard viewing window.*

Enter: 2nd DRAW 7 2 X,T,θ,n

   3 , 10) ENTER

Since both the x- and y-intercepts of the graph and the origin are within the current viewing window, the graph of the inequality is complete.

When finished, press 2nd DRAW *1 to clear the screen.*

Example **Graph $y - x \leq 1$ in the standard viewing window.**

First solve the inequality for y: $y \leq x + 1$. This inequality refers to points where y is *less than or equal to* $x + 1$. This means we want to shade below the graph of $y = x + 1$. We can use the bottom of the screen, Ymin or -10, as the lower boundary and $x + 1$ as the upper boundary.

Enter: 2nd DRAW 7 (–) 10 ,

X,T,θ,n + 1) ENTER

Don't forget to clear the screen when finished.

EXERCISES

Use a graphing calculator to graph each inequality. Sketch each graph on a sheet of paper.

1. $y \geq x + 2$
2. $y \leq -2x - 4$
3. $y + 1 \leq 0.5x$
4. $y \geq 4x$
5. $x + y \leq 0$
6. $2y + x \geq 4$
7. $3x + y \leq 18$
8. $y \geq 3$
9. $0.2x + 0.1y \leq 1$

Graphing Inequalities in Two Variables

What YOU'LL LEARN

• To graph inequalities in the coordinate plane.

Why IT'S IMPORTANT

You can graph inequalities to solve problems involving manufacturing and health.

APPLICATION
Manufacturing

Rapid Cycle, Inc. is a manufacturer and distributor of racing bicycles. It takes 3 hours to assemble a bicycle and 1 hour to road test a bicycle. Each technician in the company works no more than 45 hours a week. How many racing bikes can one technician assemble, and how many can he or she road test in one week?

Let x represent the number of bikes that are assembled in a week, and let y represent the number of bikes that are road-tested. Then the following inequality can be used to represent the solution.

Total time to assemble x bikes	plus	Total time to road test y bikes	is no more than	45 hours.
$3x$	$+$	y	$\leq$	45

There are an infinite number of ordered pairs that are solutions to this inequality. The easiest way to show all of these solutions is to draw a *graph* of the inequality. Before doing this, let's consider some simpler inequalities. *This problem will be solved in Example 3.*

Example ① **From the set {(3, 4), (0, 1), (1, 4), (1, 1)}, which ordered pairs are part of the solution set for $4x + 2y < 8$?**

Let's use a table to substitute the x and y values of each ordered pair into the inequality.

x	y	$4x + 2y < 8$	True or False?
3	4	$4(3) + 2(4) < 8$ $20 < 8$	false
0	1	$4(0) + 2(1) < 8$ $2 < 8$	true
1	4	$4(1) + 2(4) < 8$ $12 < 8$	false
1	1	$4(1) + 2(1) < 8$ $6 < 8$	true

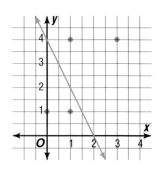

The ordered pairs {(0, 1), (1, 1)} are part of the solution set of $4x + 2y < 8$. The graph above shows the four ordered pairs of the replacement set and the equation $4x + 2y = 8$. Notice the location of the two ordered pairs that are solutions for $4x + 2y < 8$ in relation to the graph of the line.

You can use the following graphing calculator program to find out if a given ordered pair (x, y) is a solution for the inequality $5x - 3y \geq 15$.

```
PROGRAM: XYTEST
: Disp "IS (X, Y) A ","SOLUTION?"
: Prompt X,Y
: If 5X-3Y ≥ 15
: Then
: Disp "YES"
: Else
: Disp "NO"
```

To run the program for other ordered pairs, simply press ENTER and the program will begin again.

Your Turn

a. Try the program for ten ordered pairs (x, y). Keep a list of which ordered pairs you tried and which ones were solutions.

b. How do you think you could change this program to test the inequality $2x + y \geq 2y$?

c. Use your changed program to find the solution set if $x = \{-1, 0, 1\}$ and $y = \{-2, -1, 0, 1\}$.

The solution set for an inequality contains many ordered pairs when the domain and range are the set of real numbers. The graphs of all of these ordered pairs fill an area on the coordinate plane called a **half-plane.** An equation defines the **boundary** or edge for each half-plane. For example, suppose you wanted to graph the inequality $y > 5$ on the coordinate plane.

First determine the boundary by graphing $y = 5$.

Since the inequality involves only $>$, the line should be dashed. The boundary divides the coordinate plane into two half-planes.

If the inequality contains $\geq$ or $\leq$, the graph of the boundary equation would be drawn as a solid line.

To determine which half-plane contains the solution, choose a point from each half-plane and test it in the inequality.

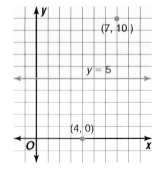

Try (7, 10).	Try (4, 0).
$y > 5$ $y = 10$	$y > 5$ $y = 0$
$10 > 5$ true	$0 > 5$ false

This graph is called an open half-plane because the boundary is not part of the graph.

The half-plane that contains (7, 10) contains the solution. Shade that half-plane.

Example **Graph $y + 2x \leq 3$.**

First solve for y in terms of x.

$$y + 2x \leq 3$$
$$y + 2x - 2x \leq 3 - 2x \quad \text{Subtract 2x from each side.}$$
$$y \leq 3 - 2x$$

(continued on the next page)

Graph $y = 3 - 2x$. Since $y \le 3 - 2x$ means $y < 3 - 2x$ or $y = 3 - 2x$, the boundary is included in the graph and should be drawn as a solid line.

Select a point in one of the half-planes and test it. For example, use the origin $(0, 0)$.

The origin is often used as a test point because the values are easy to substitute into the inequality.

The half-plane that contains the origin should be shaded.

Check: Test a point in the other half-plane, for example $(3, 3)$.

Since the statement is false, the half-plane containing $(3, 3)$ is not part of the solution.

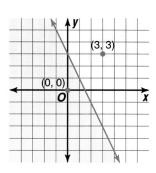

This graph is called a <u>closed half-plane</u> because the boundary line is included.

When solving real-life inequalities, the domain and range of the inequality are often restricted to nonnegative numbers or whole numbers.

Example ❸ **Refer to the application at the beginning of the lesson. How many racing bikes will the technician be able to assemble and road test?**

First, solve for y in terms of x.

$3x + y \le 45$

$3x - 3x + y \le 45 - 3x$

$y \le 45 - 3x$

Since the open sentence includes the equation, graph $y = 45 - 3x$ as a solid line. Test a point in one of the half-planes, for example $(0, 0)$. The half-plane containing $(0, 0)$ represents the solution since $3(0) + 0 \le 45$ is true.

Let's examine what the solution means. The technician cannot complete negative numbers of racing bikes. So, any point in the half-plane whose x- and y-coordinates are whole numbers is a possible solution. That is, only the portion of the shading in the first quadrant is the solution for this problem. One solution is $(5, 30)$. This represents 5 bicycles assembled and 30 road tested by the technician in a 45-hour week.

Communicating Mathematics

Study the lesson. Then complete the following.

1. **a. Graph** $y \geq x + 1$.
   b. Identify the boundary and indicate whether it is included or not.
   c. Identify the half-plane that is part of the graph.
   d. Write the coordinates of a point not on the boundary that satisfy the inequality.

2. **Explain** how you would check whether a point is part of the graph of an inequality.

3. **Assess Yourself** What do you think was the most challenging concept you learned in this chapter? Give an example of that concept and tell why you thought it was challenging.

Guided Practice

Match each inequality with its graph.

4. $y \geq \frac{1}{2}x - 2$

5. $y \leq 0.5x - 2$

6. $y \geq \frac{2}{3}x + 2$

7. $y \leq \frac{2}{3}x + 2$

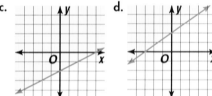

Determine which ordered pairs are solutions to the inequality. State whether the boundary is included in the graph.

8. $y \leq x$ **a.** $(-3, 2)$ **b.** $(1, -2)$ **c.** $(0, -1)$

9. $y > x - 1$ **a.** $(0, 0)$ **b.** $(2, 0)$ **c.** $(1, 3)$

10. Find which ordered pairs from the set $\{(-2, 2), (-2, 3), (2, 2), (2, 3)\}$ are part of the solution set for $a + b < 1$.

Graph each inequality.

11. $y > 3$ 12. $x + y > 1$ 13. $2x + 3y \geq -2$ 14. $-x < -y$

Practice

Copy each graph. Shade the appropriate half-plane to complete the graph of the inequality.

15. $x > 4$ 16. $3y < x$ 17. $3x + y > 4$ 18. $2x - y \leq -2$

Find which ordered pairs from the given set are part of the solution set for each inequality.

19. $y < 3x$, $\{(-3, 1), (-3, 2), (1, 1), (1, 2)\}$

20. $y - x > 0$, $\{(1, 1), (1, 2), (4, 1), (4, 2)\}$

21. $2y + x \geq 4$, $\{(-1, -3), (-1, 0), (-2, -3), (-2, 0)\}$

Graph each inequality.

22. $x > -5$	**23.** $y < -3$	**24.** $3y + 6 > 0$
25. $4x + 8 < 0$	**26.** $y \leq x + 1$	**27.** $x + y > 2$
28. $x + y < -4$	**29.** $3x - 1 \geq y$	**30.** $3x + y < 1$
31. $x - y \geq -1$	**32.** $x < y$	**33.** $-y > x$
34. $2x - 5y \leq -10$	**35.** $8y + 3x < 16$	**36.** $|y| \geq 2$
37. $y > |x + 2|$	**38.** $y > 2$ and $x < 3$	**39.** $y \leq -x$ and $x \geq -3$

Graphing Calculator

Use a graphing calculator to graph each inequality. Make a sketch of the graph.

40. $y > x - 1$ **41.** $4y + x < 16$ **42.** $x - 2y < 4$

Programming

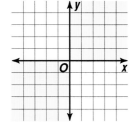

Use the program in the Exploration to determine which pairs are solutions for each inequality.

43. $x + 2y \geq 3$ **a.** $(-2, 2)$ **b.** $(4, -1)$ **c.** $(3, 1)$ **d.** $(0, 0)$

44. $2x - 3y \leq 1$ **a.** $(2, 1)$ **b.** $(5, -1)$ **c.** $(1, 1)$ **d.** $(0, 0)$

45. $-2x < 8 - y$ **a.** $(5, 10)$ **b.** $(3, 6)$ **c.** $(-4, 0)$ **d.** $(0, 0)$

Critical Thinking

46. What compound inequality is described by the graph at the right? Find a simple inequality that also describes this graph.

Applications and Problem Solving

47. Health The graph below shows the effective heart rate ranges for each type of exercise goal.

Workout Goals for Exercise

Goals

Boost performance as a competitive athlete

Improve cardiovascular conditioning

Lose weight

Improve overall health and reduce risk of heart attack

40%   50%   60%   70%   80%   90% 100%

Target Heart Rate Range

Source: *Vitality*, May 1994

Arrio is 35 years old and just beginning a bench-stepping aerobics class. In the orientation at the beginning of the first class, he learned that during exercise an effective minimum heart rate (beats/minute) should be 70% of the difference of 220 and his age. A maximum heart rate should be 80% of the difference of 220 and his age.

440   *Chapter 7   Solving Linear Inequalities*

a. Write a compound inequality that expresses the effective rate zone for a person *a* years of age.

b. In class, the participants take a break and count their heart beats for 15 seconds. What should be Arrio's effective heart rate zone for that 15-second count?

c. According to the graph and the heart rate range given, what is the goal of the bench-stepping class?

Reported Snowmobile Accidents in Minnesota

Year	Fatal	Water-related	Nonfatal	Total
1987–1988	14	3	261	278
1988–1989	8	1	313	322
1989–1990	10	4	246	260
1990–1991	11	1	354	366
1991–1992	15	0	386	401
1992–1993	19	4	546	569
1993–1994	21	1	531	553

Source: Minnesota Department of Natural Resources

48. Snowmobiling Although snowmobiling is an exhilarating sport, it can be very dangerous. The chart below shows the reported accidents in Minnesota involving snowmobiles.

a. A study suggests that the average number of nonfatal snowmobile accidents per year nationwide is about 350. Suppose *x* represents the total number of fatal snowmobile accidents in Minnesota and *y* represents the total number of snowmobile accidents in Minnesota. In what years is $x + 350 > y$?

b. Graph $x + 350 > y$.

c. The snowmobiling season in Minnesota goes from November to March. The lack of snow during the 1994–1995 season caused many to resort to riding their snowmobiles on icy lakes instead of on land. As of December 1, there had been 7 fatal accidents, 1 water-related fatal accident, and 81 nonfatal accidents. Do you think that the 1994–1995 season's accident count was greater or less than the 1993–1994 season? Explain your answer.

Mixed Review

49. Statistics Make a box-and-whisker plot of the total number of snowmobiling accidents shown in the chart for Exercise 48. (Lesson 7–7)

50. Solve $5 - |2x - 7| > 2$. (Lesson 7–6)

51. Write an equation in slope-intercept of the line that is parallel to the graph of $8x - 2y = 7$ and whose *y*-intercept is the same as the line whose equation is $2x - 9y = 18$. (Lesson 6–4)

52. Determine the value of *r* so that the line passing through $(r, 4)$ and $(-4, r)$ has a slope of 4. (Lesson 6–1)

53. Graph $-y + \frac{2}{7}x = 1$. (Lesson 5–4)

54. State the inverse of the relation $\{(4, -1), (3, 2), (-4, 0), (17, 9)\}$. (Lesson 5–2)

55. What is 98.5% of $140.32? (Lesson 4–4)

56. Find three consecutive integers whose sum is 87. (Lesson 3–3)

In·ves·ti·ga·tion

Smoke Gets In Your Eyes

Refer to the Investigation on pages 320–321. Add the results of your work below to your Investigation Folder.

The EPA's most recent long-term study on the effects of secondhand smoke shows that nonsmokers married to smokers have a 19% increased risk of having lung cancer. Lung cancer is not the only danger of secondhand smoke. Twelve studies show that heart disease is another danger. Nonsmokers who are exposed to their spouses' smoke have a 30% increased chance of death from heart disease than do other nonsmokers. After reviewing a number of studies, the EPA's risk analysis has also concluded that secondhand smoke causes an extra 150,000 to 300,000 respiratory infections a year among the nations 5.5 million children under the age of 18 months.

Katharine Hammond, an environmental-health expert at the University of California, Berkeley, has also conducted a study on the *carcinogenic* components of secondhand smoke. The carcinogenic components are the parts of smoke that are known to cause cancer in humans. She found that "in the same room, at the same time, the nonsmoker is getting as much benzene (a chemical that is known to cause cancer in humans) as a smoker gets smoking six cigarettes."

James Repace and Alfred Lowery, two statistical researchers who study the effects of secondhand smoke, have concluded that a lifetime increase in lung-cancer risk of 1 in 1000 could be caused by long-term exposure to air containing more than 6.8 micrograms of nicotine per cubic meter of air.

Analyze

You have conducted experiments and organized your data in various ways. It is now time to analyze your findings and state your conclusions.

PORTFOLIO ASSESSMENT

You may want to keep your work on this Investigation in your portfolio.

1 True secondhand smoke consists mostly of sidestream smoke. This is the smoke that comes from the smoldering cigarette. This smoke is much more toxic than inhaled smoke. How does this information affect your conclusions about the amount of smoke inhaled by nonsmokers in a room?

2 If you are a nonsmoker and live with a smoker, what are the cost factors involved? Explain your calculations.

3 Describe your personal experience with secondhand smoke.

Write

You want to inform people of the effects of secondhand smoke. You decide to write a letter to the editor of a local paper describing your investigation on the effects of secondhand smoke.

4 Use the information above and the results of your experiments and explorations to write a paper regarding the health risks of secondhand smoke.

5 You may want to do further research. The American Cancer Society and other agencies have information regarding the health risks of smoking and secondhand smoke.

6 Use data, charts, and graphs to justify your position. Use mathematics to help convince your readers of the conclusions you drew from this Investigation.

VOCABULARY

After completing this chapter, you should be able to define each term, property, or phrase and give an example or two of each.

Algebra

addition property for inequality   (p. 385)

boundary   (p. 437)

compound inequality   (p. 405)

division property for inequality   (p. 393)

half-plane   (p. 437)

intersection   (p. 406)

multiplication property for inequality
  (p. 393)

set-builder notation   (p. 385)

subtraction property for inequality   (p. 385)

union   (p. 408)

Statistics

box-and-whisker plot   (p. 427)

extreme values   (p. 427)

whiskers   (p. 428)

Probability

compound event   (p. 414)

outcomes   (p. 413)

simple events   (p. 414)

tree diagram   (p. 413)

Problem Solving

draw a diagram   (p. 406)

UNDERSTANDING AND USING THE VOCABULARY

Choose the letter of the term that best matches each statement, algebraic expression, or algebraic sentence.

1. If $\frac{1}{2}x \leq -5$, then $x \leq -10$.

2. If $8 > 4$, then $8 + 5 > 4 + 5$.

3. $\{h \mid h > 43\}$

4. $x \geq -3$ or $x < -10$

5. $x \geq -4$ and $x < 2$

6. If $4x - 1 < 7$, then $4x - 4 < 4$.

7. If $-3x < 9$, then $x > -3$.

8. $>$

9. $<$

10. $7 > x > 1$

11. $|x + 6| > 12$ means $x + 6 > 12$ or $-(x + 6) > 12$.

a. absolute value inequality

b. addition property for inequality

c. compound inequality

d. division property for inequality

e. greater than

f. intersection

g. less than

h. multiplication property for inequality

i. set builder notation

j. subtraction property for inequality

k. union

OBJECTIVES AND EXAMPLES	REVIEW EXERCISES

Upon completing this chapter, you should be able to:

- solve inequalities by using addition and subtraction   (Lesson 7–1)

$$56 > m + 16$$
$$56 - 16 > m + 16 - 16$$
$$40 > m$$
$$\{m \mid m < 40\}$$

- solve inequalities by using multiplication and division   (Lesson 7–2)

$$\frac{-5}{6}m > 25$$
$$\frac{-6}{5}\left(\frac{-5}{6}m\right) > \frac{-6}{5}(25)$$
$$m < -30$$
$$\{m \mid m < -30\}$$

- solve linear inequalities involving more than one operation   (Lesson 7–3)

$$15b - 12 > 7b + 60$$
$$15b - 7b - 12 > 7b - 7b + 60$$
$$8b - 12 > 60$$
$$8b - 12 + 12 > 60 + 12$$
$$8b > 72$$
$$\frac{8b}{8} > \frac{72}{8}$$
$$b > 9$$
$$\{b \mid b > 9\}$$

Use these exercises to review and prepare for the chapter test.

Solve each inequality. Then check your solution.

12. $r + 7 > -5$ **13.** $-35 + 6n < 7n$

14. $2t - 0.3 \leq 5.7 + t$ **15.** $-14 + p \geq 4 - (-2p)$

Define a variable, write an inequality, and solve each problem. Then check your solution.

16. The difference of a number and 3 is at least 2.

17. Three times a number is greater than four times the number less eight.

Solve each inequality. Then check your solution.

18. $7x \geq -56$ **19.** $90 \leq -6w$

20. $\frac{2}{3}k \geq \frac{2}{15}$ **21.** $9.6 < 0.3x$

Define a variable, write an inequality, and solve each problem. Then check your solution.

22. Six times a number is at most 32.4.

23. Negative three fourths of a number is no more than 30.

Find the solution set of each inequality if the replacement set for each variable is {−5, −4, −3, . . . 3, 4, 5}.

24. $\frac{x - 5}{3} > -3$

25. $3 \leq -4x + 7$

Solve each inequality. Then check your solution.

26. $2r - 3.1 > 0.5$ **27.** $4y - 11 \geq 8y + 7$

28. $-3(m - 2) > 12$

29. $-5x + 3 < 3x + 23$

30. $4(n - 1) < 7n + 8$

31. $0.3(z - 4) \leq 0.8(0.2z + 2)$

OBJECTIVES AND EXAMPLES

- solve compound inequalities and graph their solution sets (Lesson 7–4)

$$2a > a - 3 \qquad \text{and} \qquad 3a < a + 6$$
$$2a - a > a - a - 3 \qquad 3a - a < a - a + 6$$
$$a > -3 \qquad\qquad 2a < 6$$
$$\frac{2a}{2} < \frac{6}{2}$$
$$a < 3$$

$$\{a \,|\, -3 < a < 3\}$$

```
←+—●+++++++●+→
  -4 -3 -2 -1 0  1  2  3  4
```

- find the probability of a compound event (Lesson 7–5)

Draw a tree diagram to show the possibilities for boys and girls in a family of 3 children. Assume that the probabilities for girls and boys being born are the same.

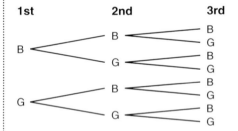

The probability that the family has exactly 3 girls is $\frac{1}{8}$ or 0.125, because there is 1 way out of 8 for this to happen. The probability that the family has exactly 2 boys and 1 girl is $\frac{3}{8}$ or 0.375, because there are 3 ways out of 8 for this to happen.

- solve open sentences involving absolute value and graph the solutions (Lesson 7–6)

$$|2x + 1| > 1$$

$$2x + 1 > 1 \qquad \text{or} \qquad 2x + 1 < -1$$
$$2x + 1 - 1 > 1 - 1 \qquad 2x + 1 - 1 < -1 - 1$$
$$\frac{2x}{2} > \frac{0}{2} \qquad\qquad \frac{2x}{2} < \frac{-2}{2}$$
$$x > 0 \qquad\qquad\qquad x < -1$$

$$\{x \,|\, x > 0 \text{ or } x < -1\}$$

```
←+++++●+●+++++→
  -4 -3 -2 -1 0  1  2  3
```

REVIEW EXERCISES

Solve each compound inequality. Then graph the solution set.

32. $x - 5 < -2$ and $x - 5 > 2$

33. $2a + 5 \le 7$ or $2a \ge a - 3$

34. $4r \ge 3r + 7$ and $3r + 7 < r + 29$

35. $-2b - 4 \ge 7$ or $-5 + 3b \le 10$

36. $a \ne 6$ and $3a + 1 > 10$

37. With each shrimp, salmon, or crab dinner at the Seafood Palace, you may have soup or salad. With shrimp, you may have broccoli or a baked potato. With salmon, you may have rice or broccoli. With crab, you may have rice, broccoli, or a potato. If all combinations are equally likely, find the probability of an order containing each item.

   a. salmon   **b.** soup

   c. rice   **d.** shrimp and rice

   e. salad and broccoli

   f. crab, soup, and rice

38. Matthew has 2 brown and 4 black socks in his dresser. While dressing one morning, he pulled out 2 socks without looking. What is the probability that he chose a matching pair?

Solve each open sentence. Then graph the solution set.

39. $|y + 5| > 0$

40. $|1 - n| \le 5$

41. $|4k + 2| \le 14$

42. $|3x - 12| < 12$

43. $|13 - 5y| \ge 8$

44. $\left|2p - \frac{1}{2}\right| > \frac{9}{2}$

OBJECTIVES AND EXAMPLES	REVIEW EXERCISES

OBJECTIVES AND EXAMPLES

• display and interpret data on box-and-whisker plots   (Lesson 7–7)

The following high temperatures were recorded during a two-week cold spell in St. Louis. Make a box-and-whisker plot of the temperatures.

20°   2°   12°   5°   4°   16°   17°
7°   6°   16°   5°   0°   5°   30°

0 5 10 15 20 25 30 35

REVIEW EXERCISES

The number of calories in a serving of french fries at 13 restaurants are 250, 240, 220, 348, 199, 200, 125, 230, 274, 239, 212, 240, and 327.

45. Make a box-and-whisker plot of these data.

46. Are there any outliers? If so, name them.

• graph inequalities in the coordinate plane
(Lesson 7–8)

Graph $2x + 3y < 9$.

Find which ordered pairs from the given set are part of the solution set for each inequality.

47. $3x + 4y < 7$, $\{(1, 1), (2, -1), (-1, 1), (-2, 4)\}$

48. $4y - 8 \geq 0$, $\{(5, -1), (0, 2), (2, 5), (-2, 0)\}$

49. $-2x < 8 - y$, $\{(5, 10), (3, 6), (-4, 0), (-3, 6)\}$

Graph each inequality.

50. $x + 2y > 5$ **51.** $4x - y \leq 8$

52. $\frac{1}{2}y \geq x + 4$ **53.** $3x - 2y < 6$

APPLICATIONS AND PROBLEM SOLVING

54. Number Theory   The sum of three consecutive integers is less than 100. Find the three integers with the greatest sum.
(Lesson 7–3)

55. Shipping   An empty book crate weighs 30 pounds. The weight of a book is 1.5 pounds. For shipping, the crate must weigh at least 55 pounds and no more than 60 pounds. What is the acceptable number of books that can be packed in the crate?   (Lesson 7–4)

56. Automobiles   An automobile dealer has cars available painted red or blue, with 4-cylinder or 6-cylinder engines, and with manual or automatic transmissions.   (Lesson 7–5)

a. What is the probability of selecting a car with manual transmission?

b. What is the probability of selecting a car with a 4-cylinder engine and a manual transmission?

c. What is the probability of selecting a blue car with a 6-cylinder engine and an automatic transmission?

A practice test for Chapter 7 is provided on page 793.

ALTERNATIVE ASSESSMENT

COOPERATIVE LEARNING PROJECT

Statistics   In this chapter, you learned how to make and interpret a box-and-whisker plot. Making a box-and-whisker plot and interpreting the data from a box-and-whisker plot are two different skills, however. One can go through the routine of drawing the plot but not be able to use the data portrayed by the plot to answer appropriate questions.

For this project, suppose there are two companies that manufacture window glass. They have each submitted a bid to a contractor who is building a library. Since glass that varies in thickness can cause distortions, the contractor has decided to measure the thickness of panes of glass from each factory, at several locations on each pane. The table below shows measurements for the two panes, one from each manufacturer.

Glass Thickness (mm)	
Company A	Company B
10.2	9.4
12.0	13.0
11.6	8.2
10.1	14.9
11.2	12.6
9.7	7.7
10.7	13.2
11.6	12.2
10.4	10.2
9.8	9.5
10.6	9.9
10.3	9.7
8.5	11.5
10.2	11.5
9.7	10.5
9.2	10.6
8.6	6.4
11.3	13.5

Prepare a stem-and-leaf plot and a box-and-whisker plot to organize and compare the two sets of data.

Follow these steps to organize your data.

- Determine how to set up the stem-and-leaf plot using decimals.

- Determine if using a common stem would be useful in comparing the two sets of data.

- Compare the ranges of the two sets of data.

- For which company is there "bunching" or "spreading out evenly" of the data?

- Compare the stem-and-leaf plot shape with the box-and-whisker plot shape.

- Compare the middle half of the data for each company.

- Write a comparative description of the sets of data and determine, with support, which company's glass should have less distortion.

THINKING CRITICALLY

- Why are multiplication and division the only two out of the four operations for which it is necessary to distinguish between positive and negative numbers when solving linear inequalities?

- Under what conditions will the compound sentence $x < a$ and $-a < x$ have no solutions?

PORTFOLIO

Select one of the assignments from this chapter for which you felt organization of the problem and reevaluation of the answer were important in order to get an accurate answer. Revise your work as necessary and place it in your portfolio. Explain why organization and reevaluation were important.

SELF EVALUATION

Do you look beyond the obvious in your math answers? Many times math students will work through a math problem rather routinely and not evaluate or check their answer. An answer must make sense and be accurate.

Assess yourself. Do you take the obvious solution as the whole answer or do you evaluate your answers for accuracy and rationalness? List two problems in mathematics and/or your daily life whereby the obvious answer was incorrect, so you needed to evaluate your solution for accuracy.

EXTRA PRACTICE

Pages 448–755 are contained in Volume Two.

Lesson 1-1   Write an algebraic expression for each verbal expression.

1. the product of x and 7

2. the quotient of r and s

3. the sum of b and 21

4. a number t decreased by 6

5. a number a to the third power

6. sixteen squared

Write a verbal expression for each algebraic expression.

7. $n - 7$

8. xy

9. m^5

10. 8^4

11. $6r^2$

12. $z^7 + 2$

Write each expression as an expression with exponents.

13. $5 \cdot 5 \cdot 5$

14. $7 \cdot a \cdot a \cdot a \cdot a$

15. $2(m)(m)(m)$

16. $5 \cdot 5 \cdot 5 \cdot x \cdot x \cdot y$

17. $p \cdot p \cdot p \cdot p \cdot p \cdot p$

18. $4 \cdot 4 \cdot 4 \cdot t \cdot t$

Evaluate each expression.

19. 2^4

20. 8^2

21. 7^3

22. 10^4

23. 3^6

24. 4^5

Lesson 1-2   Give the next two items for each pattern.

1.

2.

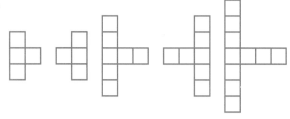

3. 12, 23, 34, 45, ...

4. 39, 33, 27, 21, ...

5. 6, 7.2, 8.4, 9.6, ...

6. 86, 81.5, 77, 72.5, ...

7. 4, 8, 16, 32, ...

8. 3125, 625, 125, 25, ...

9. 15, 16, 18, 21, 25, 30, ...

10. $w - 2, w - 4, w - 6, w - 8, ...$

11. 13, 10, 11, 8, 9, 6, ...

Lesson 1-3   Evaluate each expression.

1. $3 + 8 \div 2 - 5$

2. $4 + 7 \cdot 2 + 8$

3. $5(9 + 3) - 3 \cdot 4$

4. $4(11 + 7) - 9 \cdot 8$

5. $5^3 + 6^3 - 5^2$

6. $16 \div 2 \cdot 5 \cdot 3 \div 6$

7. $7(5^3 + 3^2)$

8. $\dfrac{9 \cdot 4 + 2 \cdot 6}{7 \cdot 7}$

9. $25 - \frac{1}{3}(18 + 9)$

Evaluate each expression when $a = 2$, $b = 5$, $x = 4$, and $n = 10$.

10. $8a + b$

11. $12x + ab$

12. $a(6 - 3n)$

13. $bx + an$

14. $x^2 - 4n$

15. $3b + 16a - 9n$

16. $n^2 + 3(a + 4)$

17. $(2x)^2 + an - 5b$

18. $[a + 8(b - 2)]^2 \div 4$

Lesson 1-4 Suppose the number 16,782 is rounded to 16,800 and plotted using stem 16 and leaf 8. Write the stem and leaf for each number below if the numbers are part of the same set of data.

1. 24,640

2. 35,788

3. 4239

4. 5865

5. 611

6. 17,903

7. The stem-and-leaf plot at the right gives the average weekly earnings for various occupations in 1993. Use the plot to answer each question.

  a. What were the highest weekly earnings?

  b. What were the lowest weekly earnings?

  c. How many occupations have weekly earnings of at least $500?

  d. What does $5 \mid 1$ represent?

Average Weekly Earnings

Stem	Leaf
2	6 7
3	1 2 8 9
4	1 3 6 8 8 8 9
5	0 1 6
6	
7	1 2 4

$3 \mid 8 = \$380-\$389$

Lesson 1-5 State whether each equation is *true* or *false* for the value of the variable given.

1. $b + \frac{2}{3} = \frac{3}{4} + \frac{1}{3}, b = \frac{1}{2}$

2. $\frac{2 + 13}{y} = \frac{3}{5}y, y = 5$

3. $x^8 = 9^4, x = 3$

4. $4t^2 - 5(3) = 9, t = 7$

5. $\frac{3^2 - 5x}{3^2 - 1} \leq 2, x = 4$

6. $a^6 \div 4 \div a^3 \div a < 3, a = 2$

Find the solution set for each inequality if the replacement set for x is {4, 5, 6, 7, 8} and for y is {10, 12, 14, 16}.

7. $x + 2 > 7$

8. $x - 1 > 3$

9. $2y - 15 \leq 17$

10. $y + 12 < 25$

11. $\frac{y + 12}{7} \geq 4$

12. $\frac{2(x - 2)}{3} < \frac{4}{7 - 5}$

13. $x - 4 > \frac{x + 2}{3}$

14. $y^2 - 100 \geq 4y$

15. $9x - 20 \geq x^2$

16. $0.3(x + 4) \leq 0.4(2x + 3)$

17. $1.3x - 12 < 0.9x + 4$

18. $1.2y - 8 \leq 0.7y - 3$

Solve each equation.

19. $x = \frac{17 + 9}{2}$

20. $3(8) + 4 = b$

21. $\frac{18 - 7}{13 - 2} = y$

22. $28 - (-14) = z$

23. $20.4 - 5.67 = t$

24. $t = 91.8 \div 27$

25. $-\frac{5}{8}\left(-\frac{4}{5}\right) = c$

26. $8\frac{1}{12} - 5\frac{5}{12} = e$

27. $\frac{3}{4} - \frac{9}{16} = s$

28. $\frac{5}{8} + \frac{1}{4} = y$

29. $n = \frac{84 \div 7}{18 \div 9}$

30. $d = 3\frac{1}{2} \div 2$

Lesson 1-6 Name the property or properties illustrated by each statement.

1. If $8 \cdot 3 = 24$, then $24 = 8 \cdot 3$.

2. $6 + (4 + 1) = 6 + 5$

3. $(12 - 3)(7) = 9(7)$

4. $qrs = 1qrs$

5. $\left(\frac{8}{9}\right)\left(\frac{9}{8}\right) = 1$

6. $4\left(6^2 \cdot \frac{1}{36}\right) = 4$

7. $0 + 45 = 45$

8. If $5 = 9 - 4$, then $9 - 4 = 5$.

9. $2(0) = 0$

10. $16 + 37 = 16 + 37$

11. $1(57) = 57$

12. $0 + h = h$

13. If $9 + 1 = 10$ and $10 = 5(2)$, then $9 + 1 = 5(2)$.

Lesson 1-7 Use the distributive property to rewrite each expression without parentheses.

1. $3(5 + w)$

2. $(h - 8)7$

3. $6(y + 4)$

4. $9(3n + 5)$

5. $32\left(x - \frac{1}{8}\right)$

6. $c(7 - d)$

Use the distributive property to find each product.

7. $6 \cdot 55$

8. $\left(4\frac{1}{18}\right) \times 18$

9. $15(108)$

10. $14(3.7)$

11. $689 \cdot 5$

12. 7×314

Simplify each expression, if possible. If not possible, write *in simplest form.*

13. $13a + 5a$

14. $21x - 10x$

15. $8(3x + 7)$

16. $4m - 4n$

17. $3(5am - 4)$

18. $15x^2 + 7x^2$

19. $9y^2 + 13y^2 + 3$

20. $11a^2 - 11a^2 + 12a^2$

21. $6a + 7a + 12b + 8b$

Lesson 1-8 Name the property illustrated by each statement.

1. $1 \cdot a^2 = a^2$

2. $x^2 + (y + z) = x^2 + (z + y)$

3. $ax + 2b = xa + 2b$

4. $29 + 0 = 29$

5. $5(a + 3b) = 5a + 15b$

6. $5a + 3b = 3b + 5a$

7. $(4 \cdot c) \cdot d = 4 \cdot (c \cdot d)$

8. $(6x^3) \cdot 0 = 0$

9. $(4 + 1)x + 2 = 5x + 2$

10. $(a + b) + 3 = a + (b + 3)$

11. $5(ab) = (5a)b$

12. $5a + \left(\frac{1}{2}b + c\right) = \left(5a + \frac{1}{2}b\right) + c$

Simplify.

13. $5a + 6b + 7a$

14. $8x + 4y + 9x$

15. $3a + 5b + 2c + 8b$

16. $\frac{2}{3}x^2 + 5x + x^2$

17. $(4p - 7q) + (5q - 8p)$

18. $8q + 5r - 7q - 6r$

19. $4(2x + y) + 5x$

20. $9r^5 + 2r^2 + r^5$

21. $12b^3 + 12 + 12b^3$

22. $7 + 3(uv - 6) + u$

23. $3(x + 2y) + 4(3x + y)$

24. $6.2(a + b) + 2.6(a + b) + 3a$

25. $3 + 8(st + 3w) + 3st$

26. $5.4(s - 3t) + 3.6(s - 4)$

27. $3[4 + 5(2x + 3y)]$

Lesson 1-9 Match each description with the most appropriate graph.

1. Jeremy picks up his speed while jogging down a hill. Then he slows down until he comes to a stop.

2. Ilene jogs up a hill at a steady speed. Then she runs down the hill and picks up her speed.

3. Casey runs down a hill and picks up his speed. Then he continues to jog at a steady pace.

4. Luisa jogs along a road at a steady speed. Then she slows down until she comes to a stop.

Lesson 2-1 Name the set of numbers graphed.

1.
   -6 -5 -4 -3 -2 -1 0 1 2 3 4 5 6

2.
   -4 -3 -2 -1 0 1 2 3 4 5 6 7 8

3.
   -3 -2 -1 0 1 2 3 4 5 6 7 8 9

4.
   4 5 6 7 8 9 10 11 12 13 14 15 16

5.
   -10 -9 -8 -7 -6 -5 -4 -3 -2 -1 0 1 2

6.
   -5 -4 -3 -2 -1 0 1 2 3 4 5 6 7

Graph each set of numbers on a number line.

7. $\{-2, -4, -6\}$

8. $\{\dots, -3, -2, -1, 0\}$

9. {integers greater than -1}

10. {integers less than -5 and greater than -10}

11. {integers less than or equal to 3}

12. {integers less than 0 and greater than or equal to -6}

Lesson 2-2 Use the line plot below to answer each question.

1. What was the highest score on the test?

2. What was the lowest score on the test?

3. How many students took the test?

4. How many students scored in the 40s?

5. What score was received by the most students?

Make a line plot for each set of data.

6. 134, 167, 137, 138, 120, 134, 145, 155, 152, 159, 164, 135, 144, 156

7. 19, 12, 11, 11, 7, 7, 8, 13, 12, 12, 9, 9, 8, 15, 11, 4, 12, 7, 7, 6

8. 66, 74, 72, 78, 68, 75, 80, 69, 62, 65, 63, 78, 81, 78, 76, 87, 80, 69, 81, 76, 79, 70, 62, 73, 85, 87, 70

9. 152, 156, 133, 154, 129, 146, 174, 138, 185, 141, 169, 176, 179, 168, 185, 154, 199, 200

Lesson 2-3 Find each sum or difference.

1. $-3 + 16$

2. $27 - 19$

3. $8 - 13$

4. $14 + (-9)$

5. $-18 + (-11)$

6. $-25 + 47$

7. $19m - 12m$

8. $8h - 23h$

9. $24b - (-9b)$

10. $97 + (-79)$

11. $4 + (-12) + (-18)$

12. $7 + (-11) + 32$

13. $|-28 + (-67)|$

14. $|-89 + 46|$

15. $|-285 + (-641)|$

16. $-35 - (-12)$

17. $24 + (-15)$

18. $-15 + (-13)$

19. $-7 + (-21)$

20. $8 - 17 + (-3)$

21. $27 - 14 - (-19)$

22. $|-9 + 15|$

23. $\begin{bmatrix} 5 & -4 \\ 0 & 3 \end{bmatrix} + \begin{bmatrix} -3 & 4 \\ -2 & -4 \end{bmatrix}$

24. $\begin{bmatrix} 1 & -4 \\ 5 & -6 \end{bmatrix} - \begin{bmatrix} 4 & -3 \\ 7 & -1 \end{bmatrix}$

Lesson 2-4 Replace each __?__ with <, >, or = to make each sentence true.

1. 6 __?__ −4

2. 12 __?__ −21

3. −4 __?__ −10

4. 4 __?__ 14

5. −13 __?__ −8

6. −5 + 2 __?__ −3

7. 7 __?__ 13 − (−6)

8. 3.4 − 5.7 __?__ −2

9. $\frac{18}{-6}$ __?__ −3

10. $\frac{8}{13}$ __?__ $\frac{9}{14}$

11. $\frac{25}{-8}$ __?__ $\frac{-28}{7}$

12. $6\left(\frac{5}{3}\right)$ __?__ $\left(\frac{3}{2}\right)6$

13. $\frac{0.6}{7}$ __?__ $\frac{1.8}{12}$

14. 24.6 __?__ 13.8 − (−12.8)

15. −54 + 26.5 __?__ 27.5

16. $\frac{5.4}{18}$ __?__ −4 + 1

17. (4.1)(0.2) __?__ 8.4

18. $-\frac{12}{17}$ __?__ $-\frac{9}{14}$

Lesson 2-5 Find each sum or difference.

1. $-\frac{11}{9} + \left(-\frac{7}{9}\right)$

2. $\frac{5}{11} - \frac{6}{11}$

3. $\frac{2}{7} - \frac{3}{14}$

4. −4.8 + 3.2

5. −1.7 − 3.9

6. −72.5 − 81.3

7. $-\frac{3}{5} + \frac{5}{6}$

8. $\frac{3}{8} + \left(-\frac{7}{12}\right)$

9. $-\frac{7}{15} + \left(-\frac{5}{12}\right)$

10. −4.5 − 8.6

11. 89.3 − (−14.2)

12. −0.007 + 0.06

13. $-\frac{2}{7} + \frac{3}{14} + \frac{3}{7}$

14. $-\frac{3}{5} + \frac{6}{7} + \left(-\frac{2}{35}\right)$

15. $\frac{7}{3} + \left(-\frac{5}{6}\right) + \left(-\frac{2}{3}\right)$

16. −4.13 + (−5.18) + 9.63

17. 6.7 + (−8.1) + (−7.3)

18. $\frac{3}{4} + \left(-\frac{5}{8}\right) + \frac{3}{32}$

19. 1.9 − (−7)

20. −1.8 − 3.7

21. −18 − (−1.3)

Lesson 2-6 Find each product.

1. 5(12)

2. (−6)(11)

3. (−7)(−5)

4. $\left(-\frac{7}{8}\right)\left(-\frac{1}{3}\right)$

5. $(-5)\left(-\frac{2}{5}\right)$

6. (−6)(4)(−3)

7. (4)(−2)(−1)(−3)

8. (−6.8)(−5.415)(3.1)

9. (−5.34)(3.2)

10. $\left(\frac{3}{5}\right)\left(-\frac{5}{7}\right)$

11. $-\frac{7}{15}\left(\frac{9}{14}\right)$

12. (4.2)(−5.1)(3.6)

13. $-6\left(\frac{5}{3}\right)\left(\frac{9}{10}\right)$

14. (3)(−6)(0)(−1)

15. (−21)(−2)(−1)

EXTRA PRACTICE

Lesson 2-7 Simplify.

1. $\dfrac{-48}{8}$

2. $-49 \div (-7)$

3. $-64 \div 8$

4. $-\dfrac{3}{4} \div 9$

5. $-9 \div \left(-\dfrac{10}{17}\right)$

6. $\dfrac{-450n}{10}$

7. $\dfrac{-36a}{-6}$

8. $\dfrac{63a}{-9}$

9. $8 \div \left(-\dfrac{5}{4}\right)$

10. $\dfrac{\frac{7}{8}}{-10}$

11. $\dfrac{12}{-\frac{8}{5}}$

12. $\dfrac{6a + 24}{6}$

13. $\dfrac{20a + 30b}{-2}$

14. $\dfrac{\frac{11}{5}}{-6}$

15. $\dfrac{70a - 42b}{-14}$

16. $\dfrac{-32x + 12y}{-4}$

17. $-\dfrac{7}{12} \div \dfrac{1}{18}$

18. $\dfrac{5}{\frac{15}{-7}}$

Lesson 2-8 Find each square root. Use a calculator if necessary. Round to the nearest hundredth if the result is not a whole number.

1. $-\sqrt{81}$

2. $\sqrt{0.0016}$

3. $\pm\sqrt{206}$

4. $\pm\sqrt{\dfrac{81}{64}}$

5. $\sqrt{85}$

6. $-\sqrt{\dfrac{36}{196}}$

7. $-\sqrt{149}$

8. $\pm\sqrt{961}$

9. $\sqrt{10.24}$

Evaluate each expression. Use a calculator if necessary. Round to the nearest hundredth if the result is not a whole number.

10. $\sqrt{m}$, if $m = 529$

11. $-\sqrt{c - d}$, if $c = 1.097$ and $d = 1.0171$

12. $-\sqrt{ab}$, if $a = 1.2$ and $b = 2.7$

13. $\pm\sqrt{\dfrac{x}{y}}$, if $x = 144$ and $y = 1521$

Lesson 2-9 Translate each sentence into an equation, inequality, or formula.

1. The square of a decreased by the cube of b is equal to c.

2. Twenty-nine decreased by the product of x and y is less than z.

3. The perimeter P of a parallelogram is twice the sum of the lengths of two adjacent sides a and b.

4. Four-fifths of the product of m, n, and the square of p is greater than 26.

5. Thirty increased by the quotient of s and t is equal to v.

6. The area A of a trapezoid is half the product of the height h and the sum of the two parallel bases a and b.

Lesson 3-1   Solve each equation. Then check your solution.

1. $-2 + g = 7$

2. $9 + s = -5$

3. $-4 + y = -9$

4. $m + 6 = 2$

5. $t + (-4) = 10$

6. $v - 7 = -4$

7. $a - (-6) = -5$

8. $-2 - x = -8$

9. $d + (-44) = -61$

10. $e - (-26) = 41$

11. $p - 47 = 22$

12. $-63 - f = -82$

13. $c + 5.4 = -11.33$

14. $-6.11 + b = 14.321$

15. $-5 = y - 22.7$

16. $-5 - q = 1.19$

17. $n + (-4.361) = 59.78$

18. $t - (-46.1) = -3.673$

19. $\frac{7}{10} - a = \frac{1}{2}$

20. $f - \left(-\frac{1}{8}\right) = \frac{3}{10}$

21. $-4\frac{5}{12} = t - \left(-10\frac{1}{36}\right)$

22. $x + \frac{3}{8} = \frac{1}{4}$

23. $1\frac{7}{16} + s = \frac{9}{8}$

24. $17\frac{8}{9} = d + \left(-2\frac{5}{6}\right)$

Lesson 3-2   Solve each equation. Then check your solution.

1. $-5p = 35$

2. $-3x = -24$

3. $62y = -2356$

4. $\frac{a}{-6} = -2$

5. $\frac{c}{-59} = -7$

6. $\frac{f}{14} = -63$

7. $84 = \frac{x}{97}$

8. $\frac{w}{5} = 3$

9. $\frac{q}{9} = -3$

10. $\frac{2}{5}x = \frac{4}{7}$

11. $\frac{z}{6} = -\frac{5}{12}$

12. $-\frac{5}{9}r = 7\frac{1}{2}$

13. $2\frac{1}{6}j = 5\frac{1}{5}$

14. $3 = 1\frac{7}{11}q$

15. $-1\frac{3}{4}p = -\frac{5}{8}$

16. $57k = 0.1824$

17. $0.0022b = 0.1958$

18. $5j = -32.15$

19. $\frac{w}{-2} = -2.48$

20. $\frac{z}{2.8} = -6.2$

21. $\frac{x}{-0.063} = 0.015$

22. $15\frac{3}{8} = -5.125p$

23. $-7.25 = -3\frac{5}{8}g$

24. $-18\frac{1}{4} = 2.50x$

Lesson 3-3   Solve each equation. Then check your solution.

1. $2x - 5 = 3$

2. $4t + 5 = 37$

3. $7a + 6 = -36$

4. $47 = -8g + 7$

5. $-3c - 9 = -24$

6. $5k - 7 = -52$

7. $5s + 4s = -72$

8. $3x - 7 = 2$

9. $8 + 3x = 5$

10. $-3y + 7.569 = 24.069$

11. $7 - 9.1f = 137.585$

12. $6.5 = 2.4m - 4.9$

13. $\frac{e}{5} + 6 = -2$

14. $\frac{d}{4} - 8 = -5$

15. $-\frac{4}{13}y - 7 = 6$

16. $\frac{p + 10}{3} = 4$

17. $\frac{h - 7}{6} = 1$

18. $\frac{5f + 1}{8} = -3$

19. $\frac{4n - 8}{-2} = 12$

20. $\frac{2a}{7} + 9 = 3$

21. $\frac{-3t - 4}{2} = 8$

Lesson 3-4   Find the complement of each angle measure.

1. $15°$
2. $79°$
3. $88°$
4. $a°$
5. $(3c)°$
6. $(b - 15)°$

Find the supplement of each angle measure.

7. $156°$
8. $94°$
9. $21°$
10. $a°$
11. $(3c)°$
12. $(b - 15)°$

Find the measure of the third angle of each triangle in which the measures of two angles of the triangle are given.

13. $90°, 2°$
14. $34°, 132°$
15. $111°, 28°$
16. $a°, b°$
17. $a°, (a - 15)°$
18. $b°, (3b - 2)°$

Lesson 3-5   Solve each equation. Then check your solution.

1. $6(y - 5) = 18$
2. $-21 = 7(p - 10)$
3. $3(h + 2) = 12$
4. $-3(x + 2) = -18$
5. $11.2n + 6 = 5.2n$
6. $2m + 5 - 6m = 25$
7. $3z - 1 = 23 - 3z$
8. $5a - 5 = 7a - 19$
9. $5b + 12 = 3b - 6$
10. $3x - 5 = 7x + 7$
11. $1.9s + 6 = 3.1 - s$
12. $2.85y - 7 = 12.85y - 2$
13. $2.9m + 1.7 = 3.5 + 2.3m$
14. $3(x + 1) - 5 = 3x - 2$
15. $4(2y - 1) = -10(y - 5)$
16. $\frac{6v - 9}{3} = v$
17. $\frac{3t + 1}{4} = \frac{3}{4}t - 5$
18. $\frac{2}{5}y + \frac{y}{2} = 9$
19. $3y - \frac{4}{5} = \frac{1}{3}y$
20. $\frac{3}{4}x - 4 = 7 + \frac{1}{2}x$
21. $\frac{x}{2} - \frac{1}{3} = \frac{x}{3} - \frac{1}{2}$

Lesson 3-6   Solve each equation for x.

1. $x + r = q$
2. $ax + 4 = 7$
3. $2bx - b = -5$
4. $\frac{x - c}{c + a} = a$
5. $\frac{x + y}{c} = d$
6. $\frac{ax + 1}{2} = b$
7. $\frac{x + t}{4} = d$
8. $6x - 7 = -r$
9. $kx + 4y = 5z$
10. $ax - 6 = t$
11. $\frac{2}{3}x + a = b$
12. $q(x + 1) = 5$
13. $\frac{x - y}{z} = 8$
14. $\frac{7a + b}{x} = 1$
15. $\frac{4cx + t}{7} = 2$
16. $\frac{9x - 4c}{z} = z$
17. $cx + a = bx$
18. $\frac{12q - x}{5} = t$

Lesson 3-7

1. **Geography** The areas in square miles of the 20 largest natural U.S. lakes are given below. Find the mean, median, and mode of the areas.

   | 31,700 | 1697 | 242 | 700 | 22,300 | 451 | 374 | 432 | 23,000 | 207 |
   | 1000 | 1361 | 315 | 625 | 9910 | 215 | 458 | 360 | 7550 | 435 |

2. **Work** Each number below represents the number of days that each employee of the Cole Corporation was absent during 1996. Find the mean, median, and mode of the number of days.

0	10	8	5	8	9	3	3	2	9	7	0	4	2	4	6	2
9	13	3	1	5	5	7	2	6	5	3	4	7	1	1	5	3
3	1	4	5	1	2

3. **Football** Tailback Michael Anderson, of the West High Bears, averaged 137.6 yards rushing per game for the first five games of the season. He rushes for 155 yards in the sixth game. What is his new rushing average?

4. **Olympic Games** One of the events in the Winter Olympics is the men's 500-meter speed skating. The winning times for this event are shown at the right. Find the mean, median, and mode of the times.

Year	Time(s)	Year	Time(s)
1932	43.4	1968	40.3
1936	43.4	1972	39.4
1948	43.1	1976	39.2
1952	43.2	1980	38.0
1956	40.2	1984	38.2
1960	40.2	1988	36.5
1964	40.1	1992	37.1

Lesson 4-1 Solve each proportion.

1. $\dfrac{4}{5} = \dfrac{x}{20}$

2. $\dfrac{b}{63} = \dfrac{3}{7}$

3. $\dfrac{y}{5} = \dfrac{3}{4}$

4. $\dfrac{7}{4} = \dfrac{3}{a}$

5. $\dfrac{t-5}{4} = \dfrac{3}{2}$

6. $\dfrac{x}{9} = \dfrac{0.24}{3}$

7. $\dfrac{n}{3} = \dfrac{n+4}{7}$

8. $\dfrac{12q}{-7} = \dfrac{30}{14}$

9. $\dfrac{1}{y-3} = \dfrac{3}{y-5}$

10. $\dfrac{r-1}{r+1} = \dfrac{3}{5}$

11. $\dfrac{a-3}{8} = \dfrac{3}{4}$

12. $\dfrac{6p-2}{7} = \dfrac{5p+7}{8}$

13. $\dfrac{2}{9} = \dfrac{k+3}{2}$

14. $\dfrac{5m-3}{4} = \dfrac{5m+3}{6}$

15. $\dfrac{w-5}{4} = \dfrac{w+3}{3}$

16. $\dfrac{96.8}{t} = \dfrac{12.1}{7}$

17. $\dfrac{x}{6.03} = \dfrac{4}{17.42}$

18. $\dfrac{4n+5}{5} = \dfrac{2n+7}{7}$

Lesson 4-2 Determine whether each pair of triangles is similar. Justify your answer.

1.

2.

3.

4.

5.

6.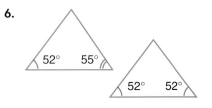

Lesson 4-3 For each triangle, find sin N, cos N, and tan N to the nearest thousandth.

1.

2.

3.

Use a calculator to find the value of each trigonometric ratio to the nearest ten thousandth.

4. cos 25°

5. tan 31°

6. sin 71°

7. cos 64°

8. tan 9°

9. sin 2°

Use a calculator to find the measure of each angle to the nearest degree.

10. tan $B = 0.5427$

11. cos $A = 0.8480$

12. sin $J = 0.9654$

13. cos $Q = 0.3645$

14. sin $R = 0.2104$

15. tan $V = 0.956$

Lesson 4-4 Write each ratio as a percent and then as a decimal.

1. $\frac{1}{4}$

2. $\frac{34}{100}$

3. $\frac{4}{25}$

4. $\frac{3}{20}$

5. $\frac{7}{8}$

6. $\frac{9}{10}$

7. $\frac{24}{40}$

8. $\frac{4}{50}$

9. $\frac{7}{15}$

10. $\frac{4}{9}$

11. $\frac{36}{15}$

12. $\frac{18}{4}$

Use a proportion to answer each question.

13. Twenty-four is what percent of 48?

14. What percent of 70 is 14?

15. Nine is what percent of 72?

16. Fourteen is 17.5% of what number?

17. What percent of 16 is 5.12?

18. What number is 25% of 64?

19. What percent of 80 is 2?

20. Forty-five is what percent of 112.5?

Lesson 4-5 State whether each percent of change is a percent of increase or a percent of decrease. Then find the percent of increase or decrease. Round to the nearest whole percent.

1. original: $100
   new: $67

2. original: 62 acres
   new: 98 acres

3. original: 322 people
   new: 289 people

4. original: 78 pennies
   new: 36 pennies

5. original: $212
   new: $230

6. original: 35 mph
   new: 65 mph

Find the final price of each item. When there is a discount and sales tax, first compute the discount price and then compute the sales tax and final price.

7. television: $299
   sales tax: 4.5%

8. boots: $49.99
   discount: 15%
   sales tax: 3.5%

9. backpack: $28.95
   discount: 10%
   sales tax: 5%

10. software: $36.99
   sales tax: 6.25%

11. jacket: $65
   discount: 30%
   sales tax: 4%

12. book: $15.95
   sales tax: 7%

Lesson 4–6   Determine the probability of each event.

1. a coin will land tails up

2. There is a December 1 this year.

3. A baby will be a girl.

4. Next year will have 400 days.

5. This is an algebra book.

6. Today is Wednesday.

Find the probability of each outcome if a computer randomly chooses a letter in the word "success."

7. the letter e

8. P(not c)

9. the letter s

10. the letter b

11. P(vowel)

12. the letters u or c

Find the odds of each outcome if a die is rolled.

13. a 4

14. a number greater than 3

15. a multiple of 3

16. a number less than 5

17. an odd number

18. not a 6

Lesson 4-7

1. **Advertising**   An advertisement for an orange drink claims that the drink contains 10% orange juice. How much pure orange juice would have to be added to 5 quarts of the drink to obtain a mixture containing 40% orange juice?

2. **Finance**   Jane Pham is investing $6000 in two accounts, part at 4.5% and the remainder at 6%. If the total annual interest earned from the two accounts is $279, how much did Jane deposit at each rate?

3. **Entertainment**   At the Golden Oldies Theater, tickets for adults cost $5.50 and tickets for children cost $3.50. How many of each kind of ticket was purchased if 21 tickets were bought for $83.50?

4. **Automotives**   A car radiator has a capacity of 14 quarts and is filled with a 20% antifreeze solution. How much must be drained off and replaced with pure antifreeze to obtain a 40% antifreeze solution?

5. **Nutrition**   A liter of cream has 9.2% butterfat. How much skim milk containing 2% butterfat should be added to the cream to obtain a mixture with 6.4% butterfat?

Lesson 4-8   Determine which equations represent inverse variations and which represent direct variations. Then find the constant of variation.

1. $ab = 6$

2. $\dfrac{50}{y} = x$

3. $\dfrac{1}{5}a = d$

4. $s = 3t$

5. $14 = cd$

6. $2x = y$

Solve. Assume that y varies directly as x.

7. If $y = 45$ when $x = 9$, find y when $x = 7$.

8. If $y = 18$ when $x = 27$, find x when $y = 8$.

9. If $y = 450$ when $x = 6$, find y when $x = 10$.

10. If $y = 6$ when $x = 48$, find y when $x = 20$.

11. If $y = 25$ when $x = 20$, find x when $y = 35$.

12. If $y = 100$ when $x = 40$, find y when $x = 16$.

13. If $y = -7$ when $x = -1$, find x when $y = -84$.

14. If $y = 5$ when $x = -10$, find y when $x = 50$.

15. If $y = 24$ when $x = 6$, find y when $x = 14$.

16. If $y = -10$ when $x = -4$, find x when $y = -15$.

Solve. Assume that y varies inversely as x.

17. If $y = 54$ when $x = 4$, find x when $y = 27$.

18. If $y = 18$ when $x = 6$, find x when $y = 12$.

19. If $y = 2$ when $x = 26$, find y when $x = 4$.

20. If $y = 3$ when $x = 8$, find x when $y = 4$.

21. If $y = 12$ when $x = 24$, find x when $y = 9$.

22. If $y = 8$ when $x = -8$, find y when $x = -16$.

23. If $y = 3$ when $x = -8$, find y when $x = 4$.

24. If $y = 27$ when $x = \dfrac{1}{3}$, find y when $x = \dfrac{3}{4}$.

25. If $y = 19.5$ when $x = 6.3$, find x when $y = 10.5$.

26. If $y = 4.8$ when $x = 10$, find y when $x = 19.2$.

Lesson 5-1   Refer to the coordinate plane below. Write the ordered pair for each point. Name the quadrant in which the point is located.

1. B 2. T 3. P

4. Q 5. A 6. K

7. J 8. L 9. S

10. D 11. M 12. N

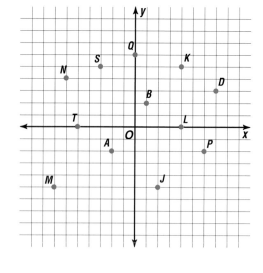

Graph each point.

13. $A(2, -4)$ 14. $B(3, 5)$

15. $C(-4, 0)$ 16. $D(-4, 3)$

17. $E(-5, -5)$ 18. $F(-1, 1)$

19. $G(0, -3)$ 20. $H(2, 3)$

Lesson 5-2   State the domain and range of each relation.

1. $\{(5, 2), (0, 0), (-9, -1)\}$

2. $\{(-4, 2), (-2, 0), (0, 2), (2, 4)\}$

3. $\{(7, 5), (-2, -3), (4, 0), (5, -7), (-9, 2)\}$

4. $\{(3.1, -1), (-4.7, 3.9), (2.4, -3.6), (-9, 12.12)\}$

Express the relation shown in each table, mapping, or graph as a set of ordered pairs. Then state the domain, range, and inverse of the relation.

5.

x	y
1	3
2	4
3	5
4	6
5	7

6.

x	y
-4	1
-2	3
0	1
2	3
4	1

7.

8.

9.

10.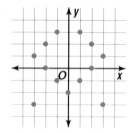

Lesson 5-3   Which ordered pairs are solutions of each equation?

1. $3r = 8s - 4$ a. $\left(\frac{2}{3}, \frac{3}{4}\right)$ b. $\left(0, \frac{1}{2}\right)$ c. $(4, 2)$ d. $(2, 4)$

2. $3y = x + 7$ a. $(2, 4)$ b. $(2, -1)$ c. $(2, 3)$ d. $(-1, 2)$

3. $4x = 8 - 2y$ a. $(2, 0)$ b. $(0, 2)$ c. $(0.5, -3)$ d. $(1, -2)$

4. $3n = 10 - 4m$ a. $(0, 3)$ b. $(-2, 6)$ c. $(1, 2)$ d. $(2, 1)$

Solve each equation if the range is $\{-3, -1, 0, 2, 3\}$.

5. $y = 2x$ 6. $y = 5x + 1$ 7. $2a + b = 4$

8. $4r + 3s = 13$ 9. $5b = 8 - 4a$ 10. $6m - n = -3$

Lesson 5-4 Determine whether each equation is a linear equation. If an equation is linear, rewrite it in the form $Ax + By = C$.

1. $3x = 2y$

2. $2x - 3 = y^2$

3. $3x - 2y = 8$

4. $5x - 7y = 2x - 7$

5. $2x + 5x = 7y$

6. $\frac{1}{x} + \frac{5}{y} = -4$

Graph each equation.

7. $3x + y = 4$

8. $y = 3x + 1$

9. $3x - 2y = 12$

10. $2x - y = 6$

11. $3x - 2y = 8$

12. $y = \frac{3}{4}$

13. $y = 5x - 7$

14. $x + \frac{1}{3}y = 6$

15. $x = -\frac{5}{2}$

16. $5x - 2y = 8$

17. $4x + 2y = 9$

18. $4x + 3y = 12$

Lesson 5-5 Determine whether each relation is a function.

1.

x	y
1	3
2	5
1	-7
2	9
3	3

2.

3.

4.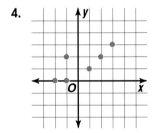

5.

a	b
1	-2
3	-4
5	-6
9	-4
10	1

6.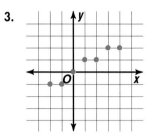

7. $\{(-2, 4), (1, 3), (5, 2), (1, 4)\}$

8. $\{(5, 4), (-6, 5), (4, 5), (0, 4)\}$

9. $\{(3, 1), (5, 1), (7, 1)\}$

10. $\{(3, -2), (4, 7), (-2, 7), (4, 5)\}$

11. $y = 2$

12. $x^2 + y = 11$

Lesson 5-6 Write an equation for each relation.

1.

x	2	4	6	8	10
f(x)	-4	-3	-2	-1	0

2.

x	-2	-1	0	1	2
f(x)	0	-3	-4	-3	0

3.

x	1	2	3	4	5
f(x)	7	11	15	19	23

4.

5.

6.

7. $\{(3, 12), (4, 14), (5, 16), (6, 18), (7, 20)\}$

8. $\{(1, 4), (2, 8), (3, 12), (-2, -8), (-3, -12)\}$

9. $\{(-6, -8), (-3, -6), (0, -4), (3, -2), (6, 0)\}$

10. $\{(-3, 9), (-2, 7), (1, 1), (4, -5), (6, -9)\}$

Lesson 5-7 Find the range, median, upper quartile, lower quartile, and interquartile range for each set of data.

1. 56, 45, 37, 43, 10, 34

2. 77, 78, 68, 96, 99, 84, 65

3. 30, 90, 40, 70, 50, 100, 80, 60

4. 4, 5.2, 1, 3, 2.4, 6, 3.7, 8, 1.3, 7.1, 9

5. 25°, 56°, 13°, 44°, 0°, 31°, 73°, 66°, 4°, 29°, 37°

6. 234, 648, 369, 112, 527, 775, 406, 268, 400

7.

Stem	Leaf
0	0 2 3
1	1 7 9
2	2 3 5 6
3	3 4 4 5 9
4	0 7 8 8

$2\,|\,2 = 22$

8.

Stem	Leaf
7	3 4 7 8
8	0 0 3 5 7
9	4 6 8
10	0 1 8
11	1 9

$9\,|\,4 = 9.4$

9.

Stem	Leaf
25	0 3 7 9
26	1 3 4 5 5 6
27	1 5 6 6 9
28	1 2 3 5 8
29	2 5 6 9

$27\,|\,5 = 2750$

Lesson 6-1 Determine the slope of each line.

1. a

2. b

3. c

4. d

5. e

6. f

7. g

8. h

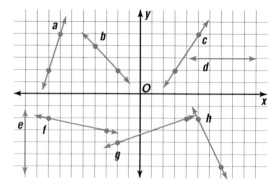

Determine the slope of the line that passes through each pair of points.

9. $(-2, 2), (3, -3)$

10. $(-2, -8), (1, 4)$

11. $(3, 4), (4, 6)$

12. $(-5, 4), (-1, 11)$

13. $(18, -4), (6, -10)$

14. $(-4, -6), (-4, -8)$

Determine the value of r so the line that passes through each pair of points has the given slope.

15. $(-1, r), (1, -4), m = -5$

16. $(-2, 1), (r, 4), m = \dfrac{3}{5}$

17. $(-1, 3), (-3, r), m = -3$

18. $(3, r), (7, -2), m = \dfrac{1}{2}$

19. $(r, -2), (-7, -1), m = -\dfrac{1}{4}$

20. $(-3, 2), (7, r), m = \dfrac{2}{3}$

Lesson 6-2 Write the point-slope form of an equation of the line that passes through the given point and has the given slope.

1. $(5, -2), m = 3$

2. $(5, 4), m = -5$

3. $(-2, -4), m = \dfrac{3}{4}$

4. $(-3, 1), m = 0$

5. $(-1, 0), m = \dfrac{2}{3}$

6. $(0, 6), m = -2$

Write the standard form of an equation of the line that passes through the given point and has the given slope.

7. $(-6, -3), m = -\dfrac{1}{2}$

8. $(4, -3), m = 2$

9. $(5, 4), m = -\dfrac{2}{3}$

10. $(1, 3), m = $ undefined

11. $(-2, 6), m = 0$

12. $(6, -2), m = \dfrac{4}{3}$

Lesson 6-3 Explain whether a scatter plot for each pair of variables would probably show a *positive*, *negative*, or *no* correlation between the variables.

1. playing time of a basketball game and points scored

2. age of a car and its value

3. heights of mothers and sons

4. temperature of the water in an outdoor swimming pool and the temperature outside

5. the weight of a person's car and the amount they pay to use a turnpike

6. the number of guests at a birthday party and the amount of food remaining after the party

7. The scatter plot at the right compares the number of hours per week people watched television to the number of hours per week they spent doing some physical activity.

   a. As people watched more television, what happened to the number of hours they spend doing some physical activity?

   b. Is there a correlation between the variables? Is it positive or negative?

Lesson 6-4 Find the *x*- and *y*-intercepts of the graph of each equation.

1. $3x + 2y = 6$

2. $5x + y = 10$

3. $2x + 5y = -11$

4. $3y = 12$

5. $y - 6x = 5$

6. $x = -2$

Write an equation in slope-intercept form of a line with the given slope and *y*-intercept. Then write the equation in standard form.

7. $m = -\frac{2}{5}, b = 2$

8. $m = 5, b = -15$

9. $m = -\frac{7}{4}, b = 2$

10. $m = -\frac{4}{3}, b = \frac{5}{3}$

11. $m = -6, b = 15$

12. $m = 12, b = -24$

Find the slope and *y*-intercept of the graph of each equation.

13. $y - \frac{3}{5}x = -\frac{1}{4}$

14. $y = 3x - 7$

15. $\frac{2}{3}x + \frac{1}{6}y = 2$

16. $2x + 3y = 5$

17. $3y = 8x + 2$

18. $5y = -8x - 2$

Write an equation in standard form for a line that passes through each pair of points.

19. $(-1, 7), (8, -2)$

20. $(6, 0), (0, 4)$

21. $(8, -1), (7, -1)$

22. $(1, 0), (0, 1)$

23. $(5, 7), (-1, 6)$

24. $(-3, -5), (3, -15)$

Lesson 6-5 Graph each equation.

1. $4x + y = 8$

2. $2x - y = 8$

3. $3x - 2y = 6$

4. $6x - 3y = 6$

5. $x + \frac{1}{2}y = 4$

6. $4x + 5y = 20$

7. $y + 3 = -2(x + 4)$

8. $y - 1 = 3(x - 5)$

9. $y + 6 = -\frac{2}{3}(x + 1)$

10. $y - 5 = 4(x + 6)$

11. $y - 2 = (x + 7)$

12. $3(x - 1) = y + \frac{4}{5}$

13. $y = \frac{3}{4}x + 4$

14. $y = 4x - 1$

15. $-4x + y = 6$

16. $-2x + y = 3$

17. $5y - 6 = 3x$

18. $y = \frac{3}{2}x - 5$

Lesson 6-6 Determine whether the graphs of each pair of equations are *parallel*, *perpendicular*, or *neither*.

1. $2x + 3y = -12$
   $2x + 3y = 6$

2. $-4x + 3y = 12$
   $x + 3y = 12$

3. $y = -3x + 9$
   $y + 3x = 14$

4. $y = 0.5x + 8$
   $2y = -8x - 3$

5. $y = 7x + 2$
   $y = 2x + 7$

6. $y + 5 = -9$
   $y + x = y - 6$

Write an equation in slope-intercept form of the line that passes through the given point and is parallel to the graph of each equation.

7. $(1, 6), y = 4x - 2$

8. $(4, 6), y = 2x - 7$

9. $(-3, 0), y = \frac{2}{3}x + 1$

10. $(2, 3), x - 5y = 7$

11. $(0, 4), 3x + 8y = 4$

12. $(5, -2), y = -3x - 7$

Write an equation in slope-intercept form of the line that passes through the given point and is perpendicular to the graph of each equation.

13. $(0, -1), y = -\frac{3}{5}x + 4$

14. $(-2, 3), 6x + y = 4$

15. $(0, 0), y = \frac{3}{4}x - 1$

16. $(4, 0), 4x - 3y = 2$

17. $(6, 7), 3x - 5y = 1$

18. $(5, -1), 8x + 4y = 15$

Lesson 6-7 Find the coordinates of the midpoint of a segment with each pair of endpoints.

1. $L(12, 2), M(8, 4)$

2. $S(9, 5), T(17, 3)$

3. $D(17, 9), E(11, -3)$

4. $F(4, 2), G(8, -6)$

5. $M(19, -3), N(11, 5)$

6. $B(-6, 5), C(8, -11)$

7. $T(-11, 6), U(13, 4)$

8. $A(-6, 1), B(8, 9)$

9. $J(6.4, -3), K(1.8, -3)$

10. $R(19, 5), S(7, 4)$

11. $G(8, 10), H(16, -6)$

12. $C(7.6, 8.3), D(-5, 6.1)$

Find the coordinates of the other endpoint of a segment given one endpoint and the midpoint M.

13. $P(9, 3), M(1, 2)$

14. $C(3, 5), M(5, -7)$

15. $G(5, -9), M\left(8, -\frac{15}{2}\right)$

16. $J(4, -7), M(-2, -3)$

17. $A(-3, 8), M(3, -5)$

18. $F(5, 7), M(5, 6)$

19. $T(-6, 12), M(4, 1)$

20. $D(-8, 5), M\left(-\frac{1}{2}, 2\right)$

21. $S(16, -9), M\left(\frac{3}{2}, -\frac{13}{2}\right)$

22. $U(-9, 14), M(0, 3)$

23. $F(21, 18), M(19, 11)$

24. $X\left(\frac{1}{4}, \frac{1}{3}\right), M\left(\frac{3}{16}, \frac{1}{3}\right)$

Lesson 7-1 Solve each inequality. Then check your solution.

1. $c + 9 \leq 3$

2. $d - (-3) < 13$

3. $z - 4 > 20$

4. $h - (-7) > -2$

5. $-11 > d - 4$

6. $2x > x - 3$

7. $2x - 3 \geq x$

8. $16 + w < -20$

9. $14p > 5 + 13p$

10. $-7 < 16 - z$

11. $-5 + 14b \leq -4 + 15b$

12. $2s - 6.5 \geq -11.4 + s$

13. $1.1v - 1 > 2.1v - 3$

14. $\frac{1}{2}t + \frac{1}{4} \geq \frac{3}{2}t - \frac{2}{3}$

15. $9x < 8x - 2$

16. $-2 + 9n \leq 10n$

17. $a - 2.3 \geq -7.8$

18. $5z - 6 > 4z$

Lesson 7-2 Solve each inequality. Then check your solution.

1. $7b \geq -49$

2. $-5j < -60$

3. $\frac{w}{3} > -12$

4. $\frac{p}{5} < 8$

5. $-8f < 48$

6. $\frac{t}{-4} \geq -10$

7. $\frac{128}{-g} < 4$

8. $-4.3x < -2.58$

9. $4c \geq -6$

10. $6 \leq 0.8n$

11. $\frac{2}{3}m \geq -22$

12. $-25 > \frac{a}{-6}$

13. $-15a < -28$

14. $-\frac{7}{9}x < 42$

15. $\frac{3y}{8} \leq 32$

16. $-7y \geq 91$

17. $0.8t > 0.96$

18. $\frac{4}{7}z \leq -\frac{2}{5}$

Lesson 7-3 Solve each inequality. Then check your solution.

1. $3y - 4 > -37$

2. $7s - 12 < 13$

3. $-5e + 9 > 24$

4. $-6v - 3 \geq -33$

5. $-2k + 12 < 30$

6. $-2x + 1 < 16 - x$

7. $15t - 4 > 11t - 16$

8. $13 - y \leq 29 + 2y$

9. $5q + 7 \leq 3(q + 1)$

10. $2(w + 4) \geq 7(w - 1)$

11. $-4t - 5 > 2t + 13$

12. $\frac{2t + 5}{3} < -9$

13. $\frac{z}{4} + 7 \geq -5$

14. $13r - 11 > 7r + 37$

15. $8c - (c - 5) > c + 17$

16. $-5(k + 4) \geq 3(k - 4)$

17. $9m + 7 < 2(4m - 1)$

18. $3(3y + 1) < 13y - 8$

19. $5x \leq 10(3x + 4)$

20. $3\left(a + \frac{2}{3}\right) \geq a - 1$

21. $0.7(n - 3) \leq n - 0.6(n + 5)$

Lesson 7-4 Solve each compound inequality. Then graph the solution set.

1. $2 + x < -5$ or $2 + x > 5$

2. $-4 + t > -5$ or $-4 + t < 7$

3. $3 \leq 2g + 7$ and $2g + 7 \leq 15$

4. $2v - 2 \leq 3v$ and $4v - 1 \geq 3v$

5. $3b - 4 \leq 7b + 12$ and $8b - 7 \leq 25$

6. $-9 < 2z + 7 < 10$

7. $5m - 8 \geq 10 - m$ or $5m + 11 < -9$

8. $12c - 4 \leq 5c + 10$ or $-4c - 1 \leq c + 24$

9. $2h - 2 \leq 3h \leq 4h - 1$

10. $3p + 6 < 8 - p$ and $5p + 8 \geq p + 6$

11. $2r + 8 > 16 - 2r$ and $7r + 21 < r - 9$

12. $4j + 3 < j + 22$ and $j - 3 < 2j - 15$

13. $2(q - 4) \leq 3(q + 2)$ or $q - 8 \leq 4 - q$

14. $\frac{1}{2}w + 5 \geq w + 2 \geq \frac{1}{2}w + 9$

Lesson 7-5

1. **Food** For breakfast at Paul's Place, you can select one item from each of the following categories for $1.99.

Meat	Potato	Bread	Beverage
ham sausage bacon	hash browns country potatoes	toast muffin bagel biscuit	juice coffee

   a. What is the probability that a customer will have ham for breakfast?

   b. What is the probability of selecting a biscuit and hash browns?

   c. What is the probability of having a bagel, country potatoes, and coffee?

2. **Music** In order to raise money for a trip to the opera, the music club has set up a lottery using two-digit numbers. The first digit will be a numeral from 1 to 4. The second digit will be a numeral from 3 to 8. The first digit in Trudy's lottery number is 2, but she can't remember the second digit. If only one two-digit lottery number is drawn, and that number has 2 as the first digit, what is the probability that Trudy will win?

3. **Law** A three-judge panel is being used to settle a dispute. Both sides in the dispute have decided that a majority decision will be upheld. If each judge will render a favorable decision based on the evidence presented two-thirds of the time, what is the probability that the correct side will win the dispute?

Lesson 7-6 Solve each open sentence. Then graph the solution set.

1. $|a - 5| = -3$
2. $|g + 6| > 8$
3. $|t - 5| \leq 3$
4. $|a + 5| \geq 0$
5. $|14 - 2z| = 16$
6. $|y - 9| < 19$
7. $|2m - 5| > 13$
8. $|14 - w| \geq 20$
9. $|13 - 5y| = 8$
10. $|3p + 5| \leq 23$
11. $|6b - 12| \leq 36$
12. $|25 - 3x| < 5$
13. $|7 + 8x| > 39$
14. $|4c + 5| \geq 25$
15. $|4 - 5s| > 46$

Lesson 7-7

1. **Travel** Speeds of the fastest train runs in the U.S. and Canada are given below in miles per hour. Make a box-and-whisker plot of the data.

   93.5   82.5   89.3   83.8   81.8   86.8
   90.8   84.9   95.0   83.1   83.2   88.2

2. **Basketball** The numbers below represent the 20 highest points-scored-per-game averages for a season in the NBA from 1947 to 1990. Make a box-and-whisker plot of this data.

   35.0   33.5   32.5   37.9   31.2   38.4   34.5   34.7   32.9   44.8
   31.7   37.1   36.5   34.0   50.4   32.3   33.6   34.8   33.1   35.6

3. **Baseball** The stem-and-leaf plot at the right shows the number of home runs hit by the home run leaders in the National League in 1990.

   a. Find the median, upper quartile, lower quartile, and interquartile range.

   b. Are there any outliers? If so, name them.

   c. Draw a box-and-whisker plot of the data.

Stem	Leaf
2	2 3 3 4 4 4 4 5 5 6 7 7 8
3	2 2 3 3 5 7
4	0

$3|3 = 33$

Lesson 7-8 **Graph each inequality.**

1. $y \leq -2$

2. $x < 4$

3. $x + y < -2$

4. $x + y > -4$

5. $y > 4x - 1$

6. $3x + y > 1$

7. $3y - 2x \leq 2$

8. $x < y$

9. $3x + y > 4$

10. $5x - y < 5$

11. $-4x + 3y \geq 12$

12. $-x + 3y \leq 9$

13. $y > -3x + 7$

14. $3x + 8y \leq 4$

15. $5x - 2y \geq 6$

EXTRA PRACTICE

Write an algebraic expression for each verbal expression.

1. the sum of a number x and 13

2. the reciprocal of a number x squared

3. the cube of a number x decreased by 7

4. the product of 5 and a number x squared

Find the next two items for each pattern.

5.

6. 4, 7, 10, 13, . . .

7. 2, 5, 10, 17, . . .

Evaluate each expression.

8. $5^2 - 12$

9. $(0.5)^3 + 2 \cdot 7$

10. $\frac{2}{5}(16 - 9)$

Evaluate each expression when $a = 2$, $b = 0.5$, $c = 3$, and $d = \frac{4}{3}$.

11. $a^2b + c$

12. $(cd)^3$

13. $(a + d)c$

Solve each equation.

14. $y = (4.5 + 0.8) - 3.2$

15. $4^2 - 3(4 - 2) = y$

16. $\frac{2^3 - 1^3}{2 + 1} = y$

Name the property illustrated by each statement.

17. $a = a + 0$

18. $\frac{1}{a} \cdot a = 1$

19. If $a = b$, and $b = c$, then $a = c$.

20. $a(bc) = (ab)c$

21. $8(st) = 8(ts)$

22. $7y + 5x - 4y = 5x + 7y - 4y$

Simplify each expression.

23. $2m + 3m$

24. $4x + 2y - 2x + y$

25. $3(2a + b) - 1.5a - 1.5b$

Use the stem-and-leaf plot below to complete Exercises 26–27.

Stem	Leaf
1	1 4 6 8 8
2	0 3 3 3 5 7 9
3	0 0 2 6 $2|4 = 24$

26. List the set of data represented by the plot on the left.

27. Which number is used most frequently?

Sketch a reasonable graph for each situation.

28. A basketball is shot from the free throw line and falls through the net.

29. A slammer is dropped on a stack of pogs and bounces off.

30. An infant grows through adulthood.

Solve.

31. Travel   If a car travels at an average speed of 50 miles per hour, how far can the car travel in 5 hours? Use the distance formula $d = r \cdot t$.

32. Geometry   If the area of a circle is given by the equation $A = \pi r^2$, find the area if the radius is 4 in. (Use 3.14 for π.)

33. If a farmer wants to put a fence around his rectangular garden that measures 11 yards by 14 yards, how many yards of fencing will he need?

CHAPTER 2 TEST

Find each sum or difference.

1. $12 - 19$

2. $-21 + (-34)$

3. $1.654 + (-2.367)$

4. $-\dfrac{7}{16} - \dfrac{3}{8}$

5. $18b + 13xy - 46b$

6. $6.32 - (-7.41)$

7. $\dfrac{5}{8} + \left(-\dfrac{3}{16}\right) + \left(-\dfrac{3}{4}\right)$

8. $32y + (-73y)$

9. $\left| -28 + (-13) \right|$

10. $\begin{bmatrix} -8 & 5 \\ 2 & -3 \end{bmatrix} + \begin{bmatrix} 3 & 6 \\ -4 & -7 \end{bmatrix}$

11. $\begin{bmatrix} 1 & 0 \\ -9 & 3 \end{bmatrix} - \begin{bmatrix} 5 & 8 \\ -4 & 2 \end{bmatrix}$

Evaluate each expression.

12. $-x - 38$, if $x = -2$

13. $\left| -\dfrac{1}{2} + z \right|$, if $z = \dfrac{1}{4}$

14. $mp - k$, if $m = -12$, $p = 1.5$, and $k = -8$

15. $w^2 - 15$, if $w = 5$

Replace each ? with <, >, or = to make each sentence true.

16. -14 _?_ -15

17. $\dfrac{9}{20}$ _?_ $\dfrac{7}{15}$

18. -4.65 _?_ -4.45

Find a number between the given numbers.

19. $-\dfrac{2}{3}$ and $-\dfrac{9}{14}$

20. $\dfrac{4}{7}$ and $\dfrac{9}{4}$

21. $\dfrac{12}{7}$ and $\dfrac{15}{8}$

Simplify.

22. $\dfrac{8(-3)}{2}$

23. $(-5)(-2)(-2) - (-6)(-3)$

24. $\dfrac{2}{3}\left(\dfrac{1}{2}\right) - \left(-\dfrac{3}{2}\right)\left(-\dfrac{2}{3}\right)$

25. $\dfrac{70x - 30y}{-5}$

26. $\dfrac{7}{\frac{-2}{5}}$

27. $\dfrac{3}{4}(8x + 12y) - \dfrac{5}{7}(21x - 35y)$

Find each square root. Use a calculator if necessary. Round to the nearest hundredth if the result is not a whole number.

28. $\pm\sqrt{\dfrac{16}{81}}$

29. $\sqrt{40}$

30. $\sqrt{2.89}$

31. Statistics The height, in inches, of the students in a health class are 65, 63, 68, 66, 72, 61, 62, 63, 59, 58, 61, 74, 65, 63, 71, 60, 62, 63, 71, 70, 59, 66, 61, 62, 68, 69, 64, 63, 70, 61, 68, and 67.

    a. Make a line plot of the data on the heights of the students in health class.

    b. What was the most common height for the students in this class?

32. Define a variable and write an equation for the following problem. Do *not* solve.

    Each week for several weeks, Save-a-Buck stores reduced the price of a sofa by $18.25. The original price was $380.25. The final reduced price was $252.50. For how many weeks was the sofa on sale?

33. Translate $r = (a - b)^3$ into a verbal sentence.

Solve each equation. Then check your solution.

1. $-15 - k = 8$

2. $-1.2\,x = 7.2$

3. $\frac{3}{4}y = -27$

4. $\frac{t-7}{4} = 11$

5. $-12 = 7 - \frac{y}{3}$

6. $k - 16 = -21$

7. $t - (-3.4) = -5.3$

8. $-3(x + 5) = 8x + 18$

9. $2 - \frac{1}{4}(b - 12) = 9$

10. $\frac{r}{5} - 3 = \frac{2r}{5} + 16$

11. $25 - 7w = 46$

12. $-w + 11 = 4.6$

Find the mean, median, and mode for each set of data.

13. 67, 31, 15, 49, 31, 35, 42, 27

14.

Stem	Leaf
18	0 5 8
19	3 3 4 4
20	8 8 9
21	4 5 5 5 9 $19|4 = 194$

Define a variable, write an equation, and solve each problem.

15. The sum of two integers is -23. One integer is -84. Find the other integer.

16. Negative two thirds of a number is eight fifths. What is the number?

17. What number decreased by 37 is -65?

18. The measures of two angles of a triangle are $23°$ and $121°$. Find the measure of the third angle.

19. Find two consecutive odd integers whose sum is 172.

Solve each equation or formula for the variable specified.

20. $h = at - 0.25vt^2$, for a

21. $A = \frac{1}{2}(b + B)h$, for h

Solve.

22. Geometry   One of two supplementary angles measures $37°$ less than four times the other. Find the measure of each angle.

23. Grades   James had a cumulative score of 338 points after his first four tests, each worth 100 points. What is the least that he would need to score on his next chapter test in order to have a mean score of 82%?

24. Merchandise Stock   A store has 49 cartons of yogurt, some plain, some with blueberry flavoring. There are six times as many cartons of plain yogurt as flavored yogurt. How many cartons of plain yogurt are there?

25. Consumerism   Teri went to SuperValue to buy some groceries. Her total bill was $8.51. She bought 2 pounds of grapes at $0.99 per pound, a gallon of milk for $2.59, and three cans of orange juice. She used a coupon worth 50¢ for the three cans of orange juice. How much was each can of orange juice?

CHAPTER 4 TEST

Solve each proportion.

1. $\dfrac{2}{5} = \dfrac{x-3}{-2}$

2. $\dfrac{n}{4} = \dfrac{3.25}{52}$

3. $\dfrac{x-3}{x+5} = \dfrac{9}{11}$

4. $\dfrac{x+1}{-3} = \dfrac{x-4}{5}$

Solve.

5. Find 6.5% of 80.

6. 42 is what percent of 126?

7. 84 is 60% of what number?

8. What number decreased by 20% is 16?

9. 24 is what percent of 8?

10. 54 is 20% more than what number?

11. A price decreased from $60 to $45. Find the percent of decrease.

12. The price in dollars p minus a 15% discount is $3.40. Find p.

△ABC and △JKH are similar. For each set of measures given, find the measures of the remaining sides.

13. $c = 20, h = 15, k = 16, j = 12$

14. $c = 12, b = 13, a = 6, h = 10$

15. $k = 5, c = 6.5, b = 7.5, a = 4.5$

16. $h = 1\frac{1}{2}, c = 4\frac{1}{2}, k = 2\frac{1}{4}, a = 3$

Solve each right triangle. State the side lengths to the nearest tenth and the angle measures to the nearest degree.

17.

18.

19.

20.

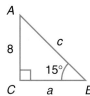

Solve.

21. Music   During a 20-song sequence on a popular radio station, 8 soft-rock, 7 hard-rock, and 5 rap songs are played at random. You tune in to the station.

   a. What is the probability that a hard-rock song is playing?

   b. What are the odds that a rap song is playing?

   c. What is the probability that a soft-rock song or a hard-rock song is playing?

22. Two buildings are separated by an alley. Joe is looking out of a window 60 feet above the ground in one building. He observes the measurement of the angle of depression of the base of the second building to be 50° and the angle of elevation of the top to be 40°. How high is the second building?

23. Finance   Gladys deposited an amount of money in the bank at 6.5% annual interest. After 6 months, she received $7.80 interest. How much money had Gladys deposited in the bank?

24. Stereo   Larry is buying a stereo that costs $399. Since he is an employee of the store, he receives a 15% discount. He also has to pay 6% sales tax. If the discount is computed first, what is the total cost of Larry's stereo?

25. Travel   At the same time Kris leaves Washington, D.C., for Detroit, Michigan, Amy leaves Detroit for Washington, D.C. The distance between the cities is 510 miles. Amy's average speed is 5 miles per hour faster than Kris's. How fast is Kris's average speed if they pass each other in 6 hours?

1. Graph $K(0, -5)$, $M(3, -5)$, and $N(-2, -3)$.

2. Name the quadrant in which $P(-5, 1)$ is located.

Express the relations shown in each table, mapping, or graph as a set of ordered pairs. Then state the domain, range, and inverse of the relation.

3.

x	f(x)
0	−1
2	4
4	5
6	10

4.

5.

Solve each equation if the domain is {−2, −1, 0, 2, 4}.

6. $y = -4x + 10$

7. $4 - 2x = 5y$

8. $-x + 3y = 1$

Graph each equation.

9. $x + 2y = -1$

10. $-3x = 5 - y$

11. $-4 = x - \frac{1}{2}y$

Determine whether each relation is a function.

12. $\{(2, 4), (3, 2), (4, 6), (5, 4)\}$

13. $8y = 7 + 3x$

If $f(x) = -2x + 5$ and $g(x) = x^2 - 4x + 1$, find each value.

14. $f\left(\frac{1}{2}\right)$

15. $g(-2)$

16. $-2g(3)$

Write an equation for each relation.

17.

x	1	2	3	4	7
y	3	8	13	18	33

18.

x	1	3	5	7	9	11	13
y	5	17	29	41	53	65	77

Solve.

19. **School** Art and Gina's test scores in algebra are given below.
    Art: 87, 54, 78, 97, 65, 82, 75, 68, 82, 73, 66, 75
    Gina: 70, 80, 57, 100, 73, 74, 65, 77, 91, 69, 71, 76
    a. Find the range and interquartile range for each set of scores.
    b. Identify any outliers.
    c. Which one had the more consistent scores?

20. **Sales** When you use Jay's Taxi Service, a two-mile trip costs $6.30, a five-mile trip costs $11.25, and a ten-mile trip costs $19.50. Write an equation to describe this relationship and use it to find the cost of a one-mile trip.

Determine the slope of the line that passes through each pair of points.

1. $(5, 8), (-3, 7)$

2. $(-2, 5), (2, 9)$

Find the slope and *x*- and *y*-intercepts of the graph of each equation.

3. $x - 8y = 3$

4. $3x - 2y = 9$

5. $y = 7$

Graph each equation.

6. $4x - 3y = 24$

7. $2x + 7y = 16$

8. $y = \frac{2}{3}x + 3$

Determine whether the graphs of each pair of equations are *parallel, perpendicular,* or *neither.*

9. $y = 4x - 11$
$2y + 1 = 8x$

10. $-7y = 4x + 14$
$7x - 4y = -12$

Write an equation in standard form of the line that satisfies the given conditions.

11. passes through $(2, 5)$ and $(8, -3)$

12. passes through $(-2, -1)$ and $(6, -4)$

13. has slope of 2 and *y*-intercept $= 3$

14. has *y*-intercept $= -4$ and passes through $(5, -3)$

15. slope $= \frac{3}{4}$ and passes through $(6, -2)$

16. parallel to $6x - y = 7$ and passes through $(-2, 8)$

Write an equation in slope-intercept form of the line that satisfies the given conditions.

17. passes through $(4, -2)$ and the origin

18. passes through $(-2, -5)$ and $(8, -3)$

19. passes through $(6, 4)$ with *y*-intercept $= -2$

20. slope $= -\frac{2}{3}$ and *y*-intercept $= 5$

21. slope $= 6$ and passes through $(-3, -4)$

22. perpendicular to $5x - 3y = 9$ and passes through the origin

23. parallel to $3x + 7y = 4$ and passes through $(5, -2)$

24. perpendicular to $x + 3y = 7$ and passes through $(5, 2)$

25. Find the coordinates of the other endpoint of segment *AB* given $A(-2, -7)$ and midpoint $M(6, -5)$.

26. The table below shows the number of students per computer in American classrooms since 1983.

School Year	Students per Computer	School Year	Students per Computer
1983–84	125	1989–90	22
1984–85	75	1990–91	20
1985–86	50	1991–92	18
1986–87	37	1992–93	16
1987–88	32	1993–94	14
1988–89	25	1994–95	12

a. Make a scatter plot of the data.

b. Describe the correlation between the variables.

c. Are these data best modeled by a line? Explain your answer.

Solve each inequality. Then check your solution.

1. $-12 \leq d + 7$

2. $7x < 6x - 11$

3. $z - 1 \geq 2z - 3$

4. $5 - 4b > -23$

5. $-\frac{2}{3}r \leq \frac{7}{12}$

6. $8y + 3 < 13y - 9$

7. $8(1 - 2z) \leq 25 + z$

8. $0.3(m + 4) > 0.5(m - 4)$

9. $\frac{2n - 3}{-7} \leq 5$

10. $y + \frac{5}{8} > \frac{11}{24}$

Solve each compound inequality. Then graph the solution set.

11. $x + 1 > -2$ and $3x < 6$

12. $2n + 1 \geq 15$ or $2n + 1 \leq -1$

13. $8 + 3t > 2$ and $-12 > 11t - 1$

14. $|2x - 1| < 5$

15. $|5 - 3b| \geq 1$

16. $|3 - 5y| < 8$

Define a variable, write an inequality, and solve each problem. Then check your solution.

17. Twice a number subtracted from 12 is no less than the number increased by 27.

18. Seven less than twice a number is between 71 and 83.

19. The product of two integers is no less than 30. One of the integers is 6. What is the other integer?

20. The average of four consecutive odd integers is less than 20. What are the greatest integers that satisfy this condition?

Graph each inequality.

21. $y \geq 5x + 1$

22. $x - 2y > 8$

23. $3x - 2y < 6$

Solve.

24. **Business**   Two men and three women are each waiting for a job interview. There is only enough time to interview two people before lunch. Two people are chosen at random.

   a. What is the probability that both people are women?

   b. What is the probability that at least one person is a woman?

   c. Which is more likely, one of the people is a woman and the other is a man, or both people are either men or women?

25. **Fire Safety**   The city council of McBride is investigating the efficiency of the fire department. The time taken by the fire department to respond to a fire alarm was surveyed. It was found that the response times in minutes for 17 alarms were as follows.

   1, 3, 2, 2, 1, 9, 4, 6, 1, 10, 1, 4, 5, 10, 1, 3, 6

   a. Draw a box-and-whisker plot of the data.

   b. Between what two values of the data is the middle 50% of the data?

GLOSSARY

This glossary contains terms from Volumes One and Two. Pages 1–447 are contained in Volume One. Pages 382–753 are contained in Volume Two.

A

absolute value (85) The absolute value of a number is its distance from zero on a number line.

acute triangle (165) In an acute triangle, all of the angles measure less than 90 degrees.

adding integers (86)

1. To add integers with the *same* sign, add their absolute values. Give the result the same sign as the integers.

2. To add integers with *different* signs, subtract the lesser absolute value from the greater absolute value. Give the result the same sign as the integer with the greater absolute value.

addition property for inequality (385) For all numbers a, b, and c, the following are true:
1. if $a > b$, then $a + c > b + c$.

2. if $a < b$, then $a + c < b + c$.

addition property of equality (144) For any numbers a, b, and c, if $a = b$, then $a + c = b + c$.

additive identity (37) For any number a, $a + 0 = 0 + a = a$.

additive inverse property (87) For any number a, $a + (-a) = 0$.

algebraic expression (6) An expression consisting of one or more numbers and variables along with one or more arithmetic operations.

angle of depression (208) An angle of depression is formed by a horizontal line and a line of sight below it.

angle of elevation (208) An angle of elevation is formed by a horizontal line and a line of sight above it.

associative property (51) For any numbers a, b, and c, $(a + b) + c = a + (b + c)$ and $(ab) c = a(bc)$.

axes (254) Two perpendicular number lines that are used to locate points on a coordinate plane.

axis of symmetry (612) The equation of the axis of symmetry for the graph of $y = ax^2 + bx + c$, where $a \neq 0$, is $x = -\frac{b}{2a}$.

B

back-to-back stem-and-leaf plot (27) A back-to-back stem-and-leaf plot is used to compare two sets of data. The same stem is used for the leaves of both plots.

base **1.** (7) In an expression of the form x^n, the base is x. **2.** (215) The number that is divided into the percentage in the percent proportion.

best-fit line (341) A line drawn on a scatter plot that passes close to most of the data points.

binomial (514) The sum of two monomials.

boundary (437) A boundary of an inequality is a line that separates the coordinate plane into half-planes.

box-and-whisker plot (427) A type of diagram or graph that shows the quartiles and extreme values of data.

C

coefficient (47) The numerical factor in a term.

commutative property (51) For any numbers a and b, $a + b = b + a$ and $ab = ba$.

comparison property (94) For any two numbers a and b, exactly one of the following sentences is true.
$$a < b \qquad a = b \qquad a > b$$

complementary angles (163) Two angles are complementary if the sum of their measures is 90 degrees.

complete graph (278) A complete graph shows the origin, the points at which the graph crosses the x- and y-axes, and other important characteristics of the graph.

completeness property (121) Each real number corresponds to exactly one point on the number line. Each point on the number line corresponds to exactly one real number.

completeness property for points in the plane (256)

1. Exactly one point in the plane is named by a given ordered pair of numbers.

2. Exactly one ordered pair of numbers names a given point in the plane.

completing the square (742, 743)   To add a constant term to a binomial of the form $x^2 + bx$ so that the resulting trinomial is a perfect square.

complex fraction (114)   If a fraction has one or more fractions in the numerator or denominator, it is called a complex fraction.

composite numbers (558)   A whole number, greater than 1, that is not prime.

compound event (414)   A compound event consists of two or more simple events.

compound inequality (405)   Two inequalities connected by *and* or *or*.

congruent angles (164)   Angles that have the same measure.

conjugates (722)   Two binomials of the form $a\sqrt{b} + c\sqrt{d}$ and $a\sqrt{b} - c\sqrt{d}$.

consecutive integers (158)   Consecutive integers are integers in counting order.

consistent (455)   A system of equations is said to be consistent when it has at least one ordered pair that satisfies both equations.

constant of variation (239)   The number k in equations of the form $y = kx$ and $xy = k$.

constants (496)   Monomials that are real numbers.

coordinate (73)   The number that corresponds to a point on a number line.

coordinate plane (254)   The plane containing the x- and y-axes.

corresponding angles (201)   Matching angles in similar triangles, which have equal measures.

corresponding sides (201)   The sides opposite the corresponding angles in similar triangles.

cosine (206)   In a right triangle with acute angle A, the cosine of angle $A =$

$$\frac{\text{measure of leg adjacent to angle } A}{\text{measure of hypotenuse}}.$$

cross products (94)   When two fractions are compared, the cross products are the products of the terms on the diagonals.

---

data (25)   Numerical information.

defining the variable (127)   Choosing a variable to represent one of the unspecified numbers in a problem.

degree   1. (515) The degree of a monomial is the sum of the exponents of its variables. **2.** (516) The degree of a polynomial is the degree of the term of the greatest degree.

density property (96)   Between every pair of distinct rational numbers, there are infinitely many rational numbers.

dependent (456)   A system of equations that has an infinite number of solutions.

dependent variable (58)   The variable in a function whose value is determined by the independent variable.

difference of squares (545)   Two perfect squares separated by a subtraction sign, $a^2 - b^2 = (a + b)(a - b)$.

dimensional analysis (174)   The process of carrying units throughout a computation.

direct variation (239)   A direct variation is described by an equation of the form $y = kx$, where $k \neq 0$.

discrete mathematics (88)   A branch of mathematics that deals with finite or discontinuous quantities.

discriminant (632)   In the quadratic formula, the expression $b^2 - 4ac$.

distance formula (737)   The distance d between any two points with coordinates (x_1, y_1) and (x_2, y_2) is given by the following formula.

$$d = \sqrt{(x_2 - x_1)^2 + (y_2 - y_1)^2}$$

distributive property (46)   For any numbers a, b, and c:

1. $a(b + c) = ab + ac$ and $(b + c)a = ba + ca$.

2. $a(b - c) = ab - ac$ and $(b - c)a = ba - ca$.

dividing rational numbers (112)   The quotient of two rational numbers having the same sign is positive. The quotient of two rational numbers having different signs is negative.

division property for inequality (393)   For all numbers a, b, and c, the following are true:

1. If c is positive and $a < b$, then $\frac{a}{c} < \frac{b}{c}$, and if c is positive and $a > b$, then $\frac{a}{c} > \frac{b}{c}$.

2. If c is negative and $a < b$, then $\frac{a}{c} > \frac{b}{c}$, and if c is negative and $a > b$, then $\frac{a}{c} < \frac{b}{c}$.

division property of equality (151)   For any numbers a, b, and c, with $c \neq 0$, if $a = b$, then $\frac{a}{c} = \frac{b}{c}$.

domain (263)   The set of all first coordinates from the ordered pairs in a relation.

draw a diagram (406)   A problem-solving strategy that is often used as an organizational tool.

E

element (33)   A member of a set.

elimination (469)   The elimination method of solving a system of equations is a method that uses addition or subtraction to eliminate one of the variables to solve for the other variable.

equally likely (229)   Outcomes that have an equal chance of occurring.

equation (33)   A mathematical sentence that contains an equals sign, $=$.

equation in two variables (271)   An equation in two variables contains two unknown values.

equilateral triangle (164)   A triangle in which all the sides have the same length and all the angles have the same measure.

equivalent equation (144)   Equations that have the same solution.

equivalent expressions (47)   Expressions that denote the same number.

evaluate (8)   To find the value of an expression when the values of the variables are known.

excluded value (660)   A value is excluded from the domain of a variable because if that value were substituted for the variable, the result would have a denominator of zero.

exponent (7)   In an expression of the form x^n, the exponent is n.

exponential function (634, 635)   A function that can be described by an equation of the form $y = a^x$, where $a > 0$ and $a \neq 1$.

extraneous solutions (697)   Solutions derived from an equation that are not solutions of the original equation.

extreme values (427)   The least value and the greatest value in a set of data.

extremes (196)   *See* proportion.

F

factored form (559)   A monomial is written in factored form when it is expressed as the product of prime numbers and variables where no variable has an exponent greater than 1.

factoring (564)   To express a polynomial as the product of monomials and polynomials.

factoring by grouping (567)   A method of factoring polynomials with four or more terms.

factors (6)   In a multiplication expression, the quantities being multiplied are called factors.

family of graphs (354)   A family of graphs includes graphs and equations of graphs that have at least one characteristic in common.

FOIL method (537)   To multiply two binomials, find the sum of the products of

F   the first terms,
O   the outside terms,
I   the inside terms, and
L   the last terms.

formula (128)   An equation that states a rule for the relationship between certain quantities.

function   **1.** (56) A relationship between input and output in which the output depends on the input.   **2.** (287) A relation in which each element of the domain is paired with exactly one element of the range.

functional notation (289)   In functional notation, the equation $y = x + 5$ is written as $f(x) = x + 5$.

G

general equation for exponential decay (644)   The general equation for exponential decay is represented by the formula $A = C(1 - r)^t$.

general equation for exponential growth (643)   The general equation for exponential growth is represented by the formula $A = C(1 + r)^t$.

graph (73, 255)   To draw, or plot, the points named by certain numbers or ordered pairs on a number line or coordinate plane, respectively.

greatest common factor (GCF) (559)   The greatest common factor of two or more integers is the greatest number that is a factor of all the integers.

guess and check (574)   A problem-solving strategy in which several values or combinations of values are tried in order to find a solution to a problem.

half-life (638)   The half-life of an element is defined as the time it takes for one-half a quantity of a radioactive element to decay.

half-plane (437)   The region of a graph on one side of a boundary is called a half-plane.

horizontal axis (56)   The horizontal line in a graph that represents the independent variable.

hypotenuse (206)   The side of a right triangle opposite the right angle.

identify subgoals (745)   A problem-solving strategy that uses a series of small steps, or subgoals.

identity (170)   An equation that is true for every value of the variable.

inconsistent (456)   A system of equations is said to be inconsistent when it has no ordered pair that satisfies both equations.

independent (456)   A system of equations is said to be independent if the system has exactly one solution.

independent variable (58)   The variable in a function whose value is subject to choice is the independent variable. The independent variable affects the value of the dependent variable.

inequality (33)   A mathematical sentence having the symbols $<$, $\leq$, $>$, or $\geq$.

integers (73)   The set of numbers represented as $\{..., -3, -2, -1, 0, 1, 2, 3, ...\}$.

interquartile range (304, 306)   The difference between the upper quartile and the lower quartile of a set of data. It represents the middle half, or 50%, of the data in the set.

intersection (406)   The intersection of two sets A and B is the set of elements common to both A and B.

inverse of a relation (264)   The inverse of any relation is obtained by switching the coordinates in each ordered pair.

inverse variation (241)   An inverse variation is described by an equation of the form $xy = k$, where $k \neq 0$.

irrational numbers (120)   A number that cannot be expressed in the form $\frac{a}{b}$, where a and b are integers and $b \neq 0$.

isosceles triangle (164)   In an isosceles triangle, at least two angles have the same measure and at least two sides have the same length.

least common denominator (LCD) (686)   The least common denominator is the least common multiple of the denominators of two or more fractions.

least common multiple (LCM) (685)   The least common multiple of two or more integers is the least positive integer that is divisible by each of the integers.

legs (206)   The sides of a right triangle that are not the hypotenuse.

like terms (47)   Terms that contain the same variables, with corresponding variables with the same power.

linear equation (280)   An equation whose graph is a line.

linear function (278)   An equation whose graph is a nonvertical line.

line plot (78)   Numerical data displayed on a number line.

look for a pattern (13, 497, 637)   A problem-solving strategy often involving the use of tables to organize information so that a pattern may be determined.

lower quartile (306)   The lower quartile divides the lower half of a set of data into two equal parts.

make an organized list (685)   A problem-solving strategy that uses an organized list to arrange and evaluate data in order to determine a solution.

mapping (263)   A mapping pairs one element in the domain with one element in the range.

matrix (88)   A matrix is a rectangular arrangement of elements in rows and columns.

maximum (611)   The highest point on the graph of a curve, such as a the vertex of parabola that opens downward.

mean (178)   The mean of a set of data is the sum of the numbers in the set divided by the number of numbers in the set.

means (196) *See* proportion.

measures of central tendency (178) Numbers known as measures of central tendency are often used to describe sets of data because they represent a centralized, or middle, value.

measures of variation (306) Measures of variation are used to describe the distribution of data.

median (178) The median is the middle number of a set of data when the numbers are arranged in numerical order.

midpoint (369) A point that is halfway between the endpoints of a segment.

minimum (611) The lowest point on the graph of a curve, such as a the vertex of parabola that opens upward.

mixed expression (690) An algebraic expression that contains a monomial and a rational expression.

mode (178) The mode of a set of data is the number that occurs most often in the set.

monomial (496) A monomial is a number, a variable, or a product of a number and one or more variables.

multi-step equations (157) Multi-step equations are equations that need more than one operation to solve them.

multiplicative identity (38) For any number a, $a \cdot 1 = 1 \cdot a = a$.

multiplicative inverse (38) For every nonzero number $\frac{a}{b}$, where $a, b \neq 0$, there is exactly one number $\frac{b}{a}$ such that $a \cdot b = 1$.

multiplication property for inequality (393) For all numbers a, b, and c, the following are true.

1. If c is positive and $a < b$, then $ac < bc$, $c \neq 0$, and if c is positive and $a > b$, then $ac > bc$, $c \neq 0$.

2. If c is negative and $a < b$, then $ac > bc$, $c \neq 0$, and if c is negative and $a > b$, then $ac < bc$, $c \neq 0$.

multiplicative property of -1 (107) The product of any number and -1 is its additive inverse.

$$-1(a) = -a \text{ and } a(-1) = -a$$

multiplicative property of equality (150) For any numbers a, b, and c, if $a = b$, then $a \cdot c = b \cdot c$.

multiplicative property of zero (38) For any number a, $a \cdot 0 = 0 \cdot a = 0$.

negative correlation (340) There is a negative correlation between x and y if the values are related in opposite ways.

negative exponent (503) For any nonzero number a and any integer n, $a^{-n} = \frac{1}{a^n}$.

negative number (73) Any number that is less than zero.

number line (72) A line with equal distances marked off to represent numbers.

number theory (158) The study of numbers and the relationships between them.

obtuse triangle (165) An obtuse triangle has one angle with measure greater than 90 degrees.

odds (229) The odds of an event occurring is the ratio of the number of ways the event can occur (successes) to the number of ways the event cannot occur (failures).

open sentences (32) Mathematical statements with one or more variables, or unknown numbers.

opposites (87) The opposite of a number is its additive inverse.

order of operations (19)

1. Simplify the expressions inside grouping symbols, such as parentheses, brackets, and braces, and as indicated by fraction bars.

2. Evaluate all powers.

3. Do all multiplications and divisions from left to right.

4. Do all additions and subtractions from left to right.

ordered pair (57) Pairs of numbers used to locate points in the coordinate plane.

organize data (464) Organizing data is useful before solving a problem. Some ways to organize data are to use tables, charts, different types of graphs, or diagrams.

origin (254) The point of intersection of the two axes in the coordinate plane.

outcomes (413) Outcomes are all possible combinations of a counting problem.

outlier (307) In a set of data, a value that is much greater or much less than the rest of the data can be called a outlier.

parabola (610, 611)   The general shape of the graph of a quadratic function.

parallel lines (362)   Lines in the plane that never intersect. Nonvertical parallel lines have the same slope.

parallelogram (362)   A quadrilateral in which opposite sides are parallel.

parent graph (354)   The simplest of the graphs in a family of graphs

percent (215)   A percent is a ratio that compares a number to 100.

percentage (215)   The number that is divided by the base in a percent proportion.

percent of decrease (222)   The ratio of an amount of decrease to the previous amount, expressed as a percent.

percent of increase (222)   The ratio of an amount of increase to the previous amount, expressed as a percent.

percent proportion (215) $\dfrac{\text{Percentage}}{\text{Base}} = \dfrac{r}{100}$

perfect square (119)   A rational number whose square root is a rational number.

perfect square trinomial (587)   A trinomial which, when factored, has the form $(a + b)^2 = (a + b)(a + b)$ or $(a - b)^2 = (a - b)(a - b)$.

perpendicular lines (362)   Lines that meet to form right angles.

point-slope form (333)   For any point (x_1, y_1) on a nonvertical line having slope m, the point-slope form of a linear equation is as follows:
$$y - y_1 = m(x - x_1).$$

polynomial (513, 514)   A polynomial is a monomial or a sum of monomials.

positive correlation (340)   There is a positive correlation between x and y if the values are related in the same way.

power (7)   An expression of the form x^n is known as a power.

power of a monomial (498)   For any numbers a and b, and any integers m, n, and p,
$$(a^m b^n)^p = a^{mp} b^{np}.$$

power of a power (498)   For any number a, and all integers m and n, $(a^m)^n = a^{mn}$.

power of a product (498)   For all numbers a and b, and any integer m, $(ab)^m = a^m b^m$.

prime factorization (558)   A whole number expressed as a product of factors that are all prime numbers.

prime number (558)   A prime number is a whole number, greater than 1, whose only factors are 1 and itself.

prime polynomial (577)   A polynomial that cannot be written as a product of two polynomials with integral coefficients is called a prime polynomial.

principal square root (119)   The nonnegative square root of an expression.

probability (228)   The ratio that tells how likely it is that an event will take place.
$$P(\text{event}) = \frac{\text{number of favorable outcomes}}{\text{total number of possible outcomes}}$$

problem-solving plan (126)
  1. Explore the problem.
  2. Plan the solution.
  3. Solve the problem.
  4. Examine the solution.

product (6)   The result of multiplication.

product of powers (497)   For any number a, and all integers m and n, $a^m \cdot a^n = a^{m + n}$.

product property of square roots (719)   For any number a and b, where $a \geq 0$ and $b \geq 0$, $\sqrt{ab} = \sqrt{a} \cdot \sqrt{b}$.

proportion (195)   In a proportion, the product of the extremes is equal to the product of the means. If $\dfrac{a}{b} = \dfrac{c}{d}$, then $ad = bc$.

Pythagorean theorem (713)   If a and b are the measures of the legs of a right triangle and c is the measure of the hypotenuse, then $c^2 = a^2 + b^2$.

Pythagorean triple (583)   Three whole numbers a, b, and c such that $a^2 + b^2 = c^2$.

quadrant (254)   One of the four regions into which the x- and y-axes separate the coordinate plane.

quadratic equation (620)   A quadratic equation is one in which the value of the related quadratic function is 0.

quadratic formula (628)   The roots of a quadratic equation in the form $ax^2 + bx + c = 0$, where $a \neq 0$, are given by the formula $x = \dfrac{-b \pm \sqrt{b^2 - 4ac}}{2a}$.

quadratic function (610, 611)   A quadratic function is a function that can be described by an equation of the form $y = ax^2 + bx + c$, where $a \neq 0$.

quartiles (304, 306)   In a set of data, the quartiles are values that divide the data into four equal parts.

quotient of powers (501)   For all integers m and n and any nonzero number a, $\dfrac{a^m}{a^n} = a^{m-n}$.

quotient property of square roots (721)   For any numbers a and b, where $a > 0$ and $b > 0$, $\sqrt{\dfrac{a}{b}} = \dfrac{\sqrt{a}}{\sqrt{b}}$.

R

radical equations (732)   Equations that contain radicals with variables in the radicand.

radical sign (119)   The symbol $\sqrt{}$, indicating the principal or nonnegative root of an expression.

radicand (719)   The radicand is the expression under the radical sign.

random (229)   When an outcome is chosen without any preference, the outcome occurs at random.

range   **1.** (263) The set of all second coordinates from the ordered pairs in the relation. **2.** (306) The difference between the greatest and the least values of a set of data.

rate   **1.** (197) The ratio of two measurements having different units of measure.   **2.** (215) In the percent proportion, the rate is the fraction with a denominator of 100.

ratio (195)   A ratio is a comparison of two numbers by division.

rational equation (696)   A rational equation is an equation that contains rational expressions.

rational expression (660)   A rational expression is an algebraic fraction whose numerator and denominator are polynomials.

rational numbers (93)   A rational number is a number that can be expressed in the form $\dfrac{a}{b}$, where a and b are integers and $b \neq 0$.

rationalizing the denominator (722)   Rationalizing the denominator of a radical expression is a method used to remove or eliminate the radicals from the denominator of a fraction.

real numbers (121)   The set of rational numbers and the set of irrational numbers together form the set of real numbers.

reciprocal (38)   The multiplicative inverse of a number.

reflexive property of equality (39)   For any number a, $a = a$.

regression line (342)   The most accurate best-fit line for a set of data, and can be determined with a graphing calculator or computer.

relation (260, 263)   A relation is a set of ordered pairs.

replacement set (33)   A set of numbers from which replacements for a variable may be chosen.

right triangle (165)   A right triangle has one angle with a measure of 90 degrees.

rise (325)   The vertical change in a line.

roots (620)   The solutions of a quadratic equation.

run (325)   The horizontal change in a line.

S

scalar multiplication (108)   In scalar multiplication, each element of a matrix is multiplied by a constant.

scale (197)   A ratio called a scale is used when making a model to represent something that is too large or too small to be conveniently drawn at actual size.

scatter plot (339)   In a scatter plot, the two sets of data are plotted as ordered pairs in the coordinate plane.

scientific notation (506)   A number is expressed in scientific notation when it is in the form $a \times 10^n$, where $1 \leq a < 10$ and n is an integer.

set (33)   A collection of objects or numbers.

set-builder notation (385)   A notation used to describe the members of a set. For example, $\{y \mid y < 17\}$ represents the set of all numbers y such that y is less than 17.

sequence (13)   A set of numbers in a specific order.

similar triangles (201) If two triangles are similar, the measures of their corresponding sides are proportional, and the measures of their corresponding angles are equal.

simple events (414) A single event in a probability problem.

simple interest (217) The amount paid or earned for the use of money. The formula $I = prt$ is used to solve simple interest problems.

simplest form (47) An expression is in simplest form when it is replaced by an equivalent expression having no like terms and no parentheses.

simplest radical form (723) A radical expression is in simplest radical form when the following three conditions have been met.
  1. No radicands have perfect square factors other than one.
  2. No radicands contain fractions.
  3. No radicals appear in the denominator of a fraction.

sine (206) In a right triangle with acute angle A, sine of angle $A = \dfrac{\text{measure of leg opposite angle } A}{\text{measure of the hypotenuse}}$.

slope (324, 325) The ratio of the rise to the run as you move from one point to another along a line.

slope-intercept form (347) An equation of the form $y = mx + b$, where m is the slope and b is the y-intercept of a given line.

solution (32) A replacement for the variable in an open sentence that results in a true sentence.

solution of an equation in two variables (271) If a true statement results when the numbers in an ordered pair are substituted into an equation in two variables, then the ordered pair is a solution of the equation.

solution set (33) The set of all replacements for the variable in an open sentence that result in a true sentence.

solve an equation (145) To solve an equation means to isolate the variable having a coefficient of 1 on one side of the equation.

solving an open sentence (32) Finding a replacement for the variable that results in a true sentence.

solving a triangle (208) Finding the measures of all sides and angles of a right triangle.

square of a difference (544) If a and b are any numbers, $(a - b)^2 = (a - b)(a - b) = a^2 - 2ab + b^2$.

square of a sum (543) If a and b are any numbers, $(a + b)^2 = (a + b)(a + b) = a^2 + 2ab + b^2$.

square root (118, 119) One of two identical factors of a number.

standard form (333) The standard form of a linear equation is $Ax + By = C$, where A, B, and C are integers, $A \geq 0$, and A and B are not both zero.

statistics (25) A branch of mathematics concerned with methods of collecting, organizing, and interpreting data.

stem-and-leaf plot (26) In a stem-and-leaf plot, each piece of data is separated into two numbers that are used to form a stem and a leaf. The data are organized into two columns. The column on the left contains the stem and the column on the right contains the leaves.

substitution (462) The substitution method of solving a system of equations is a method that uses substitution of one equation into the other equation to solve for the other variable.

substitution property of equality (39) If $a = b$, then a may be replaced by b in any expression.

subtracting integers (87) To subtract a number, add its additive inverse. For any numbers a and b, $a - b = a + (-b)$.

subtraction property for inequality (385) For all numbers a, b, and c, the following are true:
  1. if $a > b$, then $a - c > b - c$.
  2. if $a < b$, then $a - c < b - c$.

subtraction property of equality (146) For any numbers a, b, and c, if $a = b$, then $a - c = b - c$.

supplementary angles (162) Two angles are supplementary if the sum of their measures is 180 degrees.

symmetric property of equality (39) For any numbers a and b, if $a = b$, then $b = a$.

symmetry (612) Symmetrical figures are those in which the figure can be folded and each half matches the other exactly.

system of equations (455) A set of equations with the same variables.

system of inequalities (482) A set of inequalities with the same variables.

tangent (206)   In a right triangle, the tangent of angle $A = \dfrac{\text{measure of leg opposite angle } A}{\text{measure of leg adjacent to angle } A}$.

term   **1.** (13) A number in a sequence.   **2.** (47) A number, a variable, or a product or quotient of numbers and variables.

transitive property of equality (39)   For any numbers a, b, and c, if $a = b$ and $b = c$, then $a = c$.

tree diagram (413)   A tree diagram is a diagram used to show the total number of possible outcomes.

triangle (163)   A triangle is a polygon with three sides and three angles.

trigonometric ratios (206) $m \angle C = 90$

$$\sin A = \frac{a}{c} \qquad \cos A = \frac{b}{c} \qquad \tan A = \frac{a}{b}$$

trinomials (514)   A trinomial is the sum of three monomials.

uniform motion (235)   When an object moves at a constant speed, or rate, it is said to be in uniform motion.

union (408)   The union of two sets A and B is the set of elements contained in both A or B.

unique factorization theorem (558)   The prime factorization of every number is unique except for the order in which the factors are written.

unit cost (95)   The cost of one unit of something.

upper quartile (306)   The upper quartile divides the upper half of a set of data into two equal parts.

use a model (339)   A problem-solving strategy that uses models, or simulations, of mathematical situations that are difficult to solve directly.

use a table (255)   A problem-solving strategy that uses tables to organize and solve problems.

variable (6)   Variables are symbols that are used to represent unspecified numbers.

Venn diagrams (73)   Venn diagrams are diagrams that use circles or ovals inside a rectangle to show relationships of sets.

vertex (610, 611)   The maximum or minimum point of a parabola.

vertical axis (56)   The vertical line in a graph that represents the dependent variable.

vertical line test (289)   If any vertical line passes through no more than one point of the graph of a relation, then the relation is a function.

weighted average (233)   The weighted average M of a set of data is the sum of the product of each number in the set and its weight divided by the sum of all the weights.

whiskers (428)   The whiskers of a box-and-whisker plot are the segments that are drawn from the lower quartile to the least value and from the upper quartile to the greatest value.

whole numbers (72)   The set of whole numbers is represented by {0, 1, 2, 3, ...}.

work backward (156)   A problem-solving strategy that uses inverse operations to determine an original value.

x-axis (254)   The horizontal number line.

x-coordinate (254)   The first number in an ordered pair.

x-intercept (346)   The coordinate at which a graph intersects the x-axis.

y-axis (254)   The vertical number line.

y-coordinate (254)   The second number in an ordered pair.

y-intercept (346)   The coordinate at which a graph intersects the y-axis.

Z

zero exponent (502)   For any nonzero number a, $a^0 = 1$.

zero product property (594)   For all numbers a and b, if $ab = 0$, then $a = 0$, $b = 0$, or both a and b equal 0.

zeros (620)   The zeros of a function are the roots, or x-intercepts, of the function.

This glossary contains terms from Volumes One and Two. Pages 1–447 are contained in Volume One. Pages 382–753 are contained in Volume Two.

absolute value/valor absoluto (85)   El valor absoluto de un número equivale al número de unidades que dicho número dista de cero en la recta numérica.

acute triangle/triángulo agudo (165)   En un triángulo agudo, todos los ángulos miden menos de 90°.

adding integers/suma de enteros (86)

1. Para sumar enteros del *mismo* signo, suma los valores absolutos de los números. Da al resultado el mismo signo de los números.

2. Para sumar enteros de *distinto* signo, resta el valor menor del valor mayor. Da al resultado el mismo signo que el número con el mayor valor absoluto.

addition property of equality/propiedad de adición de la igualdad (144)   Para cualquiera de los números a, b y c, si $a = b$, entonces $a + c = b + c$.

addition property of inequality/propiedad de adición de la desigualdad (385)   Para todos los números a, b y c:

1. si $a > b$, entonces $a + c > b + c$;

2. si $a < b$, entonces $a + c < b + c$.

additive identity/identidad aditiva (37)   Para cualquier número a, $a + 0 = 0 + a = a$.

additive inverse property/propiedad del inverso de la adición (87)   Para cualquier número a, $a + (-a) = 0$.

algebraic expression/expresión algebraica (6)   Una expresión que consiste de uno o más números y variables, además de una o más operaciones aritméticas.

angle of depression/ángulo de depresión (208)   El ángulo de depresión se forma por una línea horizontal y una línea visual por debajo de la misma.

angle of elevation/ángulo de elevación (208)   Un ángulo de elevación se forma por una línea horizontal y una línea visual por encima de la misma.

associative property/propiedad asociativa (51)   Para cualquiera de los números a, b, y c, $(a + b) + c = a + (b + c)$ y $(ab)c = a(bc)$.

axes/ejes (254)   Dos rectas numéricas perpendiculares que se usan para ubicar puntos en un plano de coordenadas.

axis of symmetry/eje de simetría (612)   La ecuación para el eje de simetría de la gráfica $y = ax^2 + bx + c$, en la cual $a \neq 0$ y $x = -\frac{b}{2a}$.

back-to-back stem-and-leaf plot/diagrama de tallo y hojas consecutivo (27)   Un diagrama de tallo y hojas consecutivo se usa para comparar dos conjuntos de datos. El mismo tallo se usa para las hojas de ambos diagramas.

base/base   1. (7) En una expresión de la forma x^n, la base es x.   **2.** (215) El número que se divide entre el porcentaje en una proporción de porcentaje.

best-fit line/línea de mejor encaje (341)   Es una línea que se dibuja en un diagrama de dispersión y que pasa cerca de la mayoría de los puntos de datos.

binomial/binomio (514)   La suma de dos monomios.

boundary/frontera (437)   La frontera de una desigualdad es una línea que separa el plano de coordenadas en dos mitades de planos.

box-and-whisker plot/diagrama de caja y patillas (427)   Un tipo de diagrama o gráfica que muestra los valores cuartílicos y los extremos de los datos.

coefficient/coeficiente (47)   El factor numérico de un término.

commutative property/propiedad conmutativa (51)   Para cualquiera de los números a y b, $a + b = b + a$ y $ab = ba$.

comparison property/propiedad de comparación (94)   Para cualquier par de números a y b, exactamente una de las siguientes operaciones es válida.

$$a < b \quad a = b \quad a > b$$

complementary angles/ángulos complementarios (163)   Dos ángulos son complementarios si la suma de sus medidas es 90 grados.

complete graph/gráfica completa (278)   Una gráfica completa muestra el origen, los puntos en donde la gráfica cruza el eje de coordenadas x y el eje de coordenadas y, además de otras características importantes en la gráfica.

completeness property/propiedad del completo (121)   Cada número real corresponde exactamente a un punto en la recta numérica. Cada punto en la recta numérica corresponde exactamente a un número real.

completeness property for points in the plane/propiedad del completo de los puntos en el plano (256)

1. Exactamente un punto en el plano es nombrado por un par ordenado de números dado.

2. Exactamente un par ordenado de números nombra un punto dado en el plano.

completing the square/completar el cuadrado (742, 743) Sumar un término constante a un binomio de la forma $x^2 = bx$, de modo que el trinomio resultante sea un cuadrado perfecto.

complex fraction/fracción compleja (114) Si una fracción tiene uno o más fracciones en el numerador o denominador, entonces se llama una fracción compleja.

composite number/número compuesto (558) Un número entero, mayor de 1, que no es un número primo.

compound event/evento compuesto (414) Un evento compuesto consiste de dos o más eventos simples.

compound inequality/desigualdad compuesta (405) Dos desigualdades conectadas por *y* u *o*.

congruent angles/ángulos congruentes (164) Ángulos que tienen la misma medida.

conjugates/conjugados (722) Dos binomios de la forma $a\sqrt{b} + c\sqrt{d}$ y $a\sqrt{b} - c\sqrt{d}$.

consecutive integers/números enteros consecutivos (158) Son los números enteros en el orden de contar.

consistent/consistente (455) Se dice que un sistema de ecuaciones es consistente cuando tiene por lo menos un par ordenado que satisface ambas ecuaciones.

constant of variation/constante de variación (239) El número *k* en ecuaciones de la forma $y = kx$ y $xy = k$.

constants/constantes (469) Los monomios que son números reales.

coordinate/coordenada (73) El número que corresponde a un punto en una recta numérica.

coordinate plane/plano de coordenadas (254) El plano que contiene el eje de coordenadas *x* y el eje de coordenadas *y*.

corresponding angles/ángulos correspondientes (201) Ángulos que encajan en triángulos semejantes y que tienen medidas iguales.

corresponding sides/lados correspondientes (201) Los lados opuestos a los ángulos correspondientes en triángulo semejantes.

cosine/coseno (206) En un triángulo rectángulo con ángulo agudo *A*, el coseno del

$$\text{ángulo } A = \frac{\text{medida del cateto adyacente al ángulo } A}{\text{medida de la hipotenusa}}.$$

cross products/productos cruzados (94) Cuando se comparan dos fracciones, los productos cruzados son los productos de los términos en diagonales.

D

data/datos (25) La información numérica.

defining the variable/definir la variable (127) El escoger una variable para representar uno de los números no especificados en un problema.

degree/grado **1.** (515) El grado de un monomio es la suma de los exponentes de sus variables. **2.** (516) El grado de un polinomio es el grado del término con el grado más alto.

density property/propiedad de la densidad (96) Entre cada par de números racionales distintos, existe una infinidad de números racionales.

dependent/dependiente (456) Un sistema de ecuaciones que tiene un número infinito de soluciones.

dependent variable/variable dependiente (58) La variable en una función cuyo valor lo determina la variable independiente.

difference of squares/diferencia de cuadrados (545) Dos cuadrados perfectos separados por un signo de sustracción, $a^2 - b^2 = (a + b)(a - b)$.

dimensional analysis/análisis dimensional (174) El proceso de llevar unidades a lo largo de un cómputo.

direct variation/variación directa (239) Una función lineal descrita por una ecuación de la forma $y = kx$, en que $k \neq 0$.

discrete mathematics/matemáticas de números discretos (88) Una rama de las matemáticas que estudia los conjuntos de números finitos o interrumpidos.

discriminant/discriminante (632) En la fórmula cuadrática, la expresión $b^2 - 4ac$ se denomina la discriminante.

distance formula/fórmula de distancia (737) La distancia *d* entre cualquier par de puntos con coordenadas (x_1, y_1) y (x_2, y_2) es dada por la siguiente fórmula.
$$d = \sqrt{(x_2 - x_1)^2 + (y_2 - y_1)^2}.$$

distributive property/propiedad distributiva (46) Para cualquiera de los números *a*, *b* y *c*:

1. $a(b + c) = ab + ac$ y $(b + c)a = ba + ca$.

2. $a(b - c) = ab - ac$ y $(b - c)a = ba - ca$.

dividing rational numbers/división de números racionales (112) El cociente de dos números racionales que tienen el mismo signo es positivo. El cociente de dos números racionales que tienen diferente signo es negativo.

division property of equality/propiedad de división de la igualdad (151) Para cualquiera de los números *a*, *b* y *c*, en que $c \neq 0$, si $a = b$, entonces $\frac{a}{c} = \frac{b}{c}$.

division property of inequality/propiedad de división de la desigualdad (393) Para cualquiera de los números reales *a*, *b* y *c*:

1. si c es positivo y $a < b$, entonces $\dfrac{a}{c} < \dfrac{b}{c}$, y

si c es positivo y $a > b$, entonces $\dfrac{a}{c} > \dfrac{b}{c}$.

2. si c es negativo y $a < b$, entonces $\dfrac{a}{c} > \dfrac{b}{c}$, y

si c es negativo y $a > b$, entonces $\dfrac{a}{c} < \dfrac{b}{c}$.

domain/dominio (263) El conjunto de todas las primeras coordenadas de los pares ordenados de una relación.

draw a diagram/trazar un diagrama (406) Una estrategia para resolver problemas que a menudo se usa como una herramienta organizadora.

element/elemento (33) Un miembro de un conjunto.

elimination/eliminación (469) El método de eliminación para resolver un sistema de ecuaciones es un método que usa adición o sustracción para eliminar una de las variables y así despejar la otra variable.

equally likely/igualmente verosímil (229) Respuestas que tienen igual posibilidad de ocurrir.

equation/ecuación (33) Un enunciado matemático que contiene un signo de igualdad, $=$.

equation in two variables/ecuación en dos variables (271) Una ecuación en dos variables contiene dos valores desconocidos.

equilateral triangle/triángulo equilátero (164) Un triángulo en el cual todos los lados y todos los ángulos tienen la misma medida.

equivalent equation/ecuación equivalente (144) Ecuaciones que tienen la misma solución.

equivalente expressions/expresiones equivalentes (47) Expresiones que denotan el mismo número.

evaluate/evaluar (8) El método de hallar el valor de una expresión cuando se conocen los valores de las variables.

excluded value/valor excluido (660) Se excluye un valor del dominio de una variable porque si se sustituyera ese valor por la variable, el resultado tendría un cero en el denominador.

exponent/exponente (7) En una expresión de la forma x^n, el exponente es n.

exponential function/función exponencial (634, 635) Una función que se puede describir por una ecuación de la forma $y = a^x$, en que $a > 0$ y $a \neq 1$.

extraneous solutions/soluciones extrañas (697) Soluciones obtenidas de una ecuación y las cuales no son soluciones aceptables para la ecuación original.

extreme values/valores extremos (427) Los valores extremos son el menor valor y el mayor valor en un conjunto de datos.

extremes/extremos (196) *Ver* proporción.

factored form/forma factorial (559) Un monomio está escrito en forma factorial cuando está expresado como el producto de números primos y variables y las variables no tienen un exponente mayor de 1.

factoring/factorizar (564) Expresar un polinomio como el producto de monomios o polinomios.

factoring by grouping/factorizando por grupos (567) Un método de factorización de polinomios de cuatro o más términos.

factors/factores (6) En una expresión de multiplicación, los factores son las cantidades que se multiplican.

family of graphs/familia de gráficas (354) Una familia de gráficas incluye gráficas y ecuaciones de gráficas que tienen por lo menos una característica en común.

FOIL method/método FOIL (537) Para multiplicar dos binomios, halla la suma de los productos de los primeros términos, los términos de afuera, los términos de adentro y los últimos términos.

formula/fórmula (128) Una ecuación que enuncia una regla para la relación entre ciertas cantidades.

function/función **1.** (56) Una relación entre los datos de entrada y de salida, en que los datos de salida dependen de los datos de entrada. **2.** (287) Una relación en que cada elemento del dominio se aparea exactamente con un elemento de la amplitud.

functional notation/notación funcional (289) En notación funcional, la ecuación $y = x + 5$ se escribe $f(x) = x + 5$.

general equation for exponential decay/ ecuación general para la disminución exponencial (644) La ecuación general para la disminución exponencial está representada por $A = C(1 - r)^t$.

general equation for exponential growth/ ecuación general para el crecimiento exponencial (643) La ecuación general para el crecimiento exponencial está representada por $A = C(1 + r)^t$.

graph/gráfica (73, 255) Consiste en dibujar, o trazar sobre una recta numérica o plano de coordenadas, los puntos nombrados por ciertos números o pares ordenados.

greatest common factor (GCF)/máximo común divisor (MCD) (559) El máximo común divisor de dos o más números enteros es el número mayor que es un factor de todos los números.

guess and check/conjetura y cotejo (574)   Una estrategia para resolver problemas en la cual se prueban varios valores o combinaciones de valores para hallar una solución al problema.

half-life/media vida (638)   La media vida de un elemento radioactivo es el tiempo que tarda en desintegrarse la mitad del elemento.

half-plane/medio plano (437)   La región de un plano en un lado de una recta en el plano.

horizontal axis/eje horizontal (56)   La recta horizontal en una gráfica que representa la variable independiente.

hypotenuse/hipotenusa (206)   El lado de un triángulo rectángulo opuesto al ángulo recto.

identify subgoals/identificación de submetas (745)   Una estrategia para resolver problemas que utiliza una serie de pasos secundarios o submetas.

identity/identidad (170)   Una ecuación que es cierta para cualquier valor de la variable.

inconsistent/inconsistente (456)   Se dice que un sistema de ecuaciones es inconsistente cuando no tiene pares ordenados que satisfacen ambas ecuaciones.

independent/independiente (456)   Se dice que un sistema de ecuaciones es independiente si el sistema tiene exactamente una solución.

independent variable/variable independiente (58)   La variable en una función cuyo valor está sujeto a elección se dice que es una variable independiente. La variable independiente determina el valor de la variable dependiente.

inequality/desigualdad (33)   Un enunciado matemático que contiene los símbolos $<$, $\leq$, $>$, o $\geq$.

integers/enteros (73)   El conjunto de enteros representados por $\{\ldots, -3, -2, -1, 0, 1, 2, 3, \ldots\}$.

interquartile range/amplitud intercuartílica (304, 306)   La diferencia entre el cuartil superior y el inferior de un conjunto de datos. Representa la mitad inferior, ó 50%, de los datos en un conjunto.

intersection/intersección (406)   Para dos conjuntos A y B, es el conjunto de elementos comunes a ambos A y B.

inverse of a relation/inverso de una relación (264)   El inverso de cualquier relación se obtiene intercambiando las coordenadas en cada par ordenado.

inverse variation/variación inversa (241)   Una variación inversa se describe por una ecuación de la forma $xy = k$, en que $k \neq 0$.

irrational number/números irracional (120)   Un número que no se puede expresar en la forma $\frac{a}{b}$, en que a y b son números enteros y $b \neq 0$.

isosceles triangle/triángulo isósceles (164)   En un triángulo isósceles, por lo menos dos ángulos tienen la misma medida y por lo menos dos lados tienen la misma longitud.

least common denominator (LCD)/mínimo común denominador (MCD) (686)   El mínimo común múltiplo de los denominadores de dos o más fracciones.

least common multiple (LCM)/mínimo común múltiplo (MCM) (685)   Para dos o más enteros, el MCM es el menor número entero positivo divisible entre cada uno de los enteros.

legs/catetos (206)   Los lados de un triángulo rectángulo que forman el ángulo recto.

like terms/términos semejantes (47)   Términos que contienen las mismas variables, con variables correspondientes que tienen la misma potencia.

line plot/esquema lineal (78)   Datos numéricos desplegados sobre una recta numérica.

linear equation/ecuación lineal (280)   Una ecuación que puede escribirse de la forma $Ax + By = C$, en que A y B no son ambos iguales a cero. La gráfica de una ecuación lineal es una recta.

linear function/función lineal (278)   Ecuación cuya gráfica es una recta no vertical.

look for a pattern/busca un patrón (13, 497, 637)   Una estrategia para resolver problemas que a menudo involucra el uso de tablas para organizar la información de manera que se pueda determinar un patrón.

lower quartile/cuartil inferior (306)   El cuartil inferior divide la mitad inferior de un conjunto de datos en dos partes iguales.

make an organized list/haz una lista organizada (685)   Una estrategia para resolver problemas que utiliza una lista organizada para arreglar y evaluar los datos y determinar una solución.

mapping/relación (263)   Una relación aparea un elemento en el dominio con un elemento en la amplitud.

matrix/matriz (88)   Una matriz es un arreglo rectangular de elementos en hileras y columnas.

maximum/máximo (611)   El punto más alto en la gráfica de una curva, tal como el vértice de la parábola que se abre hacia abajo.

mean/media (178)   La media de un conjunto de datos es la suma de los números en el conjunto dividida entre el número de números en el conjunto.

means/media proporcional (196)   *Ver* proporción.

measures of central tendency/medidas de tendencia central (178)   Los números que se usan a menudo para describir conjuntos de datos porque estos representan un valor centralizado o en el medio.

measures of variation/medidas de variación (306) Números usados para describir la amplitud o distribución de los datos.

median/mediana (178) La mediana es el número en el centro de un conjunto de datos cuando los números se organizan en orden numérico.

midpoint/punto medio (369) El punto medio de un segmento es el punto equidistante entre los extremos del segmento.

minimum/mínimo (611) El punto más bajo en la gráfica de una curva, tal como el vértice de una parábola que se abre hacia arriba.

mixed expression/expresión mixta (690) Una expresión algebraica que contiene un monomio y una expresión racional.

mode/modal (178) El número que ocurre con más frecuencia en un conjunto.

monomial/monomio (496) Un número, una variable o el producto de un número y una o más variables.

multi-step equations/ecuaciones múltiples (157) Ecuaciones que requieren más de una operación para resolverlas.

multiplicative identity/identidad de multiplicación (38) Para cualquier número a, $a \cdot 1 = 1 \cdot a = a$.

multiplicative inverse/inverso multiplicativo (38) Para cualquier número no cero a, hay exactamente un número $\frac{1}{a}$, tal que $a \cdot \frac{1}{a} = \frac{1}{a} \cdot a = 1$.

multiplicative property of −1/propiedad multiplicativa de −1 (107) El producto de cualquier número y −1 es el inverso aditivo del número.

$$-1(a) = -a \text{ y } a(-1) = -a.$$

multiplicative property of equality/propiedad multiplicativa de la igualdad (150) Para cualquiera de los números a, b y c, si $a = b$, entonces $a \cdot c = b \cdot c$.

multiplication property of inequality/propiedad de multiplicación de la desigualdad (393) Para todos los números a, b y c, lo siguiente es cierto:

1. Si c es positivo y $a < b$, entonces $ac < bc$, y si c es positivo y $a > b$, entonces $ac > bc$.

2. Si c es negativo y $a < b$, entonces $ac > bc$, y si c es negativo y $a > b$, entonces $ac < bc$.

multiplicative property of zero/propiedad multiplicativa de cero (38) Para cualquier número a, $a \cdot 0 = 0 \cdot a = 0$.

negative correlation/correlación negativa (340) Existe una correlación negativa entre x y y si los valores están relacionados de maneras opuestas.

negative exponent/exponente negativo (503) Para cualquiera de los números no cero a y cualquier número entero n, $a^{-n} = \frac{1}{a^n}$

negative number/número negativo (73) Cualquier número que es menos de cero.

number line/recta numérica (72) Una recta con marcas equidistantes que se usa para representar números.

number theory/teoría de números (158) El estudio de los números y de las relaciones entre los mismos.

obtuse triangle/triángulo obtuso (165) Un triángulo obtuso tiene un ángulo cuya medida es mayor de 90 grados.

odds/posibilidades (229) Las posibilidades de que un evento ocurra son la proporción del número de formas en que el evento puede ocurrir (éxitos) comparada con el número de formas en que puede no ocurrir (fracasos).

open sentences/enunciados abiertos (32) Enunciado matemático que contiene una o más variables, o incógnitas.

opposites/opuestos (87) El opuesto de un número es su inverso aditivo.

order of operations/orden de operaciones (19)

1. Simplifica las expresiones dentro de los símbolos de agrupación, tales como paréntesis, corchetes y como lo indiquen las barras de fracción.
2. Evalúa todas las potencias.
3. Realiza todas las multiplicaciones y las divisiones de izquierda a derecha.
4. Realiza todas las sumas y las restas de izquierda a derecha.

ordered pair/par ordenado (57) Pares de números que se usan para ubicar puntos en el plano de coordenadas.

organize data/organizar datos (464) Una estrategia útil antes de resolver un problema. Algunas formas de organizar los datos son el uso de tablas, esquemas, diferentes tipos de gráficas o diagramas.

origin/origen (254) El punto de intersección de los dos ejes en el plano de coordenadas.

outcomes/resultados (413) Todas las maneras en que puede ocurrir un evento.

outlier/valor atípico (307) Cualquier elemento en un conjunto de datos que es por lo menos 1.5 veces el valor de la amplitud intercuartílica mayor que el cuartil superior, o menor que el cuartil inferior.

parabola/parábola (610, 611) La forma general de la gráfica de una función cuadrática.

parallel lines/rectas paralelas (362)   Rectas en el plano que nunca se intersecan. Las rectas paralelas no verticales tienen la misma pendiente.

parallelogram/paralelogramo (362)   Un cuadrilátero cuyos lados opuestos son paralelos.

parent graph/gráfica principal (354)   La gráfica más simple en una familia de gráficas.

percent/por ciento (215)   Un por ciento es una proporción que compara un número con 100.

percentage/porcentaje (215)   El número que se divide entre la base en un por ciento de proporción.

percent of decrease/porcentaje de disminución (222)   La proporción de una cantidad de disminución comparada con una cantidad previa, expresada en forma de por ciento.

percent of increase/porcentaje de aumento (222)   La proporción de una cantidad de aumento comparada con una cantidad previa, expresada en forma de por ciento.

percent proportion/proporción de porcentaje (215)
$$\frac{\text{Percentaje}}{\text{Base}} = \frac{r}{100}$$

perfect square/cuadrado perfecto (119)   Un número racional cuya raíz cuadrada es un número racional.

perfect square trinomial/cuadrado perfecto trinómico (587)   Un trinomio que al factorizarse tiene la forma $(a + b)^2 = (a + b)(a + b)$ o $(a - b)^2 = (a - b)(a - b)$.

perpendicular lines/rectas perpendiculares (362)   Rectas que se encuentran para formar ángulos rectos.

point-slope form/forma punto–pendiente (333)   Para cualquier punto (x_1, y_1) sobre una recta no vertical cuya pendiente es m, la forma punto–pendiente de una ecuación lineal es la siguiente:
$$y - y_1 = m(x - x_1).$$

polynomial/polinomio (513, 514)   Un polinomio es un monomio o la suma de monomios.

positive correlation/correlación positiva (340)   Existe una correlación positiva entre x y y si los valores están relacionados de la misma forma.

power/potencia (7)   Una expresión de la forma x^n se conoce como una potencia.

power of a monomial/potencia de un monomio (498)   Para cualquiera de los números a y b, y para todos los números enteros m, n y p, $(a^m b^n)^p = a^{mp} b^{np}$.

power of a power/potencia de una potencia (498)   Para cualquier número a y todos los números enteros m y n, $(a^m)^n = a^{mn}$.

power of a product/potencia de un producto (498)   Para todos los números a y b y cualquier número entero m, $(ab)^m = a^m b^m$.

prime factorization/factorización prima (558)   Un número entero expresado como un producto de factores que son todos números primos.

prime number/número primo (558)   Un número primo es un número entero mayor que 1 cuyos únicos factores son 1 y el número mismo.

prime polynomial/polinomio primo (577)   Un polinomio que no se puede escribir como el producto de dos polinomios con coeficientes integrales se llama un polinomio primo.

principal square root/raíz cuadrada principal (119)   La raíz cuadrada no negativa de una expresión.

probability/probabilidad (228)   Una razón que expresa la posibilidad de que algún evento suceda.
$$P(\text{evento}) = \frac{\text{número de resultados favorables}}{\text{número de resultados posibles}}$$

problem-solving plan/plan para solucionar problemas (126)

1. Explorar el problema.
2. Planificar la solución.
3. Resolver el problema.
4. Examinar la solución.

product/producto (6)   El resultado de la multiplicación.

product of powers/producto de potencias (497)   Para cualquiera de los números a y todos los números enteros m y n, $a^m \cdot a^n = a^{m + n}$.

product property of square roots/propiedad del producto de raíces cuadradas (719)   Para cualquiera de los números a y b, en que $a \geq 0$ y $b \geq 0$, $\sqrt{ab} = \sqrt{a} \cdot \sqrt{b}$.

proportion/proporción (195)   En una proporción, el producto de los extremos es igual al producto de las medias. Si $\frac{a}{b} = \frac{c}{d}$, entonces $ad = bc$.

Pythagorean theorem/teorema de Pitágoras (713)   Si a y b son las medidas de los catetos de un triángulo rectángulo y c es la medida de la hipotenusa, entonces, $c^2 = a^2 + b^2$.

Pythagorean triple/triplete de Pitágoras (583)   Tres números enteros a, b, y c, tales que $a^2 + b^2 = c^2$.

quadrant/cuadrante (254)   Una de las cuatro regiones en que el eje x y el eje y separan el plano de coordenadas.

quadratic equation/ecuación cuadrática (620)   Una ecuación cuadrática es una en que el valor de la función cuadrática relacionada es 0.

quadratic formula/fórmula cuadrática (628)   Las raíces de una ecuación cuadrática en la forma $ax^2 + bx + c = 0$, en la cual $a \neq 0$, son dadas por la fórmula
$$x = \frac{-b \pm \sqrt{b^2 - 4ac}}{2a}.$$

quadratic function/función cuadrática (610, 611) Función que se puede describir con una equación de la forma $ax^2 + bx + c = 0$, en la cual $a \neq 0$.

quartiles/cuartiles (304, 306) En un conjunto de datos, los cuartiles son valores que dividen los datos en cuatro partes iguales.

quotient of powers/cociente de potencias (501) Para todos los números enteros m y n y todo número no cero a, $\frac{a^m}{a^n} = a^{m-n}$.

quotient property of square roots/propiedad del cociente de raíces cuadradas (721) Para cualquiera de los números a y b, en que $a > 0$ y $b > 0$, $\sqrt{\frac{a}{b}} = \frac{\sqrt{a}}{\sqrt{b}}$.

radical equations/ecuaciones radicales (732) Ecuaciones que contienen radicales con variables en el radicando.

radical sign/signo radical (119) El símbolo $\sqrt{}$ que se usa para indicar la raíz cuadrada principal no negativa de una expresión.

radicand/radicando (719) El radicando es la expresión debajo del signo radical.

random/al azar (229) Cuando se escoge un resultado sin ninguna preferencia, el resultado ocurre al azar.

range/amplitud 1. (263) El conjunto de todas las segundas coordenadas de los pares ordenados de una relación. **2.** (306) La diferencia entre los valores mayor y menor en un conjunto de datos.

rate/razón (197) La proporción de dos medidas que se dan en diferentes unidades de medida.

rate/tasa (215) En una proporción de por ciento, la tasa es la fracción con 100 como denominador.

ratio/razón (195) Una razón es una comparación de dos números mediante división.

rational equation/ecuación racional (696) Una ecuación racional es una ecuación que contiene expresiones racionales.

rational expression/expresión racional (660) Una expresión racional es una fracción algebraica cuyo numerador y cuyo denominador son polinomios.

rational number/número racional (93) Un número racional es un número que se puede expresar en la forma $\frac{a}{b}$, en que a y b son números enteros y $b \neq 0$.

rationalizing the denominator/racionalizando el denominador (722) Un proceso que se usa para quitar o eliminar radicales del denominador de una fracción, en una expresión radical.

real numbers/números reales (121) El conjunto de números irracionales junto con los números racionales.

reciprocal/recíproco (38) El inverso multiplicativo de un número.

reflexive property of equality/propiedad reflexiva de la igualdad (39) Para cualquier número a, $a = a$.

regression line/línea de regresión (342) La línea más exacta de mejor ajuste para un conjunto de datos. Se puede determinar con una calculadora de gráficar o una computadora.

relation/relación (260, 263) Una relación es un conjunto de pares ordenados.

replacement set/conjunto de substitución (33) Un conjunto de números de los cuales se pueden escoger números para reemplazar una variable.

right triangle/triángulo rectángulo (165) Un triángulo que tiene un ángulo con una medida de 90 grados.

rise/altura (325) El cambio vertical en una recta.

roots/raíces (620) Las soluciones para una ecuación cuadrática.

run/carrera (325) El cambio horizontal en una recta.

scalar multiplication/multiplicación escalar (108) En multiplicación escalar, cada elemento de una matriz se multiplica por una constante.

scale/escala (197) Una razón llamada una escala se usa en la construcción de un modelo para representar algo que es muy grande o muy pequeño para ser dibujado a su tamaño real.

scatter plot/diagrama de dispersión (339) Gráfica que muestra dos conjuntos de datos trazados como puntos (pares ordenados) en el plano de coordenadas.

scientific notation/notación científica (506) Un número está expresado en notación científica cuando está en la forma de $a \times 10^n$, en que $1 \leq a < 10$ y n es un número entero.

set/conjunto (33) Una colección de objetos o números.

set-builder notation/notación de construcción de conjuntos (385) Una notación que se usa para describir los miembros de un conjunto. Por ejemplo, $\{y \mid y < 17\}$ representa el conjunto de todos los números y de modo que y es menor que 17.

sequence/sucesión (13) Un conjunto de números en un orden específico.

similar triangles/triángulos semejantes (201) Si dos triángulos son semejantes, las medidas de sus lados correspondientes son proporcionales y las medidas de sus ángulos correspondientes son iguales.

simple event/evento simple (414) Evento sencillo en un problema de probabilidad.

simple interest/interés simple (217)   La cantidad pagada o ganada por el uso de una cantidad de dinero. La fórmula $I = prt$ se usa para resolver problemas de interés simple.

simplest form/forma reducida (47)   Una expresión está en su forma reducida cuando ha sido reemplazada por una expresión similar que no tiene términos semejantes ni paréntesis.

simplest radical form/forma radical reducida (723)   Una expresión radical se encuentra en forma reducida cuando se satisfacen las siguientes condiciones:

1. Ningún radicando tiene factores cuadrados perfectos además de uno.

2. Ningún radicando contiene fracciones.

3. No aparece ningún radical en el denominador de una fracción.

sine/seno (206)   En un triángulo rectángulo con ángulo agudo A, el seno del ángulo

$$A = \frac{\text{medida del cateto opuesto al ángulo } A}{\text{medida de la hipotenusa}}.$$

slope/pendiente (324, 325)   El cambio vertical (altura) al cambio horizontal (carrera) a medida que te mueves de un punto a otro a lo largo de la recta.

slope-intercept form/forma pendiente-intersección (347)   Una ecuación de la forma $y = mx + b$, en que m es la pendiente y b es la intersección en y de una recta dada.

solution/solución (32)   Una sustitución por una variable en una ecuación que resulta en una ecuación válida.

solution of an equation in two variables/ solución de una ecuación de dos variables (271)   Si al sustituir los números en un par ordenado se satisface una ecuación de dos variables, entonces el par ordenado es una solución de la ecuación.

solution set/conjunto de solución (33)   El conjunto de todos los sustitutos para una variable en un enunciado abierto que satisfacen el enunciado.

solve an equation/resuelve una ecuación (145)   Resolver una ecuación quiere decir aislar la variable cuyo coeficiente es 1, en un lado de la ecuación.

solving an open sentence/resolviendo un enunciado abierto (32)   Hallar un sustituto que satisface la variable.

solving a triangle/resolviendo un triángulo (208)   El proceso de hallar las medidas de todos los lados y ángulos de un triángulo rectángulo.

square of a difference/cuadrado de una diferencia (544)   Si a y b son cualquier par de números, $(a - b)^2 = (a - b)(a - b) = a^2 - 2ab + b^2$.

square of a sum/cuadrado de una suma (543)   Si a y b son cualquier par de números, $(a + b)^2 = (a + b)(a + b) = a^2 + 2ab + b^2$.

square root/raíz cuadrada (118, 119)   Uno de los dos factores idénticos de un número.

standard form/forma estándar (333)   La forma estándar de una ecuación lineal es $Ax + By = C$, en la cual A, B y C son números enteros, $A \geq 0$ y A y B no son ceros ambos.

statistics/estadística (25)   Una rama de las matemáticas que tiene que ver con los métodos de recolección, organización e interpretación de datos.

stem-and-leaf plot/gráfica de tallo y hojas (26)   En una gráfica de tallo y hojas, cada dato se separa en dos números que se usan para formar un tallo y las hojas. Los datos se organizan en dos columnas. La columna de la izquierda contiene el tallo y la columna de la derecha las hojas.

substitution/sustitución (462)   El método de sustitución para resolver un sistema de ecuaciones es un método que sustituye una ecuación en la otra ecuación para despejar la otra variable.

substitution property of equality/propiedad de sustitución de la igualdad (39)   Si $a = b$, entonces, a se puede reemplazar por b en cualquier expresión.

subtracting integers/sustracción de enteros (87)   Para restar un entero, suma su inverso aditivo. Para cualquiera de los enteros a y b, $a - b = a + (-b)$.

subtraction property of equality/propiedad de sustracción de la igualdad (146)   Para cualquiera de los números a, b y c, si $a = b$, entonces $a - c = b - c$.

subtraction property of inequality/propiedad de sustracción de la desigualdad (385)   Para todos los números a, b y c los siguientes son ciertos:

1. si $a > b$, entonces $a - c > b - c$;
2. si $a < b$, entonces $a - c < b - c$.

supplementary angles/ángulos suplementarios (162)   Dos ángulos son suplementarios si la suma de sus medidas es 180 grados.

symmetric property of equality/propiedad simétrica de la igualdad (39)   Para cualquiera de los números a y b, si $a = b$, entonces $b = a$.

symmetry/simetría (612)   Las figuras simétricas son aquellas en que la figura se puede doblar y cada mitad es exactamente igual a la otra.

system of equations/sistema de ecuaciones (455)   Un conjunto de ecuaciones con las mismas variables.

system of inequalities/sistema de desigualdades (482)   Un conjunto de desigualdades con las mismas variables.

T

tangent/tangente (206)   En un triángulo rectángulo con ángulo agudo A, la tangente del ángulo $A = \frac{\text{medida del cateto opuesto al ángulo } A}{\text{medida del cateto adyacente al ángulo } A}$.

term/término   1. (13) Un número en una sucesión. **2.** (47) Un número, una variable o un producto o cociente de números y variables.

transitive property of equality/propiedad transitiva de la igualdad (39)   Para cualquiera de los números a, b y c, si $a = b$ y $b = c$, entonces $a = c$.

tree diagram/diagrama de árbol (413)   Un diagrama de árbol es un diagrama que se usa para mostrar el número total de posibles resultados.

triangle/triángulo (163)   Un triángulo es un polígono con tres lados y tres ángulos.

trigonometric ratios/razones trigonométricas (206)   Para el triángulo rectángulo ABC con ángulo agudo A, el seno $A = \frac{a}{c}$, coseno $A = \frac{b}{c}$, tangente $A = \frac{a}{b}$.

trinomial/trinomio (514)   La suma de tres monomios.

uniform motion/movimiento uniforme (235)   Cuando un objeto se mueve a una velocidad, o a un ritmo constante, se dice que se mueve en movimiento uniforme.

union/unión (408)   Para dos conjuntos A y B, el conjunto de los elementos contenidos en ambos A o B o A y B.

unique factorization theorem/teorema de la factorización única (558)   La factorización prima de cada número es única excepto por el orden en que se escriben los factores.

unit cost/costo unitario (95)   El costo de una unidad de un artículo.

upper quartile/cuartil superior (306)   El cuartil superior divide la parte superior de un conjunto de datos en dos partes iguales.

use a model/usa un modelo (339)   Una estrategia para resolver problemas que usa modelos, o simulaciones, de situaciones matemáticas difíciles de resolver directamente.

use a table/usa una tabla (255)   Una estrategia para resolver problemas que usa tablas para organizar y resolver problemas.

variable/variable (6)   Las variables son símbolos que se usan para representar números desconocidos.

Venn diagrams/diagramas de Venn (73)   Los diagramas de Venn son diagramas que usan círculos u óvalos dentro de un rectángulo para mostrar relaciones de conjuntos.

vertex/vértice (610, 611)   El punto máximo o mínimo de una parábola.

vertical axis/eje vertical (56)   La recta vertical en una gráfica que representa la variable dependiente.

vertical line test/prueba de recta vertical (289)   Si cualquier recta vertical pasa por un solo punto de la gráfica de una relación, entonces la relación es una función.

weighted average/promedio ponderado (233)   El promedio ponderado M de un conjunto de datos es la suma del producto de cada número en el conjunto y su peso divididos entre la suma de todos los pesos.

whiskers/patillas (428)   Las patillas de un diagrama de caja y patillas son los segmentos que se dibujan desde el cuartil inferior hasta el mínimo valor y desde el cuartil superior hasta el máximo valor.

whole numbers/números enteros (72)   El conjunto de números enteros se representa por {0,1,2,3,...}.

work backward/trabaja al revés (156)   Una estrategia para resolver problemas que usa operaciones inversas para determinar un valor inicial.

x-axis/eje x (254)   La recta numérica horizontal en el plano de coordenadas.

x-coordinate/coordenada x (254)   El primer número en un par ordenado.

x-intercept/intersección con el eje x (346)   La coordenada x de un punto donde una gráfica interseca el eje x.

y-axis/eje y (254)   La recta numérica vertical en el plano de coordenadas.

y-coordinate/coordenada y (254)   El segundo número en un par ordenado.

y-intercept/intersección con el eje y (346)   La coordenada y de un punto donde una gráfica interseca el eje y.

zero exponent/exponente cero (502)   Para cualquier número no cero a, $a^0 = 1$.

zero product property/propiedad del producto cero (594)   Para todos los números a y b, si $ab = 0$, entonces $a = 0$, $b = 0$, o ambos a y b son iguales a cero.

zeros/ceros (620)   Las raíces, o intersecciones con el eje x de la gráfica de una función.

Page A3   Pretest

1. D   3. E   5. B   7. C   9. C   11. E   13. A   15. D   17. BH

19. B   21. C   23. E   25. C

Page A7   Prerequisite Concept 1

1. line graph to show how values changed over time

3. circle graph to show how parts relate to the whole

5. circle graph to show how parts relate to the whole

7. 60 women

9.

Page A9   Prerequisite Concept 2

1.

3. $\frac{2}{15}$ 5. $\frac{7}{15}$ 7. $\frac{1}{3}$ 9. 24   11. $2\frac{5}{9}$ 13. $7\frac{1}{4}$ 15. $3\frac{23}{32}$ 17. 35

19. $8\frac{63}{64}$ 21. $6\frac{1}{4}$ 23a. 78 kids   23b. 24 kids   23c. 198 kids

25. $46

Page A11   Prerequisite Concept 3

1.

3. $\frac{1}{16}$ 5. $1\frac{3}{4}$ 7. 5   9. $\frac{25}{512}$ 11. $\frac{11}{6}$ 13. 3   15. 6   17. $\frac{7}{20}$

19. 104 sheets   21. $6\frac{1}{2}$ times

Page A13   Prerequisite Concept 4

1. $\frac{3}{4} + \frac{3}{4} = 1\frac{1}{2}$

3.

5. $\frac{2}{3}$ 7. $\frac{2}{3}$ 9. 1   11. $\frac{5}{6}$ 13. $1\frac{5}{12}$ 15. $2\frac{1}{8}$ 17. $5\frac{1}{8}$ 19. $8\frac{7}{12}$

21. $12\frac{1}{8}$ 23. $\frac{7}{12}$ cup   25a. $\frac{2}{5}$ 25b. $\frac{7}{50}$ 25c. $\frac{23}{50}$

Page A15   Prerequisite Concept 5

1. $\frac{7}{8} - \frac{5}{8} = \frac{1}{4}$

3.

5. $\frac{1}{2}$ 7. $\frac{1}{5}$ 9. $\frac{1}{2}$ 11. $\frac{3}{8}$ 13. $\frac{5}{12}$ 15. $\frac{1}{12}$ 17. 3   19. $1\frac{1}{8}$

21. $1\frac{3}{4}$ 23. $19\frac{5}{16}$ 25a. $1\frac{1}{8}$ 25b. $4\frac{3}{8}$ between Thursday

and Friday

Page A17   Prerequisite Concept 6

1. Sample answer: Smaller to larger means divide. Larger to smaller means multiply.   3. 2500   5. 0.34

7. 7000   9. 8000   11. 6   13. $3\frac{3}{4}$ 15. cm, m; in., ft

17. mg; oz   19. 135 tons

Page A19   Prerequisite Concept 7

1. $1 \cdot 8 = 2 \cdot x$ 3. 2   5. 10   7. 21   9. $3\frac{1}{3}$ 11. 20   13. 20

15. $1.47   17. 12 m   19. 75 hits

Page A21   Prerequisite Concept 8

1. • A

3.

5.

7.

9.

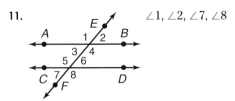

11.   ∠1, ∠2, ∠7, ∠8

13. 49°   15. 49°   17. 131°   19a. line segment

19b. 4 pairs

Page A23   Prerequisite Concept 9

1. See students' work.   3. 68 in.; 273 in^2 5. 12 ft; 6 ft^2

7. 51 mi^2 9. $278.64

Page A25   Prerequisite Concept 10

1.

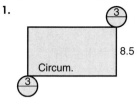

3. 248 in^2; 240 in^3 5. 266.9 cm^2; 274.8 cm^3 7. 376.8 cm^2; 552.6 cm^3 **9a.** about 71,315.68 in^2 or about 495.25 ft^2
9b. about 1,088,952 in^3 or about 630.18 ft^3

Page A27   Prerequisite Concept 11

1. $6^2 + 8^2 = 10^2$ 3. 12 yd   5. 12.4 in.   7. 17.0   9. 15
11. 13.2   13. 41.3   15. 13 yards   17. $450^2 = 200^2 + b^2$; about 403.1 ft

Page A29   Prerequisite Concept 12

1.

3. No section on the spinner is white.   5. 36   7. 25
9. $\frac{3}{8}$ 11. $\frac{1}{4}$ 13. $\frac{1}{2}$ 15. $\frac{1}{2}$ 17. $\frac{5}{12}$ 19. 60%

Page A30   Posttest

1. C   3. E   5. D   7. A   9. B   11. B   13. A   15. C   17. E
19. B   21. D   23. B   25. A

CHAPTER 1 EXPLORING EXPRESSIONS, EQUATIONS, AND FUNCTIONS

Pages 9–11   Lesson 1–1

7. $3y^2 - 6$ **9.** 3 times x squared increased by 4   **11.** a^7
13. 32   **15.** $k + 20$ **17.** a^7 **19.** $\frac{2x^2}{3}$ or $\frac{2}{3}x^2$ **21.** $b + 8$
23. 4 times m to the fifth power   **25.** c squared plus 23
27. 2 times 4 times 5 squared   **29.** 8^2 **31.** 4^7 **33.** z^5
35. 49   **37.** 64   **39.** 16   **41.** $3(55 - w^3)$ **43.** $a + b + \frac{a}{b}$
45. $y + 10x$ **47.** 16; 16   **47a.** They are equal; no; no.
47b. no; for example, $2^3 \neq 3^2$ **49a.** $3.5x$ **49b.** $3.5y$
49c. $3.5x + 3.5y$

Pages 15–18   Lesson 1–2

5.

   7. $5x + 1, 6x + 1$

9a.

4^1	4^2	4^3	4^4	4^5
4	16	64	256	1024

9b. 6; $4^6 = 4096$ **9c.** 4; When the exponent is odd, 4 is in the ones place.   **11.**

13. 48, 96   **15.** 25, 36   **17.** $a + 7, a + 9$
19a.
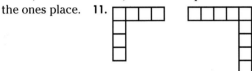

19b. White; even-numbered figures are white.
19c. 12 sides; The shapes come in pairs. Since $19 \div 2 = 9.5$, the 19th figure will be part of the 10th pair. The 10th pair will have $10 + 2$, or 12 sides.   **21a.** 4; 9; 16; 25
21b. The sums are perfect square numbers; that is 1^2, $2^2, 3^2, 4^2, 5^2, \ldots$ **21c.** 10,000   **21d.** x^2 **23a.** 1,999,998; 2,999,997; 3,999,996; 4,999,995   **23b.** 8,999,991
25. 4, 7, 10, 13, 16   **27.** 1:13 P.M.   **29a.** 100 cards
31. x cubed divided by 9   **33.** 4^3; 64 cubes
35. $x + \frac{1}{11}x$

Pages 22–24   Lesson 1–3

5. 173   **7.** 4   **9.** 25   **11.** $a^2 - a$ **13.** 14   **15.** 14
17. 9   **19.** 60   **21.** $\frac{11}{18}$ **23.** 6   **25.** 126   **27.** 147
29. 126   **31.** $r^2 + 3s$; 19   **33.** $(r + s)t^2$; $\frac{7}{4}$ **35.** 19.5 mm
37. 22 in.   **39.** 16.62   **41.** 3.77   **43a.** Sample answer:
$(4 - 2)5 \div (3 + 2)$ **43b.** $4 \times 2 \times 5 \times 3 \times 2$ **45a.** $\frac{1}{3}Bh$
45b. 198,450 m^3 **47.** 16, 19.5   **49.** June 18   **51.** $t + 3$
53. 9 more than two times y

Pages 28–31   Lesson 1–4

5. stem 12, leaf 2   **7.** stem 126, leaf 9   **9a.** 35 tickets sold one day   **9b.** 31 tickets   **9c.** 75 tickets   **9d.** 1043 tickets
11. stem 13, leaf 3   **13.** stem 44, leaf 3   **15.** stem 111, leaf 3   **17.** stem 14, leaf 3   **19.** stem 111, leaf 4
21. rounded to the nearest hundred: 9, 12, 24, 27, 38, 39, 40   **23a.** Possible answer: Both teens and young adults had similar distribution of responses.   **23b.** Sample answer: The market research shows that the new game is equally appealing to teens and young adults. Therefore, we should concentrate our marketing efforts towards both teens and young adults.

25a.

1980	Stem	1993
4	1	3
6	2	4
8 5 1	3	3 7 7
	4	6
1	5	

$3 \mid 7 = 3700$

25b. 3000 to 3900; 3000 to 3900   **25c.** Sample answer: Farm sizes seem to be shrinking.   **27.** 15   **29a.** $\frac{s}{5}$
29b. 2 miles; yes
31.

Pages 34–36   Lesson 1–5

7. false   **9.** $\{1, 3, 5\}$ **11.** 75   **13.** false   **15.** true   **17.** true
19. $\{5\}$ **21.** $\left\{\frac{1}{2}, \frac{3}{4}\right\}$ **23.** $\{10, 15, 20\}$ **25.** 3   **27.** 2
29. $4\frac{5}{6}$ **31.** Sample answer: $p = 1$ and $q = 2$, $p = 2$ and $q = 10$, $p = 3$ and $q = 8$, $p = 4$ and $q = 20$, and $p = 5$ and $q = 15$ **33a.** $C = \frac{3500 \cdot 4}{14}$ **33b.** 1000 Calories
35. 2, 5, 6, 7, 9   **37.** 8   **39.** 5 less than x to the fifth power

Page 36   Self Test

1. $3a + b^2$ **3.** 11:04, 11:08, 11:51, 11:55   **5.** 408
7.

Stem	Leaf
4	8
5	4
6	7
7	7
8	5 9

$6 \mid 7 = 67$

9. $\{6, 7, 8\}$

Pages 41–43   Lesson 1–6

7. $\frac{2}{9}$ **9.** c   **11.** e   **13.** d   **15.** f

17. $6(12 - 48 \div 4) + 9 \cdot 1$
$= 6(12 - 12) + 9 \cdot 1$ *Substitution* $(=)$
$= 6(0) + 9 \cdot 1$ *Substitution* $(=)$
$= 0 + 9 \cdot 1$ *Multiplicative property of 0*
$= 0 + 9$ *Multiplicative identity*
$= 9$ *Additive identity*
19a. $4(20) + 7$
19b. $4(20) + 7$
$= 80 + 7$ *Substitution* $(=)$
$= 87$ *Substitution* $(=)$

19c. 87 years **21.** 9 **23.** $\frac{1}{p}$ **25.** $\frac{2}{3}$ **27.** substitution ($=$)
29. multiplicative identity **31.** multiplicative inverse, multiplicative identity **33.** symmetric ($=$) **35.** reflexive ($=$) **37.** substitution ($=$); substitution ($=$); multiplicative identity; multiplicative inverse; substitution ($=$) **39.** substitution ($=$); substitution ($=$); substitution ($=$); multiplicative identity

41. $(15 - 8) \div 7 \cdot 25$
$\quad = 7 \div 7 \cdot 25 \quad$ *Substitution($=$)*
$\quad = 1 \cdot 25 \quad$ *Substitution ($=$)*
$\quad = 25 \quad$ *Multiplicative identity*

43. $(2^5 - 5^2) + (4^2 - 2^4)$
$\quad = (32 - 25) + (16 - 16) \quad$ *Substitution ($=$)*
$\quad = 7 + 0 \quad$ *Substitution ($=$)*
$\quad = 7 \quad$ *Additive identity*

45. $5^3 + 9\left(\frac{1}{3}\right)^2$
$\quad = 125 + 9\left(\frac{1}{9}\right) \quad$ *Substitution ($=$)*
$\quad = 125 + 1 \quad$ *Multiplicative inverse*
$\quad = 126 \quad$ *Substitution ($=$)*

47a. $[21(12 \cdot 2)] + [23(15 \cdot 2)] + [67(10 \cdot 2)]$
47b. $[21(12 \cdot 2)] + [23(15 \cdot 2)] + [67(10 \cdot 2)]$
$\quad = [21(24)] + [23(30)] + 67(20)] \quad$ *Substitution ($=$)*
$\quad = 504 + 690 + 1340 \quad$ *Substitution ($=$)*
$\quad = 2534 \quad$ *Substitution ($=$)*

47c. $25.34 **49.** true **51.** false **55.** 36 **57.** $12y$

Page 44   Lesson 1–7A
1. $2x + 2$ **3.** $4x + 2$ **5.** false;

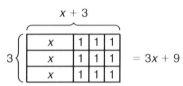

7a. Adita **7b.** Answers will vary. Answers should include the concept that multiplication by 3 is distributed over both terms in parentheses.

Pages 48–50   Lesson 1–7
5. b **7.** a **9.** c **11.** $2a - 2b$ **13.** 60 **15.** 7 **17.** $4y^4, y^4$
19. $3t^2 + 4t$ **21.** $23a^2b + 3ab^2$ **23.** $5.35(24) + 5.35(32)$; $5.35(24 + 32)$ **25.** $5g - 45$ **27.** $24m + 48$ **29.** $5a - ab$
31. 52 **33.** 60 **35.** 645 **37.** in simplest form **39.** $12a + 15b$
41. in simplest form **43.** $3x + 4y$ **45.** $1\frac{3}{5}a$ **47.** $9x + 5y$
49. no; sample counterexample: $2 + (4 \cdot 5) \neq (2 + 4)(2 + 5)$
51a. $2[x + (x + 14)] = 4x + 28$ **51b.** 96 **51c.** 527 ft^2
53. multiplicative property of 0 **55a.** $d = (1129)(2)$
55b. 2258 ft **57.** 2 **59.** 8 years

Pages 53–55   Lesson 1–8
7. commutative ($+$) **9.** associative ($\times$) **11.** $5a + 2b$
13. $14x + 3y$
15. $6z^2 + (7 + z^2 + 6)$
$\quad = 6z^2 + (z^2 + 7 + 6) \quad$ *Commutative ($+$)*
$\quad = (6z^2 + z^2) + (7 + 6) \quad$ *Associative ($+$)*
$\quad = (6 + 1)z^2 + (7 + 6) \quad$ *Distributive property*
$\quad = 7z^2 + 13 \quad$ *Substitution ($=$)*
17. multiplicative identity **19.** associative ($\times$)
21. distributive property **23.** commutative ($\times$)
25. associative ($+$) **27.** $10x + 5y$ **29.** $7x + 10y$
31. $10x + 2y$ **33.** $32a^2 + 16$ **35.** $5x$ **37.** $\frac{3}{4} + \frac{5}{3}m + \frac{4}{3}n$

39. $2(s + t) - s$
$\quad = 2s + 2t - s \quad$ *Distributive property*
$\quad = 2t + 2s - s \quad$ *Commutative ($+$)*
$\quad = 2t + (2s - s) \quad$ *Associative ($+$)*
$\quad = 2t + (2s - 1s) \quad$ *Multiplicative identity*
$\quad = 2t + (2 - 1)s \quad$ *Distributive property*
$\quad = 2t + 1s \quad$ *Substitution ($=$)*
$\quad = 2t + s \quad$ *Multiplicative identity*

41. $5xy + 3xy$
$\quad = (5 + 3)xy \quad$ *Distributive property*
$\quad = 8xy \quad$ *Substitution ($=$)*

43. $\frac{1}{100}$; Each denominator and the following numerator represent the number 1. The resulting expression is 1 in the numerator and 100 in the denominator multiplied by numerous 1s. Since 1 is the multiplicative identity, the product is $\frac{1}{100}$. **45.** no; Sample example: Let $a = 1$ and b $= 2$, then $1 * 2 = 1 + 2(2)$ or 5 and $2 * 1 = 2 + 2(1)$ or 4.
47a. $G = 3.73$ **49.** $100d + 80d + 8d, 188d$ **51.** $\frac{3}{2}$

53.

Stem	Leaf
10	0 0 0 1 1 1 1 3
9	5 7 7 9 9 9

$10|3 = 103$

55.

Pages 59–62   Lesson 1–9
5a. Graph 3 **5b.** Graph 4 **5c.** Graph 2 **5d.** Graph 1
7a. False, A is the younger player, but B runs the mile in less time. **7b.** False, A made more 3-point shots, but B made more 2-point shots. **7c.** True, B is the older player and B made more 2-point shots. **7d.** True, A is the younger player and A made more free-throw shots.
9. Graph a; An average person makes no money as a child, then his or her income rises for several years, and finally levels off. **11a.** Graph 5 **11b.** Graph 3 **11c.** Graph 1 **11d.** Graph 4 **11e.** Graph 6 **11f.** Graph 2

13.

15.

Height vs. Distance graph

17a. horizontal axis: time in years; vertical axis: millions of dollars **17b.** In 1991 the telethon raised 45 million dollars. **17c.** Money raised in the telethon decreased, then increased steadily. **17d.** The money raised increased. **19.** $7p + 11q + 9$ **21.** 14 **23.** $\frac{1}{3}Bh$

Page 63   Chapter 1 Highlights
1. a **3.** e **5.** f **7.** g **9.** b

Pages 64–66   Chapter 1 Study Guide and Assessment
11. x^5 **13.** $5x^2$ **15.** three times a number m to the fifth power **17.** the difference of four times m squared and twice m **19.** 32, 64 **21.** $4x + y, 5x + y$ **23.** $10^1, 10^2, 10^3, 10^4$ **25.** 11 **27.** 9 **29.** 26 **31.** 2.4 **33.** 9.2 **35.** 90

37. No; some former presidents are still living. **39.** false
41. true **43.** $13\frac{1}{2}$ **45.** $\frac{1}{3}$ **47.** 5 **49.** additive identity

51. $2 \times 4 + 2 \times 7$ **53.** $1 - 3p$ **55.** 294 **57.** $8m + 8n$
59. commutative $(+)$ **61.** commutative $(\times)$ **63.** $5x + 5y$
65. $3pq$ **67.** Graph c **69a.** $80s$ **69b.** 320 **69c.** 640
71a. $3 + 4 > 5, 4 + 5 > 3, 3 + 5 > 4$ **71b.** $x < 9$ ft and $x > 3$ ft

CHAPTER 2 EXPLORING RATIONAL NUMBERS

Pages 75–77 Lesson 2–1
7. $\{-1, 0, 1, 2, \ldots\}$
9.

$-4\ -3\ -2\ -1\ \ 0\ \ 1\ \ 2\ \ 3\ \ 4$

11. $-4 + (-3) = -7$ **13.** -5 **15.** 9-yard gain
17. $\{-4, -3, -2, -1\}$ **19.** $\{-7, -3\}$
21. $\{\ldots, -5, -4, -3, -2, -1, 0\}$
23.
$-3\ -2\ -1\ \ 0\ \ 1\ \ 2\ \ 3\ \ 4\ \ 5$
25.
$-4\ -3\ -2\ -1\ \ 0\ \ 1\ \ 2\ \ 3\ \ 4$
27.
$-7\ -6\ -5\ -4\ -3\ -2\ -1\ \ 0\ \ 1$
29.
$-7\ -6\ -5\ -4\ -3\ -2\ -1\ \ 0\ \ 1$

31. 13 **33.** -5 **35.** -12 **37.** 6 **39.** -23
41. $-17 + 82 = 65; 65°F$ **43a.** $4°F$
43b. $-31°F$ **43c.** $-15°F$ **43d.** $9°F$

45.

47. commutative $(\times)$ **49.** substitution $(=)$
51. 13.9

Pages 80–83 Lesson 2–2
5. from 35 to 80;

$35\ \ 40\ \ 45\ \ 50\ \ 55\ \ 60\ \ 65\ \ 70\ \ 75\ \ 80$

7a.

$30\ \ 35\ \ 40\ \ 45\ \ 50\ \ 55\ \ 60\ \ 65\ \ 70$

7b. 70 mph; cheetah **7c.** 30 mph **7d.** 30 mph **7e.** 12 **7f.** 4
9a. from 20 to 75 by 5;

$20\ \ 25\ \ 30\ \ 35\ \ 40\ \ 45\ \ 50\ \ 55\ \ 60\ \ 65\ \ 70\ \ 75$

9b. yes; 25 and 42 **11a.** Ms. Martinez's **11b.** No; in Ms. Martinez's class, the hours that the students talked on the phone were close to each other (4 ± 2 hours). In Mr. Thomas' class, the students had a wider range of hours spent talking on the phone. The range was from 0 to 8 hours, with no hours clustered around a certain number.

13. Mr. Thomas', 3.6 hours; Ms. Martinez's, 3.7 hours; yes **15.** associative $(\times)$ **17.** 2

Page 84 Lesson 2–3A
1. 6 **3.** -2 **5.** 2 **7.** 6 **9.** false **11.** true **13.** Answers should include using the number line.

Pages 90–92 Lesson 2–3
7. $-7, 7$ **9.** $0, 0$ **11.** -4 **13.** -11 **15.** 0 **17.** $40c$
19. -4 **21.** 3 **23.** $\begin{bmatrix} -1 & 4 \\ 6 & -6 \end{bmatrix}$ **25.** $-12, 12$ **27.** $302, 302$
29. 0 **31.** -32 **33.** -16 **35.** -70 **37.** -15 **39.** -5
41. $40b$ **43.** $-29p$ **45.** $26d$ **47.** -4 **49.** -15 **51.** 5
53. 8 **55.** 99 **57.** -59 **59.** $\begin{bmatrix} 0 & 3 \\ 2 & 2 \end{bmatrix}$ **61.** $\begin{bmatrix} -4 & -5 \\ -4 & -5 \\ 10 & -3 \end{bmatrix}$
63. 24th floor

65a.
Monday	Sesame	Poppy	Blue	Plain
East Store	120	80	64	75
West Store	65	105	77	53

Tuesday	Sesame	Poppy	Blue	Plain
East Store	112	76	56	74
West Store	69	95	82	50

65b.
Monday + Tuesday	Sesame	Poppy	Blue	Plain
East Store	232	156	120	149
West Store	134	200	159	103

65c.
Monday − Tuesday	Sesame	Poppy	Blue	Plain
East Store	8	4	8	1
West Store	−4	10	−5	3

This matrix represents the difference between Monday's and Tuesday's sales in each category.
67a. from 100 to 146

67b.
$100\ 105\ 110\ 115\ 120\ 125\ 130\ 135\ 140\ 145\ 150$

67c. yes; 114 **67d.** Siberian Elm; American Elm

69.
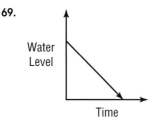

71. $55y^2$ **73.** symmetric $(=)$ **75.** 14, 20, 29, 34, 37, 38, 43, 59, 64, 74, 84 **77.** 24

Pages 97–99 Lesson 2–4
7. $<$ **9.** $=$ **11.** $-0.5, \frac{3}{4}, \frac{7}{8}, 2.5$ **15.** $>$ **17.** $<$ **19.** $<$
21. $>$ **23.** $\frac{3}{8}, \frac{2}{3}, \frac{6}{7}$ **25.** $\frac{3}{23}, \frac{8}{42}, \frac{4}{14}$ **27.** $-\frac{2}{5}, -0.2, 0.2$
29. a 16-ounce drink for $0.59 **31.** a package of 75 paper plates for $3.29 **33.** Sample answer: $\frac{2}{3}$ **35.** Sample answer: $\frac{7}{20}$ **37.** Sample answer: $\frac{1}{6}$ **39.** $E = \frac{4}{14}$ or $\frac{2}{7}$; $G = \frac{10}{14}$ or $\frac{5}{7}$; $H = \frac{13}{14}$ **41.** 0.375 inch **43.** $-9, 9$

45.

47. $m + 2n + \frac{3}{2}$ **49.** two times x squared plus six

Page 99   Self Test

1.

3. -17 **5.** 55

7a.

7b. yes; $270   **9.** $<$

Pages 102–104   Lesson 2–5

5. $-\frac{1}{9}$ **7.** $-2\frac{5}{8}$ **9.** 0.88 **11.** 5.75 **13a.** $+\frac{1}{2}$

13b.

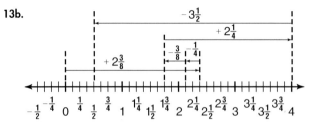

15. $-\frac{11}{16}$ **17.** -1.3 **19.** $\frac{4}{9}$ **21.** -0.2007 **23.** $-8\frac{5}{8}$

25. 0.0485 **27.** -22.94 **29.** -2.17 **31.** 1 **33.** -3.5

35. -16.7 **37.** $-\frac{52}{21}$ **39.** $\begin{bmatrix} -3.8 & -0.1 \\ 1.7 & 2.9 \end{bmatrix}$ **41.** $\begin{bmatrix} \frac{1}{4} & 11 \\ -5 & 8 \\ \frac{1}{2} & -12 \end{bmatrix}$

43. Answers will vary. Sample answer: $\begin{bmatrix} 5 & 3 \\ 1 & 7 \end{bmatrix}, \begin{bmatrix} 4 & -3 \\ 1 & 2 \end{bmatrix}$,

$\begin{bmatrix} 9 & 0 \\ 2 & 9 \end{bmatrix}, \begin{bmatrix} 18 & 0 \\ 4 & 18 \end{bmatrix}$; $\begin{bmatrix} 5 & 3 \\ 1 & 7 \end{bmatrix} + \begin{bmatrix} 4 & -3 \\ 1 & 2 \end{bmatrix} = \begin{bmatrix} 9 & 0 \\ 2 & 9 \end{bmatrix} = \begin{bmatrix} 4 & -3 \\ 1 & 2 \end{bmatrix} +$

$\begin{bmatrix} 5 & 3 \\ 1 & 7 \end{bmatrix}; \left(\begin{bmatrix} 5 & 3 \\ 1 & 7 \end{bmatrix} + \begin{bmatrix} 4 & -3 \\ 1 & 2 \end{bmatrix} \right) + \begin{bmatrix} 9 & 0 \\ 2 & 9 \end{bmatrix} = \begin{bmatrix} 18 & 0 \\ 4 & 18 \end{bmatrix} =$

$\begin{bmatrix} 5 & 3 \\ 1 & 7 \end{bmatrix} + \left(\begin{bmatrix} 4 & -3 \\ 1 & 2 \end{bmatrix} + \begin{bmatrix} 9 & 0 \\ 2 & 9 \end{bmatrix} \right)$ **45a.** $9.01, 10.13, 11.25$

45b. 1.12 **45c.** $-2, -\frac{5}{4}, -\frac{1}{2}, \frac{1}{4}, 1; -2\frac{1}{2}$ **47.** $<$

49a.

49b. no   **51.** multiplicative property of 0   **53.** $\frac{7}{24}$

Page 105   Lesson 2–6A

3. -10 **5.** -10 **7.** 10

Pages 109–111   Lesson 2–6

5. -18 **7.** 24 **9.** $-\frac{4}{5}$ **11.** -4 **13.** $-46st$ **15.** $\begin{bmatrix} -6 & 12 \\ -3 & 15 \end{bmatrix}$

17. $28.65   **19.** -60 **21.** -1 **23.** 2 **25.** 0.00879 **27.** $-\frac{6}{5}$

29. $-\frac{9}{17}$ **31.** 85.7095 **33.** 3 **35.** -6 **37.** $\frac{13}{12}$ **39.** $-\frac{179}{24}$

41. $\frac{25}{24}$ **43.** $-30rt + 4s$ **45.** $21x$ **47.** $16.48x - 5.3y$

49. $\begin{bmatrix} 2 & 6 & 3 \\ \frac{5}{2} & 5 & 1 \end{bmatrix}$ **51.** $\begin{bmatrix} -9 & 22.68 \\ -22.4 & -10 \\ 28.8 & 11.12 \end{bmatrix}$ **53.** $\begin{bmatrix} 6 & 18 & 4 \\ 0 & 2 & \frac{8}{3} \end{bmatrix}$

55. It is negative.   **57a.** No; it meets the requirement that no dimensions can be less than 20 feet, but the minimum of 1250 square feet is not met because this yard would have 1216 square feet.   **57b.** No; it meets the requirement that yards must have a minimum of 1250 square feet because it would have 1330 square feet, but the requirement that no dimension can be less than 20 feet is not met because it has a side of length 19 feet.

57c. Answers will vary. Sample answer: 30 feet.

59. -2.2 **61.** -20 **63.** -7 **65.** $\{6, 7\}$ **67.** 408

Pages 115–117   Lesson 2–7

5. -4 **7.** $-\frac{3}{32}$ **9.** $-\frac{5}{48}$ **11.** $-6x$ **13.** about 2 minutes

15. 6 **17.** $-\frac{1}{18}$ **19.** $-\frac{1}{16}$ **21.** -6 **23.** $-\frac{1}{12}$ **25.** $\frac{243}{10}$

27. $-\frac{35}{3}$ **29.** $-65m$ **31.** $r + 3$ **33.** $-20a - 25b$

35. $-14c + 6d$ **37.** $-a - 4b$ **39.** -1.2 **41.** $-0.8\overline{3}$ **43.** 4

45a.

45b. $5\frac{9}{16}$ inches   **47a.** 17.4 ± 0.161 cm   **47b.** 0.161 cm

49. $\frac{13}{12}$ **51a.** $\begin{bmatrix} -2 & 4 \\ 3 & 12 \end{bmatrix}$ **51b.** $\begin{bmatrix} 4 & 4 \\ 7 & 2 \end{bmatrix}$ **53.** $20b + 24$

55a.

Stem	Leaf
13	0
15	4
17	2 3
19	4
20	0
22	0 2
24	5
25	1 4
27	9
29	0
30	4 7
33	0
34	6
52	5

13|0 = 13,000

55b. $13,000   **55c.** $52,500   **55d.** 9

Page 118   Lesson 2–8A

1. 4–5 **3.** 13–14 **5.** 1–2

Pages 123–125   Lesson 2–8

7. 8 **9.** 11.05 **11.** 16 **13.** Q   **15.** Q

17. (number line 2 to 10, open circle near 7.5)

19. (number line −2 to 6)

21. 13 **23.** $\frac{2}{3}$ **25.** 20.49 **27.** 15 **29.** −25 **31.** $\frac{3}{5}$ **33.** 9.33
35. −47 **37.** −20 **39.** Z, Q **41.** Q **43.** Q **45.** Q **47.** I
49. Q **51.** (number line −6 to 2)

53. (number line −7 to 1)

55. (number line −15 to −7)

57. (number line −5.6 to −4.8)

59. (number line $4\frac{1}{2}$ to $6\frac{1}{2}$)

61. Yes; it lies between 27 and 28. **63a.** $d = 1.4\sqrt{1275}$; Estimates should be close to 50. **63b.** 49.989999
65. Fibonacci **67.** −12 **69.** a 1-pound package of lunch meat for $1.95 **71.** from 500 to 520 by 5s;

73. b **75a.** to the nearest ten thousand **75b.** 10
75c. $690,000 **77.** 26, 32

Pages 130–132   Lesson 2–9
5a. 4 points **5b.** 15 questions **5c.** 86 **5d.** 6 points
5e. 15 − n **5f.** 4n **7.** n + 5 ≥ 48 **9.** Let s = average speed; $s = \frac{189}{3} = 63$. **11.** The product of a times the sum of y and 1 is b. **13a.** 640 feet **13b.** y − 80 **13c.** ℓ + ℓ + w + w or 2ℓ + 2w **13d.** 200 ft **13e.** 24,000 ft² **13f.** Is there enough space to build the playground as designed? **15.** $a^2 + b^3 = 25$ **17.** $x < (a − 4)^2$ **19.** $\frac{7}{8}(a + b + c^2) = 48$

21. Sample answer: x = first number
x + 25 = second number
25 + 2x = 106
23. Sample answer:
y = how much money Hector saves
y = 2(7)(50)
25. V is equal to the quotient of a times h and three.
27. Sample answer: Three times the number of miles home is from school is 36. How many miles away from home is the school? **29.** Sample answer: The number of minutes spent waiting for a doctor in the waiting room is 5 minutes more than 6 times the number of appointments she has. How long would you have to wait for a doctor with d appointments?
31. ab + cd + 2d − 8
33. Sample answer: 4.5 + x + x = 24 + x
x = 19.5
35a. $A = \frac{1}{2}h(a + b)$ **35b.** $A = 108\frac{1}{2}$ ft² **37.** −8
39. $\begin{vmatrix} 8 & −1.6 \\ 12 & −20 \end{vmatrix}$ **41.** −25, 25 **43.** {2, 3, 4, 5}

Page 133   Chapter 2 Highlights
1. true **3.** true **5.** false, sample answer: $\frac{1}{2}$ **7.** true
9. true

Pages 134–136   Chapter 2 Study Guide and Assessment
11. (number line −4 to 6)

13. (number line −3 to 5)

15. −5 **17.** −4 **19.** −5 **21.** 56% **23.** 8 **25.** −2
27. 4 **29.** −5 **31.** > **33.** < **35.** Sample answer: 0
37. −1 **39.** 12.37 **41.** −13.26 **43.** −99 **45.** $−\frac{3}{7}$
47. −90 **49.** −9 **51.** $\frac{-4}{35}$ **53.** −109 **55.** 14 **57.** −30

59a. (line plot 5 to 30)

59b. 14 detergents **61.** 1.25 liters of soda for $1.31
63a. They both saw a gain of $\frac{1}{4}$ for the week. **63b.** CompNet
63c. It changed $−\frac{3}{4}$ from Wednesday to Thursday.
65. Let n = the number; 3n − 21 = 57.

CHAPTER 3   SOLVING LINEAR EQUATIONS

Page 143   Lesson 3–1A
1. 1 **3.** −11 **5.** 8 **7.** 4 **9.** no **11.** yes

Pages 148–149   Lesson 3–1
7. −17 **9.** −12 **11.** −14 **13.** n + (−56) = −82; −26
15. −9 **17.** −1.3 **19.** 1.4 **21.** −3 **23.** −6 **25.** −7 **27.** 24
29. $1\frac{1}{4}$ **31.** $−\frac{1}{10}$ **33.** n + 5 = 34; 29 **35.** n − (−23) = 35; 12 **37.** n + (−35) = 98; 133 **39a.** 203,600 + s = 19,300,000; 19,096,400 subscribers **39b.** Sample answer: 25 million **41a.** 24 years old **41b.** 24 years **41c.** 78 years **41d.** 34 years old **41e.** 24 years **43.** −5 **45.** −0.73x **47.** 48

Pages 152–154   Lesson 3–2
7. 17 **9.** 48 **11.** $\frac{11}{15}$ **13.** −7n = 1.477; −0.211 **15.** −9
17. −17.33 or $−17\frac{1}{3}$ **19.** −14 **21.** 12.81 **23.** 600
25. −275 **27.** −25 **29.** $8\frac{6}{13}$ **31.** −8 **33.** −12n = −156; 13
35. $\frac{4}{3}n = 4.82$; 3.615 **37.** 45 **39.** 30 **41.** 12 **43a.** 7ℓ = 350, 50 people; 7ℓ = 583, ≈ 83 people **43b.** $\frac{p}{7} = 65$; 455 people **45a.** 0.10m = 2.30; 23 minutes **45b.** 0.10(18) = c; $1.80 **47.** $17 + 1\frac{1}{2}y = 33\frac{1}{2}$; 11 years **49.** $\frac{19}{28}$ **51a.** The number of bags increased during the day. **51b.** The machine was refilled. **53.** 31

Page 155   Lesson 3–3A
1. 5 **3.** −2 **5.** 2 **7.** 4 **9.** −2

Pages 159–161   Lesson 3–3
9. −6 **11.** 35 **13.** −17 **15.** n + (n + 2) + (n + 4) = 21;

5, 7, 9   **17.** -1 **19.** -2 **21.** -42.72 **23.** $12\frac{2}{3}$ **25.** -6
27. -48 **29.** 30   **31.** 57   **33.** $4\frac{3}{7}$ **35.** $12 - 2n = -7; 9\frac{1}{2}$
37. $n + (n + 1) + (n + 2) + (n + 3) = 86; 20, 21, 22, 23$
39. $n + (n + 2) + (n + 4) = 39; 11$ m, 13 m, 15 m   **41.** 28
43. -38 **45.** $(3n - 1) + (3n + 1) + (3n + 3)$ **47.** $2a +$
$2 = 26; 12$ letters   **51.** I   **53.** -5 **55.** 5

Page 161   Self Test
1. 44   **3.** 6   **5.** $-8\frac{1}{3}$ **7.** $23 - n = 42; -19$ **9.** $n + (n + 2) =$
$126; 62, 64$

Pages 165–167   Lesson 3–4
7. $79°, 169°$ **9.** $(90 - 3x)°, (180 - 3x)°$ **11.** $(110 - x)°,$
$(200 - x)°$ **13.** $85°$ **15.** $(c - 38) + c = 90; 26°, 64°$
17. $48°, 138°$ **19.** none, $55°$ **21.** $69°, 159°$ **23.** none, $81°$
25. $(90 - 3a)°, (180 - 3a)°$ **27.** $(128 - b)°, (218 - b)°$
29. $70°$ **31.** $105°$ **33.** $138°$ **35.** $(190 - 2p)°$ **37.** $45°$
39. $x + (x - 30) = 180; 75°$ **41.** $x + (30 + 3x) = 90; 15°,$
$75°$ **43.** $x + 3x + 4x = 180; 22.5°, 67.5°, 90°$ **45.** $60°$
47. $40°$ **49a.** $\$80$ **49b.** $\$74.75$ **51.** $7t$

53a.

53b. yes; 51, 54, 55, and 57

55.

Stem	Leaf
4	2 3 6 6 7 8 9 9
5	0 0 1 1 1 1 2 4
	4 4 4 5 5 5 5 6
	6 6 7 7 7 7 8
6	0 1 1 1 2 2 4 4
	5 8 9

$5 \,|\, 2 = 52$

Pages 170–172   Lesson 3–5
5. $-\frac{1}{2}$; Add $4x$ to each side, subtract 10 from each side,
then divide each side by 14.   **7.** -25; Subtract $\frac{1}{5}$ from
each side, subtract 3 from each side, then multiply by $\frac{5}{2}$.
9. all numbers   **11.** $2(n + 2) = 3n - 13; 17, 19$ **13.** $\frac{1}{2}n +$
$16 = \frac{2}{3}n - 4; 120$ **15.** no solution   **17.** -2 **19.** 5.6
21. -16 **23.** no solution   **25.** 4   **27.** 2.6 or $2\frac{3}{5}$ **29.** 8
31. 2   **33.** all numbers   **35.** $\frac{1}{5}n + 5n = 7n - 18; 10$
37. $n + (n + 2) + (n + 4) = 180; 58°, 60°, 62°$ **39a.** $4x +$
$6 = 4x + 6$; identity   **39b.** $5x - 7 = x + 3; 2.5$ **39c.** $-3x +$
$6 = 3x - 6; 2$ **39d.** $5.4x + 6.8 = 4.6x + 2.8; -5$ **39e.** $2x -$
$8 = 2x - 6$; no solution   **41a.** 6.827 years   **41b.** Air
conditioner sales are decreasing while fan sales are
increasing.   **43.** $148°$ **45.** -2.7 **47.** $\frac{11}{9}$ **49.** 11.05

Pages 175–177   Lesson 3–6
5. $\frac{3x - 7}{4}$ **7.** $\frac{c - b}{2}$ **9.** $\frac{2S - nt}{n}$ **11.** $\frac{b}{9}$ **13.** $3c - a$
15. $\frac{v - r}{t}$ **17.** $\frac{I}{pt}$ **19.** $\frac{H}{0.24I^2t}$ **21.** $\frac{-t + 5}{4}$ **23.** $\frac{4}{3}(c - b)$

25. $\frac{5x + y}{2}$ **27.** $2x + 12 = 3y - 31; \frac{3y - 43}{2}$ **29.** $\frac{2}{3}x + 5 =$
$\frac{1}{2}y - 3; \frac{3}{4}y - 12$ **31.** $\$900$ **33a.** 36 words per minute
33b. Clarence, with 74 words per minute   **35.** $5°$
37. $x \neq 2$ **39.** $\{8\}$

Pages 181–183   Lesson 3–7
11. $8.2; 8; 8$ **13.** Sample answer: 20, 20, 30, 50, 90, 90
15a. $3.857; 3; 3$ **15b.** mean   **17.** $96.8; 50$; none   **19.** $9; 9$;
none   **21.** $5.69; 4.56$; none   **23.** $212.94; 218; 219$ **25.** 14
27a. mean   **27b.** Yes, because the article indicates that the
median value is in the middle of the data set.   **27c.** No, he
should have used the term *mean* since the mean can be
affected by extremely high values, causing the majority to
be below the mean value.   **29.** $1660\frac{3}{5}; 1076\frac{1}{2}$; none
31. 37.5 mph   **33.** 16   **35.** -20 **37.** $5 + 9ac + 14b$

Page 185   Chapter 3 Highlights
1. b   **3.** i   **5.** d   **7.** k   **9.** a   **11.** g

Pages 186–188   Chapter 3 Study Guide and Assessment
13. -16 **15.** 21   **17.** -10 **19.** -32 **21.** -116 **23.** 7
25. $83\frac{1}{3}$ **27.** 10   **29.** -16 **31.** 3   **33.** -153 **35.** 11
37. 50   **39.** 16, 17   **41.** $21°; 111°$ **43.** $(70 - y)°; (160 - y)°$
45. $30°$ **47.** $15°$ **49.** $180° - (z + z - 30°)$ or $(210° - 2z)$
51. -30 **53.** 2   **55.** $\frac{y}{5}$ **57.** $\frac{a}{y - c}$ **59.** $21; 21; 21$

61a. 25.5 years   **63.** $13\frac{3}{4}$ years

CHAPTER 4   USING PROPORTIONAL REASONING

Pages 198–200   Lesson 4–1
5. $=$ **7.** 12   **9.** 5   **11.** 4.62   **13.** 9.5 gallons   **15.** $=$ **17.** $\neq$
19. $=$ **21.** 9   **23.** $\frac{8}{5}; 1.6$ **25.** 0.84   **27.** $-\frac{149}{6}; -24.8$
29. 11   **31.** 2.28   **33.** 1.251

35a.

Louis' age	1	2	3	6	10	20	30
Mariah's age	9	10	11	14	18	28	38

35b. $9, 5, 3.\overline{6}, 2.\overline{3}, 1.8, 1.4, 1.2\overline{6}$ **35c.** $r = \frac{y + 8}{y}$ **35d.** The
ratio gets smaller.   **35e.** No; if the ratio equaled 1, Mariah
and Louis would be the same age.   **37.** 85 movies
39. $23; 19; 18$ **41.** $4x - 2x = 100; 50$ **43a.** 93   **43b.** $\$9695$
43c. more than 40   **43d.** $93 - p$ **43e.** yes   **45.** 10

Pages 203–205   Lesson 4–2
5. $\triangle DEF$ **7.** yes   **9.** $\ell = 12, m = 6$ **11.** 27 feet high
13. $\triangle DFE$ **15.** no   **17.** yes   **19.** no   **21.** $a = \frac{55}{6}, b = \frac{22}{3}$
23. $d = \frac{51}{5}, e = 9$ **25.** $a = 2.78, c = 4.24$ **27.** $b = 16.2,$
$d = 6.3$ **29.** $c = \frac{7}{2}, d = \frac{17}{8}$

31. Sample answer:

33a. $\frac{3}{9} = \frac{d}{d + 39}; d = 19.5$ ft; yes   **33b.** When the ball is served from 8 feet, $d = 23.4$ ft, and the serve is a fault. When the ball is served from 10 feet, $d = 16.7$ ft. Therefore, tall players have an easier time serving. **35.** $16\frac{1}{2}$ feet by 21 feet **37.** 160 **39.** 60 million; 420 million

41. 23;

$(19 - 12) \div 7 \cdot 23$

$= 7 \div 7 \cdot 23$ *Substitution (=)*

$= 1 \cdot 23$ *Substitution (=)*

$= 23$ *Multiplicative identity*

Pages 211–214 Lesson 4–3

7. $\sin Y = 0.600$, $\cos Y = 0.800$, $\tan Y = 0.750$ **9.** 0.8192
11. 0.1228 **13.** 46° **15.** 36° **17.** $m\angle B = 50°$, $AC = 12.3$ m, $BC = 10.3$ m **19.** $m\angle A = 30°$, $AC = 13.9$ m, $BC = 8$ m
21. $\sin G = 0.600$, $\cos G = 0.800$, $\tan G = 0.750$ **23.** $\sin G = 0.471$, $\cos G = 0.882$, $\tan G = 0.533$ **25.** $\sin G = 0.923$, $\cos G = 0.385$, $\tan G = 2.4$ **27.** 0.3584 **29.** 0.9703
31. 0.3746 **33.** 33° **35.** 22° **37.** 62° **39.** 77° **41.** 58°
43. 18° **45.** $m\angle B = 60°$, $AC = 12.1$ m, $BC = 7$ m **47.** $m\angle A = 45°$, $AC = 6$ ft, $AB = 8.5$ ft **49.** $m\angle A = 63°$, $AC = 9.1$ in., $BC = 17.8$ in. **51.** $m\angle A = 23°$, $m\angle B = 67°$, $AB \approx 12.8$ ft
53. $m\angle A = 30°$, $m\angle B = 60°$, $AC = 5.2$ cm **55a.** true; $\sin C = \frac{c}{a}$, $\cos B = \frac{c}{a}$ **55b.** $\cos B = \frac{c}{a}$, $\frac{1}{\sin B} = \frac{a}{b}$ **55c.** $\tan C = \frac{c}{b}$, $\frac{\cos C}{\sin C} = \frac{b}{a} \div \frac{c}{a} = \frac{b}{c}$ **55d.** true; $\tan C = \frac{c}{b}$, $\frac{\sin C}{\cos C} = \frac{c}{a} \div \frac{b}{a} = \frac{c}{b}$ **55e.** true; $\sin B = \frac{b}{a}$, $(\tan B)(\cos B) = \frac{b}{c} \cdot \frac{c}{a} = \frac{b}{a}$ **57.** 2229 ft **59.** 3 **61.** 10 **63.** N, W, Z, Q **65.** $\frac{2}{15}$
67. $9x + 2y$ **69.** 125

Pages 218–221 Lesson 4–4

7. 43%, 0.43 **9.** 55% **11.** 15 **13.** $52.50 **15.** $1200 at 10%, $6000 at 14% **17.** 30%, 0.30 **19.** 70%, 0.70
21. 180%, 1.8 **23.** $62\frac{1}{2}$%; 0.625 **25.** 50% **27.** $12\frac{1}{2}$%, 12.5% **29.** 400% **31.** 2.5% **33.** 72.3 **35.** 702.4 **37.** 28%
39. 25.92 **41.** 12% **43.** $3125 **45.** $2400 **47.** 39%, 34%, 15%, 12% **49.** 11.5% **51.** $5.45 **53a.** about 54 people; about 23 people **53b.** Respondents could choose more than one entree. **55.** about 277 feet **57.** $\frac{77}{4}$; 19.25
59. −4.7 **61.** −28y **63.** 2

Page 221 Self Test

1. 5 **3.** 5 **5.** $m\angle B = 34°$, $b \approx 11.5$, $c \approx 20.5$ **7.** no
9. $62\frac{1}{2}$%, 62.5%

Pages 225–227 Lesson 4–5

5. I; 40% **7.** D; 50% **9.** $85.85 **11.** $196 **13.** I; 69%
15. D; 55% **17.** D; 27% **19.** I; 2% **21.** $233.24 **23.** $19.90
25. $85.39 **27.** $16.59 **29.** $30.09 **31.** no **33.** 133%
35. 8.6% **37.** an increase of more than 22% **39.** 11%
41. 65 ft **43.** 11 **45.** heavy, 571; light, 286

Pages 230–232 Lesson 4–6

7. $\frac{1}{2}$ **9.** $\frac{2}{3}$ **11.** 5:1 **13.** $\frac{1}{2}$ **15.** $\frac{1}{2}$ **17.** $\frac{1}{2}$ **19.** $\frac{2}{11}$ **21.** 0
23. $\frac{8}{11}$ **25.** 1:5 **27.** 4:11 **29.** 23:7 **31.** $\frac{1}{2}$ **33.** 1:3
35. $\frac{8}{13}$ **37.** $\frac{1}{7}$ **39a.** 1:14 **39b.** $\frac{1}{27}$ **41a.** $\frac{3}{26}$ **41b.** $\frac{4}{13}$
41c. 3:23 **43.** 9° **45.** $50 **47.** $\frac{4}{21}$ **49.** 75

Pages 236–238 Lesson 4–7

5. 226 dozen chocolate chip; 311 dozen peanut butter
7. 11:30 A.M. **9.** 5 quarters **11.** 3 hours **13.** 2.5 hours
15. 46 mph **17.** 67 mph **19a.** 7 **19b.** 16 **21.** Sample answer: How much pure antifreeze must be mixed with a 20% solution to produce 40 quarts of a 28% solution?
23. 0 **25.** $11,000 **27.** 49 **29.** associative (+)

Pages 243–244 Lesson 4–8

5. I, 5 **7.** 10 **9.** 99 **11.** 20 inches from the 8-ounce weight, 16 inches from the 10-ounce weight **13.** I, 15
15. I, 9 **17.** D, 4 **19.** $26\frac{1}{4}$ **21.** −8 **23.** 12 **25.** 6.075
27. 8.3875 **29.** $\frac{1}{4}$ **31.** 18 pounds **33.** 1.6 feet **35.** 2:1
37. $4300 at 5%, $7400 at 7% **39.** −12

Page 245 Chapter 4 Highlights

1. angle of elevation **3.** equal, proportional **5.** legs
7. odds **9.** inverse variation

Pages 246–248 Chapter 4 Study Guide and Assessment

11. = **13.** ≠ **15.** 18 **17.** 16 **19.** $d = \frac{45}{8}$, $e = \frac{27}{4}$
21. $b = \frac{44}{3}$, $d = 6$ **23.** 0.528 **25.** 0.849 **27.** 1.607
29. 39° **31.** 80° **33.** $m\angle B = 45°$, $AB = 8.5$, $BC = 6$
35. $m\angle A \approx 66°$, $AB \approx 9.8$, $m\angle B \approx 24°$ **37.** 48 **39.** 87.5%
41. 0.1881 **43.** decrease, 13% **45.** increase, 6%
47. decrease, 10% **49.** $85.66 **51.** $9272.23 **53.** $\frac{1}{12}$
55. $\frac{5}{6}$ **57.** 10:39 **59.** 34:15 **61.** 21 **63.** 6 **65.** 21
67. 48 **69.** 450 mph and 530 mph

CHAPTER 5 GRAPHING RELATIONS AND FUNCTIONS

Pages 257–259 Lesson 5–1

5. $(-3, -1)$; III **7.** $(0, -2)$; none
8–11.

13. $(-1, -3)$; III **15.** $(0, 3)$; none **17.** $(0, 0)$; none
19. $(3, -2)$; IV **21.** $(2, 2)$; I **23.** $(-5, 4)$; II
25–36.

37.

41a. New Orleans **41b.** Oregon **41c.** Sample answer: $(75°, 40°)$ **41d.** Honolulu, Hawaii **41f.** Sample answer: The longitude lines are not the same distance apart; they meet at the poles. **45.** $3\frac{1}{2}$ hours **47.** 4; 3; none **49.** $-a - 5$ **51.** $31a + 21b$

Page 261   Lesson 5–2A

1–6.

7. They are points on the axes.

Pages 266–269   Lesson 5–2
5. $D = \{0, 1, 2\}$; $R = \{2, -2, 4\}$ **7.** $\{(1, 3), (2, 4), (3, 5), (5, 7)\}$; $D = \{1, 2, 3, 5\}$; $R = \{3, 4, 5, 7\}$; Inv $= \{(3, 1), (4, 2), (5, 3), (7, 5)\}$ **9.** $\{(1, 3), (2, 2), (4, 9), (6, 5)\}$; $D = \{1, 2, 4, 6\}$; $R = \{2, 3, 5, 9\}$; Inv $= \{(3, 1), (2, 2), (9, 4), (5, 6)\}$
11. $\{(-2, 2), (-1, 1), (0, 1), (1, 1), (1, -1), (2, -1), (3, 1)\}$; $D = \{-2, -1, 0, 1, 2, 3\}$; $R = \{-1, 1, 2\}$; Inv $= \{(2, -2), (1, -1), (1, 0), (1, 1), (-1, 1), (-1, 2), (1, 3)\}$

13.

15b. 5.6% **15c.** Except for the first 6 months of 1992, the unemployment rate seems to be decreasing.
17. $D = \{-5, -2, 1, 3\}$; $R = \{7\}$ **19.** $D = \left\{-5\frac{1}{4}, -3, \frac{1}{2}, 1\frac{1}{2}\right\}$; $R = \left\{-6\frac{2}{7}, -\frac{2}{3}, \frac{1}{4}, \frac{2}{5}\right\}$ **21.** $\{(6, 4), (4, -2), (3, 4), (1, -2)\}$; $D = \{1, 3, 4, 6\}$; $R = \{-2, 4\}$; Inv $= \{(4, 6), (-2, 4), (4, 3), (-2, 1)\}$ **23.** $\{(6, 0), (-3, 5), (2, -2), (-3, 3)\}$; $D = \{-3, 2, 6\}$, $R = \{-2, 0, 3, 5\}$; Inv $= \{(0, 6), (5, -3), (-2, 2), (3, -3)\}$ **25.** $(3, 4), (3, 2), (2, 9), (5, 4), (5, 8), (-7, 2)\}$; $D = \{-7, 2, 3, 5\}$; $R = \{2, 4, 8, 9\}$; Inv $= \{(4, 3), (2, 3), (9, 2), (4, 5), (8, 5), (2, -7)\}$ **27.** $\{(0, 25), (1, 50), (2, 75), (3, 100)\}$; $D = \{0, 1, 2, 3\}$; $R = \{25, 50, 75, 100\}$; Inv $= \{(25, 0), (50, 1), (75, 2), (100, 3)\}$ **29.** $\{(-3, 4), (-2, 2), (-1, -2), (2, 2)\}$; $D = \{-3, -2, -1, 2\}$; $R = \{-2, 2, 4\}$; Inv $= \{(4, -3), (2, -2), (-2, -1), (2, 2)\}$ **31.** $\{(-3, 3), (-3, -3), (3, 3), (3, -3), (0, 0)\}$; $D = \{-3, 0, 3\}$; $R = \{-3, 0, 3\}$; Inv $= \{(3, -3), (-3, -3), (3, 3), (-3, 3), (0, 0)\}$ **33.** $\{(-3, 1), (-1, 1), (2, 1), (3, 1), (4, 1)\}$; $D = \{-3, -1, 2, 3, 4\}$; $R = \{1\}$; Inv $= \{(1, -3), (1, -1), (1, 2), (1, 3), (1, 4)\}$

35.

37.

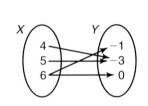

39a. $[-10, 10]$ by $[-10, 10]$

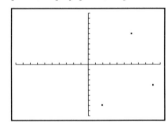

39b. $\{(10, 0), (-8, 2), (6, 6), (-4, 9)\}$

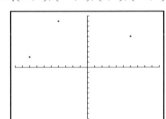

39c.

(x, y)	Quadrant	Inverse's Quadrant
(0, 10)	I	I
(2, −8)	IV	II
(6, 6)	I	I
(9, −4)	IV	II

41. If a point lies in Quadrants I or III, its inverse will lie in the same quadrant as the point. If a point lies in Quadrant II, its inverse lies in Quadrant IV, and vice versa. If a point lies on the x-axis, its inverse lies on the y-axis and vice versa. **43.** Sample answers: **a.** 157 billion, 191 billion

b. Retail sales have increased from 1992 to 1994.
c. As unemployment decreases, retail sales increase, because people have more money to spend.

45a.

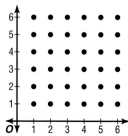

D = R = {1, 2, 3, 4, 5, 6} **45b.** D = R = {1, 2, 3, 4, 5, 6};
relation = inverse
45c. 11 possible sums   **45d.**

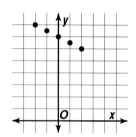

45e. $\frac{6}{36}$ or $\frac{1}{6}$; There are 6 out of 36 ways to roll a sum of 7.

47. 32   **49.** $480   **51.** 2214, 2290   **53.** $4 + 80x + 32y$

Page 270   Lesson 5–3A

1. {(−3, −19), (−2, −15), (−1, −11), (0, −7), (1, −3), (2, −1), (3, 5)}   **3.** {(−3, −10.4), (−2, −9.2), (−1, −8), (0, −6.8), (1, −5.6), (2, −4.4), (3, −3.2)}

Pages 274–277   Lesson 5–3

7. {(−2, −1), (−1, 1), (0, 3), (1, 5), (2, 7), (3, 9)}   **9.** a, d
11. {(−2, 8), (−1, 7.5), (0, 7), (1, 6.5), (2, 6)}

13. a, c   **15.** a, b   **17.** c   **19.** {(−3, −12), (−2, −8), (0, 0), (3, 12), (6, 24)}   **21.** {(−3, 10), (−2, 9), (0, 7), (3, 4), (6, 1)}   **23.** {(−3, 11), (−2, 9.5), (0, 6.5), (3, 2), (6, −2.5)}   **25.** {(−3, −12), (−2, −7), (0, 3), (3, 18), (6, 33)}   **27.** {(−3, 1.8), (−2, 1.4), (0, 0.6), (3, −0.6), (6, −1.8)}

29.

x	y
−5	−9
−3	−5
0	1
1	3
3	7
6	13

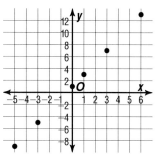

31.

a	b
−2	4.5
−1	3.25
0	2.00
1	0.75
3	−1.75
4	−3.00
5	−4.25

33a. $3x + 4y = 180$ **33b.** $y = 45 - \frac{3}{4}x$ **33c.** Sample answer:
(1, 44.25), (2, 43.5), (3, 42.75), (4, 42), (5, 41.25)

35. $\left\{ -\frac{2}{3}, -\frac{1}{3}, 0, \frac{2}{3}, 1 \right\}$ **37.** $\left\{ -\frac{7}{4}, -\frac{1}{2}, 2, \frac{13}{4}, \frac{9}{2} \right\}$

39.

x	y
−2	−8
−1	−1
0	0
1	1
2	8

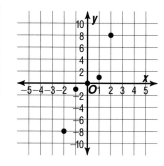

41. {(−2.5, −4.26), (−1.75, −3.21), (0, −0.76), (1.25, 0.99), (3.333, 3.902)}   **43.** {(−100, 350), (−30, 116.$\overline{6}$), (0, 16.$\overline{6}$), (120, −383.$\overline{3}$), (360, −1183.$\overline{3}$), (720, −2383.$\overline{3}$)}
45a. {−6, −4, 0, 4, 6}   **45b.** {−13, −8, −4, 4, 8, 13}
45c. {−5, 0, 4, 8, 13}   **47a.** $D = \frac{m}{V}$ **47b.** silver and gasoline

49.   Women   Men

L	S	(L, S)		L	S	(L, S)
$9\frac{1}{3}$	6	$\left(9\frac{1}{3}, 6\right)$		$11\frac{1}{3}$	8	$\left(11\frac{1}{3}, 8\right)$
$9\frac{5}{6}$	$7\frac{1}{2}$	$\left(9\frac{5}{6}, 7\frac{1}{2}\right)$		$11\frac{5}{6}$	$9\frac{1}{2}$	$\left(11\frac{5}{6}, 9\frac{1}{2}\right)$
$10\frac{1}{6}$	$8\frac{1}{2}$	$\left(10\frac{1}{6}, 8\frac{1}{2}\right)$		$12\frac{1}{3}$	11	$\left(12\frac{1}{3}, 11\right)$
$10\frac{2}{3}$	10	$\left(10\frac{2}{3}, 10\right)$		$12\frac{5}{6}$	$12\frac{1}{2}$	$\left(12\frac{5}{6}, 12\frac{1}{2}\right)$

51a.

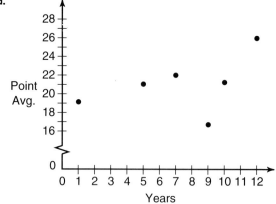

51b. The more years played, the higher the point per game average of the player.   **53.** $467.50   **55.** $18,000
57a. $2w + 2\ell = 148$ **57b.** w, 14.25 in.; ℓ, 59.75 in.
59. −7.976

Page 280   Lesson 5–4A

1.

3.

5.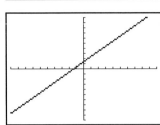

7a. Sample answer: [−10, 110] by [−5, 15]

7b.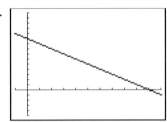

7c. Sample answer: (0, 10), (100, 0), (10, 9)

9a. Sample answer: [−2.5, 0.5] by [−0.05, 0.05]

9b.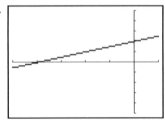

9c. Sample answer: (−2, 0), (0, 0.02), (1, 0.03)

Pages 283–286   Lesson 5–4

7. yes; $2x + y = 6$ **9.** yes; $3x + 2y = 7$

11.

$4x + 3y = 12$

13.

$x = 6$

15.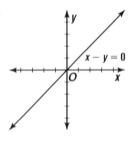

$x - y = 0$

17. yes; $\frac{3}{5}x - \frac{2}{3}y = 5$ **19.** no   **21.** yes; $3x - 2y = 8$ **23.** yes; $7x - 7y = 0$ **25.** yes; $3m - 2n = 0$ **27.** yes; $6a - 7b = -5$

29.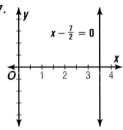

$y = 3x + 1$

33.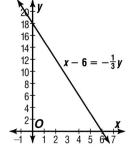

$x - 6 = -\frac{1}{3}y$

37.

$x - \frac{7}{2} = 0$

41.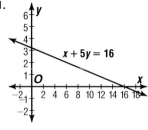

$x + 5y = 16$

45.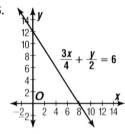

$\frac{3x}{4} + \frac{y}{2} = 6$

47. −5, 7.5   **49.** 6, 9, 10

51. [−5, 15] by [−10, 10], Xscl: 1, Yscl: 1

53. [−10, 10] by [−2, 2], Xscl: 1, Yscl: 0.25

55. [−5, 25] by [−20, 5], Xscl: 5, Yscl: 5

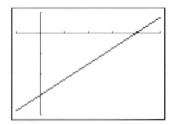

57a. Sample answer: Parallel lines that slant upward and intersect the *x*-axis at −7, −2.5, 0, and 4.5.

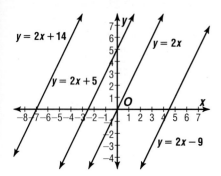

57b. Sample answer: Parallel lines that slant downward and intersect the *x*-axis at 0, $1\frac{1}{3}$, $2\frac{1}{3}$, and $-3\frac{1}{3}$.

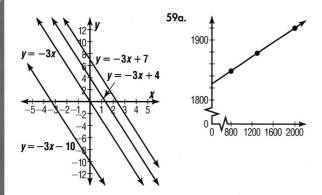

59a.

59b. yes, but only if her sales are $1300 or $2000 over target **61.** $y = 3 - 4x$

63a.

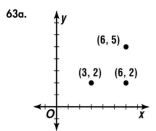

63b. (3, 5) **63c.** 12 inches **65.** 15% **67.** 120° **69.** $\frac{1}{48}$

Page 286 Self Test

1.

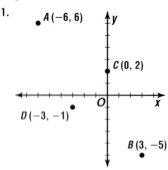

3. {(−5, −3), (−1, 4), (4, 4), (4, −3)}, D = {−5, −1, 4}, R = {−3, 4}, Inv = {(−3, −5), (4, −1), (4, 4), (−3, 4)}

5. $b = 3 - \frac{2}{3}a$: $\left\{\left(-2, 4\frac{1}{3}\right), \left(-1, 3\frac{2}{3}\right), (0, 3), \left(1, 2\frac{1}{3}\right), (3, 1)\right\}$

7.

9.

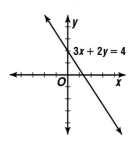

Pages 291–294 Lesson 5–5

7. no **9.** yes **11.** yes **13.** no **15.** −10 **17.** $3w + 2$
19. yes **21.** no **23.** yes **25.** yes **27.** yes **29.** yes
31. yes **33.** no **35.** −14 **37.** $-\frac{9}{25}$ **39.** 5.25 **41.** −18
43. $9b^2 - 6b$ **45.** 3 **47.** $5a^4 - 10a^2$ **49.** $24p - 36$
51a. Sample answer: $f(x) = x$. **51b.** Sample answer: $f(x) = x^2$ **53.** a; sample answer: because they have a single price per shirt, and since you cannot have a fraction of a shirt, you must use points instead of lines. It is a function.

55a. D: $0 \le k \le 24$; R: $0 \le g \le 100$

55b.

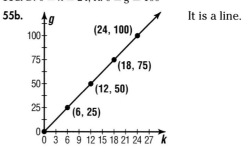

It is a line.

55c. 24 karats **57a.** $260.87 **57b.** Sample equation:
$B = \frac{3}{23}P$ **59.** {−16, −13, 5, 11} **61.** $\frac{3}{13}$ **63.** 45° **65.** −48

67a.

67b. 149 **67c.** 91 **67d.** 121 and 125 **67e.** 13 players

Pages 299–302 Lesson 5–6

5. $f(x) = 2x + 6$ **7.** $y = \frac{1}{2}x - \frac{3}{2}$ **9.** −6 **11.** 48, 60
13. 8, −2 **15.** $f(x) = 5x$ **17.** $g(x) = 11 - x$ **19.** $h(x) = \frac{1}{3}x - 2$ **21.** $y = -3x$ **23.** $y = \frac{1}{2}x$ **25.** $y = 2x - 10$
27. $xy = -24$ **29.** $y = x^3$ **31.** *y*-intercept: $f(0)$, *x*-intercept: $f(x) = 0$ **33a.** $f(x) = 34x - 34$ **33b.** You must go deeper in fresh water to get the same pressure as in ocean water.
35a. 1514 C **35b.** 11.04 C/min

35c.

Minutes	Calories
1	11.04
2	22.08
3	33.12
4	44.16
5	55.2
6	66.24
7	77.28
8	88.32
9	99.36
10	110.4

35d. $C(t) = 11.04t$; yes　**35e.** 1324.8 C burned; 189.2 C remaining　**37.** D = {1, 3, 5}; R = {2, 4, 6}　**39.** 5

41.

Page 304　Lesson 5–7A

1. Q1, 14; Med, 17; Q3, 20.5; R, 11; IQR, 6.5
3. Q1, 68; Med. 78; Q3, 96; R, 34; IQR, 28
5. Q1, 3.4; Med, 5.3, Q3, 21; R, 77; IQR, 17.6
7. At least 50% of the data is clustered around the median.

Pages 309–313　Lesson 5–7

7. 45, 40, 45, 34, 11; 11　**9.** 48, 26, 39, 17, 22
11. 34, 78, 96, 68, 28　**13.** 10, 5, 8, 2, 6　**15.** 1.1, 30.6, 30.9, 30.05, 0.85　**17.** 340, 1075, 1125, 1025, 100　**19.** 39, 218, 221, 202, 19　**21a.** 9,198,630; 11,750,000; 5,700,000; 24,000,000; 6,050,000　**21b.** No; the libraries would have accumulated books throughout the years.　**23.** 3760, 3224, 4201.5, 2708.5, 1421　**25.** 21,674; 9790; 12,194; 5475; 6719　**27a.** males: 36, 29, 37.5, 44, 15; females: 23, 29, 33, 39, 10　**27b.** There are no outliers.　**27c.** The ages of the top female golfers are less varied than those of the top male golfers.　**31.** c, d　**33a.** $6.2 + p = 9.4$; about 3.2 million people　**33b.** $6.0 + p = 6.9$; about 0.9 million people

Page 315　Chapter 5 Highlights

1. e　**3.** d　**5.** j　**7.** c　**9.** b

Pages 316–318　Chapter 5 Study Guide and Assessment

10–13.

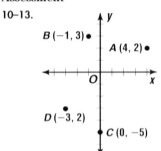

15. (2, −1); IV　**17.** (1, 1); I　**19.** D = {−3, 4}, R = {5, 6}
21. D = {−3, −2, −1, 0}, R = {0, 1, 2}　**23.** {(2, 0), (−1, 3), (2, 2), (−1, −2)}　**25.** {(−4, −13), (−2, −11), (0, −9), (2, −7), (4, −5)}　**27.** $\left\{\left(-4, -5\frac{1}{3}\right), \left(-2, -2\frac{2}{3}\right), (0, 0), \left(2, 2\frac{2}{3}\right), \left(4, 5\frac{1}{3}\right)\right\}$

29.

x	y
−2	−7
0	3
2	13
4	23
6	33

31.　**33.**

35.

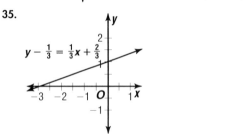

37. yes　**39.** no　**41.** yes　**43.** 3　**45.** $a^2 + a + 1$
47. $2a^2 - 14a + 26$　**49.** $y = x - 4$　**51.** 70, 65, 85, 45, 40
53. 37, 73, 77, 62, 15

55a.

55b. $56　**57a.** 9.6, 6, 7.05, 8.3, 2.3　**57b.** 12.6, 14.7

CHAPTER 6　ANALYZING LINEAR EQUATIONS

Page 324　Lesson 6–1A

1. $-\frac{1}{2}$　**3.** Sample answer: (3, 3)　**5.** Yes; sample answer: suppose the endpoints are $C(-4, -1)$ and $D(-2, -2)$. Let the upper right peg represent (0, 0). To go from C to D, the y value decreases by 1 and the x value increased by 2. The ratio is $\frac{-1}{2}$. If you use the rule from Exercise 4, $\frac{-2-(-1)}{-2-(-4)} = \frac{-1}{2}$. The results are the same.

Pages 329–331　Lesson 6–1

7. $\frac{5}{3}$　**9.** $\frac{3}{2}$　**11.** undefined　**13.** 2　**15.** $-\frac{1}{5}$　**17.** 0　**19.** 1

21. $\frac{4}{7}$ **23.** $-\frac{2}{3}$ **25.** undefined **27.** undefined **29.** $\frac{9}{5}$

31. -5 **33.** -1 **35.** 7

37.

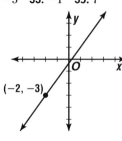

39. Sample answer: (7, 6). The slope of the line containing A and B is -2. Use the slope to go 2 units down and 1 unit to the right of either point.
41. about 1478 feet
43. 11,160 feet **45.** 77; 5; 8; 3.2; 4.8 **47.** 12% **49.** -6

51a. 12 animals
51b.

51c. 15 years **51d.** 4 animals

Pages 336–338   Lesson 6–2
5. 4; (2, -3) **7.** 3; (-7, 1) **9.** $y - 4 = -3(x + 2)$
11. $3x + 4y = -9$ **13.** $2x - y = -6$ **15.** $y + 2 = -\frac{4}{7}(x + 1)$
or $y - 2 = -\frac{4}{7}(x + 8)$; $4x + 7y = -18$ **17.** $3x - 5y = -2$
19. $y - 5 = 3(x - 4)$ **21.** $y - 1 = -4(x + 6)$ **23.** $y - 3 = -2(x - 1)$ **25.** $y + 3 = \frac{3}{4}(x - 8)$ **27.** $4x - y = -5$
29. $3x - 2y = -24$ **31.** $2x + 5y = 26$ **33.** $y - 2 = -\frac{1}{3}(x + 5)$ or $y + 1 = -\frac{1}{3}(x - 4)$ **35.** $y + 1 = \frac{3}{7}(x + 8)$ or $y - 5 = \frac{3}{7}(x - 6)$ **37.** $y + 2 = 0(x - 4)$ or $y + 2 = 0(x - 8)$
39. $4x + 3y = 39$ **41.** $11x + 8y = 17$ **43.** $36x - 102y = 61$
45. $\frac{5 - 1}{5 - 9} = \frac{4}{-4}$ or -1; An equation of the line is $(y - 1) = -1(x - 9)$. Let $y = 0$ in the equation and see if $x = 10$.
$$0 - 1 = -x + 9$$
$$-10 = -x$$
$$10 = x$$
(10, 0) lies on the line. Since (10, 0) is a point on the x-axis, the line intersects the x-axis at (10, 0).
47a. No; for a rise of 30 inches the ramp must be 30 feet long, but there only 18 feet available.
47b.

sidewalk

49a. $0.26, $0.09 **49b.** $0.32 **51.** 28 **53.** No; because there could be 2 pink and 1 white or 1 pink and 2 white.
55. associative property of addition

Pages 343–345   Lesson 6–3
7. positive **9a.** It gets better. **9b.** yes; negative
9c. (7, 18) and (8, 16) **11.** positive **13.** no **15.** no

17. positive **19.** negative **21.** Yes; sample reason: Dots are grouped in an upward diagonal pattern. **23.** c; If 1 is correct then 19 are wrong, if 2 are correct, then 18 are wrong, and so on. Graph c shows these pairs of numbers. **25a.** Sample answer: The more taxation increases, the more in debt the government becomes.
25b. Sample answer: You work harder and your grades go up. **25c.** Sample answer: As more money is spent on research, fewer people die of cancer. **25d.** Sample answer: Comparing the number of professional golfers with the number of holes-in-one. **27a.** The correlation shown by the graph shows a slightly positive correlation between SAT scores and graduation rate. **27b.** Sample equation: $y = 0.89x + 1142.17$

29a., c.

29b. positive **29c.** Sample answer: $15x - 13y = 129$
31a. $7500x - y = 120,000$ **31b.** 15,000 feet **31c.** No; it only describes the plane's path in that part of the flight.
33. $75°$ **35.** 1.45

Pages 350–353   Lesson 6–4
7a. 2 **7b.** 2, -4 **7c.** $y = 2x - 4$ **9.** $-\frac{28}{3}, \frac{7}{2}$
11. $y = \frac{2}{3}x - 10, 2x - 3y = 30$ **13.** $-2, -4$ **15.** $x + 5y = 13$
17. $y = \frac{11}{3}x; 44$ **19.** $-\frac{3}{2}; 0, 0; y = -\frac{3}{2}x$ **21.** $\frac{2}{5}, 5, -2, y = \frac{2}{5}x - 2$ **23.** 2, $\frac{8}{7}$ **25.** $-\frac{5}{6}, 5$ **27.** none, 6
29. $y = 7x - 2, 7x - y = 2$ **31.** $y = -1.5x + 3.75, 6x + 4y = 15$ **33.** $y = -7, y = -7$ **35.** $-\frac{5}{4}, \frac{5}{2}$ **37.** $-4, 12$
39. $\frac{4}{5}, \frac{4}{15}$ **41.** $y = -2$ **43.** $x = 3$ **45.** $y = 9$
47. $y = \frac{11}{24}x, \frac{33}{2}$ **49.** $y = \frac{2}{3}x - \frac{8}{3}$ **51.** ($-3, -1$)
53. $y = 2.04x - 21.27$ **55a.** 42.24 ft^3 **55b.** 270 K
57a. $y = 0.1x + 3$ **57b.** Go to the other bank, since this one would charge you $5.50. **59.** 0

61.

63. 13 ft 11 in.
65. -9
67. $14x + 14$

A (5, -2)

Page 353   Self Test
1. -1 **3.** undefined **5.** $y - 4 = \frac{1}{2}(x + 6)$ **7.** $x + 5y = 17$
9. $-6, -14$

Page 355   Lesson 6–5A

1. All graphs are of the family $y = -ax + 0$, where a represents different negative slopes.

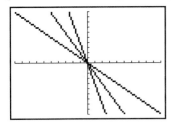

3. All graphs have the same slope, -1, but have different y-intercepts.

5.

7.

9. $y = -x + 2.5$

Pages 359–360   Lesson 6–5

7.

$y + 5 = -2(x + 1)$

9.

$y = 2x - 3$

11.

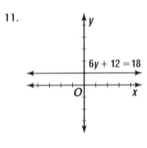

$6y + 12 = 18$

13a.

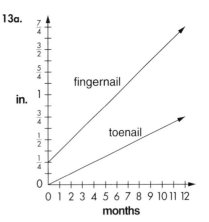

fingernail

toenail

months

13b. $1\frac{3}{4}$ inches   **13c.** $\frac{1}{16}$ inch   **13d.** See 13a.

13e. the rate of growth per month

15.

$6x - y = 9$

19.

$2(x - 3) = y + \frac{3}{2}$

23.

$y = 3x - 5$

27.

$5x + 2y = 20$

31.

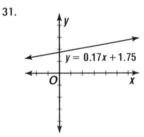

$y = 0.17x + 1.75$

35.

All lines have the same slope 4.

37a.

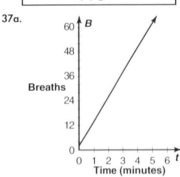

Breaths

Time (minutes)

37b. 122 breaths

39. $y = \frac{2}{5}x + 12$

41. 4:30 P.M.   **43.** 1.8

45. a number m minus 1

Pages 366–368 Lesson 6–6

7. $\frac{2}{3}, -\frac{3}{2}$ **9.** perpendicular   **11.** $y = \frac{5}{6}x - \frac{21}{2}$

13. $y = x - 9$ **15.** $y = \frac{9}{5}x$ **17.** perpendicular

19. parallel   **21.** perpendicular   **23.** perpendicular

25. $y = x + 1$ **27.** $y = \frac{8}{7}x + \frac{2}{7}$ **29.** $y = 2.5x - 5$

31. $y = \frac{2}{3}x + 4$ **33.** $y = -\frac{9}{2}x + 14$ **35.** $y = 3x - 19$

37. $y = -\frac{1}{5}x - 1$ **39.** $y = -\frac{7}{2}x + 11$ **41.** $y = -3$

43. $y = \frac{5}{4}x$ **45.** $y = \frac{1}{3}x - 6$ **47.** No, because the slope

of $\overline{AC}$ is $\frac{6}{7}$ and the slope of $\overline{BC}$ is $-\frac{2}{3}$. These slopes are

not negative reciprocals of each other, so the lines are

not perpendicular and the figure is not a rhombus.

49a. $y = \frac{1}{2}x + \frac{7}{2}, y = -2x + 11$ **49b.** right or 90° angle

51.

53. \$242.80   **55a.** $\frac{5}{12}; \frac{7}{12}$ **55b.** 16-karat gold

57. multiplicative identity

Pages 372–374 Lesson 6–7

5. (1.5, 6)   **7.** (−6, 1)   **9.** (−11, 7)   **11.** (8, 9.8)

13. (12.5, 6)   **15.** (1, 5)   **17.** (3, 2)   **19.** $\left(\frac{1}{2}, 1\right)$

21. (0.8, 2.7)   **23.** (4x, 9y)   **25.** (−7, 0)   **27.** (21, −6)

29. (9, 10)   **31.** $\left(\frac{5}{6}, \frac{1}{3}\right)$ **33.** B(2.3, 6.8)   **35.** P(6.65, −1.85)

37. (−1, 5)   **39.** $\left(1, \frac{5}{2}\right)$ **41a.** N(6, 3), M(10, 3)

41b. parallel, $MN = \frac{1}{2}AB$ **43a.** P(−4, 1), Q(10, −1),

R(2, 9)   **43b.** 62 square units; Sample answer: The area of

the smaller triangle is $\frac{1}{2}bh$. Since the base of the larger

triangle is twice that of the smaller one and the height is

also twice the length of the small one, the area of the

larger is $\frac{1}{2}(2b)(2h)$, or $2bh$. This is 4 times the area of the

small one.   **45.** $y = -\frac{7}{9}x - \frac{8}{3}$ **47.** yes; $9x - 6y = 7$

49. −20   **51.** $3.1x + 1.54$

Page 375 Chapter 6 Highlights

1. parallel   3. midpoint   5. perpendicular

7. slope-intercept   9. slope

Pages 376–378 Chapter 6 Study Guide and Assessment

11. $-\frac{1}{3}$ **13.** $\frac{2}{5}$ **15.** $\frac{25}{3}$ **17.** $y + 3 = -2(x - 4)$

19. $y - 7 = 0$ **21.** $y - 3 = \frac{3}{5}x$ **23.** $y - 1 = -\frac{6}{7}(x - 4)$

25. $3x - y = 18$ **27.** $3x - 4y = 22$ **29.** $y = 5$ **31.** $x = -2$

33a. Yes; it is positive   **33b.** Sample answer: 35 stories

33c. $y = 0.03x + 21$

35. $y = -3x$
37. $y = 5.5$
39. −8, 4
41. −2, 12

43.   **45.**

47.   **49.**

51. $y = -\frac{7}{2}x - 14$ **53.** $y = -\frac{3}{8}x + \frac{13}{2}$ **55.** $y = 5x - 15$

57. $\left(1, -\frac{5}{2}\right)$ **59.** $\left(5, \frac{11}{2}\right)$ **61.** $\left(\frac{7}{2}, -\frac{3}{2}\right)$ **63.** (13, 11)

65. (−5, 20)   **67.** $d = 45t - 10$

CHAPTER 7   SOLVING LINEAR EQUATIONS

Pages 388–390 Lesson 7–1

7. c   **9.** d   **11.** $\{x \mid x > -5\}$ **13.** $\{y \mid y < -5\}$

15. $x - 17 < -13, \{x \mid x < 4\}$

17. $\{a \mid a < 18\}$

19. $\{x \mid x \le 1\}$

21. $\left\{x \mid x > \frac{11}{3}\right\}$

23. $\{x \mid x > 2\}$

25. $\left\{x \mid x < \frac{3}{8}\right\}$ **27.** $\{x \mid x \le 15\}$ **29.** $\{x \mid x < 0.98\}$

31. $\{r \mid r < 10\}$ **33.** $x - (-4) \ge 9, \{x \mid x \ge 5\}$ **35.** $3x < 2x + 8, \{x \mid x < 8\}$ **37.** $20 + x < 53, \{x \mid x < 33\}$ **39.** $2x > x - 6, \{x \mid x > -6\}$ **41.** 12   **43.** −2   **45a.** no   **45b.** yes

45c. yes   **45d.** yes   **47.** The value of x falls between −2.4

and 3.6.   **49a.** $x \le \$12.88$ **49b.** Sample answer: There

may be sales tax on his purchases.   **51.** $y = -3x + 3$

53. 42, 131, 145, 159, 28   **55.** 12   **57.** <

Page 391   Lesson 7–2A

1. Sample answer: The variable remains on the left, but the inequality symbol is reversed.   **3.** When the coefficient of x is positive, you can solve the inequality like an equation and retain the same inequality symbol. If the coefficient of x is negative, you can solve like an equation, but the symbol must be reversed.

Pages 396–398   Lesson 7–2

9. multiply by $-\frac{1}{6}$ or divide by -6; yes; $\{y \mid y \le 4\}$

11. multiply by 4; no; $\{x \mid x < -20\}$ **13.** $\{x \mid x < 30\}$

15. $\{t \mid t \le -30\}$ **17.** $\frac{1}{5}x \le 4.025$; $\{x \mid x \le 20.125\}$ **19.** $s \ge 12$

21. $\{b \mid b > -12\}$ **23.** $\{x \mid x \ge -44\}$ **25.** $\{r \mid r < -6\}$

27. $\{t \mid t < 169\}$ **29.** $\{g \mid g \ge 7.5\}$ **31.** $\{x \mid x \ge -0.7\}$

33. $\left\{r \mid r < -\frac{1}{20}\right\}$ **35.** $\{x \mid x < -27\}$ **37.** $\{m \mid m \ge -24\}$

39. $36 \ge \frac{1}{2}x$; $\{x \mid x \le 72\}$ **41.** $\frac{3}{4}x \le -24$; $\{x \mid x \le -32\}$

43. $-8x \le 144$; -18 or greater   **45.** $y < 7.14$ meters   **47.** $\ge$ **49.** $<$ **51.** up to 416 miles   **53.** at least 5883 signatures   **55.** $(-1, 1)$ **57.** -2 **59.** \$155.64   **61.** 65 yd by 120 yd

Pages 402–404   Lesson 7–3

7. c   **9.** $\{x \mid x > 2\}$ **11.** $\{d \mid d > -125\}$ **13.** $\{2, 3\}$

15a. $x + (x + 2) > 75$ **15b.** $x > 36.5$ **15c.** Sample answer: 38 and 40.   **17.** $\{-10, -9, \ldots, 2, 3\}$ **19.** $\{-10, -9, \ldots, -5, -4\}$ **21.** $\{t \mid t > 3\}$ **23.** $\{w \mid w \le 15\}$ **25.** $\{n \mid n > -9\}$ **27.** $\{m \mid m < 15\}$ **29.** $\{x \mid x < -15\}$ **31.** $\left\{p \mid p \le \frac{14}{3}\right\}$ **33.** $\{x \mid x > -10\}$ **35.** $\{k \mid k \le -1\}$ **37.** $\{y \mid y < -1\}$ **39.** $3(x + 7) > 5x - 13$; $\{x \mid x < 17\}$ **41.** $2x + 2 \le 18$ for $x > 0$; 7 and 9; 5 and 7; 3 and 5; 1 and 3   **43.** no solution $\{\varnothing\}$ **45a.** $x \le -8$ **45b.** $x > 8$ **45c.** $x > 2$ **45d.** $x \le -1$ **47.** $x + 0.04x + 0.15(x + 0.04x) \le \$50$, $x \le \$41.80$ **49.** at least \$571,428.57   **51a.** at most 2.9 weeks   **51b.** no change   **51c.** at most 4.1 weeks   **53.** $\{y \mid y > 10\}$ **55.** $3x + 2y = 14$ **57.** $\{-5, -3, -2, 4, 16\}$ **59.** 25.1; 23.5; no mode

Pages 409–412   Lesson 7–4

7. $0 \le x \le 9$

9. $-3 < x \le 1$ **11.** The solution is the empty set. There are no numbers greater than 5 but less than -3.

13. $\{h \mid h \le -7 \text{ or } h \ge 1\}$

15. $\{w \mid 1 > w \ge -5\}$

17. Drawings will vary; 16 pieces.   **19.** $\varnothing$

21.
23.

25. $-4 \le x \le 5$ **27.** $x \le -2$ or $x > 1$

29. $\{x \mid -1 < x < 5\}$

31. $\{x \mid x < -2 \text{ or } x > 3\}$

33. $\{c \mid c < 7\}$

35. $\varnothing$

37. $\{x \mid x \text{ is a real number.}\}$

39. $\{y \mid y > 3 \text{ and } y \ne 6\}$

41. $\{x \mid x \text{ is a real number.}\}$

43. $\{w \mid w < 4\}$

45. Sample answer: $x > 5$ and $x < -4$ **47.** $n + 2 \ge 6$ or $n + 2 \ge 10$; $\{n \mid n \le 4 \text{ or } n \ge 8\}$ **49.** $31 \le 6n - 5 \le 37$; $\{n \mid 6 \le n \le 7\}$

51. $\{m \mid -4 < m < 1\}$

53a. $\{x \mid x < -7 \text{ or } x > 1\}$ **53b.** $\{x \mid -5 \le x < 1\}$ **55.** $-4 \le x \le -1.5$ or $x \ge 2$ **57.** $4.4 < x < 6.7$ **59.** $\left\{m \mid m \ge \frac{44}{3}\right\}$ **61.** -5

63.

65. a little more than half a mile   **67.** $18px - 15bg$

Page 412   Self Test

1. $\{y \mid y \ge -17\}$ **3.** $\{n \mid n < 4\}$ **5.** $\{g \mid g < -5\}$ **7.** c   **9.** more than 17 points

Pages 415–419   Lesson 7–5

7a. outcomes from tree diagram:
burger, soup, lemonade; burger, soup, soft drink; burger, salad, lemonade; burger, salad, soft drink; burger, french fries, lemonade; burger, french fries, soft drink; sandwich, soup, lemonade; sandwich, soup, soft drink; sandwich, salad, lemonade; sandwich, salad, soft drink; sandwich, french fries, lemonade; sandwich, french fries, soft drink; taco, soup, lemonade; taco, soup, soft drink; taco, salad, lemonade; taco, salad, soft drink; taco, french fries, lemonade; taco, french fries, soft drink; pizza, soup, lemonade; pizza, soup, soft drink; pizza, salad, lemonade; pizza, salad, soft drink; pizza, french fries, lemonade; pizza, french fries, soft drink   **7b.** $\frac{1}{3}$ or $0.\overline{3}$ **7c.** $\frac{1}{12}$ or $0.08\overline{3}$ **7d.** $\frac{1}{24}$ or $0.041\overline{6}$ **9a.** 15   **9b.** $\frac{1}{5}$ or 0.2   **11.** $\frac{1}{3}$ or $0.\overline{3}$

13a. R3-G5, R3-R10, R3-B10, R3-G1, R3-Y14, B3-G5, B3-R10, B3-B10, B3-G1, B3-Y14, R5-G5, R5-R10, R5-B10, R5-G1, R5-Y14, R14-G5, R14-R10, R14-B10, R14-G1, R14-Y14, Y10-G5, Y10-R10, Y10-B10, Y10-G1, Y10-Y14   **13b.** $\frac{3}{25}$ or 0.12

13c. $\frac{2}{25}$ or 0.08   **13d.** 0   **13e.** $\frac{14}{25}$ or 0.56   **15a.** about 5.6%   **15b.** about 26.3%   **17.** 32%   **19.** between 83 and 99, inclusive   **21.** 3; -9 **23a.** $50°$ **23b.** $130°$ **23c.** yes

Pages 423–426   Lesson 7–6

5. c   **7.** c   **9.** d

11. $\{m \mid m \le -5 \text{ or } m \ge 5\}$

13. $\{r \mid -9 < r < 3\}$

15. $|x| = 2$

17. {−2, 6}

−6 −4 −2 0 2 4 6 8 10

19. ∅

21. {y | 1 ≤ y ≤ 3}

−2 −1 0 1 2 3 4 5 6

23. ∅

25. $\left\{e \mid \frac{5}{3} < e < 3\right\}$

1 2 3 4

27. {y | y is a real number.}

−4 −3 −2 −1 0 1 2 3 4

29. {w | 0 ≤ w ≤ 18}

0 2 4 6 8 10 12 14 16 18

31. {−2, 3}

−4 −3 −2 −1 0 1 2 3 4

33. $\left\{x \mid x \le -\frac{8}{3} \text{ or } x \ge 4\right\}$

−4 −3 −2 −1 0 1 2 3 4 5 6

35. $|p - 1| \le 0.01$ **37.** $|t - 50| > 50$
39. $|x + 1| = 3$ **41.** $|x - 1| \le 1$ **43.** $|x - 8| \ge 3$
45. {−2, −1, 0, 1, 2} **47.** $2a + 1$ **49.** $a \ne 0$; never
51. $\frac{8}{13}$ or 0.61 **53.** no; $52 \le s \le 66$ **55.** $16,500 \le p \le$

$18,000 **57a.** Outcomes from tree diagram: BBBB,
BBBG, BBGB, BBGG, BGBB, BGBG, BGGB, BGGG, GBBB,
GBBG, GBGB, GBGG, GGBB, GGBG, GGGB, GGGG
57b. $\frac{1}{16}$ or 0.0625, regardless of gender **57c.** $\frac{3}{8}$ or 0.375
59. {x | x ≥ −1} **61.** {k | k ≥ −15}

63.

$2x - 9 = 2y$

65. $m = -6 - \frac{n}{2}$

Pages 430–432 Lesson 7-7
7a. A; 25, 65, 30, 60, 40; B; 20, 70, 40, 60, 45 **7b.** B **7c.** A
7d. B **9a.** Q2 = 6.5, Q3 = 16, Q1 = 5, IQR = 11 **9b.** no

9c.

0 5 10 15 20 25 30 35

11a.

0 5 10 15 20 25 30 35

11b. clustered with lots of outliers **11c.** There are four
western states that have more American Indian people
than other states. **11d.** It is greater than the median.
13a.

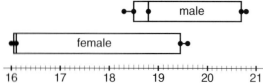

male

female

16 17 18 19 20 21

13b. 1990 **15.** Sample answer: Class A appears to be a
more difficult class than B because the students don't do
as well. **17.** {m | m > 1}

834 *Selected Answers*

19.

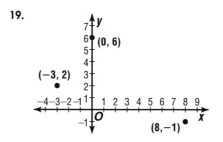

(0, 6)

(−3, 2)

−4 −3 −2 −1   1 2 3 4 5 6 7 8 9

(8, −1)

Page 434   Lesson 7-7B
1. women identifying objects with their left hands
3. The left hand data are more clustered. **5.** males

Page 435   Lesson 7-8A
1.

3.

5.

7.

9.

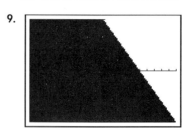

Page 439–441   Lesson 7-8
5. a **7.** b **9.** a, c; no

11.

13.

15.

17.

19. $\{1, 1), (1, 2)\}$ **21.** $\varnothing$

23.

25.

27.

29.

31.

33.

35.

37.

39.

41.

43. c   **45.** a, b, d
47a. $0.7(220 - a) \leq z \leq 0.8(220 - a)$
47b. $32 \leq z \leq 37$
47c. improve cardiovascular conditioning

49.

51. $y = 4x - 2$

53.

55. $138.22

Page 443   Chapter 7 Highlights
1. h   **3.** i   **5.** c, f   **7.** d   **9.** g   **11.** a

Pages 444–446   Chapter 7 Study Guide and Assessment
13. $\{n \mid n > -35\}$ **15.** $\{p \mid p \leq -18\}$ **17.** $3n > 4n - 8$, $\{n \mid n < 8\}$ **19.** $\{w \mid w \leq -15\}$ **21.** $\{x \mid x > 32\}$
23. $-\frac{3}{4}n \leq 30$, $\{n \mid n \geq -40\}$ **25.** $\{-5, -4, \ldots 0, 1\}$
27. $\left\{y \mid y \leq -\frac{9}{2}\right\}$ **29.** $\left\{x \mid x > -\frac{5}{2}\right\}$ **31.** $\{z \mid z \leq 20\}$
33. $\{a \mid a$ is a real number$\}$

35. $\{b \mid b \leq 5\}$

37a. $\frac{2}{7}$ **37b.** $\frac{1}{2}$ **37c.** $\frac{2}{7}$ **37d.** 0 **37e.** $\frac{3}{14}$ **37f.** $\frac{1}{14}$

39. $\{y \mid y > -5$ or $y < -5\}$

41. $\{k \mid 3 \geq k \geq -4\}$

43. $\left\{y \mid y \geq \frac{21}{5}$ or $y \leq 1\right\}$

45.

47. $\{(2, -1), (-1, 1)\}$ **49.** $\{(5, 10), (3, 6)\}$

51. 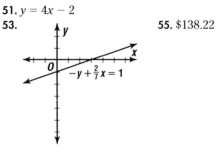   **53.**

55. 17 to 20 books

Martin; **392** The Gordon Hart Collection, Bluffton, IN; **394** Life Images; **395** Kenji Kerins; **397** (l)courtesy the office of Congresswoman Patsy Mink, (r)Doug Martin; **398** (l)Duomo/Paul Sutton, (r)Mark Lawrence/The Stock Market; **399** Aaron Haupt; **404** (t)Kenji Kerins, (b)Doug Martin; **405** (t)Alain Evrard/Photo Researchers, (c)Nikolas/Konstantinou/Tony Stone Images, (b)Norbert Wu/Tony Stone Images; **407** Aaron Haupt; **411** Pal Hermansen/Tony Stone Images; **413** (t)Suzanne Murphy-Larronde/FPG (b)Bill Ross/Westlight; **415** (bkgd)Duomo/William Sallaz,(t)KS Studio; **416** Travelpix/FPG; **417** (t)KS Studio, (b)Matt Meadows; 418 (t)Eclipse Studios, (b)Skip Comer; **419** Aaron Haupt; **420** (t)Joseph Drivas/The Image Bank, (bl)Michael Salas/The Image Bank; **422** Aaron Haupt; **423** Scott Cunningham; **425** (t)David R. Frazier/Photo Researchers, (b)MAK-1; **429** C.T. Tracy/FPG; **430** Gary Buss/FPG; **431** Lawrence Migdale; **432** courtesy The White House; **433** KS Studio; **436** Lori Adamski Peek/Tony Stone Images; **438** Court Mast/FPG; **440** Kenji Kerins; **442** (bkgd)Jane Shemilt/Science Photo Library/Photo Researchers, (t)Aaron Haupt, (b)Photo Researchers.

APPLICATIONS & CONNECTIONS INDEX

This index contains terms from Volumes One and Two. Pages 1–447 are contained in Volume One. Pages 382–753 are contained in Volume Two.

A

Academics. *See* Colleges; Education; School

Accounting, 23, 294, 468. *See also* Economics

Acoustics, 694

Advertising, 225, 404

Aeronautics, 664. *See also* Aviation; Space science

Agriculture, 31, 259, 269, 518, 616, 647, 714

Air conditioning, 172, 741

Airplanes
airline fares, 108, 287
atmospheric pressure on, 664–665
drilling holes in, 98
passenger statistics, 226, 506
pilots, 112, 507
routes, 416
slope of descent, 337, 512
speed of, 236, 248, 338, 505, 701
takeoff, 345

Anatomy, 276

Animals, 80, 172, 262, 264, 331, 341, 648

Anthropology, 702

Aquatics, 295, 300

Archaeology, 98, 213, 301, 413, 638–639

Architecture
art gallery, 626
housing development, 461
Leaning Tower of Pisa, 206, 208, 599
model home, 426
Sears Tower, 526
similar triangles, 201
slope of roof, 329, 331
tall buildings, 29, 183, 526

Art, 514, 536, 658, 659, 740

Astronomy, 508, 617, 642, 684, 685, 747. *See also* Space science

Automobiles, 330
antifreeze in, 237
braking distance for, 735
commuting in, 183, 486
depreciation of, 226
gas needed for, 198, 432
leasing vs. purchasing, 645
loans for, 505, 527, 552
maintenance schedule for, 688
MPG ratios for, 25, 26
options in purchasing, 446
popular, 311
rental of, 318, 397
repair charges for, 527
traffic and, 124, 667, 668, 670

Aviation, 98. *See also* Airplanes

Babies, 160, 418

Babysitting, 10

Balloons, 210, 490

Banking, 346, 352, 518. *See also* Finance

Birth rates, 429

Birth weight, 160

Book binding, 104

Book publishing, 4–5

Bridges, 205, 599

Budgeting, 294, 386–387, 398

Business and work
car repairs, 527
car wash, 404, 700
employment, 49, 100, 239–240, 266, 277, 285, 482, 593
keyboarding speeds, 176, 352
lawn care service, 696, 706
mileage reimbursement, 109
modular offices, 540
profit calculations, 40, 404
strikes, 404
toy evaluation, 460
T-shirt day, 416
See also Commerce; Sales

C

Carpentry, 116, 166, 330, 664, 731

Cartography, 256

Charities, 62, 178

City planning, 642

Civil engineering, 214

Climate. *See* Meteorology

Colleges, 238, 311, 382, 616
entrance exams for, 339, 345, 431, 471

Commerce, 308, 627, 697

Communication, 149, 153, 292, 475–476, 710, 739, 740

Community service, 608, 609

Commuting, 183, 486

Computers
CD-ROMs for, 620
chess and, 419
cost of, 353
graphing software for, 471, 733
keyboarding speed, 127
memory of, 495
molecular, 649
on-line services, 226
programming, 8, 49, 171
shipments for home use, 98
spreadsheets, 297, 327, 646
student use of, 418

Construction
building codes, 110
clearances, 325, 328
of deck, 566–567
of dining area with fountain, 747
excavation site, 674
of garden, 533
handrails for physically challenged persons, 422–423
ladders for, 227

lot size, 325, 328, 592
paneling, 717
ramps for people with disabilities, 150, 151
of roads, 330, 368
roof support, 353
slip forming, 730
slope of roof, 350
of swimming pool, 540, 566–567

Consumerism
cost comparisons, 95, 116, 136, 149
credit cards, 220, 302
discounting, 223, 224, 225, 226, 227
leasing vs. purchasing, 645
MPG ratios of automobiles, 25, 26
payment plans, 600
printer prices, 541
See also Sales, Economics

Contests, 228

Cooking, 481, 670

Cosmetics, 215

Cultures
Aztec agriculture, 714
Chinese calendar, 50
Chinese kite-making, 362
Egyptian pyramids, 24, 474
Gupta Empire (India), 460
Hawaiian alphabet, 160
Japanese royal wedding, 156

Currency. *See* Money

D

Decision making, 654

Demographics. *See* Population statistics

Diet, 70–71
for diabetic, 227
guidelines for, 384, 385
planning, 71
See also Food; Nutrition

Drafting, 116, 205

Driving. *See* Automobiles

E

Earthquakes, 30

Ecology, 205, 313. *See also* Environment

Economics
allowances, 50
cost comparisons, 95, 116, 136, 149
depreciation, 56–57, 226
farming, 31
feeding the world, 511
inflation, 161
money, 115, 220, 237, 641, 670
stock market, 19, 102, 103, 136, 468, 496, 500, 580
See also Finance; Consumerism

Education
ACT Program scores, 431
earnings potential and, 382
SAT scores, 339, 345, 471
See also Colleges; School

Electricity, 136, 547, 698, 701, 725

Electronics, 698, 699, 700, 701, 706

Elevators, 91

Employment, 49, 100, 239–240, 266, 277, 285, 482, 593. *See also* Business and work

Energy, 647

Engineering, 214, 399, 684, 735

Entertainment
balloons, 210, 490
carnivals, 465
cartoons, 469
fireworks, 613–614
games, 72
ice cream party, 43
movies, 197, 199, 227, 269, 312, 378, 469, 510, 557
music, 62, 77, 237, 241, 244, 252–253, 571, 674
on-line, 473
parades, 671, 672, 687
radio, 648
state fairs, 198
television, 78, 318, 425
theater, 311, 467
See also Games; Sports

Environment
energy resources, 647
hazardous waste sites, 431
plastic bottle usage, 205
recycling, 11, 93, 199
reducing pollutants, 680
water usage, 55, 239

Farming, 31, 259, 269, 518, 616, 647, 714

Finance
banking, 346, 352, 518
budgeting, 294, 386–387, 398
car loans, 505, 527, 552
checking accounts, 251, 346, 352
credit cards, 220, 302
fund raising, 124, 404, 706
income, 533
mortgage payments, 505
real estate, 404, 527
savings accounts, 515, 648, 654
savings plan, 318
spending, 390, 403, 425
stocks, 19, 102, 103, 136, 468, 496, 500, 580
trust fund, 552
See also Economics; Investments

Fireworks, 613–614

Folklore, 635

Food
calcium in, 486
Calories in, 35, 66, 220, 311, 552, 747
chocolate, 195, 196, 232
cost of, 132
drinks, 35, 234, 248
fats in, 32, 34, 161, 220, 384, 385
marshmallows, 222
probability, 417
in restaurants, 188
sales of, 104, 485
in school cafeteria, 45–46
tacos, 747
for world population, 511
See also Diet; Nutrition

Forests, 92, 220, 725, 752

Fountains, 599

Fund-raising, 124, 404, 706

Games, 403, 450–451
Battleship, 753
billiards, 204
cards, 417
chess, 419
contests, 228
hopscotch, 529, 533, 546
house of cards, 18
Jeopardy!, 231
Monopoly, 72
Olympic, 83, 407–408, 427, 429
Price Is Right, 425
puzzles, 91, 707
Rook, 674
Scrabble, 228
Twister, 417
video, 416
Yahtzee, 415
See also Sports

Gardening, 563, 570, 599

Geography
altitude, 332
areas of South American countries, 183
Great Pyramid (Egypt), 24
latitude and longitude, 258
New York City, 462
rivers, 31, 743, 744
time zones, 76, 580
Washington, D.C., 480
water on Earth's surface, 580

Geology
earthquakes, 30
underground temperature, 312

Handicrafts, 12

Hardware, 298

Health
AIDS virus, 509
bacteria, 511
birth weight, 160
blood pressure, 290
burning Calories, 280, 282–283, 552
caloric requirements, 301
CPR, 360
daily intake of fat, 161, 384, 385
exercise heart rate, 177, 440–441
fingernail growth, 359
heartbeats, 674
height differences, 27–28
ideal weight, 188, 364
life expectancy, 352
medicine, 633
muscle length, 226
nursing, 113–114
percentage of fractures by age, 219
Sun Protection Factor (SPF), 6
weight training, 285
See also Diet; Nutrition

Home economics, 188, 294

Home repair, 157

Humor, 467

Hydraulics, 631

Icebergs, 681, 684

Immigrants, 322, 323

Insurance, 519, 647

Interest
compound, 648, 654, 670
simple, 66, 217, 218, 219, 220, 238, 244, 251, 277, 461, 505, 518, 552, 593, 643–644, 647

Interior design, 675, 677

Investments
certificates of deposit (CDs), 593, 643–644, 654
interest on, 66, 217, 218, 219, 220, 238, 244, 251, 277, 461, 505, 518, 552, 593, 643–644, 647, 648, 654, 670
real estate, 404, 527
savings accounts, 515, 648, 654
stocks, 19, 102, 103, 136, 468, 496, 500, 580

Kite-making, 362, 364

Labor. *See* Business and work

Landmarks
Cincinnati parks, 254
Gateway Arch (St. Louis), 611
Great Pyramid (Egypt), 24
Leaning Tower of Pisa, 206, 208, 599
national parks, 511, 626
Royal Gorge, Colorado, 541
Statue of Liberty, 593
tall buildings in North America, 183, 526

Landscaping, 491, 656–657, 663, 665, 695

Law enforcement, 237, 425, 752

Leisure. *See* Entertainment; Games; Sports

Levers, 242, 392, 393

Life expectancy, 352

Manufacturing, 30, 293, 430, 436, 438, 519

Marketing, 52

Metallurgy, 293, 368

Meteorology
precipitation, 81, 617
temperature, 55, 74, 76, 176, 177, 305–306, 313, 430, 541, 633
thunder and lightning, 31, 284
tornadoes, 35, 82
windchill factor, 76

Military, 124

Money, 115, 220, 237, 641, 670

Motion, uniform, 235–237, 477–478, 479

Nature, 752

Networks, 51

APPLICATIONS & CONNECTIONS INDEX

Nutrition
calcium, 486
Calories, 35, 66, 220, 280, 282–283, 301, 311, 552, 747
for diabetic, 227
dietary guidelines, 384, 385
fats, 32, 34, 161, 220, 384, 385
restaurant meals, 188
See also Diet

Oceanography, 732
Oil leases, 600
On-line computer systems, 473, 481
Optics, 419

Packaging, 560
Parks, 254, 511, 626
Personal finance. *See* Finance
Pet ownership, 172
Photography, 547, 604
Physical fitness, 77, 285, 474. *See also* Sports
Political science, 397
Population statistics
on American Indians, 431
for cities, 180, 462, 644, 689
for countries, 182, 647
for local communities, 648
on population growth, 276, 374, 643
for states, 462, 643, 695
on teenagers, 153
Postal service, 43, 188, 522, 526
Predictions, 226
Printing, 338
Psychology, 702
Puzzles, 91, 707
Pyramids, 24, 474
Pyrotechnics, 613–614

Radioactivity, 638–639, 648
Railroads, 211, 238, 259, 673, 674
Real estate, 404, 527
Records, world, 154, 454–455, 600
Recreation. *See* Entertainment; Games; Sports
Recycling, 11, 93, 199
Rescue missions, 596–597
Restaurants, 104, 188, 416
Road construction, 330, 368

Sales
calculating profits from, 40, 404
clearance, 167, 250
comparisons of, 236
discounts on, 223, 224, 225, 226, 227
displays for, 18
of food, 91, 104, 485
of music, 62
price vs. quantity, 57
retail, 80, 172, 200, 244, 268, 466, 481, 702, 741
by students, 237
volume of, 394
weekend specials, 490
Sales tax, 223, 225, 505
School
cafeteria food in, 45–46
grades, 233, 238, 390
oral presentations, 232
telephone time, 82
test scores, 139, 339, 345, 417, 431, 471, 477, 730
See also Colleges
Science, 18, 116, 284, 680
biology, 36, 119, 126, 199, 222, 262, 264, 302, 345, 358, 509, 510, 511, 542, 547, 640, 642
chemistry, 54, 236, 299, 352, 405, 424, 474, 509, 533
earth science, 293, 356
physics and physical science, 50, 55, 176, 242, 243, 271, 276, 361, 392, 393, 397, 411, 509, 563, 593, 633, 654, 660, 662–663, 665, 708, 719, 721, 741, 752
Shipping, 446, 579
Sight distance, 752
Social studies
civics and government, 110, 571, 628, 630, 632
history, 81, 167, 179–180, 432, 543–544, 547, 599, 646, 681
Sociology, 153
Sound, 694, 736
Space science, 85, 90, 167, 243, 269, 420, 425, 527
Sports, 450–451
aquatics, 136
archery, 343, 372, 654
athletic field, 563
ballooning, 490
baseball, 76, 214, 238, 294, 598, 702, 716, 752
basketball, 78–79, 181, 277, 378, 412, 415, 593, 642, 746
bicycling, 237, 313, 669, 701
boating, 218, 236, 700, 713, 717, 725
diving, 221, 594, 597, 654
football, 37–38, 75, 144, 145–146, 162, 307, 309, 430, 616
golf, 91, 125, 163, 200, 303, 304, 312
hockey, 36, 232
in-line skating, 313
miniature golf, 286
mountain biking, 313
mountain climbing, 167
Olympic Games, 83, 407–408, 427, 429
racing, 725
rugby, 565, 566, 571
running, 286, 552
sailing, 218, 713, 717, 725
scuba diving, 295
skiing, 378
snowmobiling, 441
soccer, 99, 398
speed skating, 451
swimming, 349, 512
table tennis, 205, 690
tennis, 205, 527, 615
track and field, 104, 168, 481, 530–531
volleyball, 569
women athletes, 468
Stock market, 19, 102, 103, 136, 468, 496, 500, 580
Superheroes, 599
Surveying, 202, 480

Taxes, 223, 225, 226, 505
Technology. *See* Computers
Telecommunications, 149, 153, 292, 475–476, 739, 740
Time zones, 76, 580
Tourism, 259
Traffic, 124, 667, 668, 670
Trains, 211, 238, 259, 673, 674
Transportation
airplanes. *See* Airplanes
automobiles. *See* Automobiles
buses, 17, 235–236
modes of, 413–414
ships, 718, 727, 728
taxis, 533
trains, 211, 238, 259, 673, 674
Travel, 138, 689
camping trips, 736
car rental for, 318, 397
commuting, 183, 486
cost of, 533
cruises, 718, 727, 728
fuel needed for, 198, 432
rate of, 237, 244, 248, 259, 361, 378, 486, 680
speed of, 236, 238, 490, 500, 735
time needed for, 235
vacations, 483, 736
See also Transportation
Trees, 92, 220, 725, 752

Uniform motion, 235–237, 477–478, 479
Utilities, 684

Volunteering, 608, 609

Water supply, 724
Water usage, 55, 239
Waves, 563
Weather. *See* Meteorology
Wildlife, 648. *See also* Animals
Woodworking, 369
Work. *See* Business and work
World cultures. *See* Cultures
World records, 154, 454–455, 600

APPLICATIONS & CONNECTIONS INDEX

INDEX

This index contains terms from Volumes One and Two. Pages 1–447 are contained in Volume One. Pages 382–753 are contained in Volume Two.

A

Absolute value, 85–86, 87, 420–426
function, 355

Acute angles, 165, 400, 731

Acute triangle, 165

Addition
of areas, 46–47
associative property of, 51–55
commutative property of, 51–55, 92
elimination using, 469–474
of integers, 74–76, 84, 86–92
of polynomials, 520–527
of radical expressions, 727–731
of rational expressions, 681–684,
686–688
of rational numbers, 100–104
solving equations with, 144–149
solving inequalities with, 384–390

Addition property
of equality, 144–146
for inequalities, 385–390

Additive identity property, 37, 52

Additive inverse, 87, 523, 682

Algebraic expressions, 6–7, 14, 21, 23

Alternative Assessment, 67, 137, 189, 249,
319, 379, 447, 491, 553, 605, 655, 707, 753

Angles
acute, 165, 400, 731
complementary, 163, 533
congruent, 164
corresponding, 201
of depression, 208
of elevation, 208, 209
measuring, 162–163, 169
supplementary, 162–163, 500
of triangle, 163–164, 395, 525

Applications. *See* Applications & Connections Index

Area
adding, 46–47
of circle, 9, 129, 504, 518, 543–544
of rectangle, 21, 22, 44, 46–47, 130, 275,
397, 499, 517, 518, 532, 558, 562, 576,
579, 585, 586, 589, 604, 642, 679, 680,
708, 729
of square, 121, 396, 504, 585, 604, 708
surface, 11, 128, 129
of trapezoid, 132, 173, 286
of triangle, 380, 571, 627

Arithmetic sequence, 104

Assessment
Alternative, 67, 137, 189, 249, 319, 379,
447, 491, 553, 605, 655, 707, 753
Chapter Tests, 788–800
Math Journal, 28, 41, 53, 170, 175, 181,
224, 236, 274, 283, 291, 343, 350, 359,
388, 415, 439, 478, 484, 499, 504, 531,
561, 591, 624, 631, 668, 672, 699, 723,
729, 734

Portfolio, 67, 137, 184, 189, 249, 314, 319,
379, 442, 447, 491, 548, 553, 605, 650,
655, 707, 748, 753
Self Evaluation, 67, 137, 189, 249, 319,
379, 447, 491, 553, 605, 655, 707, 753
Self Tests, 36, 99, 161, 221, 286, 353, 412,
474, 527, 580, 633, 680, 731
Study Guide and Assessment, 64–67,
134–137, 186–189, 246–249, 316–319,
376–379, 444–447, 488–491, 550–553,
602–605, 652–655, 704–707, 750–753

Associative property, 51–55

Averages, 96, 149, 411, 418
mean, 178–183, 200, 221, 251, 472, 586, 747
median, 178–183, 200, 221, 251, 472, 586
mode, 178–183, 200, 221, 251, 472, 586
weighted, 233–238

Axes
of symmetry, 612, 616
x, 56, 254
y, 56, 254

B

Back-to-back stem-and-leaf plots, 27–28

Balance, 189

Banneker, Benjamin, 480

Bases
percents, 215–216
of power, 7

Base-ten tiles, 118

BASIC programming language, 8

Best-fit lines, 314, 341–345, 352

Binomials
conjugates, 722–723
defined, 514
differences of squares, 545, 581–586
dividing by, 675–680
factoring, 564, 568, 581–586
FOIL method for multiplying, 537, 574
identifying, 515
multiplying, 537–538, 574
square of, 544

Boundary, of half-plane, 437

Box-and-whisker plots, 427–434, 441, 481,
527, 571

Braces { }, 20

Brackets [ ], 20

C

Calculators. *See* Graphing calculators;
Scientific calculators

Campanus, Johannes, 608

Capture-recapture method of counting,
190–191

Carbon dioxide, 68–69, 83, 111, 154, 177,
184

Career Choices
astronomer, 685
computer programmer, 174
construction contractor, 325
ecologist, 455
entomologist, 126
land surveyors, 202
meteorologist, 720
nutritionist, 384
pastry chef, 560
pilot, 507
registered nurse, 6
urban or city planner, 643
zoologist, 264

Celsius, Anders, 177

Celsius temperature scale, 176, 177, 541,
593

Central tendency, measures of, 178–183,
200, 221, 251, 472, 586, 747

Change, percent of, 222–227

Chapter Projects
career research, 383
college expenses, 141
communication of teens and parents,
711
computer memory, 495
diet research and planning, 71
favorite movies, 557
heritage survey, 193
immigrant income, 323
income of volunteers, 609
music survey, 253
Native American art, 659
poetry, 5
wheelchair race, 451

Chapter Tests, 787–799

Charts, 111, 154, 191, 200

Chin Chiu-Shao, 608

Circle(s), 9, 129, 504, 518, 543–544

Circle graphs, 216, 218, 219

Coefficient, 47–48
leading, 158

Commutative property, 51–55, 92

Comparison property, 94–95

Compass, 216

Complementary angles, 163, 533

Complete graph, 278

Completeness properties
for points in plane, 256
for points on number line, 121

Completing the square, 742–747

Complex fractions, 114, 691–695

Composite number, 558

Compound events, 414–419

Compound inequalities, 405–412

Compound interest, 648, 654, 670

Compound sentence, 421

Computers
graphing software for, 471, 733
programming, 8, 49, 171
spreadsheets, 297, 327, 646
See also Applications & Connections Index

Cones, volume of, 62

Congruent angles, 164

Conjugates, 722–723

Connections. *See* Applications & Connections Index

Consecutive integers, 158–159, 160, 161, 402, 446, 574, 599, 604

Consistent systems of equations, 455

Constant, 496
of variation, 239

Continuous Review. *See* Reviews, Mixed

Control experiment, 68, 69, 83, 657

Cooperative Learning Projects
art and framing, 605
balances and solving equations, 189
Battleship™ game, 753
catering charges, 319
cooking, 137
fun puzzle, 707
landscaping, 491
letter occurrence, 249
operating music store, 67
saving for college, 553
sightseeing tours, 655
video rental store policies, 379
window-glass manufacturing, 447

Coordinate(s), 73–75
x, 254, 346
y, 254, 333, 346, 622

Coordinate planes, 254–259
completeness property for points in, 256
graphing systems of equations on, 455
half-plane, 437–438, 482
midpoint of line segment in, 370–374, 571
parallel lines in, 362–368
perpendicular lines in, 364–368

Correlation, 340, 341

Corresponding angles, 201

Corresponding sides, 201

Cosine, 206–214

Cost, unit, 95

Counting
capture-recapture method of, 190–191
estimation in, 200, 227, 302
fish in pond, 190–191, 200, 227, 269, 302, 314

Cousteau, Jacques-Yves, 295

Cross products, 94, 196

Cube, volume of, 501, 649

Cumulative Reviews, 138–139, 250–251, 380–381, 492–493, 606–607, 708–709

D

Data, 25
analysis of, 184, 220, 314, 321, 338, 442, 548, 650, 748
organization of, 464, 468, 480, 486, 490

Data collection
experiments, 68, 69, 83, 321, 361, 449, 461, 519, 541, 657
research, 71, 74, 190–191, 383, 718
sampling, 314
surveys, 80, 193, 253, 711
See also Investigations

Decay, exponential, 644–649

Decimals
expressing rational numbers as, 95
repeating, 95
solving equations with, 147
terminating, 95

Defining variables, 127

Degree
of monomial, 515
of polynomial, 516

Denominators
least common, 686–687, 692
like, 681–684
rationalizing, 722–723
unlike, 685–689

Density formula, 271

Density property, 96

Dependent events, 414

Dependent systems of equations, 456

Dependent variable (quantity), 58–62, 83, 150, 154, 239, 273, 280, 290, 332, 341, 358, 614, 620

Depression, angle of, 209

Diagrams, 259, 337
in problem solving, 406, 407–408, 410, 411, 412, 425
Punnett squares, 542
tree, 413–419
Venn, 73, 121, 418

Differences of squares, 545
factoring, 581–586

Dimensional analysis, 174–175, 668, 673

Diophantus of Alexandria, 172

Direct variation, 239–240, 242–244

Discounting, 223, 224, 225, 227

Discrete mathematics, 88, 90, 110
arithmetic sequence, 104
counting, 190–191, 200, 227, 269, 302, 314
data analysis, 184, 220, 314, 321, 338, 442, 548, 650, 748
geometric mean, 123, 734, 735
geometric sequence, 110
matrix addition, 103

Discriminant, 632

Distance formula, 737–741, 747

Distributive property, 44–50, 52, 89, 114, 536–537, 564–571

Division
by binomials, 675–680

by monomials, 501–505
of polynomials, 675–680
of rational expressions, 671–674
of rational numbers, 112–117
solving equations with, 151–154
solving inequalities with, 392–398

Division property
of equality, 151–152
for inequalities, 393–398

Domain, 263, 264
reasonable, 276, 284, 293, 616, 632

E

Edge of half-plane, 437

Element, 33, 263

Elevation, angle of, 209, 210

Elimination
using addition and subtraction, 469–474
using multiplication, 475–481

Ellipsis (...), 54

Equality, properties of, 39, 46, 52, 144–154, 639

Equally likely outcomes, 229

Equals sign (=), 127

Equation(s), 33
balance in, 189
with decimals, 147
equivalent, 144
for exponential decay, 644–649
for exponential growth, 643–649
functional notation, 289
linear. *See* Linear equations
multi-step, 155–161
one-step, 142–154
in problem solving, 126–132
quadratic. *See* Quadratic equations
radical, 732–736
rational, 696–702
as relations, 271–277
solving, 145, 173–177, 270, 271–277
solving by factoring, 594–600
solving with addition, 144–149
solving with division, 151–154
solving with multiplication, 150–154
solving with subtraction, 146–149
systems of. *See* Systems of equations in two variables, 271–277
with variable on both sides, 168–172
writing, 126–132
writing from patterns, 295–302, 331

Equation mat, 142, 145, 152, 155

Equation models, 142–143

Equilateral triangle, 164

Equivalent equation, 144

Equivalent expressions, 47

Estimation
in counting, 200, 227, 302
in finding roots of quadratic equations, 621
in problem solving, 621
projections, 398
of square roots, 118

Even numbers, 158, 402, 532

Events
compound, 414–419
dependent, 414
independent, 414
probability of, 228, 229
simple, 414

Excluded values, 660–661, 666

Experiments, 68, 69, 83, 321, 361, 449, 461, 519, 541, 657

Exponent(s), 7, 497, 501–505
negative, 503
in scientific notation, 506–512
zero, 502

Exponential decay, 644–649

Exponential functions, 634–642
defined, 635
modeling, 636, 640
property of equality for, 639

Exponential growth, 643–649

Expressions
algebraic, 6–7, 14, 21, 23
equivalent, 47
evaluating, 8, 20, 21, 23
mixed, 690–695
numerical, 6
quadratic, 628
radical, 719–731
rational, 660–674, 681–689
simplest form of, 47, 48, 49, 53, 54, 723
verbal, 7

Extraneous solutions, 697

Extremes, 196

Extreme values, 427–428

F

Fabulous Firsts
Elmer Simms Campbell, 469
Charles Curtis, 571
Gertrude Ederle, 349
Maya Ying Lin, 201
Patsy Takemoto Mink, 397
Dr. Antonia C. Novello, 27
Clair C. Patterson, 642
Sally Ride, 420
Wilma Rudolph, 168
Alberto Salazar, 95
Muriel Siebert, 496
Jim Thorpe, 702
Granville T. Woods, 740

Factor(s), 6, 558
greatest common (GCF), 558, 559–563
prime, 558–563, 685

Factored form, 559, 565

Factoring, 564
binomials, 564, 568, 581–586
differences of squares, 581–586
by grouping, 567
perfect squares and, 587–593
polynomials, 565–571
solving equations by, 594–600
trinomials, 572–580
using distributive property, 564–571

Fahrenheit temperature scale, 176, 177,
541, 593

Families, (See also Parent graphs)
of graphs, 354, 618
of parabolas, 618–619, 622, 626

Fibonacci, Leonard, 17, 124

FOIL method, 537, 574

Formulas, 128–132
for area of circles, 129
for area of rectangles, 21, 275, 579
for area of trapezoids, 132, 173
for area of triangles, 380, 571
for batting average, 702
for changing Celsius to Fahrenheit, 176, 177, 593
for changing Fahrenheit to Celsius, 541
for density, 271
distance, 737–741, 747
for distance, 175, 176
for g-force, 55
for height of projectile, 633, 654, 708
Hooke's Law, 411
for interest, 66
for length of shadow, 361
for mass, 271
for mechanical advantage, 392, 600, 662–663
for monthly payments, 505
for perimeter of parallelograms, 23
for perimeter of rectangles, 20, 273
for perimeter of squares, 23
for perimeter of triangles, 23
for periods of pendulums, 719, 721, 752
for population doubling time, 276
for power, 547
quadratic, 628–633
for resistance, 397, 698, 699
for sight distance, 752
for speed, 735
for surface area of rectangular solids, 128
for voltage, 725
for volume of prisms, 538, 579

Fraction(s)
adding, 100–104, 681–684, 686–688
algebraic. See Rational expressions
complex, 114, 691–695
dividing, 112–117, 671–674
expressing as decimals, 95
improper, 151
least common denominators, 686–687
multiplying, 106–111, 667–670
percents and, 215–216
simplifying, 114, 660–666, 691–695
subtracting, 101–104, 681–684
See also Rational numbers

Function(s), 56–62, 287–294, 610–617, 634–649, 660
absolute value, 355
definition, 56, 287
dependent variable (quantity), 58–62, 83, 150, 154, 239, 273, 280, 290, 332, 341, 358, 614, 620
direct variation, 239–240, 242–244, 349
domain, 58, 83, 154, 242, 263, 273, 280, 361, 611, 614, 620, 631
exponential, 634–642, 643–649
functional notation, 289
graphing technology, 278–279, 610, 634, 666
independent variable (quantity), 58–62,
83, 150, 154, 239, 273, 280, 290, 332, 341, 358, 614, 620
linear, 278–279, 358
maximum and minimum points of, 611
modeling real-world applications, 57, 77, 83, 150, 154, 172, 184, 239, 287, 290, 295, 298, 412, 611, 635, 638, 643–649, 743
parent graphs, 354–355, 362, 618–619
preview, 56–62
quadratic, 293, 610–617, 620, 746
range, 58, 83, 154, 242, 263, 273, 280, 361, 611, 614, 620, 631
rational, 660
relations, 58, 263
vertical line test for, 288

Functional notation, 289

G

GCF (greatest common factor), 558, 559–563

Geometric mean, 123, 734, 735

Geometric sequence, 110

Geometry, 8, 10, 18, 111, 124, 231
angles, 162–167, 169, 170, 171, 201, 209, 210, 227, 232, 361, 395, 400, 500, 512, 533, 731
area of circle, 9, 129, 504, 518, 543–544
area of rectangle, 21, 22, 44, 46–47, 130, 275, 397, 499, 517, 518, 532, 558, 562, 576, 579, 585, 586, 589, 604, 627, 642, 679, 680, 708, 729
area of square, 121, 396, 504, 585, 604
area of trapezoid, 132, 173, 286
area of triangle, 380, 571
diagonals in polygons, 13–14, 533
distance formula, 737–741, 747
graphing systems of equations, 457, 459
height of triangle, 599
length of sides of rectangle, 161, 251
length of sides of triangle, 66, 160, 389, 725, 747
midpoint of line segment, 369–374, 571
parallel lines, 362–368
perimeter of parallelogram, 23
perimeter of pentagon, 106
perimeter of quadrilateral, 684
perimeter of rectangle, 20, 50, 273–274, 277, 561, 570, 682, 683, 729
perimeter of square, 23, 121, 570, 600
perimeter of trapezoid, 14
perimeter of triangle, 23, 36, 160, 397, 524, 525, 752
perpendicular lines, 254, 362–368, 512, 649, 725
Pythagorean theorem, 712–718, 737
quadratic functions, 746
quadrilaterals, 13, 160, 362–363, 526, 684, 695
similar triangles, 201–205, 220, 250, 541, 593, 642
slope, 335, 337, 398, 412
surface area of prism, 11, 128
volume of cone, 62
volume of cube, 501, 649
volume of prism, 538, 540, 579
volume of rectangular solid, 124

Gerson, Levi ben, 322

Girard, Albert, 658

Global Connections, 12, 73, 151, 196, 306, 363, 392, 483, 536, 566, 611, 697, 714

Graph(s), 73
  axes on, 56, 254
  best-fit line on, 314
  box-and-whisker, 427–434, 441, 481, 527, 571
  circle, 216, 218, 219
  complete, 278
  coordinates on, 73–75
  of experiments, 184
  families of, 354, 618
  of families of parabolas, 618–619, 622, 626
  of inequalities, 435–441
  interpreting and sketching, 56–62, 731
  intersection of, 406–407
  linear, 354–355
  of linear equations, 356–361, 480
  of ordered pairs, 255, 358, 436–441
  origin, 56, 254, 438
  parent, 354–355, 362, 618–619
  plotting point on, 255
  of quadratic equations, 620–627
  of quadratic functions, 610–617, 746
  of relations, 260–261, 262–269
  of solution set, 122, 123, 124
  solving quadratic equations with, 620–627
  solving systems of equations with, 452–461
  of systems of inequalities, 482–486
  union of, 408
  x-intercept and y-intercept of, 346–349, 356, 358, 620

Graphing calculators, 2–3
  best-fit lines, 342
  box-and-whisker plots, 433–434
  discounting, 224, 225
  distance formula, 740
  dividing polynomials, 678, 679
  evaluating algebraic expressions, 21, 23
  exponential functions, 634, 641
  factoring trinomials, 578
  families of linear graphs, 354–355
  families of parabolas, 618–619, 622, 626
  graphing equations, 284, 338, 360
  greatest common factor, 562
  inequalities, 389, 401, 403, 411, 435, 437, 440, 484, 486
  linear relations and functions, 278–279
  means and medians, 180
  measures of variation, 303–304, 310
  measuring hypotenuse of triangle, 213
  measuring lengths of legs of triangle, 213, 389
  pixels, 373
  plotting points on graphs, 258
  Pythagorean theorem, 717
  quadratic equations, 622, 632
  quadratic functions, 610
  radical expressions, 720, 726
  radioactivity, 648
  rational expressions, 662, 664, 692, 694
  relations, 260–261, 268
  scatter plots, 344
  scientific notation, 507, 510

  slope, 351
  solving equations, 270, 273, 276
  solving equations involving decimals, 147
  solving multi-step equations, 157, 160
  standard form of linear equations, 480
  symmetry of parabolas, 616
  systems of equations, 452–453, 459
  systems of inequalities, 484, 486

Graphing software, 471, 733

Greater than (>) symbol, 94, 385, 386

Greater than or equal to (≥) symbol, 33, 94, 385, 386

Greatest common factor (GCF), 558, 559–563

Grouping, factoring by, 567

Grouping symbols, 20, 21

Growth, exponential, 643–649

Guess-and-check strategy, 574–575

Half-life, 638–639

Half-plane, 437–438, 482

Hands-On Activities. *See* Modeling Mathematics

Hexagon, 13

Hooke's Law, 411

Horizontal axis, 56, 254

Hypotenuse, 206, 213, 712–718

Hypsometer, 209, 211

Identity, 170

Identity properties, 37–38, 40–43
  additive, 37, 52
  multiplicative, 38, 52, 92

Inconsistent systems of equations, 456

Independent events, 414

Independent systems of equations, 456

Independent variable (quantity), 58–62, 83, 150, 154, 239, 273, 280, 290, 332, 341, 358, 614, 620

Inequalities, 33, 384–447
  addition and subtraction properties for, 385–390
  compound, 405–412
  defined, 93
  graphing, 435
  graphing in two variables, 436–441
  multiplication and division properties for, 393–398
  multi-step, 399–404
  as open sentences, 122, 420–426
  solving by using addition and subtraction, 384–390
  solving by using multiplication and division, 392–398

  systems of, 482–486

Integers, 73–77
  adding, 74–76, 84, 86–92
  consecutive, 158–159, 160, 161, 402, 446, 574, 599, 604
  multiplying, 105, 109
  as rational numbers, 93
  subtracting, 84, 87–92

Integration
  discrete mathematics, 88, 90, 103, 104, 110, 123, 220, 227, 228–232, 244, 250, 269, 285, 290–291
  geometry. *See* Geometry
  modeling, 581
  number theory. *See* Number theory
  probability, 227, 228–232, 244, 250, 269, 285, 353, 380, 413–419, 424, 481, 519, 585, 593, 736
  problem solving, 13, 56–62, 126–132
  statistics. *See* Statistics
  technology. *See* Computers; Graphing calculators; Programming; Scientific calculators; Spreadsheets
  trigonometry, 294, 302, 345, 374, 586

Intercepts
  x, 346–347, 356, 358, 620
  y, 346–349, 356, 358

Interdisciplinary connections. *See* Connections

Interest
  compound, 648, 654, 670
  simple, 66, 217, 218, 219, 220, 238, 244, 251, 277, 461, 505, 518, 552, 593, 643–644, 647

Interquartile range, 304, 306–313, 428–432

Interval, 78

Inverses
  additive, 87, 523, 682
  multiplicative, 38
  of relations, 264–265

Inverse variation, 241–244

Investigations
  The Brickyard, 554–555, 580, 600, 617, 627, 650
  Go Fish!, 190–191, 200, 227, 269, 302, 314
  The Greenhouse Effect, 68–69, 83, 111, 154, 177, 184
  A Growing Concern, 656–657, 665, 695, 718, 741, 748
  Ready, Set, Drop!, 448–449, 461, 519, 541, 548
  Smoke Gets In Your Eyes, 320–321, 338, 361, 398, 426, 442

Irrational numbers, 120, 121

Isosceles triangle, 164, 738

Journal. *See* Math Journal

Al-Khowarizmi, Muhammed ibn Musa, 141

INDEX

INDEX

LCD (least common denominator), 686–687, 692

LCM (least common multiple), 685–688

Leading coefficient, 158

Leaf, 25. *See also* Stem-and-leaf plots

Least common denominator (LCD), 686–687, 692

Least common multiple (LCM), 685–688

Legs of triangles, 206, 213, 389, 713

Less than (<) symbol, 94, 385, 386

Less than or equal to (≤) symbol, 33, 94, 384, 385, 386

Like denominators, 681–684

Like terms, 47

Line(s)
  number, 72–77, 91, 94, 100, 121, 131
  parallel, 362–368
  perpendicular, 254, 362–368, 512, 649, 725
  regression, 342
  slopes of, 324–331
  vertical, 289

Linear equations, 280–286, 332–338
  graphing, 356–361, 480
  point-slope form of, 333–338
  slope-intercept form of, 346–353, 358, 363, 365
  standard form of, 281, 333–338, 357–358, 480

Linear functions, 278–279

Linear graphs, 354–355

Line plots, 78–83, 99, 104, 111, 117, 125, 221, 227

Line segment, midpoint of, 369–374, 571

Liu Hsin, 658

Liu Hui, 124

Long-Term Projects. *See* Investigations

Lower quartile, 306–313, 427–432

Manipulatives. *See* Modeling Mathematics

Mapping, 263

Mathematical Modeling
  charts, 111, 154, 191, 200
  diagrams, 160, 166, 171, 259, 337. *See also* Diagrams
  equations, 33–34, 126–132, 142, 145, 152, 155
  graphs, 56–62, 73–75, 122, 123, 124
  in problem solving, 340
  scatter plots, 340, 345

Mathematics and Society
  Computers and Chess, 419
  High-Tech Checkout Lanes, 481
  International Mathematical Olympiad, 11
  Mayan Mathematics, 77
  Measurements Great and Small, 512
  Minimizing Computers, 649

Morgan's Conjecture, 214
Nonlinear Math, 736
Number Sieve, 563
People Around the World, 374
Reading the Labels, 689
Record Cold Temperatures, 313
Teen Talk Barbies, 183

Mathematics Exploration Toolkit (MET) **(graphing software),** 471, 733

Math history, 124, 172

Math Journal, 28, 41, 53, 170, 175, 181, 224, 236, 274, 283, 291, 343, 350, 359, 388, 415, 439, 478, 484, 499, 504, 531, 561, 591, 624, 631, 668, 672, 699, 723, 729, 734

Matrix, 88
  addition of, 103
  scalar multiplication of, 108

Maximum point of function, 611

Mayan mathematics, 77

Mean(s), 178–183, 200, 221, 251, 472, 586, 747
  geometric, 123, 734, 735
  in proportions, 196

Means-extremes property of proportions, 196

Measurement
  of angles, 162–163, 169
  great and small, 512
  of hypotenuse of triangle, 213
  of lengths of legs of triangle, 213, 389
  proportional measures, 201
  protractors, 162, 164, 216

Measures of central tendency, 178–183, 200, 221, 251, 472, 586, 747

Measures of variation, 303–313, 331, 390, 461, 702

Mechanical advantage, 392, 600, 662–663

Median, 178–183, 200, 221, 251, 472, 586

Midpoint of line segment, 369–374, 571

Minimum point of function, 611

Mitchell, Maria, 141

Mixed expressions, 690–695

Mixed Review. *See* Reviews, Mixed

Mode, 178–183, 200, 221, 251, 472, 586

Model(s)
  brick, 555
  equation, 142–143
  house, 426
  in problem solving, 340

Modeling Mathematics
  adding integers, 74, 75, 84
  adding polynomials, 520–521, 524
  angles of a triangle, 164, 165
  circle graphs, 216, 218
  completing the square, 742, 746
  distributive property, 44, 48
  dividing polynomials, 676, 678
  estimating square roots, 118
  exponential functions, 636, 640
  factoring differences of squares, 581, 584
  factoring trinomials, 572–573, 578, 579

factoring using distributive property, 564, 568
  graphs, 257
  hypsometer, 209, 211
  line plots, 79, 80
  looking for patterns, 15
  measures of variation, 309
  midpoint of line segment, 370, 371
  multiplying integers, 105, 109
  multiplying polynomials, 528, 534–535, 539, 545, 546
  number line, 102
  perpendicular lines, 365, 366
  polynomials, 513, 516, 520–521
  probability, 229, 230
  Pythagorean theorem, 712, 715
  ratios, 194, 198, 203
  relations and inverses, 265, 266
  similar triangles, 203
  slope, 324
  solving inequalities, 391, 395, 402
  solving multi-step equations, 155, 159
  solving one-step equations, 142–143, 148, 152
  subtracting integers, 84, 89
  subtracting polynomials, 520–521, 524
  surface area, 128, 129
  symmetry of parabolas, 612, 614
  systems of equations, 458, 464, 466
  See also Mathematical modeling

Monomials
  defined, 496
  degree of, 515
  dividing by, 501–505
  identifying, 515
  multiplying, 496–500
  multiplying polynomials by, 528–535
  power of, 498

Morgan's Conjecture, 214

Motion, uniform, 235–236, 477–478, 479

Multiples, least common, 685–688

Multiplication
  associative property of, 51–55
  of binomials, 537–538, 574
  elimination using, 475–481
  factors in, 6, 558
  of integers, 105, 109
  of monomials, 496–500
  of polynomials, 528–541
  products in, 6
  of rational expressions, 667–670
  of rational numbers, 106–111
  scalar, 108
  solving equations with, 150–154
  symbol for, 6

Multiplication property
  of equality, 150–151
  for inequalities, 393–398

Multiplicative identity property, 38, 52, 92

Multiplicative inverse property, 38

Multiplicative property
  of −1, 107
  of zero, 38

Multiplicative reciprocals, 38

Multi-step equations, 155–161

Multi-step inequalities, 399–404

N

Natural numbers, 73

Negative correlation, 340

Negative exponents, 503

Negative numbers, 73

Negative sign (−), 73

Nonlinear math, 736

Notation
functional, 289
scientific, 506–512
set-builder, 385

Not equal to (≠) symbol, 94

Number(s)
comparing on number line, 94
composite, 558
even, 158, 402, 532
graphing, 73–75
irrational, 120, 121
natural, 73
negative, 73
odd, 158–159, 403, 574, 599, 604
prime, 558, 562
rational, 93–117
real, 121–125
sign of, 73, 88, 112
whole, 72, 73
See also Integers

Number line, 72–77, 91, 131
adding rational numbers on, 100
comparing numbers on, 94
completeness property for points on, 121

Number sieve, 563

Number theory, 467, 470, 473, 479, 486, 490, 518, 539, 623–624, 625, 633, 654, 665, 718, 752
composite numbers, 558
consecutive numbers, 158–159, 160, 161, 402, 446, 574, 599, 604
geometric mean, 123, 734, 735
perfect square numbers, 587
prime numbers, 558, 562
Pythagorean triple, 583, 585
twin primes, 562

Numerical expressions, 6

O

Obtuse triangle, 165

Odd numbers, 158–159, 403, 574, 599, 604

Odds, 229–232, 380, 593, 736

One-step equations, 142–154

Open half-plane, 437

Open sentences, 32–36, 122
with absolute values, 420–426
solving, 32, 420–426

Opposites, 87

Ordered pairs, 57, 254–255, 358, 436–441

Order of operations, 19–24

Origin, 56, 254, 438

Outcomes, 229, 413–419

Outlier, 307–313, 428

P

Parabolas
axes of symmetry of, 612, 616
defined, 610, 611
families of, 618–619, 622, 626
graphing, 620–627
maximum points of, 611
minimum points of, 611
symmetry of, 612–613, 616
vertices of, 610–617

Parallel lines, 362–368

Parallelograms, 362–363
equations for sides of, 335
perimeter of, 23

Parent graphs
linear, 354–355, 362
quadratic, 618–619

Patterns, 12–18, 23, 36, 99
looking for, 13, 15, 92, 104, 138, 139, 221, 368, 390, 497, 500, 526, 580, 600, 617, 637, 638
in problem solving, 13, 497
writing equations from, 295–302, 331

Pendulum, period of, 719, 721, 752

Pentagons, 13, 106

People in the News
Jorge Aguilar, 323
Liz Alvarez, 609
Ayrris Layug Aunario, 193
Julie Goskowicz, 451
Tony Goskowicz, 451
Amy Gusfa, 495
Shakema Hodge, 383
Ryan Holladay, 711
Wendy Isdell, 5
Jamie Knight, 253
Joe Maktima, 659
Marianne Ragins, 141
Robert Rodriguez, 557
Danny Seo, 71
Jenny Slabaugh, 495

Percent, 215–227
of change, 222–227
of decrease, 222–227
fractions and, 215–216
of increase, 222–227
interest, 217, 218, 219, 220
solving using equations, 216
solving using proportions, 215–216

Percent proportion, 215–216

Perfect squares, 119
factoring and, 587–593
quadratic equations as, 743–747

Perfect square trinomials, 587–593

Perimeter, 139
of parallelogram, 23
of pentagon, 106
of quadrilateral, 684
of rectangle, 20, 50, 273–274, 277, 561, 570, 682, 683, 729
of square, 23, 121, 570, 600, 708
of trapezoid, 14

of triangle, 23, 36, 160, 397, 524, 525, 752

Perpendicular lines, 254, 362–368, 512, 649, 725

Pi (π), 124

Planes. *See* Coordinate planes

Points
maximum, 611
midpoint, 369–374, 571
minimum, 611
origins, 56, 254, 438

Point-slope form of linear equations, 333–338

Polygons, diagonals in, 13–14, 533

Polynomials, 513–547
adding, 520–527
ascending order, 516
defined, 513, 514
degree of, 516
descending order, 516
dividing, 675–680
factoring, 565–571. *See also* Factoring
identifying, 515
multiplying, 528–541
prime, 577
subtracting, 520–527
See also Binomials; Monomials; Trinomials

Portfolio, 67, 137, 184, 189, 249, 314, 319, 379, 442, 447, 491, 548, 553, 605, 650, 655, 707, 748, 753

Positive correlation, 340

Positive sign (+), 73

Possibilities, list, 685, 688, 706

Power(s), 7, 497
formula for, 547
of monomial, 498
power of, 498
of product, 498
product of, 497
quotient of, 501–505
zero, 502

Precision, 116

Prime factorization, 558–563, 685

Prime number, 558, 562

Prime polynomial, 577

Principal, 217

Principal square root, 119

Prisms
surface area of, 11, 128
volume of, 538, 540, 579

Probability, 227, 228–232, 244, 250, 285, 353, 380, 424, 481, 519, 585, 684
of compound events, 414–419
defined, 228
of event, 228, 229
graphing outcomes, 269
odds and, 229–232, 380, 593, 736

Problem-solving strategies, 156
drawing diagrams, 406, 407–408, 410, 411, 412, 425
guess and check, 574–575
identifying subgoals, 745, 747
listing possibilities (making organized list), 685, 688, 706

INDEX

looking for patterns, 13, 497, 637
models, 340
organizing data, 464, 468, 480, 486, 490
planning, 126–127
tables, 255, 600
working backward, 156–157, 161, 172, 177, 338
writing equations and formulas, 126–132

Product(s), 6
cross products, 94, 196
power of, 498
of powers, 497

Product property of square roots, 719–720

Programming, 49
in BASIC, 8
solving equations, 171
See also Graphing calculators

Project(s). *See* Chapter Projects; Cooperative Learning Projects; Investigations

Projections, 398

Properties
additive identity, 37, 52
additive inverse, 87, 523, 682
associative, 51–55
commutative, 51–55, 92
comparison, 94–95
completeness, 121, 256
density, 96
distributive, 44–50, 52, 89, 114, 536–537, 564–571
of equalities, 39, 46, 52, 144–154, 639
for inequalities, 385–390, 393–398
means-extremes, 196
multiplicative, of −1, 107
multiplicative identity, 38, 52, 92
multiplicative inverse, 38
product, 719–720
of proportions, 196
quotient, 721–722
reflexive, 39
of square roots, 719–722
substitution, 39, 52
symmetric, 39, 46
transitive, 39
zero product, 594–600

Proportion(s), 195–200
capture-recapture method, 190–191
cross products of, 94, 196
direct variations, 239–240, 242–244
means-extremes property of, 196
percent, 215–216
in rate problems, 197–198
similar triangles and, 201–205

Proportional measures, 201

Protractor, 162, 164, 209, 216

Punnett squares, 542

Pyramid, volume of, 24

Pythagoras, 17

Pythagorean theorem, 712–718, 737
corollary to, 714
modeling, 712, 715
statement of, 713

Pythagorean triple, 583, 585

Quadrants, 254

Quadratic equations, 620–633
defined, 620
discriminant of, 632
roots of, 620–627
solving by completing the square, 743–747
solving by graphing, 620–627
solving by using quadratic formula, 628–633

Quadratic formula, 628–633

Quadratic functions
defined, 610, 611
graphing, 610–617, 746
zeros of, 620

Quadrilaterals, 13, 362–363
length of sides of, 160
perimeter of, 684
relationship among sides of, 695
sum of angles in, 526

Quartiles, 304, 306–313
lower, 306–313, 427–432
upper, 306–313, 427–432

Quotient, 112–113
of powers, 501–505
sign of, 112

Quotient property of square roots, 721–722

Radical equations, 732–736

Radical expressions, 719–731
adding, 727–731
simplest form of, 723
simplifying, 719–726
subtracting, 727–731

Radical sign ($\sqrt{}$), 119

Radicand, 719

Random outcome, 229

Range
defined, 306
interquartile, 304, 306–313, 428–432
of relation, 263, 264
reasonable, 276, 284, 293, 616, 632

Rates
percents, 215
ratios, 197–198

Ratio(s), 194–200
trigonometric, 206–214

Rational equations, 696–702
solving, 696–702

Rational expressions
adding, 681–684, 686–688
defined, 660
dividing, 671–674
with like denominators, 681–684
multiplying, 667–670
simplifying, 660–666
with unlike denominators, 685–689

Rationalizing the denominator, 722–723

Rational numbers, 93–117
adding, 100–104
comparing, 93–99
comparison property for, 95
defined, 93
density property for, 96
dividing, 112–117
expressing as decimals, 95
multiplying, 106–111
subtracting, 101–104

Real numbers, 121–125

Reciprocals, 116, 502
multiplicative, 38
of zero, 38

Rectangles
area of, 21, 22, 44, 46–47, 130, 275, 397, 499, 517, 518, 532, 558, 562, 576, 579, 585, 586, 589, 604, 642, 679, 680, 708, 729
length of sides of, 161, 251, 598
perimeter of, 20, 50, 273–274, 277, 561, 570, 682, 683, 729
surface area of, 128
volume of, 124

Reflexive property of equality, 39

Regression line, 342

Relations, 58, 260–302
defined, 260, 263
domain of, 263, 264
equations as, 271–277
functions, 58, 263, 278–279, 287–294
graphing calculators and, 260–261, 268
graphs of, 260–261, 262–269
inverse of, 264–265
linear, 278–286
mapping, 263
range of, 263, 264

Repeating decimals, 95

Replacement set, 33, 400–401

Reviews
Chapter Tests, 788–800
Cumulative, 138–139, 250–251, 380–381, 492–493, 606–607, 708–709
Extra Practice, 756–786
Mixed, 18, 24, 31, 36, 43, 50, 55, 62, 77, 83, 92, 99, 104, 111, 117, 125, 132, 149, 154, 161, 167, 172, 177, 183, 200, 205, 214, 220–221, 227, 232, 238, 244, 259, 269, 277, 285–286, 294, 302, 312–313, 331, 337, 345, 352–353, 361, 368, 374, 390, 397–398, 404, 412, 418–419, 425–426, 432, 441, 460–461, 468, 474, 481, 486, 500, 505, 511–512, 518–519, 526–527, 533, 540–541, 547, 563, 571, 580, 585–586, 593, 600, 616–617, 627, 633, 642, 649, 665, 670, 674, 680, 684, 689, 695, 702, 718, 725, 731, 736, 741, 747
Study Guide and Assessment, 64–67, 134–137, 186–189, 246–249, 316–319, 376–379, 444–447, 488–491, 550–553, 602–605, 652–655, 704–707, 750–753

Right triangle, 165, 712–718

Rise, 325

Roots
of quadratic equations, 620–627
square, 118–125, 719–722. *See also* Square roots

Run, 325

S

Sampling, 314

Scalar multiplication, 108

Scale, 78, 80, 197

Scatter plots, 339–345, 352, 353, 368, 519, 547, 593, 642

Schwarzschild radius, formula for, 508

Scientific calculators
  rational expressions, 662, 664, 666
  reciprocals, 116
  scientific notation, 502
  sign of number, 88
  solving equations involving decimals, 147
  square roots, 120

Scientific notation, 506–512

Self Evaluation, 67, 137, 189, 249, 319, 379, 447, 491, 553, 605, 655, 707, 753

Self Tests, 36, 99, 161, 221, 286, 353, 412, 474, 527, 580, 633, 680, 731

Sentences
  compound, 421
  open, 32–36, 122, 420–426

Sequences, 12–18
  arithmetic, 104
  Fibonacci, 17
  geometric, 110

Set
  replacement, 33, 400–401
  solution, 33, 122, 123, 124, 400–401

Set-builder notation, 385

Sides, corresponding, 201

Sign of number, 73, 88, 112

Similar triangles, 201–205, 220, 250, 541, 593, 642

Simple events, 414

Simple interest, 66, 217, 218, 219, 220, 238, 244, 251, 277, 461, 505, 518, 552, 593, 643–644, 647

Simplest form of expression, 47, 48, 49, 53, 54, 723

Simplest radical form, 723

Simplifying
  complex fractions, 114, 691–695
  radical expressions, 719–726
  rational expressions, 660–666

Sine, 206–214

Slope, 324–331

Slope-intercept form of linear equations, 346–353, 358, 363, 365

Software, graphing, 471, 733

Solution
  of equation in two variables, 271–277
  extraneous, 697
  of open sentence, 32, 420–426

Solution set, 33, 122, 123, 124, 400–401

Spatial reasoning
  circles, 9, 129, 504, 518, 543–544
  cones, 62
  cubes, 501, 649

  hexagons, 13
  parallelograms, 23, 335, 362–363
  pentagons, 13, 106
  polygons, 13–14, 533
  prisms, 11, 128, 538, 540, 579
  pyramids, 24
  quadrilaterals, 13, 160, 362–363, 526, 684, 695
  squares, 23, 121, 396, 504, 544, 545, 570, 581–593, 600, 604, 708, 742–747
  trapezoids, 14, 132, 173, 286
  See also Rectangles; Triangles

Spiral Review. *See* Reviews, Mixed

Spreadsheets, 297, 327, 646

Square(s)
  area of, 121, 396, 504, 585, 604, 708
  completing, 742–747
  of a difference, 544
  differences of, 545, 581–586
  perfect, 119, 587–593, 743–747
  perimeter of, 23, 121, 570, 600, 708
  of sum, 543

Square roots, 118–125
  defined, 119
  estimating, 118
  principal, 119
  product property of, 719–720
  quotient property of, 721–722

Standard form of linear equations, 281, 333–338, 357–358, 480

Statistics
  averages. *See* Averages
  best-fit lines, 314, 341–345, 352
  box-and-whisker plots, 427–434, 441, 481, 527, 571
  on circle graphs, 219
  defined, 25
  geometric mean, 123, 734, 735
  line plots, 78–83, 99, 104, 111, 117, 125, 221, 227
  measures of central tendency, 178–183, 200, 221, 251, 472, 586
  measures of variation, 303–313, 331, 390, 461, 702
  misleading, 25
  percents, 215–227
  population. *See* Applications & Connections Index
  scatter plots, 339–345, 352, 368, 519, 547, 593
  stem-and-leaf plots, 25–31, 92, 111, 117, 138, 167, 232, 337, 431
  test scores, 139, 339, 345, 417, 431, 471, 477, 731

Stem, 25

Stem-and-leaf plots, 25–31, 92, 111, 117, 138, 167, 232, 337, 431
  back-to-back, 27–28

Study Guide and Assessment, 64–67, 134–137, 186–189, 246–249, 316–319, 376–379, 444–447, 488–491, 550–553, 602–605, 652–655, 704–707, 750–753

Subgoals, identifying, 745, 747

Substitution, 462–468

Substitution property of equality, 39, 52

Subtraction
  elimination using, 469–474

  of integers, 84, 87–92
  of polynomials, 520–527
  of radical expressions, 727–731
  of rational numbers, 101–104
  solving equations with, 146–149

Subtraction property
  of equality, 146–149
  for inequalities, 385–390

Sum, square of, 543

Supplementary angles, 162–163, 500

Surface area, 11, 128, 129

Symbols, 1
  for absolute value, 85
  equals sign, 127
  for plus or minus ($\pm$), 628
  greater than, 94, 385, 386
  greater than or equal to, 33, 94, 385, 386
  grouping, 20, 21
  less than, 94, 385, 386
  less than or equal to, 33, 94, 384, 385, 386
  not equal to, 94
  radical sign, 119

Symmetric property of equality, 39, 46

Symmetry
  axis of, 612, 616
  defined, 612
  of parabolas, 612–613, 616

Systems of equations, 452–481
  consistent, 455
  dependent, 456
  inconsistent, 456
  independent, 456
  solving by elimination using addition and subtraction, 469–474
  solving by elimination using multiplication, 475–481
  solving by graphing, 452–461
  solving by substitution, 462–468

Systems of inequalities, 482–486

T

Tables, 156, 170, 255, 268, 270, 277, 600

Tangent, 206–214

Technology. *See* Computers; Graphing calculators; Programming; Scientific calculators; Spreadsheets

Technology Tips, 88, 169, 216, 261, 282, 299, 351, 428, 613, 622, 637

Temperature scales, 176, 177, 541, 593

Term(s), 47
  coefficient of, 47–48
  like, 47
  in sequences, 13

Terminating decimals, 95

Tests. *See* Assessment

Tiles
  algebra, 44, 513, 520–521, 524, 528, 534–535, 545, 564, 572–573, 742
  base-ten, 118

Time lines, 4–5, 70–71, 140–141, 192–193, 252–253, 322–323, 382–383, 450–451, 494–495, 556–557, 608–609, 658–659, 710–711

Top Five, 31, 72, 163, 228, 290, 330, 385, 454, 514, 596, 626, 698, 727

Transitive property of equality, 39

Trapezoids
    area of, 132, 173, 286
    perimeter of, 14

Tree diagrams, 413–419

Triangles, 163–167
    acute, 165
    angles of, 163–164, 395, 525
    area of, 380, 571, 627
    diagonals in, 13
    equilateral, 164
    height of, 599
    hypotenuse of, 206, 213, 712–718
    isosceles, 164, 738
    legs of, 206, 213, 389, 713
    length of sides of, 66, 160, 389, 725, 747
    obtuse, 165
    perimeter of, 23, 36, 160, 397, 524, 525, 752
    Pythagorean theorem, 712–718, 737
    right, 165, 712–718
    similar, 201–205, 220, 250, 541, 593, 642
    solving, 208–210
    sum of angles of, 163–164, 169
    trigonometric ratios and, 206–214

Trigonometric ratios, 206–214

Trinomials
    defined, 514
    factoring, 572–580
    identifying, 515
    perfect square, 587–593

Twin primes, 562

Uniform motion, 235–236, 477–478, 479

Union, 408

Unique factorization theorem, 558

Unit cost, 95

Unlike denominators, 685–689

Upper quartile, 306–313, 427–432

Values
    absolute, 85–86, 87, 420–426
    excluded, 660–661, 666
    extreme, 427–428

Variables, 6
    defining, 127, 130
    dependent, 58, 273
    graphing inequalities in, 436–441
    independent, 58, 273
    solving equations with, 168–172, 271–277

Variation
    constant of, 239
    direct, 239–240, 242–244
    inverse, 241–244
    measures of, 303–313, 331, 390, 461, 702

Venn diagrams, 73, 121, 418

Verbal expressions, 7

Vertex, 610–617

Vertical axis, 56, 254

Vertical line test, 289

Volume
    of cone, 62
    of cube, 501, 649
    of prism, 538, 540, 579
    of pyramid, 24
    of rectangular solid, 124

Weighted average, 233–238

Whiskers, 428. *See also* Box-and-whisker plots

Whole numbers, 72, 73

Working backward, 156–157, 161, 172, 177, 338

x-axis, 254

x-coordinate, 254, 346

x-intercept, 346–347, 356, 358, 620

y-axis, 254

y-coordinate, 254, 333, 346, 622

y-intercept, 346–349, 356, 358

Zero(s)
    positive and negative and, 73
    of quadratic function, 620
    reciprocal of, 38

Zero exponent, 502

Zero product property, 594–600